Reservoir Sandstones

Robert R. Berg
Texas A&M University

Prentice-Hall, Inc., Englewood Cliffs, New Jersey 07632

Library of Congress Cataloging in Publication Data

Berg, Robert R.
Reservoir sandstones.

Includes bibliographies and index.
1. Sandstones. 2. Sandstones—United States.
3. Oil reservoir engineering. I. Title.
QE471.15.S25B47 1986 552′.5 85-6537
ISBN 0-13-774373-4

Editorial/production supervision and
interior design: Maria McColligan
Cover design: Diane Saxe
Manufacturing buyer: John Hall

Printed in the United States of America

10 9 8 7 6 5 4 3 2 1

ISBN 0-13-774373-4 01

Prentice-Hall International (UK) Limited, *London*
Prentice-Hall of Australia Pty. Limited, *Sydney*
Prentice-Hall Canada Inc., *Toronto*
Prentice-Hall Hispanoamericana, S.A., *Mexico*
Prentice-Hall of India Private Limited, *New Delhi*
Prentice-Hall of Japan, Inc., *Tokyo*
Prentice-Hall of Southeast Asia Pte. Ltd., *Singapore*
Editora Prentice-Hall do Brasil, Ltda., *Rio de Janeiro*
Whitehall Books Limited, *Wellington, New Zealand*

Contents

Preface

Sandstones are important as reservoirs for oil and gas, and it has been estimated that 50% of the world's petroleum will be produced from sandstones. Sandstones are also important as aquifers for the production of water, for underground storage of natural gas, and disposal of liquid waste. The purpose of this book is to describe the character and origins of sandstones that are reservoirs for oil and gas and, especially, to promote a better understanding of reservoir properties.

A knowledge of sandstone properties can be valuable to geologists and engineers who are employed in exploration for, and development of, subsurface fluids or other mineable resources. The morphologies of reservoirs and aquifers are of special importance. The size and shape of sandstone bodies largely determine the amount of fluid that can be stored in the subsurface; composition, texture, and sedimentary structures control the production of fluids. A knowledge of fundamental rock properties, therefore, is important to the prediction of occurrence and behavior of sandstone reservoirs.

Many of the properties of sandstone reservoirs are determined by depositional environment, and therefore, the descriptions are organized by environment. Interpretations of environment are made from rock properties in vertical sequence, and from these interpretations a prediction can be made of the local size and shape of sandstone bodies. The interpretations may be aided by a knowledge of regional stratigraphy, the nature of adjacent rocks, fossil content, and a host of other facts commonly available to the geological interpreter. The sandstones themselves, however, provide the keys to interpretation. From composition, texture, and sedimentary structures in vertical sequence, reliable interpretations can be made of depositional environment and local morphology.

This book can form the basis for a one-semester course on the origin of sandstones for upper-level undergraduate and graduate students in geology, geophysics, and petroleum engineering. Basic knowledge is presented on sandstone properties, transport mechanisms, and modern depositional environments so that this book alone provides the necessary background for advanced study. This presentation should be supplemented, of course, by readings from the technical literature, but there is no other book currently available that describes in detail those properties of sandstones most important to subsurface geological practice.

The approach used here is the study of oil and gas fields where sandstone properties and morphology are known with some assurance. With the aid of these examples, other sandstones may be interpreted by similarity or contrast with the reservoirs described. The fields illustrated herein represent only a select number of the possible morphologies for sandstones. Undoubtedly there are many more, and equally well known, examples that could be documented. The fields described here, however, represent basic morphologies that can be expected to occur most commonly in sequences of different ages. Although the field examples represent specific reservoirs and their properties, some general principles emerge that may have wider application, both in future geologic studies as well as in exploration and development of subsurface resources.

The reservoir descriptions will also be of interest to professional geologists and petroleum engineers. All the sandstones exhibit inhomogeneities, to a greater or lesser degree, depending on origin. These variations in rock properties, both lateral and vertical, pose problems for fluid flow and recovery. Variations impose a degree of antisotropy with regard to flow, and these variations must be considered in planning for future recovery of oil, especially by tertiary methods.

ACKNOWLEDGMENTS

Support for these studies has come from many sources. Foremost among them are the companies and persons that supplied the cores, and in many cases, financial assistance. Most are identified only as the operators of the wells from which the cores were recovered. Although corporate names appear prominently, these names represent many individuals who supplied materials as well as enthusiasm and encouragement for geologic interpretations of cores.

Many persons also have participated directly in these studies. Faculty members and students at Texas A&M University have provided, over a period of more than 15 years, analyses and descriptions that have contributed substantially to the final interpretations. These persons are credited individually as analysts or as authors of theses and publications, and their part in these interpretations is gratefully acknowledged. The manuscript was improved by the critical reading of David G. Kersey, and others who commented on individual sections.

R.R.B.

1

Sandstone Properties

Sandstones are common reservoirs for oil and gas, and thus they are a class of sedimentary rocks of great economic importance. A knowledge of their properties is essential in the exploration for, and the production of, subsurface fluids. The primary properties of texture, composition, sedimentary structures, and morphology are determined largely by the source materials and by the environment of deposition. These properties, in turn, control the production of fluids through a borehole. All the properties, then, have economic significance, and knowledge of sandstone properties is essential to understanding the reservoir and its capacity for oil or gas production.

CLASSES OF PROPERTIES

Many reservoir studies require that geologic variables be handled in a quantitative manner. For the purpose of analysis it is important to categorize reservoir rock properties so that interrelationships among the variables may be recognized. Three main classes of rock properties can be established (Berg 1970).

The first group of properties may be called *primary or definitive properties* (Table 1–1) because these are the fundamental characteristics of the reservoir. The definitive properties are mineral composition, texture, sedimentary structures, and morphology, or size and shape of the reservoir body. These fundamental properties exert a primary control on other reservoir properties, which may therefore be thought of as *secondary or dependent properties*. The dependent properties include porosity, permeability, fluid saturation, and bulk density.

TABLE 1–1. SELECTED RESERVOIR ROCK PROPERTIES, SYMBOLS, AND CUSTOMARY UNITS OF MEASUREMENT

Primary (definitive)	Secondary (dependent)	Tertiary (latent)
Composition (%)	Porosity, ϕ (%)	Resistivity, R (Ωm)
Texture (mm)	Permeability, k (md)	Spontaneous potential, SP (mV)
Sedimentary structures (m)	Saturation, S (%)	Radioactivity, γ (counts/sec)
Morphology (descriptive)	Bulk density, ρ_b (g/cm^3)	Sonic travel time, Δt (μsec/ft)

The dependency of secondary properties on the primary properties is illustrated by porosity. The storage capacity of a reservoir is controlled by the mineral composition, primarily the cementing materials, and also by the textural properties of grain size, sorting, and grain packing. Therefore, a knowledge of the primary properties for a given reservoir is often an indication, at least in a qualitative way, of the secondary properties.

A third group of properties may be called *tertiary or latent properties.* These are the properties measured in the borehole by geophysical logs. The most important of these properties are resistivity, spontaneous potential, radioactivity, both natural and induced, and sonic travel time. The latent properties are determined, to a great extent, by mineral composition, porosity, and the nature of the fluids in the rock. For this reason, the rock type and its fluid saturation often can be determined quite accurately from log measurements, but it is more difficult to interpret the latent properties in terms of primary rock properties.

The categories of rock properties (Table 1–1) indicate the difficulty of predicting the dependent properties in the subsurface without first knowing as much as possible about the primary properties. There are gaps in our knowledge of reservoir rocks and their contained fluids that prevent us from estimating the amount of fluids stored within the reservoir and the ability to recover these fluids. It is quite likely that further studies of the geologic variables will yield relationships that may make the prediction of reservoir properties more quantitative. The next section describes the primary rock properties, especially as they may be observed and measured in sandstone cores.

PRIMARY PROPERTIES

Composition

Sandstones are composed of mineral particles, called *grains,* which consist largely of quartz, feldspars, and rock fragments [Figure 1–1(A)]. The finer mineral material between the grains is called the *matrix* and commonly consists of clay minerals but may also contain very fine particles of quartz and feldspar. Both the grains and matrix represent the products of weathering that were carried as sediment to their

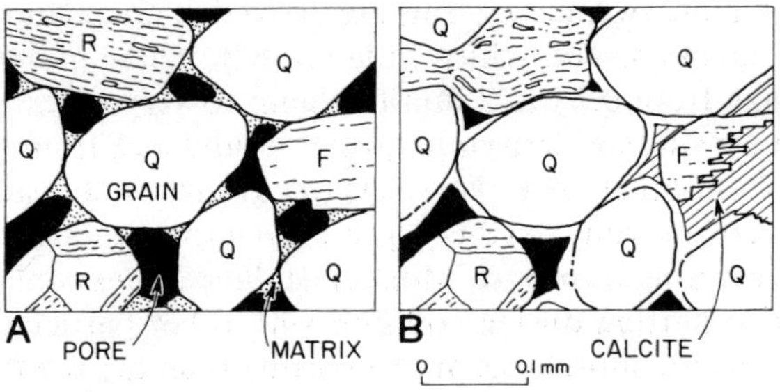

Figure 1-1 Diagrams showing relationships among some compositional and textural elements of sandstones: (A) detrital grains and matrix shortly after desposition (Q = quartz, F = feldspar, R = rock fragments); (B) altered grains and matrix including silica as overgrowths on quartz grains and calcite as intergranular cement and as a replacement of feldspar grains. Note reduction of pore size in altered rock.

site of deposition and are referred to as *detritus.* After deposition and burial, the grains and matrix are commonly altered by the physical effects of compaction and by chemical changes. These changes are broadly referred to as *diagenesis,* and early chemical changes are called *authigenesis.*

A third major component, in addition to the grains and matrix, is the cement introduced between the particles during diagenesis [Figure 1–1(B)]. The term *cement* refers to any mineral that is precipitated after the grains and matrix have been deposited. Thus, the cements are formed after deposition and burial, in contrast to detrital components that were carried to the sites by currents. As the term implies, cements are responsible for *lithification,* or joining grains into a competent mass.

This brief description of the principal compositional elements serves to emphasize that many, if not most, sandstones may be highly complex mixtures. The detrital particles give important clues to the source materials and, in some cases, indicate the degree to which the original materials have been changed by the effects of transport. Composition, then, may be important in environmental interpretations. The nature of particles also determines, to a large extent, the degree to which these materials may be altered, either physically or chemically, by diagenesis. These changes may greatly affect the reservoir properties of porosity and permeability, generally in a deleterious manner. The mineral particles and cements may be considered to be in equilibrium under subsurface conditions of temperature, pressure, and adjacent fluids. Any change of these conditions, such as by deep burial or by the withdrawal or injection of fluids, may promote some change in the character of grains and matrix. Thus, sandstones must be thought of as chemically complex mixtures that may become unstable under long- or short-term changes in the physical and chemical environment.

Quartz. Grains of quartz are the major components of many sandstones, and quartz is generally regarded as the common mineral most resistant to weathering and abrasion during transport. There are two distinct types of quartz, however, that apparently react somewhat differently during transport of sediment. *Monocrystalline grains* are those composed of part of a single quartz crystal, and *polycrystalline grains* are composed of parts of two or more crystals and may properly be considered as rock fragments. Both types of quartz grains are derived in abundance during the weathering of granites, gneisses, and schists (Blatt 1967). These grains are of large size, mostly in the range from 0.25 to 2 mm (medium to very coarse sand).

Monocrystalline quartz grains are derived in greater abundance from plutonic rocks such as granite than from other rock types. Quartz grains in sandstones are generally finer than grains derived from weathering, but even the weathered quartz grains suffer reduction in size during transport. Monocrystalline grains contain fracture planes that are subject to parting during collision with other particles, but in general, monocrystalline grains are more resistant to parting than are the polycrystalline grains. Polycrystalline grains contain numerous crystal boundaries that range from planar to sutured, and hence they may be more susceptible to breakage during transport than are the monocrystalline grains. Therefore, the two types of quartz are separately identified when composition of sandstone is determined, and polycrystalline grains are classed as rock fragments so that there may be greater insight into the nature of the parent materials and the results of transport. Both types of quartz are highly resistant to chemical alteration under reservoir conditions.

Feldspar. The feldspar minerals orthoclase and plagioclase are largely derived from the weathering of igneous and metamorphic rocks, but because these minerals are more easily degraded by both weathering and transport, their abundance in sandstones is generally much less than that of quartz. Orthoclase and microcline, the potassium feldspars, appear to be more resistant than plagioclase to weathering, but like all feldspars, orthoclase contains cleavage planes along which weathering takes place and the grains are easily parted. Plagioclase feldspars, which contain sodium or calcium, are especially susceptible to chemical weathering, and these grains are less likely to survive during transport than the orthoclase feldspars. In addition, all feldspars are subject to further chemical alteration after deposition and during burial [Figure 1-1(B)]. Because of their low resistance, the relative abundance of feldspar minerals sometimes may be a possible indication of source sediment or distance of transport.

Rock fragments. Many sand grains are composed of fragments of other rocks and show the character of the terrain from which they were derived. Common rock fragments may be the result of weathering of igneous, metamorphic, or sedimentary rocks; and despite the rigors of transport, some fragments may survive in the sediment to indicate the nature of the source terrain. Large amounts of igneous rock fragments suggest that there may have been only a short distance of sediment transport and possibly high topographic relief. Volcanic fragments, in particular, may not survive the long distance of stream travel (Davies et al. 1978). On the other hand, some fragments are highly resistant, and polycrystalline quartz, chert, and metamorphic quartzite can be present when all other rock fragments have disappeared. The nature of rock fragments, therefore, can be a useful supplement to the interpretation of depositional environments.

Many fine-grained rock fragments of contrasting origin, such as shale, schist, and volcanic glass, are unstable both physically and chemically. The fine-grained fragments of shale and schist may be deformed between the more resistant grains during compaction and flow into adjacent pores [Figure 1–1(B)]. Glass, especially, may be chemically altered during diagenesis to produce clay minerals with an increase in bulk volume and a reduction in pore space. The fragments more easily deformed are termed *labile components* and may be important in the reduction of porosity and permeability.

Matrix. Fine-grained materials between grains may be composed largely of clay minerals, some of which may be further altered during diagenesis. The common products of rock weathering are the clays—kaolinite, illite, and montmorillonite. Kaolinite is derived largely from the weathering of feldspar and is the most stable under conditions of burial. Montmorillonite is derived from a variety of minerals but especially from fine-grained volcanic materials such as ash. This mineral contains abundant interlayered water, which may be removed during burial by the influence of temperature, and above 100°C, montmorillonite may be converted rapidly to the clay mineral illite. Diagenetic changes may add significantly to the total matrix by altering other minerals. It is believed that feldspar grains can alter in place to kaolinite by reaction with ground water, and some kaolinite may form by precipitation from solution. At higher temperatures volcanic glass may alter to chlorite and to interlayered clay minerals of complex composition.

The interpretation of clay minerals as related to depositional environment is hazardous without a knowledge of diagenetic history. Nevertheless, large amounts of detrital matrix may be significant to the interpretation of source terrain, distance of transport, and velocity or constancy of transporting currents.

Accessories. Many other minerals are found in sandstones but most commonly in trace amounts of 1 percent or less. Among the accessory minerals, the micas may be most abundant. Muscovite, or potassium mica, survives weathering and transport in greater abundance than does biotite, an iron mica, and muscovite is an expected accessory mineral. The heavy, or high-density, minerals may also be present but in extremely small quantities. The common heavies include tourmaline and zircon, which are most likely to survive weathering, transportation, and resedimentation through several cycles. Other opaque minerals may be the result of diagenesis, such as iron oxides and pyrite, each indicating a different chemical environment. The least common accessory minerals are those that are easily weathered, such as hornblende, epidote, and other magnesium and iron silicates. Generally, the accessory minerals are of little environmental significance, and their small amounts limit their usefulness. In most petrographic analyses, they may be dismissed in the category of "other" minerals.

Cement. Intergranular cements are commonly silica and carbonate. Silica is precipitated as overgrowths, or layers, on quartz grains [Figure 1–1(B)] and in optical continuity with the original grains. Silica overgrowths may develop soon after burial and may continue to be deposited with increased temperature and depth. Carbonate, usually calcite, is precipitated early to form fillings of intergranular spaces and may occur in small patches or as a pervasive groundmass throughout the sandstone. Other minerals are more rare as cements, but in some cases unusual cements such as authigenic feldspar may be relatively abundant. Even the clay minerals, especially those formed by diagenesis, may help to bind the particles together into a coherent mass. Clays, however, are not now considered a cement since their ability to bind is not as great as other cement minerals.

Measurement. A quantitative estimate of composition is best made by the observation of thin sections under the petrographic microscope. A standard method is used, that of mineral identification at numerous points during traverses of the thin section in a direction normal to bedding. The points are selected at random by means of a mechanical stage that allows the traverse to proceed by uniform small steps. The number of points counted can depend on composition and grain size. For example, a very few points would be required to determine accurately the percentage of the dominant mineral grain present if the sandstone is essentially monomineralic. A very large number of grains must be counted, however, to establish accurately the percentage of minor compositional elements, or the number of each grain type in a highly mixed composition. Commonly, 100 to 200 points are sufficient to represent a fairly reliable estimate of composition. This relatively small number of points is not accurate statistically, but the degree of accuracy is not much improved unless an exceedingly large number of grains are counted. In routine analysis, the point-count method yields an estimate of considerable value despite, perhaps, a relatively modest degree of accuracy.

Analysis by X-ray diffraction methods provides the only accurate way of determining the character of clay minerals and other matrix materials. For clays, diffraction methods can be supplemented by scanning-electron microscope observations. For more accurate identification of some materials, elemental analyses may be necessary. These methods should not be substituted for petrographic analysis in thin section, however. Mineral identification by X-ray diffraction of bulk samples, for example, cannot be more than a first approximation of the composition of grains and matrix.

Classification. Sandstones are classified and named according to the composition of grains. Several classifications are in current use, and in fact a large number have been proposed (Klein 1963). However, most classifications employ three end members or groups of grains as displayed on a ternary diagram [Figure 1-2(A)]. One commonly used system employs the following groups (Folk 1968):

1. Quartz including metaquartzite.
2. Feldspar including feldspar-rich granitic and gneissic rock fragments.
3. Rock fragments other than those listed above.

The major rock classes are quartzarenite, arkose, and litharenite, with compositions as indicated by the relative percentages on the ternary diagram [Figure 1–2(A)]. Intermediate compositions are designated by the appropriate combined terms.

Another method of classification employs matrix or clays as a fourth end member, and composition is shown by means of a tetrahedral diagram [Figure 1–2(B)]. This classification is modified from one proposed by Dott (1964) and is still preferred in some textbooks (Pettijohn et al. 1973). The end members for grains remain essentially the same, but when the matrix exceeds 15 percent, the names *feldspathic graywacke* and *lithic graywacke* are used. When the matrix exceeds 50 percent the rock is a mudstone rather than a sandstone.

The advantage in considering matrix content is that for many sandstones, the matrix is an important part of the composition. Therefore it should be recognized by specific rock terms. Furthermore, its inclusion in a tetrahedral diagram emphasizes an important relationship, and that is that the composition and texture are interrelated. In general, median grain size decreases from the three-end member face of the diagram toward the matrix pole, from sand size to nearly clay size. The "clean sand" face, therefore, represents good porosity and permeability, whereas these properties decrease rapidly toward the matrix pole.

Color is an important feature of sandstones but only as far as this property is determined by composition, and therefore color itself is a secondary property. Highly quartzose sandstones are nearly white, arkoses may be either pink or white, and litharenites may be speckled gray due to the scattered dark rock fragments. Graywackes are "dirty" gray because of their high matrix. Contained residual fluids such as petroleum add a tan-to-brown cast. Red denotes oxidizing conditions, and dark gray denotes reducing conditions, but the chemical conditions can be either primary or secondary. Thus, the colors of sandstones can be important to interpretations and should always be recorded, but color alone cannot substitute for mineralogical analysis.

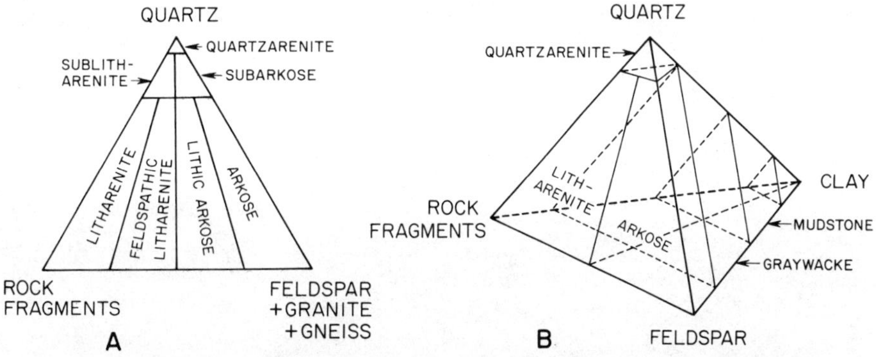

Figure 1–2 Classifications of sandstones according to different end members: (A) classification that uses three end members for grain composition [from Folk 1968]; (B) classification including grains and matrix [modified from Dott 1964].

Texture

Sandstones are composed of particles that may range in size through several orders of magnitude. The grains and matrix, therefore, represent populations of individual particles that vary greatly in size as well as in composition. Particle sizes are classed according to a standard scheme (Table 1-2), and a population of dominant "sand" may contain a wide range of particle size. The term *texture* has a broad meaning and refers to the interrelationships among the population (Krynine 1948). The principal elements of texture are grain size, sorting, packing, shape, and orientation. The first two are the most significant and commonly measured elements.

Grain size. There are several ways to express the "average size" of particles, or the central tendency of the grains-plus-matrix population, and the terms are necessarily statistical. The *mode* is the midpoint of the most abundant class when the

TABLE 1-2. GRAIN SIZE CLASSIFICATION OF CLASTIC SEDIMENT

Diameter (mm)	Phi class[a] (ϕ)	Modal class	Sediment	Rock
		Boulders		
256	−8			
		Cobbles		
64	−6		Gravel	Conglomerate
		Pebbles		
4	−2			
		Granules		
2	−1			
		Very coarse		
1	0			
		Coarse		
0.50	+1			
		Medium	Sand	Sandstone
0.25	2			
		Fine		
0.125	3			
		Very fine		
0.062	4			
		Coarse		
0.031	5			
		Medium		
0.015	6		Silt	Siltstone
		Fine		
0.007	7			
		Very fine		
0.004	8			
<0.004		Clay	Clay	Claystone

SOURCE: Wentworth 1922.

[a] $\phi = -\log_2$ (diameter, mm).

grain population is separated into groups and is probably the measure that can be estimated in a hand specimen or by low-power magnification. The *median size* is the midpoint, or fiftieth percentile, of the population and is measured by graphic methods after a sample is separated into classes, or grouped, by sieving the disaggregated rock. The classes are designated by logarithmic units, or phi classes, for statistical purposes (Table 1–2). The *mean size* is an average value determined by the measurement of individual particles as from thin-section observations, and thus the measurements are ungrouped. The expression for "average" size, therefore, should denote how this size was determined, whether by classed or unclassed data. If a "mean" value is determined by graphic methods, it is called a *graphic mean* or *median* in order to distinguish it from the *arithmetic mean* determined for unclassed data.

The most accurate measures of the total population are those derived from grouped data by sieving or by settling velocity. The data may be interpreted by graphic methods from a cumulative frequency plot of classes by weight—probably the most common method. Values may also be calculated by means of "moment" formulas (Griffiths 1967). Sieve analysis requires time and considerable labor, and so sieving is only used, if at all, for a limited number of samples.

A more efficient method is direct measurement of grains in thin section by petrographic microscope. This technique is the only practical way to examine an indurated rock, and larger number of samples may be processed in less time than by sieving. The thin section is traversed normal to bedding, and measurement is made of the long axes for individual grains at points along the traverse. The simple arithmetic mean of the individual values is used as an expression of mean grain size for the population according to the equation

$$\bar{x} = \sum_{i=1}^{n} x_i/n$$

where $\bar{x}$ is the mean size, x_i are individual grain measurements, and n is the number of grains measured. Obviously, this measure does not account for the entire population because the very fine-grained matrix particles cannot be measured accurately. Measurements are made for particles down to a size of about 0.02 to 0.03 mm, or coarse to medium silt. The division, then, between grains and matrix is at this arbitrary point in thin-section analysis. Nevertheless, reliable and reproducible estimates of grain size can be made by this method.

Grains selected for measurement may include those of any composition or only a single mineral type. Because quartz is the major grain type, measurements are only made on monocrystalline grains for most sandstones. When quartz is not a major type, then measurements are made on all grains. Measures of a single grain type tends to reduce any variation in mean values that may be introduced by compositional differences. For example, quartz and feldspar grains tend to be finer grained than rock fragments or some other minerals such as mica. The selection of quartz for measurement results in values that can be compared among samples despite changes in composition of other grains.

Measurements are made of the apparent long axis for each grain. The majority of grains in thin section are elongate, and neither axis, long or short, represents a true length because of the oblique orientation of most grains. However, long-axis measurements result in mean sizes that are close to, but slightly larger than, the median grain size determined by sieve analysis (Friedman 1958). This relationship holds for many sandstones but probably not for those that contain a large amount of matrix. In any case, apparent long-axis measurements result in values that are comparable among many samples.

The number of grains that must be counted depends on composition, grain size, and sorting. Only a small number of grains are required for accurate size determinations in a highly quartzose sandstone, whereas a larger number are required for lithic and arkosic sandstones. For most sandstones, 100 long-axis measurements are sufficient to characterize mean size, especially if a single dominant grain type is selected.

Sorting. The next most important measure of texture is the degree of dispersion of the grain population about the median or mean size. Dispersion is referred to qualitatively as *sorting* but is expressed quantitatively as *standard deviation*. Graphically, the standard deviation may be determined by one-half the difference between the 84th and 16th percentiles. Thus, standard deviation is expressed as a plus or minus value from the median, and it accounts for dispersion of 68 percent of the grain population. For ungrouped values from thin-section measurements, the standard deviation is calculated by means of the basic equation

$$\sigma = \left[\sum_{i=1}^{n} (\bar{x} - x_i)^2/(n - 1) \right]^{1/2}$$

where σ is the standard deviation, $\bar{x}$ is the mean size, x_i are the individual grain measurements, and n is the number of grains measured.

Other measures. The particle population can be further characterized by two other expressions, *skewness* and *kurtosis*. Skewness is a measure of population asymmetry about the median. The calculated skewness in phi units ranges from positive for a finely skewed population to negative for a coarsely skewed population. Kurtosis is a measure of peakedness of the frequency distribution and ranges from greater than unity for highly peaked curves to less than unity for a flattened curve. Neither of these parameters may be calculated accurately for ungrouped values obtained from thin-section measurements, primarily because of the small size of the population that is sampled. Both skewness and kurtosis have been claimed to be significant to environmental interpretation, but their value has not been firmly established. They have only been tested on relatively well-sorted populations with contradictory results. Perhaps they will prove to be of value in interpreting more poorly sorted populations. Skewness and kurtosis are not calculated for any of the reservoir sandstones described in this text.

Packing. The relation of grains to each other is an estimate of the degree of packing. This textural property is never measured in routine analyses and, in fact, is difficult to estimate. The concept of packing, however, is vital to understanding the origin of porosity. More precisely, packing describes the manner in which the grains are held in place by tangent contacts with neighboring grains. Hence, the fundamental measure of packing would be the number of contacts per grain. Packing is best visualized by assuming systematic arrangements of uniform spheres (Graton and Fraser 1935). A large number of such spheres may represent a well-sorted sandstone. Only eight spheres in two layers are required to display the symmetry of the entire systematic arrangement (Figure 1–3). The smallest unit of symmetry, in fact, is the unit cell that is formed by the intersection of planes that pass through the centers of the spheres. The unit cell contains only one sphere volume in each packing, and the remaining part of the bulk volume is void.

Each layer may be visualized as consisting of layers in either square or rhombic configurations [Figure 1–3(A) and (E)]. Four systematic packings may be considered. *Cubic* packing is formed by two square layers superimposed so that the centers of the four spheres in the upper layer are coincident with the centers in the lower layer [Figure 1–3(B)]. Each sphere has 6 contacts with neighboring spheres. It is easily calculated that the void space in such an arrangement is 47.6 percent of the unit cell volume, or, in other words, the porosity (n) is 47.6 percent. If the top layer is moved a distance of one radius and in line with connecting radii between the spheres of the lower layer, then the arrangement is more tightly packed [Figure 1–

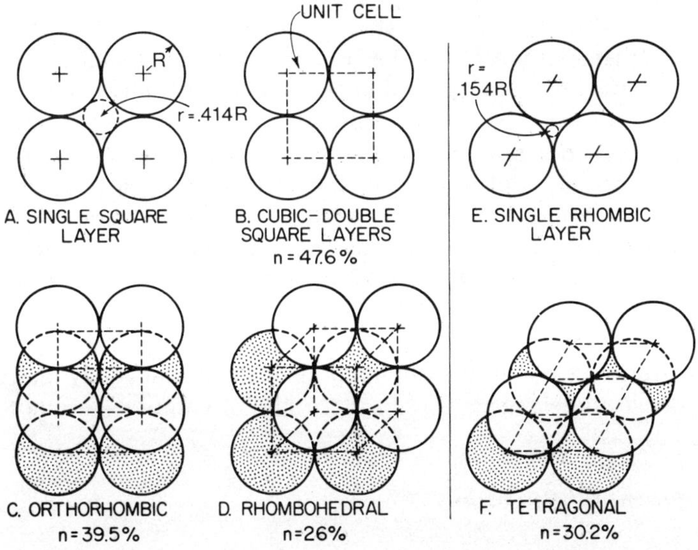

Figure 1–3 Diagrams of systematic packings of uniform spheres as described by Graton and Fraser (1935). Porosity (n) is given for the principal packings.

3(C)]. This packing is called *orthorhombic,* the number of contacts per sphere has increased to 8, and the porosity is 30.5 percent. The tightest packing is visualized by moving the top layer diagonally so that the spheres in the upper layer nestle in the opening between the four spheres of the lower layer [Figure 1–3(D)]. This packing is called *rhombohedral,* the number of contacts per sphere is 12, and the porosity is 26 percent. A packing of intermediate porosity is formed by layers in rhombic configuration [Figure 1–3(E)]. If the top layer is moved a distance of one radius and in line with connecting radii between the spheres of the lower layer, the arrangement is called *tetragonal* [Figure 1–3(F)]. The number of contacts per sphere is 10, and the porosity of the unit cell is 30.2 percent.

These systematic arrangements illustrate that as the degree of packing increases, the bulk volume of the unit cell decreases, and the amount of void space also decreases. By analogy, the mechanical compaction of well-sorted sand may be visualized as proceeding by the same process, forming successively more tightly packed arrangements, and reducing porosity in a systematic manner with depth of burial. These arrangements also illustrate that the measurement of porosity is an indirect measure of packing for many sandstones. Of course, the analogy is invalid when a large amount of cement or matrix is present between the grains, but systematic arrangements of grains may be applicable to many well-sorted sandstones that form oil reservoirs and aquifers.

It is difficult to estimate packing in a two-dimensional thin section of rock. Therefore, this measurement is seldom made, and, in fact, is usually unnecessary, especially if the important property of porosity has already been measured for the sample.

Other elements. Other textural elements include *grain shape* and *orientation.* Shape may be expressed in two ways: *sphericity* is a measure of the departure of a grain from a spherical form, and *roundness* is a measure of the angularity of grain edges. For example, a cube has a high degree of sphericity but a low roundness. Sphericity and roundness appear to be related to grain size, the finer grains having lower values. Grain shape is difficult and tedious to measure, and its value to environmental interpretations has not been demonstrated. However, new techniques of measurement and analysis (Ehrlich and Weinberg 1970) suggest that shape studies may become useful in the future.

Orientation is a measure of preferred direction for the long axes of grains, but like grain shape, this measurement is difficult and rarely made. Therefore, both of these textural elements—grain shape and orientation—are not included in reservoir studies.

Sedimentary Structures

The most characteristic features of sedimentary rocks are the layering or stratification, known as *sedimentary structures.* These internal structures are the result of either primary stratification that resulted from sedimentary processes or secondary stratification produced by biological or physical changes shortly after deposition. In either case, they reflect processes at the site of deposition and are most valuable indicators of depositional conditions.

In the most general sense, sedimentary structures are inhomogeneities in the distribution of textural and compositional elements within the rock (Krynine 1948, p. 145). This definition means that most layering is produced by regular alternations of coarser and finer particles. In the same way, other primary and secondary structures are produced by either regular or irregular distributions of grains so that a visible pattern results. If no visible structure is present, the rock is termed *massive,* which means that inhomogeneities may be present but on a large scale.

Primary sedimentary structures. Description of sedimentary structures employs three different groups of terms: (1) qualitative terms such as *stratification* and *cross-stratification* that describe the attitude of layering, (2) quantitative terms such as *thin-bedded* or *thick-bedded* that describe the thickness of layering, and (3) quantitative terms that describe the thickness of splitting or fracture parallel to layering such as shaly or flaggy (McKee and Weir 1953, p. 381). All of these terms are useful for the description of cores as well as outcrops.

A standardized terminology is desirable, and the one proposed by McKee and Weir (1953) is useful and most complete. The qualitative terms are important to the description because they denote the fundamental divisions of the stratified section. The basic unit, or single layer, is a *stratum,* or when inclined to the dip of the formation, a *cross-stratum.* A group of conformable strata is called a *set,* and a group of two or more sets is called a *coset* (Figure 1–4). A group composed of both strata and cross-strata is called a *composite set.* The utility of these terms is apparent when describing a thick sandstone unit that comprises a single sedimentary sequence. The total unit may be a composite set that is composed of a regular succession of sets and cosets. Thus, the recognition of such a succession aids in the description and also in identifying the sequence of different sedimentary events that resulted in the total unit.

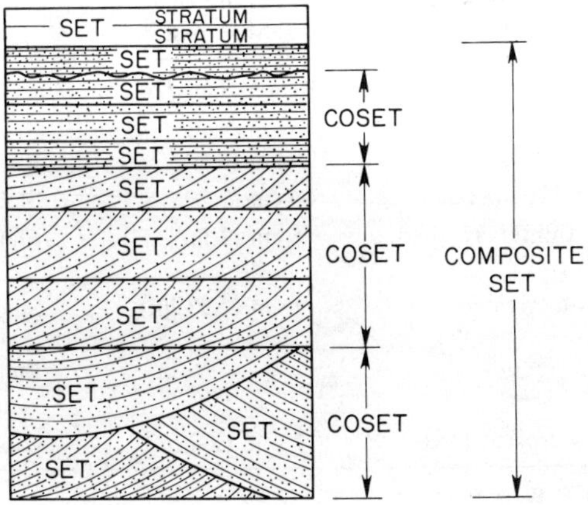

Figure 1–4 Terminology for stratified and cross-stratified units and types of crossbedding. [From McKee and Weir 1953, pp. 383–84.]

Quantitative terms for describing layered rocks were also defined by McKee and Weir (1953) (Table 1–3). *Laminae* and *laminated* refer to layers that are closely spaced, 1 cm or less apart. *Bed* and *bedding* refer to layers that are greater than 1 cm apart. Both groups of terms are further qualified by adjectives denoting relative thickness.

The terms for splitting properties include *massive* for very thick-bedded units (Table 1–3). This term is also used in a qualitative sense to denote even thin units that lack layering.

The description of inclined or cross-layering is made easier by defining the relationships between adjacent sets of cross-strata (Figure 1–5). *Simple cross-stratification* refers to sets in which the lower bounding surfaces are nonerosional. The difference in dips between sets results from the shifting current directions. *Planar cross-stratification* refers to sets in which the lower bounding surfaces show erosion. In these sets, the erosion represents local scouring of the lower surface before deposition of the succeeding set. *Trough cross-stratification* refers to sets in which the lower bounding surface is a curved surface of erosion. These sets result from the deposition and truncation of large sediment waves formed by current flow. Terms for cross-sets may be further qualified when the shape of the set can be described as either *lenticular, wedge-shaped,* or *tabular.*

Recently another type of cross-stratification has been described, called *hummocky* (Harms et al. 1975). In this type of bedding the lower bounding surfaces are erosional and slope at low angles similar to that in planar crossbedding [Figure 1–5(D)]. However, within sets the laminae are nearly parallel to the lower bounding surface, and the laminae are broadly curved, either concave or convex upward. This latter character of upward convexity distinguishes hummocky bedding from trough cross-lamination in which only the lower surfaces are curved. Hummocky bedding

TABLE 1–3. QUANTITATIVE TERMS FOR DESCRIBING LAYERED ROCKS

Terms to describe stratification	Terms to describe cross-stratification	Upper limit of thickness	Terms to describe splitting property
Very thick-bedded	Very thickly crossbedded	Greater than 120 cm	Massive
Thick-bedded	Thickly crossbedded	120 cm (4 ft)	Blocky
Thin-bedded	Thinly crossbedded	60 cm (2 ft)	Slabby
Very thin-bedded	Very thinly crossbedded	5 cm (2 in.)	Flaggy
Laminated	Cross-laminated	1 cm (0.33 in.)	Shaly (claystone, siltstone) Platy (sandstone, limestone)
Thinly laminated	Thinly crossbedded	2 mm (.08 in.)	Papery

SOURCE: Data from McKee and Weir 1953.

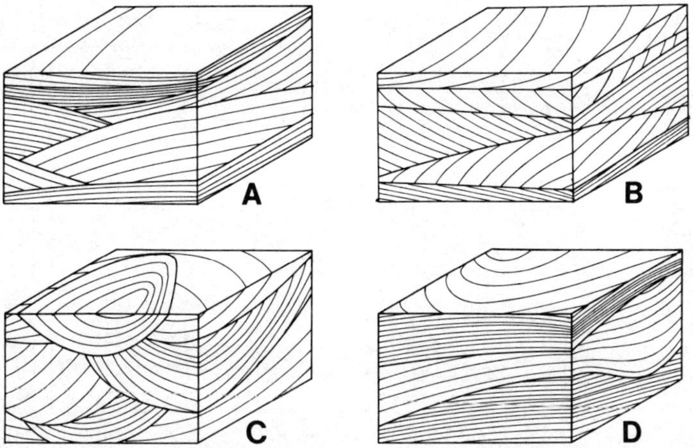

Figure 1–5 Relationships of inclined laminae within and between cosets: (A) simple cross-stratification; (B) planar cross-stratification; (C) trough cross-stratification; (D) hummocky cross-stratification. [(A), (B), and (C) from McKee and Weir 1953; (D) from Harms et al. 1975.]

is believed to have been formed by storm waves in marine shelf settings (Dott and Bourgeois 1982) in contrast to other types of cross-lamination that are the result of current flow. Thus, the recognition of hummocky bedding could be important to the interpretation of the processes that last affected the sediment.

The terminology for stratification was proposed for more detailed descriptions from outcrops where a considerable lateral extent of the rock may be observed, but most terms can be adapted readily to descriptions of stratification in cores. In all descriptions it is important to note that angle of inclination for cross-stratified units. Low-angle dips of 10° or less represent cross-stratification that is distinctly different in origin from higher-angle dips of greater than 10°.

The description of sedimentary structures in cored sections poses some special problems. Structures are best observed on a freshly cut face, but the exposed section is only 4 in. (10 cm) or less in width, which prevents a view for the shape of sets except for small-scale units. Also, the core may have been slabbed at an angle that does not permit an optimum view of the structures, and furthermore, many cores may have been broken and sampled for the measurement of reservoir properties before the geologic description is made. All of these conditions inhibit the interpretation of structures. Therefore, core descriptions should employ terms that are objective and noninterpretive as far as possible.

A systematic group of descriptive terms has been proposed for core description (Campbell 1967). Such terms emphasize the attitude and continuity of small layers of laminae within sets (Figure 1–6). These terms are preferred when making original core descriptions so that a high degree of objectivity may be maintained. Interpretation may follow easily as, for example, wavy, nonparallel laminae that are narrowly spaced at about 2 cm apart may be interpreted as ripple-form structure,

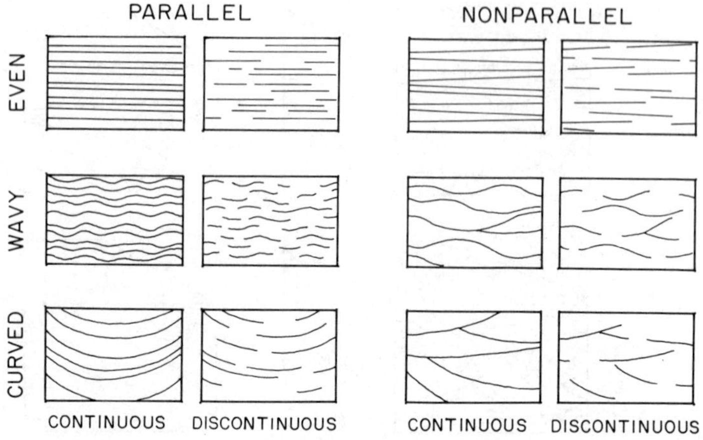

Figure 1–6 Objective terms for description of lamination in cores. [From Campbell 1967, Figure 2.]

a low-amplitude bed that results from low-velocity currents. Another interpretation that could be made, however, is that of physical deformation of soft sediment caused by loading shortly after deposition. Some interpretation should always be made, but objective terms will help to support the interpretation. All objective terms, of course, should be supplemented by measurements of spacing and dip.

Secondary sedimentary structures. An important class of secondary sedimentary structures results from the deformation of the sediment after deposition and before lithification. Physical deformation is the result of sliding, slumping, or other internal deformation that disturbs the original bedding in a manner that is obviously not a product of deposition. Physical deformation can be of two types: (1) *contorted bedding,* which is characterized by a degree of folding of the original strata, or (2) *sheared bedding,* which is characterized by movement of sediment along planar surfaces that may or may not be parallel to the original stratification.

Contorted bedding may occur at scales that range from small to very large. Wavy, parallel laminae (Figure 1–6) may form on inclined strata by short-distance movement and result in small waves that have amplitudes of only a few millimeters. In contrast, entire beds greater than 1 m in thickness may be deformed to produce an overturned fold of high amplitude but of short length. All types of contorted bedding may be found in clean sands as well as in shaly sediment, but they are most common in finer-grained sediment that apparently retains a high degree of plasticity at shallow depth of burial. Another type of contorted bedding is that formed by the upward flow of water during the earliest period of compaction and dewatering of freshly deposited sediment. Such structures commonly take the form of irregular, vertical laminae that form tubes of contorted sediment that appear to be intruding upward through the normal sedimentary structures.

Sheared bedding is characterized by a degree of foliation within the sediment and the development of parallel-to-nonparallel laminae that are commonly distinct and continuous. These laminations are the result of internal slippage along planar surfaces during deformation of thicker sets of soft sediment. In well-sorted sediment, the shear planes may seem to represent original sedimentary structures, but they differ from common cross-laminae in that they are narrowly spaced and the planes may join or bifurcate, abruptly terminating the laminae or very thin beds. More severe deformation results in extension features such as small normal faults that may be synthetic or antithetic to the direction of original bedding or to the shear planes. Displacements on such faults may range upwards from a few millimeters to many meters. Internal deformation by shearing may be so pervasive that the sediment attains a secondary structure that nearly obliterates the original character.

The recognition of physical deformation in soft sediment may be difficult, especially in cores. A useful criterion is that deformation is commonly restricted to a given interval, above and below which the section displays undeformed, primary sedimentary structures. Another criterion is that some evidence of original structures remains, even if only that of interbedded or interlaminated sediment types, so that the secondary structures are clearly seen to be superimposed on the original. In the absence of these criteria, only an "unnatural" appearance, unexpected in the section, may be the clue to soft-sediment deformation.

Another class of sedimentary structures is that produced by burrowing organisms. Especially in marine-shelf environments, the sea floor is populated by a benthic fauna of great variety ranging from wormlike organisms to crustacea and large, shelled mollusks. Many of these animals depend on the organic material in sediment as their food supply. Others bury themselves in the soft sediment but maintain communication with the water from which they draw sustenance. The action of these organisms alters the primary sedimentary structures, and collectively, this type of alteration is called *bioturbation.*

Progressive stages of alteration have been noted in modern sediments (Moore and Scruton 1957). From an original layered structure, bioturbation first disrupts the layering to produce irregular, discontinuous masses of sediment (Figure 1-7). Further disturbance may proceed to the point at which the original layers are completely separated into irregular, isolated masses of sediment that give the total sediment a mottled appearance. The mottles may be further redistributed so that the sediment has an indistinctly churned appearance, and finally, bioturbation may result in a completely homogenized sediment in contrast to the original layering of different grain size and composition.

In an opposite manner, an original, homogeneous sediment may become mottled by the feeding burrows of some animals. As sediment is passed through their bodies and the food extracted, some minerals or grains may be redistributed, leaving a burrow of somewhat different composition than the original sediment. In other cases, the animal can maintain an open burrow, the walls strengthened by its secretions, and the open burrow may be filled later by sediment of a different composition.

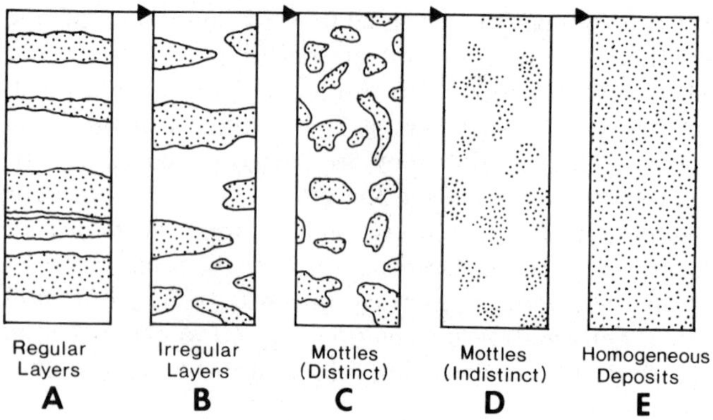

Figure 1–7 Bioturbation of modern sediment: (A) original layered sediment; (B) disrupted irregular layers; (C) and (D) mottled sediment; (E) homogenized sediments. [From Moore and Scruton 1957.]

All types of bioturbation, then, alter or destroy the primary sedimentary structures so that little evidence may be left of the stratification formed by the sedimentary processes. However, bioturbation does give some clues to the environment of deposition and indicates, at the least, that the disturbance was produced by marine organisms, probably in relatively shallow waters. In many cases, the organisms leave distinctive tracks or burrows such that the water depth may be estimated by the characteristics of the bioturbation. Such "trace fossils" are not always definitive of environment, however, and should be used with other evidence.

One distinctive type of bioturbation is that which results from the roots of plants that extend downward, sometimes for several meters, into the underlying sediment. The roots themselves may be partly preserved as thin wisps of carbonaceous material or as larger carbonaceous stringers. Rooting commonly results in the complete churning of soil and a homogenization that is itself distinctive, even without the preservation of carbonaceous material. Claystones with this type of bioturbation are the familiar underclays or "seat rocks" beneath coal seams.

The complete range of sedimentary structures is illustrated for the reservoir sandstones described in this book. The primary sedimentary structures are most important to environmental interpretations for all types of sandstones. Physical deformation is shown in diverse types from eolian to deep-marine deposits. Bioturbation of distinctive kinds is also shown ranging from the marsh through the marine environments. All are important in the final interpretations of the reservoirs.

Morphology

The size, shape, and extent of a sandstone body are referred to here as its *morphology.* This term is commonly used to denote the topology of landforms in other branches of geology and is quite appropriate for denoting the shape of a sedimentary unit as determined by original depositional processes. The importance of the morphology is that it determines the reservoir volume, and it is therefore, the primary property that most directly influences the economic potential of sandstones.

Descriptive terms for different morphologies have been defined by Krynine

(1948), mostly for thick, complex stratigraphic units of interbedded sandstone, shale, and limestone. Some of these terms such as a *blanket* deposit are still useful, but others such as a *shoestring* deposit for a single sedimentary unit are no longer used. Other terms more descriptive of single sedimentary bodies are required. Several of these terms especially useful to local description of sandstones can be defined (Table 1–4).

Blanket refers to an extensive layer in which the length and width are many times that of the thickness (Figure 1-8). The only single depositional unit that approaches a blanket deposit is an Eolian dune.

Lobate refers to deposits of more limited extent in which width and length are 100 times or more than that of the thickness and generally much less than 1000 times the thickness. The term *lobe* signifies a curved or rounded projection and is appropriate for broadly rounded sediment bodies that are composites of different types of sandstones such as alluvial fans, composite deltas, and submarine fans.

Ovate refers to deposits in which the length is somewhat greater than the width, and both are 100 times or more than that of the thickness. The term implies an isolated body, not abruptly bounded on one side as may be the case for lobate deposits. Many sands of shallow-marine origin have such a morphology.

Linear refers to a local deposit in which the length is much greater than either the width or thickness, and the width may be on the order of 300 times or more than that of the thickness. The term signifies the body is bounded by straight lines and is appropriate for description of shoreline sandstone deposited in barrier islands and some shallow-marine sandstones of longshore-bar origin.

TABLE 1–4. TERMS FOR DESCRIPTION OF MORPHOLOGIES OF SANDSTONES

Term	Description	Common dimension	Example
Blanket	Extensive layer	Length $= \text{width} > 1000 \times$ thickness	Eolian dunes
Lobate	Rounded projection	Length $= \text{width} > 100 \times$ thickness	Deltas, fans (composite bodies)
Ovate	Elliptical shape	Length $>$ width $> 100 \times$ thickness	Shallow marine
Linear	Straight, narrow	Length $> 10 \times$ width $> 300 \times$ thickness	Barrier island, longshore bar
Finger	Linear, short	Length $> 10 \times$ width $> 300 \times$ thickness	Delta distributary
Sinuous	Narrow, curved	Length $>$ width $> 100 \times$ thickness	Fluvial channel fill
Braided	Low sinuosity (S)	$S = \dfrac{\text{actual length}}{\text{direct length}} < 1.5$	High-gradient stream
Meander	High sinuosity (S)	$S = \dfrac{\text{actual length}}{\text{direct length}} \geq 1.5$	Low-gradient stream

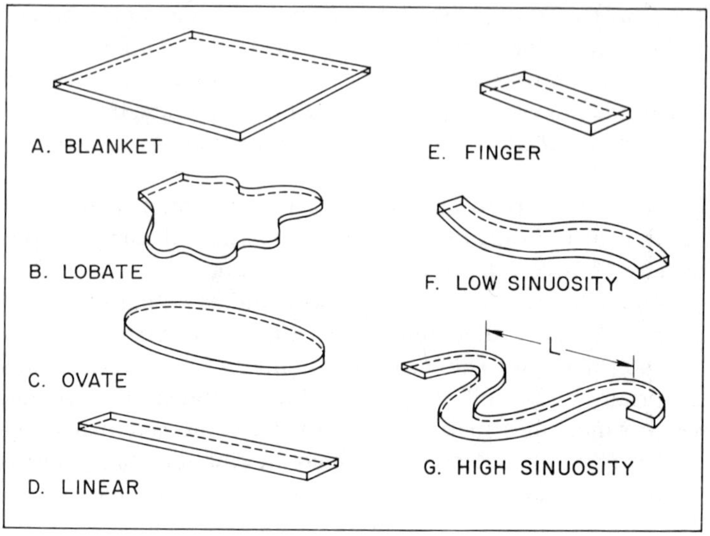

Figure 1–8 Diagrammatic morphologies for sandstone bodies. Relative dimensions are given in Table 1–4. For sinuous bodies, the actual length varies in relation to the direct length (*L*).

Finger refers to a narrow, short, linear-to-sinuous deposit that may be part of a distributary pattern of channels. The term *bar finger* has been used in this sense for distributary-mouth sands of the modern Mississippi delta (Fisk 1961), and its use has been restricted largely to this environment.

Sinuous refers to a local deposit in which the length is much greater than either the width or thickness but in which the width is generally only 100 times or more than that of the thickness. Furthermore, the body is bounded by curved lines that may be slightly wavy-to-serpentine in pattern. In other words, the body is narrow and "curvilinear." Fluvial, channel-fill sandstones have morphologies that may range from nearly straight to slightly sinuous (braided) and to highly sinuous (meandering).

Morphologies not only indicate depositional origin and the degree of internal homogeneity but also suggest trends for the long axis or length for elongate bodies. Trends in the direction of depositional dip are common for finger and sinuous sandstones, although they may be locally highly curved. In the same way, lobate bodies are more broadly curved outward in the direction of dip. Trends parallel to depositional strike are common for ovate and linear bodies. However, local trends may deviate greatly from the expected "dip" or "strike" for all sandstones.

Morphology alone, as may be mapped by interpretation of well logs, is not convincing evidence for depositional origin. For example, a broadly lobate morphology may characterize alluvial fans as well as deep-marine fans. Only the morphology is common to these bodies, but the reservoir properties can be strikingly different. The other primary properties of composition, texture, and sedimentary structures can confirm the depositional environment and, in turn, can make the interpretation of morphology more certain.

SECONDARY PROPERTIES

Porosity

Porosity of a rock is an important secondary property and is defined as the percent of void space, or nonmineral space, that occurs within or between the grains, matrix, and cement. In the subsurface, all voids are filled with fluids. Thus, porosity is a measure of the capacity of a reservoir rock to store oil, gas, and water.

Total porosity is the measure of all voids, both the separated and isolated spaces as well as the connected spaces. Effective porosity is the measure of those voids that are interconnected, and only effective porosity contributes to the flow of fluids. The amount of voids is readily determined by laboratory measurements. Porosity (ϕ) is expressed as percent of bulk volume, or

$$\phi = (V_p/V_b)\,100 = (V_b - V_m)100/V_b \qquad (1\text{–}1)$$

where V_p is pore volume, V_b is bulk volume, and V_m is mineral volume. Porosity may also be expressed as a fraction of bulk volume, in which case the symbol f is commonly used.

From Eq. (1–1) it is apparent that the measurement of porosity of a rock sample requires the determination of two values, either pore volume and bulk volume, or mineral volume and bulk volume. Because rocks are irregular in shape, bulk volume is best determined by measuring the amount of fluid that a sample displaces. The sample must be coated with paraffin so that the fluid cannot enter the pore space, or a fluid such as mercury can be used that does not penetrate the rock. Pore volume is measured directly by saturating a sample with a wetting fluid and determining the volume of fluid held in the rock by weighing the sample when dry and when saturated. In this case, only the effective porosity is determined. Mineral volume is measured by disaggregating or crushing the sample and determining the amount of fluid that the grains displace. In this case, the total porosity is accurately determined, but of course, the sample is destroyed. Mineral volume may be determined by weighing a dry specimen, and if the average mineral density can be estimated, then

$$V_m = W_m/\rho_m \qquad (1\text{–}2)$$

where W_m is the weight of the minerals and ρ_m is the average mineral density. The porosity is calculated by the equation

$$\phi = (V_b - W_m/\rho_m)100/V_b \qquad (1\text{–}3)$$

This method may be satisfactory for an essentially monomineralic sample, such as an orthoquartzite, but it is less reliable for rocks of variable composition such as those that contain a significant amount of clay and for which mineral density is not easily estimated.

Routine laboratory measurements of porosity commonly are made by placing a sample in a chamber of known volume, evacuating the chamber, and measuring the amount of air extracted. The difference in volumes of the chamber and the air removed is the mineral volume. This method determines the effective porosity and does not contaminate or disturb the sample. Details of porosity analysis are given in most textbooks of petroleum engineering (Amyx, Bass, and Whiting 1960, pp. 43–57).

The porosity of detrital rocks depends on grain size, shape, sorting, packing, and the amount of intergranular matrix or cement. Sandstones at shallow depths may have porosities of 30 to 40 percent, but sandstone porosities at greater depths may range from 10 to 25 percent. Orthoquartzites have the highest porosities, clayey sandstones have less porosity, and siltstones have the least. Because composition and grain size are interdependent, porosity may be largely a function of median grain size for most sandstones.

Although the porosity of detrital rocks depends on composition and grain size, packing is a fundamental factor. Unfortunately, there is no direct way to measure the degree of packing in clastic rocks because of the many variables in texture and composition that are possible in nature. An approach to the understanding of packing is possible, however, if some variables are held constant. For example, if grains are considered to be spheres of uniform size, then variations in shape, sorting, and composition can be eliminated, and only the relation of grains to each other remains as the most important factor. In addition, spherical grains may be thought of as assuming various geometrical arrangements in space, or systematic packings (Figure 1-3), and such idealizations permit a simple mathematical treatment of packing.

Cubic packing represents the loosest arrangement of tangent spheres, and *rhombohedral packing* represents the tightest (Figure 1–9). The volume of the unit void may be computed for each packing and is expressed as porosity or percent of the unit cell volume. For example, in cubic packing, the unit cell has a volume of $(2R)^3$ where R is the sphere radius. There are parts of eight spheres in the unit cell, each part has a volume of $\frac{1}{8}(4\pi R^3/3)$, and the total sphere volume is $4\pi R^3/3$. The volume of the unit void is the unit cell volume minus the total volume of spheres. Porosity (ϕ) is the volume of the unit voids divided by the volume of the unit cell:

$$\phi = (8R^3 - 4\pi R^3/3)100/8R^3 = (3.81R^3)100/8R^3 = 47.6\% \qquad (1\text{–}4)$$

Therefore, the theoretical maximum porosity for a well-sorted sandstone is about 48 percent.

For rhombohedral packing, the unit cell is an oblique prism that has a height equal to the layer spacing of $\sqrt{2}R$ and an area in the plane of the layers of $(2R)^2$. The volume of the unit cell is $4\sqrt{2}R^3$. The volume of spheres is the same as in cubic packing, and so porosity is

$$\phi = (4\sqrt{2}R^3 - 4\pi R^3/3)100/4\sqrt{2}R^3 = (1.47R^3)100/5.66R^3 = 25.9\% \qquad (1\text{-}5)$$

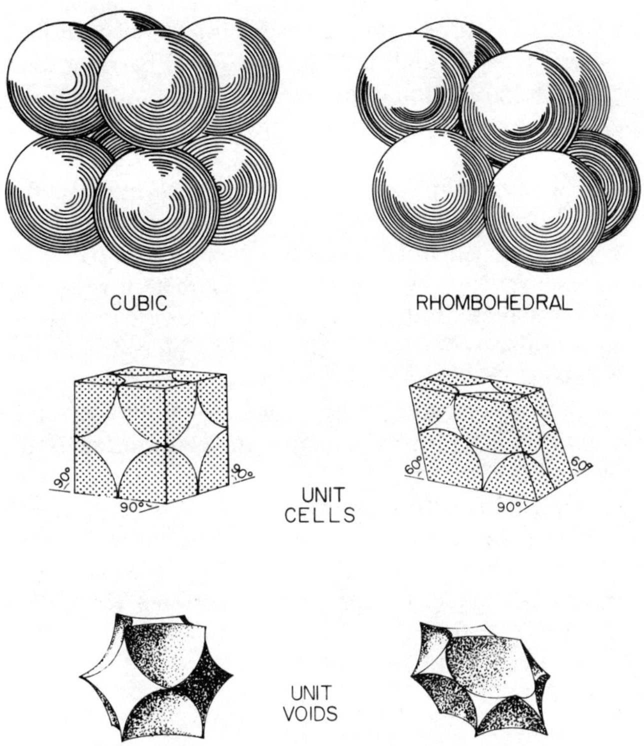

Figure 1–9 Diagrams of unit cells and unit voids for cubic and rhombohedral packings of uniform spheres.

Similar calculations are easily made for the other systematic packings. *Orthorhombic packings* have a porosity of 39.5 percent, and *tetragonal packings* have a porosity of 30.2 percent (Figure 1–3).

The porosity described so far refers to the pores between the grains, or *intergranular porosity.* There are two other types of porosity, however, that may be developed in sandstones of the subsurface. The first is *fracture porosity,* in which the amount of void space in the rock in increased by fractures that may remain open when the rock is subjected to extension stress. Such fractures also may be held open by mineral crystals deposited along the fracture planes after their initial formation. The second type of porosity is known as *secondary porosity* because the voids have been produced by leaching of cement or mineral grains after the rock has been buried at considerable depths. The interpretation of porosity of natural sandstones must consider the possible occurrence of these two types of porosity, both of which are largely independent of grain packing.

Most well-sorted sands have an initial porosity when deposited of about 40 percent, equivalent to the orthorhombic packing of spheres. Reduction of porosity takes place by the processes of diagenesis during burial. The term *diagenesis* is generally used to include all processes that take place in the sediment after deposition. Diagenesis results in the consolidation and lithification of the sediment to produce a competent, coherent rock. The processes of diagenesis result in an increase in bulk density and a corresponding decrease in porosity and permeability. The conditions of diagenesis extend to about 400°F (204°C) and to 30,000 psi (2000 bars) of total normal stress. Diagenetic processes are gradational to metamorphism, and, in fact, at greater depths of burial some low-grade metamorphic minerals begin to form in sedimentary rocks.

Diagenetic processes are of two types—physical and chemical. *Physical processes* can be thought of as compaction, or the making of a more dense or compact mass, largely by the effects of pressure. *Chemical processes* largely result in consolidation, or binding the sediment particles together to form a coherent rock.

Physical processes include an increase in degree of grain packing by mechanical adjustments among the grains (Table 1–5). Compaction also includes fracturing and shearing of soft or hard sediment. Increased packing often results in distortion of labile components such as grains of volcanic rock fragments within the sandstone, squeezing these particles until they deform to fill adjoining void spaces. Fracturing of individual grains may also take place but is not commonly observed, at least not until the sedimentary rock is buried to great depths.

Chemical processes in diagenesis include *cementation* by precipitation of minerals between grains. For example, silica may be deposited as overgrowths on quartz grains, resulting in successively thicker layers forming bonds between the grains

TABLE 1–5. COMMON DIAGENETIC CHANGES IN SANDSTONES

Process	Result	Depth	Effect
Compaction	Mechanical rearrangement of grains	Shallow	Reduces porosity from 40% to less than 30%
Cementation	Intergranular calcite, silica overgrowths in quartz	Shallow to moderate	Consolidation (lithification) and porosity reduction
Clay precipitation	Kaolinite, chlorite, and illite as grain coatings	Shallow to moderate	Slight decrease in porosity
Clay dehydration	Loss of interlayered water in expandable clays (smectite/illite)	Moderate	Slight increase in porosity
Dissolution	Leaching of calcite cement and feldspar grains	Moderate	Significant increase in porosity
Cementation	Intergranular ferroan calcite, zeolites	Moderate to deep	Significant decrease in porosity
Pressure solution	Styolitization by silica dissolution	Moderate to deep	Local decrease in porosity (along laminae)
Microfracturing	Breaking of grains	Deep	Slight increase in porosity

(Figure 1–1). Silica overgrowths tend to reduce pore size, but continuity between pores usually remains. Carbonate cements, generally calcite, may also be deposited between grains; but in this case the cement tends to fill pores, and continuity between pores is lost.

Another type of chemical diagenesis is the alteration of the mineral grains and matrix. Feldspar grains can alter in place to form clay minerals. In other cases, clay minerals such as kaolinite or chlorite may be precipitated from interstitial waters to form coatings on grains. At moderate depths of burial and under the influence of high temperatures, dehydration of some clay minerals may take place by the removal of water molecules that are loosely bound between the layers of the clay structure (Burst 1969).

A third type of chemical diagenesis is that of the *dissolution* of mineral particles. Stress concentration at points of grain contact may result in solution at these points and reprecipitation of the mineral in adjacent pores. Pressure solution of quartz grains was previously believed to be responsible for many silica overgrowths, but it may not be a common process until great depths of burial are reached, when stylolitic structures may be produced.

Another kind of dissolution is the leaching of certain mineral grains or cement to produce a secondary porosity. Secondary porosity by leaching has recently been recognized as an important process that enhances porosity in some deeply buried sandstones (Scholle and Schlueger 1979).

A common type of secondary porosity is produced by the removal of earlier deposited calcite cement. Calcite appears to be easily mobilized and dissolved by the flow of deep subsurface waters rich in carbon dioxide. The dissolved calcite may later be redeposited in other voids, or it may form a replacement of other mineral grains such as feldspar and quartz (Figure 1–1).

A well-known example of sandstone diagenesis is the Miocene section of South Louisiana (Atwater and Miller 1965; Gregory 1977). Porosity reduction appears to be a linear function of depth (Figure 1–10). Tight packing of 26 percent porosity is achieved by 12,000 ft (3658 m), and below this depth, mineral cements are found in increasing amounts. No significant fracturing or crushing of grains is observed, at least in samples as deep as 20,000 ft (6096 m). Porosity reduction in the Miocene may be accepted as a "normal" relationship in a compacting sequence of sediments because the sandstones are mostly near their maximum depth of burial and have not been affected by strong tectonic forces.

Shales in the Gulf Coast also show a characteristic compaction with depth (Figure 1–10). At shallow depths of less than 1000 ft (305 m), shales have bulk densities of less than 2.0 g/cm^3, but water is removed at a rapid rate by compaction, with only slight increases in burial depth and then more slowly with greater burial. At a depth of 12,000 ft (3658 m), bulk density has increased to about 2.4 g/cm^3. At many places, however, high fluid pressures are encountered, commonly at depths below 10,000 ft (3050 m), in which case, shale density is found to have decreased to minimum values on the order of 2.1 g/cm^3. This phenomenon is commonly attributed to clay dehydration at temperatures greater than 200°F (93°C) (Burst 1969).

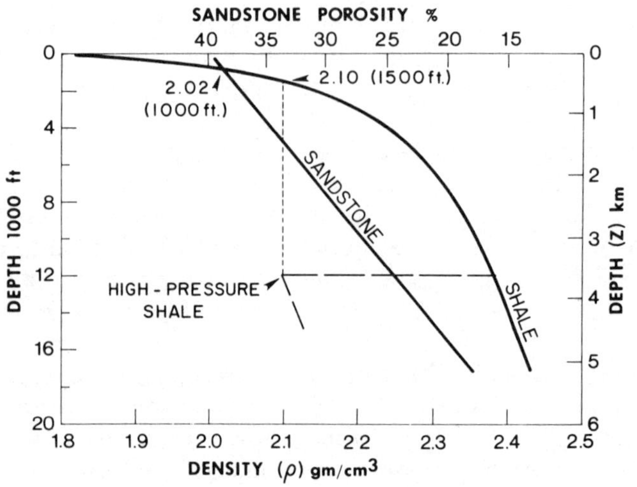

Figure 1–10 Sandstone and shale compaction in the Gulf Coast province. [Values from Gregory 1977.]

Sandstones that contain significant amounts of clay have compaction curves that are intermediate between the sandstone and shale functions (Figure 1–10). In fact, porosities of shaly sandstones may be reduced rapidly at shallow depths according to a curved function similar to that of shale. If secondary porosity by leaching has developed, the deeper parts of such curves may be nearly vertical, or even reversed. The importance of abnormally high porosity has recently been recognized in many deep sandstones of the Gulf Coast province. Leaching is believed to be responsible for much of the higher porosity at depth, whereas abnormal fluid pressures may also have prevented "normal" compaction and consolidation in some sections.

A striking example of secondary porosity is found in the Late Cretaceous Tuscaloosa sandstone of southern Louisiana (Figure 1–11). The Tuscaloosa produces gas at depths of 16,000 ft to nearly 21,000 ft (4877 to 6400 m) from sandstones that have porosities in the range of 20 percent to 30 percent. Nonreservoir sandstones at the same depths are cemented by calcite and have porosities of 8 percent to 10 percent, which are those expected by projection of porosities in shallower Cretaceous sandstones. The higher porosities in the gas reservoirs are the result of dissolution of calcite cement, probably by the updip flow of subsurface waters induced by excess pressures in the producing trend (Hudder and Tieh 1982). Thus, the secondary porosity by leaching can be significantly greater than that which would normally be anticipated for deep burial.

Permeability

Permeability is a measure of the capacity of a rock to transmit fluid. Permeability depends on the connected voids within the rock and, more importantly, on the size, shape, and arrangement of the connected pores. Because permeability depends on

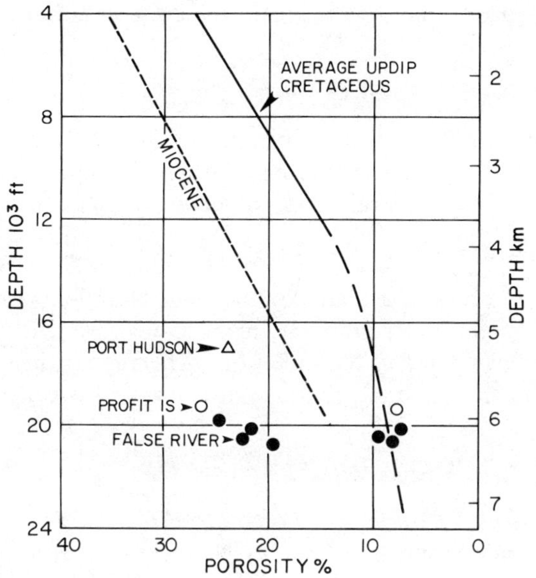

Figure 1–11 Secondary porosities in Late Cretaceous Tuscaloosa sandstones in southern Louisiana. Lower-porosity nonreservoir sandstones are tightly cemented with calcite. For comparison, the porosity-depth functions are shown for Miocene and updip Cretaceous sandstones.

the voids, it can be expected that permeability is a function of porosity, and therefore it is also dependent on the primary properties of the rock.

The first attempt to study the flow of water through sediment was made by Henri Darcy (1856), who wished to design a filtration system for the water supply of Dijon, France. His experiments established the fundamental relationship for fluid flow that applies to both sediments and rocks. A complete explanation of flow was not established until the definitive work of M. King Hubbert (1940), who placed the study of ground-water movement on a firm theoretical basis.

The Darcy experiment consisted of a vertical column of sand through which water was allowed to flow under the influence of gravity (Figure 1–12). He found that the rate of flow was proportional to the loss of head, as measured by manom-

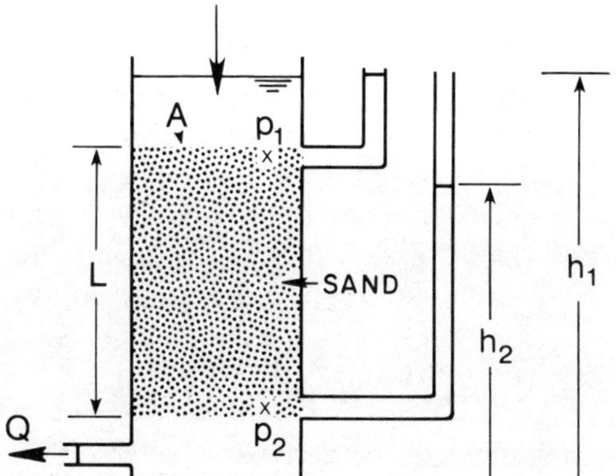

Figure 1–12 Diagram of apparatus used by Henri Darcy (1856) to investigate the flow of water in a column of sand. Symbols are defined in the text.

eter tubes, and inversely proportional to the length of the sand column. Expressed as an equation, this reads

$$Q = -KA\,(h_1 - h_2)/L \tag{1-6}$$

where Q is the volume rate of flow, K is a constant of proportionality, A is the cross-section area, and h_1 and h_2 are the heights of water in the manometers. Expressed as volume flux (v) and in general form, we have

$$v = Q/A = -K(dh/dL) \tag{1-7}$$

This is the equation often referred to as Darcy's Law. The equation states that rate of flow is proportional to hydrostatic gradient, or head loss per unit length. The minus sign indicates flow downward in the direction of decreasing head.

In the Darcy equation the constant of proportionality (K) is called *fluid conductivity* and is shown as

$$K = v(dL/dh) \tag{1-8}$$

K is also known as the *hydraulic conductivity* or the *coefficient of permeability* in ground-water studies. The measurement of K may be given as centimeters per second, cubic feet per day, or gallons per day per square foot, all of the measures being at a value of unit head loss per unit length. Thus, in hydrology the fluid conductivity of a sand may be expressed as gallons of water per day per square foot of sand area at a head loss of 1 foot of water per foot of horizontal distance and at a temperature of 68°F (20°C).

When fluids other than water are used, it is found that

$$K = k(\rho g/\mu) \tag{1-9}$$

where k is the specific conductivity for the porous medium through which a given fluid passes, and μ is the viscosity of the fluid. Thus, the Darcy equation becomes

$$v = k(\rho g/\mu)\,(dh/dL) \tag{1-10}$$

and

$$k = v(\mu/\rho g)\,(dL/dh) \tag{1-11}$$

where k is the permeability, a property of the sediment or rock. Permeabilty is measured in the laboratory on small samples of sediment or on cylindrical plugs cut from rock. The sample is placed in a closed chamber, and a pressure difference is applied to the sample. If the sample is in a horizontal position, then this is a special case of the Darcy relation, and the difference in pressure is equal to the difference in head. In other words, flow is now from high presure (p_1) to low pressure (p_2), or

$$k = (Q/A)\,(\mu)\,(dL/p_1 - p_2) \tag{1-12}$$

The dimensions of permeability can be derived by substitution of length (L), mass (M), and time (T) in Eq. (1-11), as follows:

$$k = (L^3T^{-1}L^{-2})\,(ML^{-1}T^{-1})\,(M^{-1}L^3)\,(L^{-1}T^2)\,(LL^{-1}) \tag{1-13}$$

which reduces to

$$k = L^2 \tag{1-14}$$

Thus, the dimensions of permeability are length squared or area (Hubbert 1940, p. 918). In a theoretical explanation for permeability, Hubbert (1940) showed that L could be any characteristic length for a specific porous medium such as grain size. It may be inferred that the area (L^2) represents a pore area that is a function of grain size for any given packing of well-sorted grains. Thus, permeability is most probably a function of grain size, sorting and packing.

In engineering practice the adopted unit of permeability is the darcy, which is defined as

$$1 \text{ darcy} = (Q/A)\,(\mu)\,(dL/dp)\,(1.0133 \times 10^6) \tag{1-15}$$

where

Q/A = volume flow per unit area, $cm^3/sec\ cm^2$
μ = viscosity in centipoise, dynes/cm
dp/dL = pressure gradient along unit length, atm/cm
1.0133×10^6 = conversion factor, $dynes/cm^2$ atm

In this form of the permeability equation, the pressure term (dp) is substituted for the product of head change (dh) and specific weight of the fluid (ρg) in Eq. (1–10). Unconsolidated sediments may have permeabilities of several tens or hundreds of darcys, whereas rocks commonly have permeabilities much less than a darcy, and so the smaller unit millidarcy, or 10^{-3} darcy, is commonly used.

The definition of permeability, Eq. (1–15), contains nonconsistent units since it includes atmosphere as a measure of pressure rather than as a fundamental unit. Accordingly, a more rational unit has been adopted (Lowman et al. 1972, p. 9). In the mks system, the standard dimensions of permeability are m^2, and the permeability unit is defined as

$$1 \text{ unit} = 10^{-12}\text{m}^2 = (\mu\text{m})^2$$

where (μm) is micrometer or 10^{-6} m, so the permeability values are now expressed in fundamental units. Comparison of values is facilitated by the fact that

$$1\ (\mu\text{m})^2 = 1.01 \text{ darcy} = 1.01 \times 10^3 \text{ millidarcy (md)}$$

Therefore the two units are essentially compatible and do not require the use of conversion factors. Permeabilities are commonly expressed in millidarcys (md), and this unit will not be easily displaced from the literature. Its continued use should cause no problem in understanding.

The Darcy equation as defined by the experiment in Eq. (1–6), or as used in measurement in Eq. (1–15), is based on two conditions: (1) that a single fluid is flowing, and (2) that the flow is laminar, or viscous, and not turbulent. Laboratory measurements are commonly made by the flow of air through samples. In this case, the sample is dry and saturated with air, and the applied pressure gradient is such that laminar flow is maintained. The measurement is called *absolute permeability* when only one fluid is present.

If two or more fluids are present, such as gas, oil, and water, the permeability is reduced from that of a single saturating fluid, and the measurement is called *effective permeability.* The flowing fluid must be specified. The amount of reduction in permeability depends directly on the amount of the fluid in the rock; effective permeability is lower as the amount of fluid is smaller. The reduced permeability is often expressed as the *relative permeability,* which is the ratio of effective permeability to the absolute permeability for a given fluid.

The air or gas permeabilities given in routine measurements are essentially those for water permeabilities because the viscosity difference between fluids has been accounted for in Eq. (1–15). The exception is for rocks of lower permeabilities, say 10 md or less, in which case, gas flow is somewhat greater than water flow, presumably because of a "slippage" effect first detected by Klinkenberg (1941). Permeabilities in a subsurface reservoir may be lower than those measured in the laboratory, primarily because the rock is compressed by overburden pressure.

Saturation

The amount of fluid in a rock is called the *saturation* of that fluid. The degree of saturation is expressed as a percentage of pore space. For example, 100 percent water saturation means that all of the pore space is occupied by water. This is the case for most rock volumes in the subsurface. In reservoirs, however, the rock contains other fluids (oil and gas) in variable amounts, although water is still present. Water adheres to the grain surfaces in most cases because the grains were originally wet, either because they were deposited in an aqueous environment or because they became water-saturated soon after deposition and burial. When the pores were later invaded by migrating oil or gas, the petroleum fluids did not wet the grain surfaces such as those of quartz, feldspar, and rock fragments. Only later, some types of heavier oil displaced the water from the mineral surfaces, probably under the influence of time and temperature. One mineral so affected is calcite, and many calcareous rocks tend to have a significant degree of oil wetness. Most sandstones, however, can be considered to be dominantly water-wet.

Even with large amounts of oil or gas, water remains as a thin film on grains and around the points of grain contact. When oil saturation (S_o) has a high value of, say, 80 percent, then water saturation (S_w) is still 20 percent of the pore volume. Most of this water is held at grain contacts. It is not surprising, then, that for clean, well-sorted oil-bearing sandstones, water saturation is inversely proportional to porosity. The reason for this can be explained by the systematic packings of uniform spheres where the number of contacts per unit sphere increases with tighter packing and reduced porosity. Hence, more water can be held at grain contacts in tight packings.

The configurations of oil (or gas) saturation in water-wet, well-sorted sandstones can be visualized by considering pore sizes and shapes in systematic packings of uniform spheres. In rhombohedral packings of 26 percent porosity, the pore radii can be easily calculated. The unit cell contains one net pore of diameter 0.414*D* and two pores of diameter 0.225*D* (Figure 1–13), where *D* is the sphere diameter. The narrow connections, or pore throats, between the pores have diameters of 0.154*D*. If a stringer of oil occupies a succession of rhombohedral unit cells, the surface of the oil assumes different spherical curvatures that represent the configurations for minimum interfacial pressure between the oil and water (Berg 1975, p. 943). The stringer passes alternately through pores of diameter 0.414*D* and 0.225*D*, each of which is connected by a pore throat of diameter 0.154*D*. In a well-sorted sandstone with high oil saturation, all of the pores can be imagined to be filled by such stringers connected through pore throats to form a three-dimensional network of oil.

Core Analysis

When a core is recovered from the subsurface, measurements of porosity, permeability, and fluid saturations are made on small samples, or *plugs,* taken from the rock at intervals of 1 ft (0.3 m). The laboratory procedures are designed so that the plug is not destroyed but remains intact. For example, the fluid content is determined by fractional distillation of the fluids so that the relative amounts of oil and water in the plug are determined; and the porosity and permeability are measured by using air as the saturating or flowing fluid (Amyx, Bass, and Whiting 1960). Thus, the plug is preserved for further study.

A typical core analysis does not show a total of 100 percent fluid saturation (Figure 1–14). The reason for this, of course, is that the rock has been removed from its subsurface environment of higher pressure and temperature. The greatest loss is in the more volatile components of petroleum, which bleed into the drilling mud and into the atmosphere as the core is brought to the surface. Furthermore,

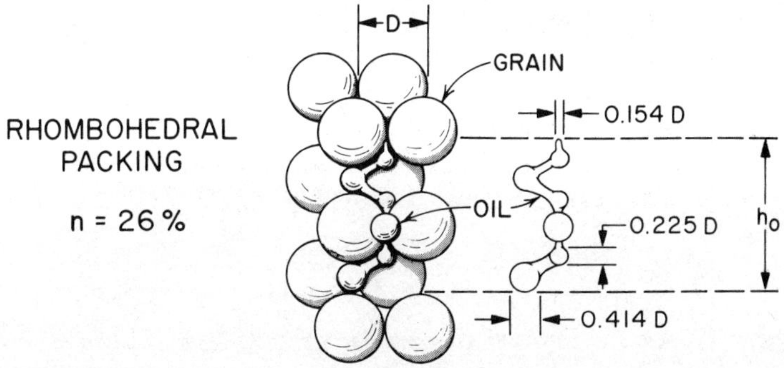

Figure 1–13 Diagram of an oil stringer connected through pores in a rhombohedral packing of uniform spherical grains where *D* is the grain diameter and the porosity is 26 percent. [From Berg 1975, p. 943.]

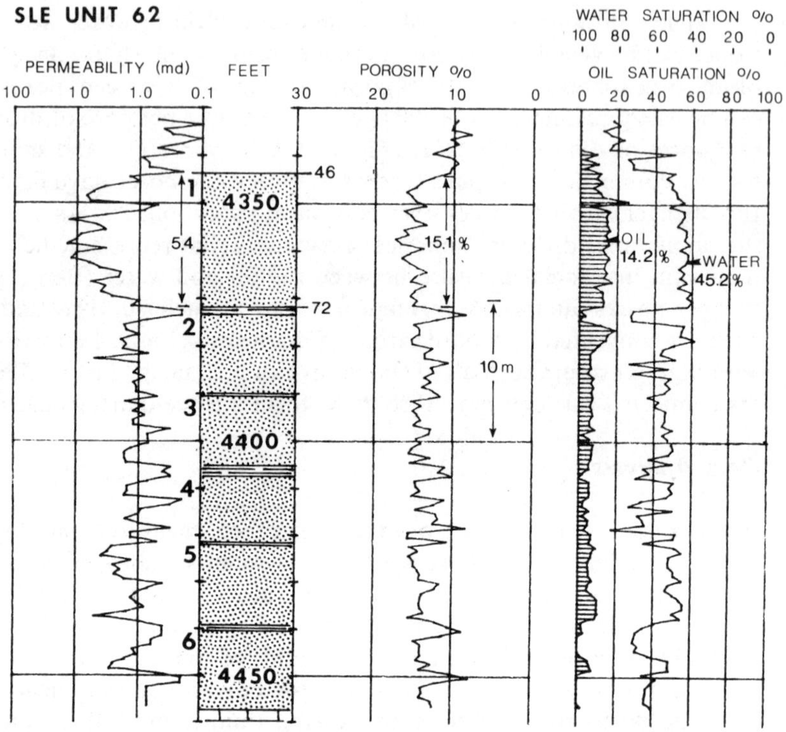

Figure 1–14 Permeabilities, porosities, and fluid saturations in a core of Pennsylvanian sandstone, Oklahoma.

the drilling mud may invade the rock pores and displace some of the original fluids, both water and oil, contaminating the core with mud. The result is that the fluid content is highly altered. It may be assumed, as a first approximation, that the remaining fluid saturations represent the irreducible or "immovable" fluids. Thus, in the example core analysis, the upper sandstone (4346–4372 ft) has an average oil saturation of 14 percent and a water saturation of 45 percent (Figure 1–14). It may be estimated, then, that the movable, or producible, oil in the reservoir amounts to the balance of fluid content or 41 percent of pore volume.

From core analysis measurements, an economic evaluation of the reservoir can be made. The total amount of movable oil, O_m, can be calculated on a per-acre basis as follows:

$$O_m = 7758\,\phi\,(1 - S_o - S_w)h \qquad (1\text{–}16)$$

where 7758 is the number of barrels per acre-ft, ϕ is porosity as a fraction, S_o is oil saturation, and S_w is water saturation from the core analysis, both as fractions, and h is the net thickness of the reservoir in feet. For the example core analysis, the upper sandstone (4346–4372 ft) can be estimated to contain the following amount of movable oil:

$$O_m = 7758\ (0.151)\ (1 - 0.142 - 0.452)\ (26) = 12{,}366 \text{ bbls/acre}$$

or a total of

$$O_m = 494{,}636 \text{ bbls/40 acres}$$

if it is assumed that the well could drain all of the movable oil from a surrounding area of 40 acres. In fact, a nearby well having essentially the same section of sandstone has produced only 145,251 barrels during a period of 17 years for a recovery of only 29 percent of the movable oil originally in the reservoir. The producing well now yields mostly water, presumably because the remaining oil is no longer in continuous connection between the pores but is in the form of isolated globules within the pores. Therefore, the well is at its economic limit for production of oil by simple pumping. The remaining oil, or 71 percent of the estimated movable oil, must be recovered by other methods such as injecting water into the reservoir in nearby wells and attempting to flush the additional oil toward the producing well. Eventually, more expensive methods may be attempted such as injecting chemicals to form an emulsion with the oil, which then may be recovered by the producing well.

The example core analysis (Figure 1–14) serves to illustrate the practical definition of the term *reservoir,* that is, a rock from which oil or gas can be produced at a profit. By this definition, only the upper sandstone (4346–4372 ft) was a reservoir for the economic conditions under which the well was originally drilled and produced. The remaining section (4372–4450 ft) contained a water saturation of more than 50 percent, a residual oil saturation of less than 10 percent, and presumably, a significant amount of "movable" oil as the balance of fluid content. Yet any attempt to pump fluid from the lower section would most likely result in an excessively large amount of water and little or no oil, and hence the lower section is not now a reservoir. However, the section may be thought of as a potential reservoir, one that might yield profitable amounts of oil by advanced technology under improved economic conditions in the future.

The example reservoir (Figure 1–14) is not exceptional but probably represents an average condition for oil recovery that has occurred in thousands of older producing wells in the United States. The success of future oil recovery from these wells will depend not only on sound engineering practice but also on more accurate predictions of reservoir properties, especially morphology, from geologic interpretations of cores.

TERTIARY PROPERTIES

Well logs record the physical properties of the rocks in place, including their electrical, radioactive, and acoustic characteristics. These properties depend not only on rock type but also on the amount or type of fluid they contain. Hence, the properties recorded or inferred by logging are tertiary properties in that they are largely dependent on both the primary and secondary properties. The tertiary properties may be thought of as *latent* properties because they are not themselves apparent but require the application of electrical or mechanical energy for their detection and measurement. Most of the logs are more accurately described as *geophysical logs* because of the nature of the measurements, but they are referred to here by the broad term *well logs.*

The most common log is the electric log, which records resistivity and spontaneous potential of the drilled section. Resistivity is measured by lowering electrodes in an instrument package, or *sonde,* into the borehole on a wire line. The sonde contains current electrodes and potential electrodes, and a continuous record or log is made of measurements as the sonde is withdrawn from the bottom of the hole. Other latent properties of the rock section are measured in a similar way by appropriate instruments. The logs so obtained represent the most complete record of nearly the entire section penetrated by the drill.

The purpose of this section is to review briefly the common log types, their qualitative interpretation, and typical response in sandstone sequences.

Types of Well Logs

The common borehole logs can be classified according to the type of log, the property measured, and the interpretation made from the measurements (Table 1–6). The first class of logs records electrical properties. The spontaneous potential (SP) log records the flow of ions between the fluid in the borehole and the more highly saline water in the formation. The effect is similar to that encountered in an electrochemical cell in which there is an exchange of ions between electrodes. In the case of the SP log, there is a natural current detected as a potential difference between an electrode in the borehole and the electrode of fixed potential at the surface. The resulting record of potential difference as a function of depth is called the *SP curve.*

The greatest potential difference is at the contact between shales and sandstones so that the shales indicate a positive potential relative to the sandstones. The potential difference at the bed contact immediately identifies the two rock types, and the log is useful to define the thickness of the sandstone beds. Thus, the interpretation is that of more porous and permeable sandstone, as opposed to the less porous and less permeable shale. In addition to its use for bed definition, the SP curve permits a calculation of resistivity of formation water. The SP curve is, therefore, an important aid to the quantitative interpretation of well logs.

Resistivity of formations may be measured by one or more types of logs. The standard electric log employs one current electrode and two measuring electrodes in the sonde with a second current electrode at a long distance from the first, either at the surface or on the logging cable itself. Within a homogeneous bed, a current of constant intensity flows through the earth from the downhole current electrode to the distant electrode, and the potential drop is measured between the remaining two measuring electrodes. The distance between the downhole current electrode and the measuring electrodes, or spacing, determines the depth of penetration of current in the formation. A short spacing between the electrodes is called the *normal* spac-

TABLE 1–6. COMMON BOREHOLE LOGS AND MEASUREMENTS

Type	Name[a]	Measurement (units)	Interpretation	Remarks[b]
Electric logs	Spontaneous potential (SP)	SP (mV)	Porosity type (qualitative), formation-water salinity	Open hole
	Resistivity log	Resistivity (ohm-m)	Fluid saturation	Open hole $R_t > R_m$
	Microlog* (minilog)**	Microresistivity (ohm-m)	Permeability (qualitative)	Open hole
	Focused log (Laterolog*)	Resistivity (ohm-m) of thin zones	Fluid saturation	Open hole $R_t >> R_m$
	Induction log	Conductivity (mmhos/m) and Resistivity (ohm-m)	Fluid saturation	Open hole $R_t > R_m$
	Dip log (Dipmeter*)	Detailed resistivity (ohm-m)	Structural and stratigraphic dip (magnitude, direction)	Open hole
Radioactivity logs	Gamma-ray log	Total gamma intensity (γ/sec)	Lithology (shale and nonshale)	Open or cased hole
	Gamma-gamma log	Gamma backscatter (γ/sec, g/cm³)	Bulk density, porosity	Open or cased hole
	Neutron-gamma	Induced gamma radiation (γ/sec)	Porosity, hydrogen density	Open or cased hole
	Neutron-neutron	Neutron capture (n/sec)	Porosity, hydrogen density	Open or cased hole
Velocity logs	Sonic (Acoustic**)	Interval transit time (μsec/ft)	Porosity	Open hole
Mechanical logs	Caliper	Hole diameter (inches)	Correction of other log measurement	Open hole

[a]Some log names are registered trademarks: *mark of Schlumberger, Ltd.; **mark of Dresser Industries, Inc.

[b]R_t = "true" resistivity of formation; R_m = resistivity of drilling mud. Cased hole measurements may be corrected to give approximate or relative rock measurements.

ing, and the curves recorded by this arrangement generally have shallow penetration that is approximately equal to about two times the electrode spacings. Thus, the shorter-spaced resistivity measurements are highly affected by any mud filtrate that invades the formation from the borehole. Longer spacing between electrodes is called *lateral* spacing, and the curves recorded have deeper penetration that is approximately equal to the electrode spacing. Commonly, resistivity measurements represent three different electrode spacings from short, intermediate, to long, which are, respectively, about 1 ft, 5 ft, and 18 ft.

The usefulness of three spacings becomes apparent when the resistivities of the mud filtrate and formation fluid are considered. The mud is generally composed of fresh water and suspended particles, whereas the formation water is generally more highly saline and less resistant to current flow. Consequently, the short-spaced electrodes measure a higher resistivity than the long-spaced electrodes when oil or gas is present in the formation beyond the zone invaded by mud filtrate. Thus, the electric log detects the presence of highly resistive fluid, either oil or gas, and in fact, the amount of water saturation may be calculated by the use of an empirical equation (Archie 1950).

Other types of resistivity measurement also reveal fluid saturations. For example, the focused log employs a double electrode system whereby the flow of current is forced to penetrate narrow porous zones within the formation by means of a "bucking" or focused current. This type of log was initially developed for use in more highly resistive formations, such as carbonate rocks, in which the resistivity of the formation (R_t) is commonly much greater than the resistivity of the mud (R_m). The log permits an accurate measurement of resistivity in thin zones.

The modern electric log is commonly the *induction log,* so called because the sonde develops a magnetic field within the formation, which in turn induces a current flow. The intensity of the induced current is measured and reflects the conductivity of the formation, which, in turn, is inverted to a resistivity measurement. This log gives accurate resistivity measurements in many sandstones in which the formation resistivity (R_t) is less than that of the mud resistivity (R_m). Again, the resistivity measurements allow a calculation of fluid saturation.

Another resistivity device is that of the *Microlog,* which is an older log developed with the specific purpose of detecting porous and permeable zones. The sonde employs closely spaced measuring electrodes that are located only 1 in. (2.5 cm) and 2 in. (5.0 cm) from the current electrode. The shorter spacing measures the resistivity of the mud cake (R_m), the solid material accumulated at the wall of the borehole, by infiltration of the mud solution into the formation. The longer spacing measures the resistivity of the formation close to the borehole where it has been invaded or flushed of original fluids by mud filtrate. The higher resistivity of the longer spacing denotes a more porous and permeable section such as a sandstone as compared with the much lower permeability of an adjacent shale, which would not be invaded by mud filtrate. Thus, separation of the two resistivity curves permits qualitative interpretation of a porous and permeable bed and an estimate of porous-sand thickness.

Dip logs represent a special class of resistivity logs in which the formation resistivity is measured by electrodes mounted on arms pressed closely against the wall of the borehole. Four resistivity curves are recorded, and small-scale variations in resistivity caused by single beds may be correlated. The geographic orientation of a reference electrode is determined magnetically so that the resistivity variations are properly oriented in space. Therefore, the resistivity variations of single dipping beds are displaced vertically, and the magnitude of dip and its direction can be calculated. For the purpose of dip calculation, the size of the borehole and its deviation from the vertical are also accounted for. The information on dip can be used for stratigraphic as well as structural interpretations.

A major class of logs depends on natural or induced radioactivity of the rock (Table 1–6). The *gamma-ray log* records natural gamma radiation. The most abundant radioactive isotope is K^{40}, which occurs in potassium-bearing minerals and is especially abundant in clay minerals. Therefore, the natural gamma-ray log distinguishes shale beds from nonshale beds by recording a high gamma radiation. Another type of radioactivity log is the *gamma-gamma log* in which the radiation is induced. The rock is bombarded by gamma rays, and the amount of backscatter is recorded. Because the more dense atoms resist the bombardment, the backscatter is greater. Therefore, the amount of backscatter is directly related to the bulk density of the rock and to the porosity.

Another type of log is the *neutron-gamma* in which radioactivity by neutron bombardment induces gamma radiation from the heavier atoms. In this reaction, hydrogen ions absorb the neutron particles, and reduced gamma radiation indicates the relative abundance of hydrogen, which may exist largely in the fluids of pores. Therefore, the induced gamma radiation is inversely proportional to the porosity of the rock. A somewhat similar device, the *neutron-neutron log,* measures neutron capture within the formation, which again is proportional to the hydrogen density and therefore to the porosity or bulk density of the rock. One of the principal advantages of the radioactivity logs is that they may be run in either open or cased holes, and furthermore these logs are relatively unaffected by the type of fluid in the hole.

The *sonic* or *Acoustic log* employs a sonde that generates a compressional sound wave that travels to the wall of the borehole, along which it propagates and generates, in turn, a reflected wave. The reflected wave is received by the sonde, and the log records the total travel time of the compressional wave from the transmitter to the detector. The travel time is inversely proportional to the density of the rock and therefore directly proportional to the porosity. The sonic log records the travel time that can be correlated empirically with porosity under most subsurface conditions. However, calculated porosity is unreliable in unconsolidated formations at shallow depths and under certan other conditions such as high pressure in the formation.

The *caliper log* measures the size of the hole by means of mechanical spring-loaded arms that press against the wall of the borehole. The borehole diameter varies depending on rock type, and in most sections the shales tend to wash out, whereas sandstones tend to maintain a more uniform hole size, only slightly enlarged. Knowledge of hole diameter may be used to correct other log readings, especially those that are greatly influenced by the volume of the mud between the instrument and the rock itself.

Logs run on older wells commonly have only the SP and standard resistivity curves because these were the original logging devices. Later the Microlog was developed and widely used because it was the first log to indicate in a reliable manner the presence of porous and permeable beds, although in a qualitative manner. Still later, as other logs were developed, most boreholes have been logged with a combination of curves. An initial logging run commonly includes the SP curve and a resistivity curve of a type best suited to provide reliable values for calculations of fluid saturation. A second logging run may record a natural gamma-ray log with one or more radiation logs that give values of both density and porosity. Less commonly, a third logging run is made to record a dip log from which valuable information may be obtained for structural and stratigraphic interpretation. Difficulties in maintaining an open hole without the drill string, or simply the time and costs involved, often prevent the recording of dip logs.

A caliper log may be recorded simultaneously with other logs such as radioactivity, sonic, or dip logs. Therefore, logs recorded on multiple runs often provide for accurate calculations of fluid saturations, porosity, and other valuable subsurface data.

Log Examples

Modern well logs and their qualitative interpretation are illustrated by logs through a section of sandstone and shale in the Gulf of Mexico, offshore Louisiana (Figure 1–15). The SP curve is recorded to the left of the depth column. Maximum departure to the right (positive) identifies shale, and response to the left (negative) denotes sandstones. Two thicker sandstones are apparent on the log at 6901–6932 ft and at 6947–6992 ft as indicated by the dot pattern in the depth column.

Resistivity is recorded in the right-hand track nearest the depth column, and the scale is logarithmic. The shallow focused log (SFL) (Figure 1–15) records resistivity of the sandstones invaded by mud filtrate. The induction log (IL) records resistivity beyond the invaded zone. In the upper sandstone, resistivity on the induction log is higher than that of the shallow focused log, which denotes a high resistivity fluid, either oil or gas, beyond the invaded zone. In the lower sandstone a similar response in the upper half also denotes oil or gas saturation. However, in the lower half of the sandstone the induction resistivity is less than that measured by the short focused log, which indicates a saturation of saline water beyond the invaded zone.

In the far right-hand track is recorded the interval transit time from the sonic log. Increased transit time to the left indicates porosity in the sandstones greater than that in the shales.

The same section was also logged with radioactivity tools (Figure 1–16). The gamma-ray curve (GR) shows higher readings in the shales (right) and lower readings in the sandstone (left). A somewhat greater gamma-ray reading in the lower sandstone suggests that it has a greater shale content than the upper sandstone.

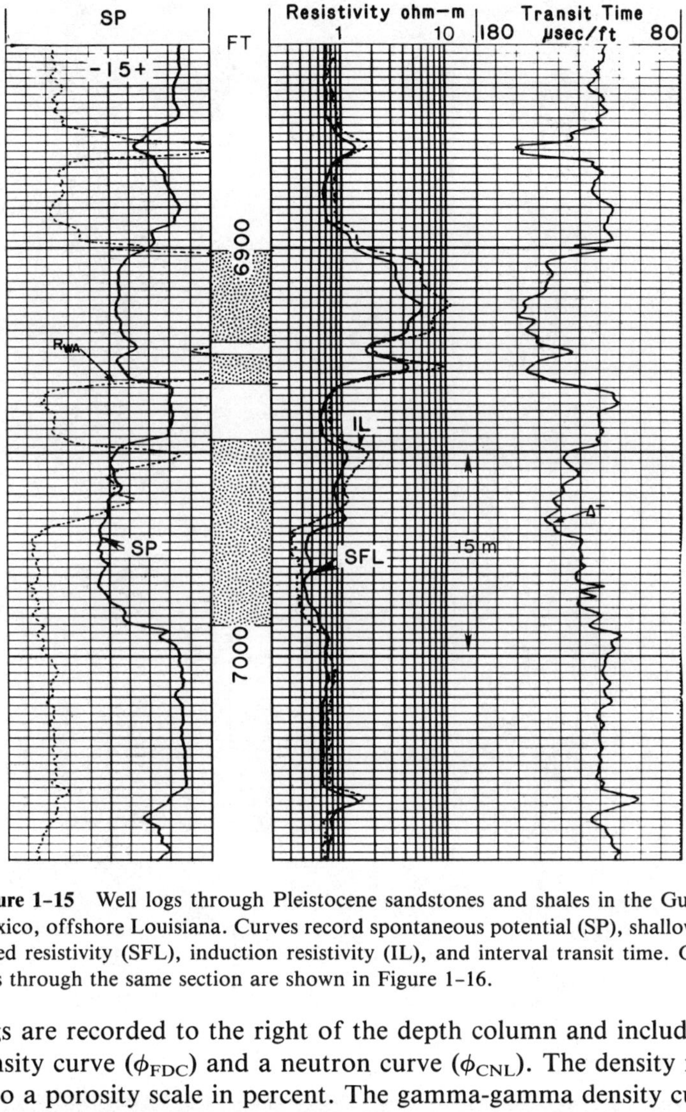

Figure 1–15 Well logs through Pleistocene sandstones and shales in the Gulf of Mexico, offshore Louisiana. Curves record spontaneous potential (SP), shallow focused resistivity (SFL), induction resistivity (IL), and interval transit time. Other logs through the same section are shown in Figure 1–16.

Density logs are recorded to the right of the depth column and include a gamma-gamma density curve (ϕ_{FDC}) and a neutron curve (ϕ_{CNL}). The density readings are converted to a porosity scale in percent. The gamma-gamma density curve records an average of 30 percent porosity in the upper sandstone and somewhat more variable porosity in the lower sandstone that ranges from about 27 percent to 33 percent. The readings from the neutron log are highly variable and are, in fact, not correct. When the formation fluid has low density, such as gas, the neutron response is spurious and indicates low porosity. This effect is seen in the upper sandstone where the neutron density and gamma-gamma density curves essentially overlap, thereby indicating that a significant saturation of gas is present in the formation.

The reservoir sandstones described in the following chapters are illustrated by

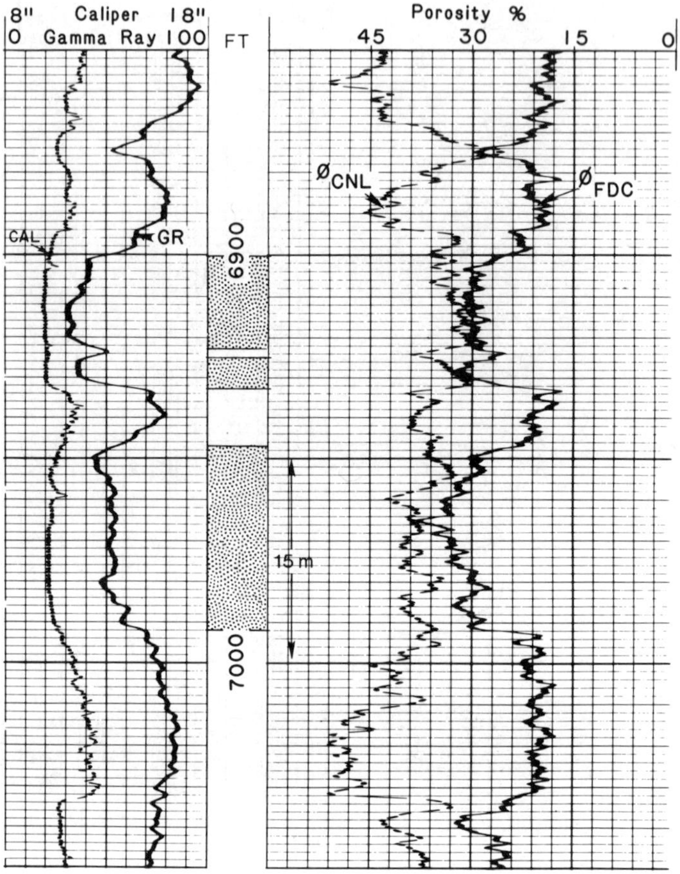

Figure 1–16 Well logs through Pleistocene sandstone and shales in the Gulf of Mexico, offshore Louisiana. Curves record natural gamma radiation (GR), hole size (caliper), porosity from the gamma-gamma density log (ϕ_{FDC}), and porosity from the neutron-density log (ϕ_{CNL}). Other logs through the same section are shown in Figure 1–15.

well logs of various types, both older logs as well as modern logs, depending on when the wells were drilled. Most logs reflect the general character of the cored sections, but no attempt is made to interpret the logs other than to illustrate their response for each sandstone type.

The interpretation of well logs has advanced to a state of great technical achievement. Quantitative methods are completely described in many industrial publications (Dresser Atlas 1974 and Schlumberger 1972, 1974) as well as in textbooks (Asquith and Gibson 1982; Lynch 1962; and Pirson 1963). The reader is referred to these publications for more complete descriptions of well logs and their use.

SUMMARY

The primary, or definitive, sandstone properties are composition, texture, sedimentary structures, and morphology. These properties are determined, in large part, by the processes of transport and deposition. Therefore, observation and measurement of these properties are important to understanding a reservoir rock. Furthermore, these properties are interdependent, so that some knowledge of only one or two properties may result in estimations of others.

Sandstones are composed of particles that are largely the more resistant products of rock weathering—quartz, feldspar, rock fragments, and clays. The relative abundance of these components determines the rock name, whether an orthoquartzite, arkose, litharenite, or graywacke. Cements are chemical precipitates, commonly calcite or silica, which bind the particles into a consolidated mass. Texture is described by several elements, the most important of which is mean grain size, or median size, and an expression for variation about the mean, or standard deviation. Another important textural element is the packing or arrangement of grains. Fine-grained detritus between grains is termed the *matrix* and may consist mostly of clay minerals.

Primary sedimentary structures are the result of the original depositional processes, but these may be altered by physical or biological disturbance after deposition. Morphology, or external configuration, is highly variable and may range from widespread "blankets" to narrow sinuous bodies of limited extent.

The secondary, or dependent, rock properties are porosity, permeability, bulk density, and fluid saturation. These properties are determined largely by the primary properties and are, therefore, also related indirectly to the original conditions of deposition. However, the primary properties are altered by burial and compaction, thereby altering the secondary properties and usually resulting in a reduction in porosity and permeability. Postdepositional changes, called *diagenesis,* also include chemical changes in composition, which commonly result in the growth of new minerals, a reduction of pore space, and a decrease in permeability. In some cases, leaching of both grains and cement may take place in the deep subsurface and may result in improved porosity and permeability with increasing depth.

The tertiary, or latent, rock properties are those commonly measured in a borehole and include electrical resistivity, spontaneous potential, and natural or induced radiation, as well as other properties. Some of these measurements are influenced by fluid content of the rocks and are, therefore, difficult to interpret in terms of primary rock properties. Nevertheless, these measurements are highly useful in reservoir interpretation, and they yield information of great value, especially from measurements that may be closely correlated with the primary rock properties by means of a cored section.

REFERENCES CITED

AMYX, J. W., D. M. BASS, JR., and R. L. WHITING, 1960, *Petroleum reservoir engineering, physical properties:* New York, McGraw-Hill, 610 p.

ARCHIE, G. E., 1950, Introduction to petrophysics of reservoir rocks: American Association of Petroleum Geologists Bulletin, v. 34, p. 943–961.

ASQUITH, G. B., and C. R. GIBSON, 1982, *Basic well log analysis for geologists:* Tulsa, Oklahoma, American Association of Petroleum Geologists, 216 p.

ATWATER, G. I., and E. E. MILLER, 1965, Effect of decrease in porosity with depth on future development of oil and gas reserves in South Louisiana (abstract): American Association of Petroleum Geologists Bulletin, v. 49, p. 334.

BERG, R. R. 1970, Identification of sedimentary environments in reservoir sandstones: Gulf Coast Association of Geological Societies Transactions, v. 20, p. 137–143.

———, 1975, Capillary pressures in stratigraphic traps: American Association of Petroleum Geologists Bulletin, v. 59, no. 6, p. 939–956.

BLATT, HARVEY, 1967, Original characteristics of clastic quartz grains: Journal of Sedimentary Petrology, v. 37, no. 2, p. 401–424.

BURST, J. F., 1969, Diagenesis of Gulf Coast clayey sediments and its possible relation to petroleum migration: American Association of Petroleum Geologists Bulletin, v. 53, p. 73–93.

CAMPBELL, C. V., 1967, Lamina, laminaset, bed, and bedset: Sedimentology, v. 8, p. 7–26.

DARCY, HENRY, 1856, Determination des lois d'ecoulement de l'eau a travers le sable, *in* Hubbert, M. K., 1969, *The theory of ground water motion and related papers:* New York, Hafner Publishing Company, p. 303–311.

DAVIES, D. K., R. K. VESSELL, R. C. MILES, M. G. FOLEY, and S. B. BONIS, 1978, Fluvial transport and downstream sediment modification in an active volcanic region, *in* A. D. Miall, editor, *Fluvial sedimentology:* Canadian Society of Petroleum Geologists Memoir 5, p. 61–84.

DOTT, R. H., JR., 1964, Wacke, graywacke and matrix—what approach to immature sandstone classification?: Journal of Sedimentary Petrology, v. 34, p. 625–632.

———, and JOANNE BOURGEOIS, 1982, Hummocky stratification: significance of its variable bedding sequences: Geological Society of America Bulletin, v. 93, p. 636–680.

DRESSER ATLAS, 1974, *Log review 1:* Houston, Texas, Dresser Atlas Division, Dresser Industries, Inc., unpaged.

EHRLICH, R., and B. WEINBERG, 1970, An exact method for the characterization of grain shape: Journal of Sedimentary Petrology, v. 40, p. 205–212.

FISK, H. N., 1961, Bar-finger sands of Mississippi Delta, *in* J. S. Peterson, and J. C. Osmond, editors, *Geometry of sandstone bodies,* Tulsa, Oklahoma, American Association of Petroleum Geologists, p. 29–52.

FOLK, R. L., 1968, *Petrology of sedimentary rocks:* Austin, Texas, Hemphills, 170 p.

FRIEDMAN, G. M., 1958, Determination of seive size distribution from thin-section data for sedimentary petrological studies: Journal of Geology, v. 66, p. 394–416.

GRATON, L. C., and H. J. FRASER, 1935, Systematic packing of spheres with particular relation to porosity and permeability: Journal of Geology, v. 43, p. 785–909.

GREGORY, A. R., 1977, Aspects of rock physics from laboratory and log data that are important to seismic interpretation: American Association of Petroleum Geologists Memoir 26, p. 15–46.

GRIFFITHS, J. C., 1967, Scientific method in analysis of sediments: New York, McGraw-Hill, 508 p.

HARMS, J. C., J. B. SOUTHARD, D. R. SPEARING, and R. G. WALKER, 1975, *Depositional environments as interpreted from primary sedimentary structures and stratification sequences:* Tulsa, Oklahoma, Society of Economic Paleontologists and Mineralogists Short Course 2, 161 p.

HUBBERT, M. K., 1940, The theory of ground-water motion: Journal of Geology, v. 48, p. 785–944.

HUDDER, K. G., and T. T. TIEH, 1983, Diagenesis of deep Tuscaloosa sandstones, Profit Island field, Louisiana (abstract): Geological Society of America Abstracts and Programs, 17th Annual Meeting, South-Central Section, p. 36.

KLEIN, G. deV., 1963, Analysis and review of sandstone classifications in the North American geological literature, 1940–1960: Geological Society of America Bulletin, v. 74, p. 555–576.

KLINKENBERG, L. J., 1941, The permeability of porous media to liquids and gases: *Drilling and Production Practices,* Washington, D.C., American Petroleum Institute, p. 200–211.

KRYNINE, P. D., 1948, The megascopic study and field classification of sedimentary rocks: Journal of Geology, v. 56, p. 130–165.

LOWMAN, S. W. and OTHERS, 1972, Definitions of selected groundwater terms-revisions and conceptual refinements: U.S. Geological Survey Water-Supply Paper 1988, 21 p.

LYNCH, E. J., 1962, *Formation evaluation:* New York, Harper & Row 422 p.

MCKEE, E. D., and G. W. WEIR, 1953, Terminology for stratification and cross-stratification in sedimentary rocks: Geological Society of America Bulletin, v. 64, p. 381–390.

MOORE, D. G., and P. C. SCRUTON, 1957, Minor internal structures of some recent unconsolidated sediments: American Association of Petroleum Geologists Bulletin, v. 41, p. 2723–2751.

PETTIJOHN, F. J., 1954, Classification of sandstones: Journal of Geology, v. 62, p. 360–365.

———, P. E. POTTER, and RAYMOND SIEVER, 1973, *Sand and sandstones:* New York, Springer-Verlag, 618 p.

PIRSON, S. J., 1963, *Handbook of well log analysis:* Englewood Cliffs, N.J., Prentice-Hall, Inc., 326 p.

SCHLUMBERGER, 1972, *Log interpretation, Vol. 1—Principles:* New York, Schlumberger Ltd., 113 p.

———, 1974, *Log interpretation, Vol. 2—Applications:* New York, Schlumberger Ltd., 116 p.

SCHOLLE, P. A., and P. R. SCHLUEGER, editors, 1979, Aspects of diagenesis: Society of Economic Paleontologists and Mineralogists Special Publication 26, 443 p.

WENTWORTH, C. K., 1922, A scale of grade and class terms for clastic sediments: Journal of Geology, v. 31, p. 228–232.

2

Sedimentary Processes

Sediment is eroded, transported, and deposited primarily by the processes of water flow, wave action, and wind. The most important process geologically is current flow, which results in transport and deposition of well-sorted sands that may become reservoirs. The erosion and transport of sand can be understood by the analysis of forces that act on water to cause flow and on grains to cause movement. The deposition of sand can be explained by the effects of decreasing flow, which produce distinctive structures in the deposits. The results of current flow are preserved in the sandstones and are an aid in the interpretation of depositional environment and in predicting the morphology of sandstones. Therefore, the causes and effects of flow are emphasized here in order to explain some fundamental properties of reservoir sandstones.

CHANNEL FLOW

Flow in open channels can be described most simply by the relation between channel discharge (Q), the velocity of the flow (U), and its cross-sectional area (A).

$$Q = UA \tag{2-1}$$

The effect of flow on sediments is easily visualized. In natural channels, the total discharge increases at times of heavy rainfall, and there are corresponding increases in the velocity and in the area, or width and depth, of flow. As velocity increases, the force of the moving water is transmitted to the bottom, and sediment is eroded and transported. When discharge and velocity decrease, the force of moving water is no longer sufficient to carry sediment, and the sediment is deposited. Therefore, large amounts of fluvial sediments are periodically moved downstream, eventually reaching the coastal region.

Flow in open channels can be expressed quantitatively if several simplifying assumptions are made. If the flow is steady, uniform, and without acceleration, then the sum of forces acting on a unit mass of water is zero:

$$\Sigma F_i = dU/dt = 0 \tag{2-2}$$

where F_i are the individual forces, U is flow velocity, and t is time. The impelling force is gravity, and the impeding force is friction along the boundaries of the channel. Consider a channel segment of width (w), depth (d), and length (l) (Figure 2-1). The impelling force of gravity (F_g) acts on the unit mass along a component in the downslope direction:

$$F_g = \rho g A l \sin\theta \tag{2-3}$$

where ρg is the specific weight of the water, A is the cross-sectional area ($A = wd$), and θ is the angle of inclination of the water surface. The resisting force (F_d) is

$$F_d = \tau l P \tag{2-4}$$

where τ is the shearing stress per unit area and P is the wetted perimeter of the channel ($w + 2d$). If it is assumed that the two forces are balanced such that there is no acceleration, then

$$\tau l P = \rho g A l \sin\theta \tag{2-5}$$

and

$$\tau = \rho g (A/P) \sin\theta = \rho g R \sin\theta \tag{2-6}$$

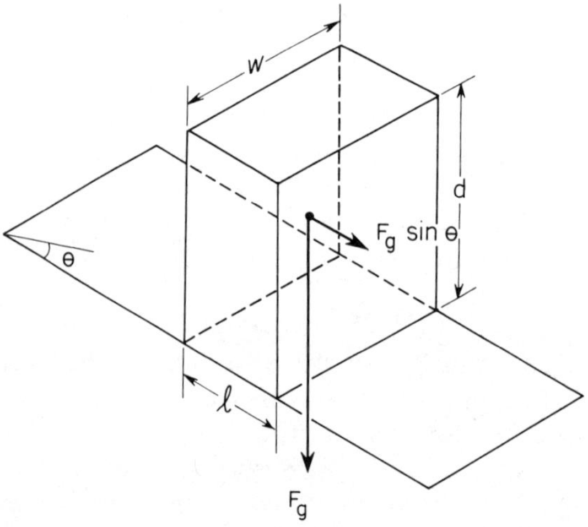

Figure 2-1 Dimensions of a water mass that has a width (w), a depth (d), and a length (ℓ), and whose base is inclined at an angle θ.

The term (A/P) is called the *hydraulic radius* (R), and $R = (wd)/(w + 2d)$, which has the dimension of length.

In the case of a large river, where width is much larger than depth, the value of R approaches that of depth. For a river in which flow is fully turbulent, the shearing stress can be assumed to be related to an average velocity (U), such that

$$\tau = \rho f U^2 \tag{2-7}$$

where f is a dimensionless friction factor. If this is true, then the two equations [Eqs. (2–6) and (2–7)] are equal:

$$\rho f U^2 = \rho g R \sin \theta \tag{2-8}$$

and solving for velocity,

$$U = (gR \sin \theta / f)^{1/2} \tag{2-9}$$

Because the angle θ is small, $\sin \theta$ is approximately equal to $\tan \theta$ or slope (S), and

$$U = (gRS/f)^{1/2} \tag{2-10}$$

This expression for velocity is called the *Chezy equation* for open-channel flow, and the factor $(g/f)^{1/2}$ is called *Chezy coefficient* (C), so that the Chezy equation is often expressed as

$$U = C(RS)^{1/2} \tag{2-11}$$

The friction factor (f) must be determined by experiment for individual channels. Nevertheless, the Chezy equation, Eq. (2–10), is widely applied in engineering studies and is useful in that it demonstrates that flow velocity is a function of slope and water depth as expressed by the hydraulic radius, and is inversely proportional to a friction factor. Erosion of bottom sediment may occur when velocity is increased, that is, when discharge, slope, and depth increase. Conversely, deposition may take place when velocity is reduced by a decrease in depth, a decrease in slope, or an increase in bottom friction. For example, when a river overflows its banks in times of flood, there is an abrupt decrease in depth for the overbank flow and an increase in bottom friction, and therefore deposition may take place laterally to a channel as in natural levees. Thus, the Chezy equation, Eq. (2–10), gives some insight as to the causes of water flow and channel processes.

EFFECTS OF FLOW

In order to understand sediment transport, a great amount of work has been expended on observations and analyses of channel flow. This work has been conducted largely by engineers (Raudkivi 1976), but important early contributions were made by geologists (Gilbert 1914; and Rubey 1933, 1938). The engineer has concentrated on relatively short-term and transient effects in channels; he has been interested in local scouring and fill, channel modification, bar formation, and bank failure. The geologist, on the other hand, is more interested in the long-term and more permanent effects of stream flow; he needs to know how sediment is eroded, transported for long distances, and deposited in thick sequences of rather uniformly sized sediment. The geologist wants to know the effects of these processes in order to extrapolate backward from the final sediment to the formative process.

An obvious relationship for study is the velocity of flow and its effect on bottom sediment, and a particularly important set of observations was compiled by Hjulström (1935, 1939). In this summary, mean current velocity was related to grain size of sediment that was eroded, transported, and deposited (Figure 2–2).

The relationships defined by Hjulström (Figure 2–2) were compiled from his own observations as well as from the experiments of others (Gilbert 1914; and Fortier and Scobey 1925). The velocities were estimated to be those at a height of 1 m above the bottom. These velocities are relatively unaffected by the friction of the stream bed, and therefore they reflect the full, undisturbed effect of water flow. Erosion velocities for fine-grained sediment refer to *aged* sediment, that is, sediment

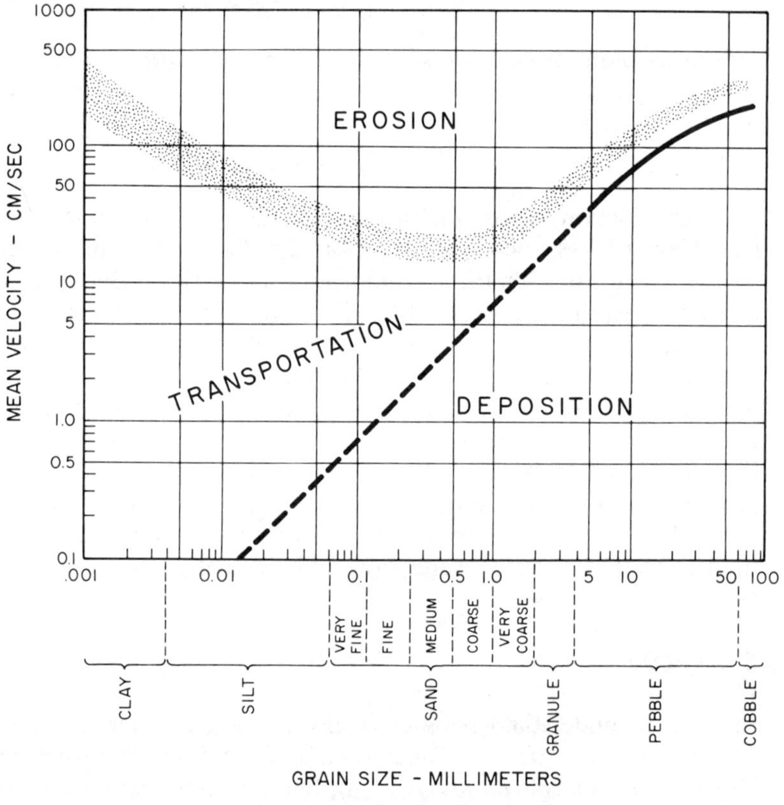

Figure 2–2 Mean current velocity required for erosion, transportation, and deposition of sediment as a function of grain size. Velocities measured 1 m above bottom. [From Hjulström 1935.]

that has been partly dried and has achieved a high degree of cohesion. Therefore, the relationships may be assumed to apply to natural fluvial channels. It has been shown that fine-grained unconsolidated sediment can be eroded at velocities less than those of the diagram (Figure 2–2), but newly deposited muds still require nearly the same velocities for erosion as do fine sand and silt (Sundborg 1956).

These observations showed that the sedimentary processes are highly selective. Relatively low velocities of about 30 cm/sec can erode and transport the entire range of sand sizes, leaving behind both coarser and finer materials. Further, the observations suggested that streams can maintain rather stable channels in fine-grained sediment; but once eroded, the finer particles will remain in suspended transport, even at low flow velocities. The sorting action of fluvial systems is well displayed because periodic fluctuations in flow can segregate and deposit the well-sorted sands, which may become aquifers and reservoirs after burial.

Definitions for the terms *erosion, transportation,* and *deposition* as these pertain to current velocity are necessary to understand the criteria used in compiling the diagram in Figure 2–2. The erosion field represents those velocities at which the grains are in continuous motion. The erosion field is bounded below by the initial erosion velocity, which is the lowest velocity at which grains of a given size first begin to move. This velocity is often called the *threshold velocity.* The difficulty in defining first movement of grains in flume experiments was recognized by Hjulström as a problem of bed roughness. He stated that erosion velocity is reached when uniform grains move over a bed of loose grains of the same size (Hjulström 1935, p. 294). Higher velocities may be required for movement of poorly sorted grains for which the bed is more irregular because of grain-size differences. Thus, the erosion velocity of the diagram (Figure 2–2) is a minimum velocity for well-sorted grains.

The transportation field represents velocities at which grains already in motion continue to move. In other words, higher velocities are required to erode grains at rest, but once moved, the loose grains continue to be moved by lower velocities. The reason for this difference is that a grain on the bed exposes only part of its surface area to flow, but after erosion, the entire grain surface is exposed to the force of flowing water. The transportation field is bounded below by the lowest velocity for loose grain movement and that at which deposition occurs. Furthermore, the transportation field is one of equilibrium in which there is neither net erosion nor deposition of grains (Hjulström 1935, p. 327).

The deposition field includes those velocities at which there is no movement of grains. It is bounded above by the lowest transport velocity, but this boundary was uncertain for the available observations. Hjulström (1935, p. 320) stated that the lowest transport velocity was extrapolated for grains finer than 5 mm, and he viewed this minimum velocity as different from the settling velocity of grains in still water.

Erosion velocity is least for sand-size sediment and has a minimum of about 15 cm/sec for sand of 0.5 mm mean size. Greater velocities are required for erosion of larger grain sizes, but surprisingly, equally large velocities are needed to erode very fine-grained sediment. Obviously, erosion velocity may not be a simple function of particle size or mass, and other sediment properties such as sorting and packing may determine the velocity for erosion of fine sediment. Erosion velocity is not a single value for a given grain size but is represented by a range of values. The reason for this lies partly in the uncertainty of observations and partly in the difficulty of determining exactly when a large amount of sediment is being actively eroded rather than merely being involved in transient movement.

Transport velocity is great for larger grain sizes and appears to decrease regularly with decrease in grain size. Again, there is some uncertainty in the velocities for medium and finer-grained sediment.

The above observations permit the definition of velocity fields for erosion, transportation, and deposition of sediment as a function of mean grain size. More precise definitions of these phenomena must be based on the physical relations between the grains and moving water.

GRAIN AND FLOW INTERACTION

First consider the forces that act on a spherical grain immersed in water (Figure 2–3). The sum of the forces gives the grain an acceleration, or

$$\Sigma F_i/m = (dV/dt) \tag{2–12}$$

where F_i are the individual forces, m is mass, V is velocity of the grain, and t is time. If the water velocity (U) is zero, the grain falls vertically through the fluid at an increasing rate until a certain velocity is reached, acceleration ceases, and the rate of fall is constant [Figure 2–3(A)]. At this point the sum of the forces acting on a unit mass is zero, or

$$\Sigma F_i = 0 \tag{2–13}$$

and the grain has attained a "steady state" of uniform rate of fall. This rate is the *terminal velocity,* or settling velocity (V_s). Because this velocity does not appear in Eq. (2–13), it is apparent that it must be found as part of one of the individual forces.

When the steady state has been reached, the forces that act on the grain [Figure 2–3(A)] are the gravitational force (F_g), which is directed downward, and the re-

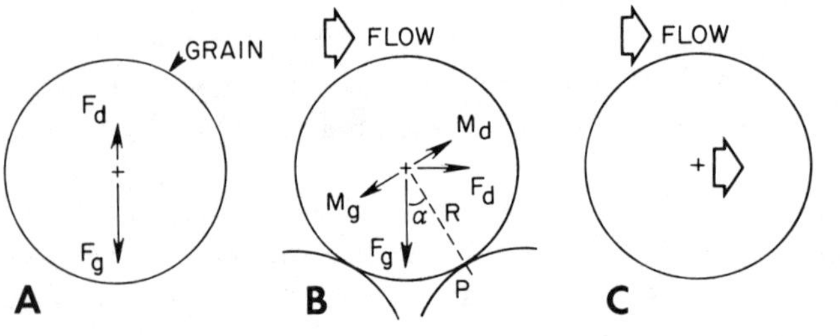

Figure 2–3 Forces that act on a spherical grain in water: (A) deposition in still water; (B) threshold of grain movement; (C) transport in suspension.

sisting force of the fluid, or drag force (F_d), which is directed upward. The sum of the forces is

$$F_g + F_d = 0 \tag{2–14}$$

and the grain falls through the water and is eventually deposited on the bottom.

If the grain rests on the bottom and is subject to horizontal flow, the water velocity may be great enough to cause the grain to move [Figure 2–3(B)]. The sum of the forces gives the grain an acceleration, as in Eq. (2–12) above, until a certain velocity is reached, acceleration ceases, and the rate of movement is constant. During initial movement, the grain rotates about a point of contact (P) with the underlying grain, and the forces that act on the grain must be determined as moments: the gravitational moment (M_g) and the drag moment (M_d) (White 1940). These moments are, respectively,

$$M_g = F_g (R) \sin \alpha \tag{2–15}$$

and

$$M_d = F_d (R) \cos \alpha \tag{2–16}$$

where the grain radius (R) is moment length. When the drag moment is equal to the gravitational movement,

$$F_d (R) \cos \alpha = F_g (R) \sin \alpha \tag{2–17}$$

and

$$F_d = F_g \tan \alpha \tag{2–18}$$

The forces are balanced, and the grain is at the threshold of movement. Any increase in water velocity will cause the drag forces to increase, and the grain will move. Again, the water velocity does not appear in Eq. (2–18) and must be found as part of the drag force.

If the water velocity is not zero and the flow is horizontal, the grain may be transported in suspension with the flow [Figure 2–3(C)]. In this case, there is no relative movement between the grain and water, the drag force is zero, and the grain moves in a horizontal direction at the same velocity as the water.

In the resolution of forces described above, water flow is considered to be laminar, or viscous, and not turbulent. For natural streams, the presence of turbulence produces transient components of drag force that act in different directions. However, the net effect of turbulent flow can be viewed as an average drag force in the direction of the average velocity of flow. Therefore, the complexities of turbulence can be ignored. Furthermore, at the threshold of grain movement [Figure 2–3(B)], a lift force acts during grain rotation, but this force is small and ceases to operate as soon as rotation ceases. Therefore, the lift force can also be ignored.

The analysis of forces is thus simplified to consider only those forces that can be easily resolved in a straightforward manner. This procedure is an idealization, but it permits quantification of the variables and an approach to better understanding of erosion, transportation, and deposition.

DEFINITION OF FORCES

Gravitational force (F_g) in the above equations is the force produced by acceleration due to gravity (g), which acts on the mass of the grain. If a spherical grain is immersed in water, the net mass is the difference in densities of the grain (ρ_s) and the water (ρ_w) times the volume of the sphere $(\pi)D^3/6$). Then, the gravitational force is

$$F_g = (\pi/6)D^3g(\rho_s - \rho_w) \qquad (2\text{–}19)$$

where D is the diameter of the grain.

Drag force is the force exerted by the water on the grain when the water and grain are in relative motion. There are three types of drag force and each depends on the type of flow around the grain (Allen 1970, p. 45): (1) *Viscous drag* acts when the viscosity of the water provides the dominant force of the fluid. (2) *Surface drag* acts when there is a transition from viscous flow to turbulent flow. (3) *Form drag* acts in fully turbulent flow when there is separation of flow around the grain.

Viscous drag is the force exerted by the water when flow is laminar, flow velocity is low, and drag is due largely to the viscosity of the fluid. The pressure on a sphere due to viscous flow was derived by Stokes (1850) as

$$p_v = 3\,\mu U/D \qquad (2\text{–}20)$$

where p_v is viscous pressure, μ is water viscosity, U is water velocity, and D is sphere diameter. This pressure acts on the entire surface of the sphere so that the total viscous drag (F_v) is the product of drag pressure and the surface area (πD^2) of the sphere; thus,

$$F_v = (3\mu U/D)\,(\pi D^2) \qquad (2\text{–}21)$$

or

$$F_v = 3\pi\mu UD \qquad (2\text{–}22)$$

Surface drag is the force exerted by the water when flow is still laminar, but velocity is greater, and drag is due largely to the mass of the moving fluid. A jet of water exerts a pressure that is equal to the product of its mass and velocity times its discharge and cross-section area, so that

$$p_s = (\rho_w U)\,(UA)/A \qquad (2\text{–}23)$$

and

$$p_s = \rho_w U^2 \qquad (2\text{–}24)$$

where p_s is surface pressure, ρ_w is water density, U is velocity, the quantity UA is discharge, and A is the cross-section area (Allen 1970, p. 29). The pressure acts on the projected area of a sphere exposed to flow, an area that is nearly that of a circle ($\pi D^2/4$). Thus, the surface drag (F_s) is

$$F_s = (\rho_w U^2)\,(\pi D^2/4) \qquad (2\text{–}25)$$

Surface drag has also been called *inertial drag* (Rubey 1933) because the mass of the moving fluid is largely responsible for the drag force as opposed to the effect of viscosity in the viscous drag [Eq. (2–22)].

Form drag is the force exerted by the water when flow is turbulent, velocity is high, and drag is due largely to the kinetic energy of the moving fluid. Kinetic energy (E_k) is defined as

$$E_k = \tfrac{1}{2}\,(mU^2) \qquad (2\text{–}26)$$

where m is mass and U is the velocity of the moving fluid. When the water contacts a solid, it exerts a pressure such that

$$p_d = \tfrac{1}{2}\,(\rho_w U^2) \qquad (2\text{–}27)$$

where p_d is the dynamic pressure and ρ_w is the water density. The dynamic pressure acts on a unit area of the solid surface, and the dynamic drag force (F_d) can be expressed as

$$F_d = \tfrac{1}{2}\,(\rho_w U^2 A) \qquad (2\text{–}28)$$

where A is the projected surface area that the water contacts. If this force acts on the surface of the sphere, the projected area is that of a circle, and

$$F_d = \tfrac{1}{2}\,(\rho_w U^2)\,(\pi D^2/4) \qquad (2\text{–}29)$$

or

$$F_d = (\rho_w U^2)\,(\pi D^2/8) \qquad (2\text{–}30)$$

Form drag acts when there is a separation of flow lines as occurs when the flow is over an irregular surface (Allen 1970, p. 45).

The three types of drag—viscous, surface, and form drag—do not act alone but act together at all times. It is the condition of flow that determines which force is dominant. As may be expected, there is a gradation from the condition of dominant viscous drag to surface drag and finally to dynamic drag during increasing flow velocity and during the transition from laminar flow to fully turbulent flow.

VELOCITY EQUATIONS

Settling Velocity

The effect of combined drag forces was first shown by Rubey (1933) to explain the settling velocities of sand grains. The problem is illustrated by the resolution of forces in which a grain falling in still water reaches a uniform rate of fall or settling velocity [Figure 2–3(a)]. In this case, the impelling force is that of gravity acting on the net weight of the grain, or as previously given in Eq. (2–19),

$$F_g = (\pi/6)\,(D^3 g)\,(\rho_s - \rho_w)$$

When the grains are small and less than about 0.1 mm in diameter, viscous drag is the dominant force. When the grains are large and greater than about 2 mm in diameter, surface drag is the dominant force, the so-called inertial drag of Rubey (1933). For intermediate grain sizes, both forces act on the grain depending on the relative size or mass of the grain. Therefore, the viscous drag equation [Eq. (2–22)] and the surface drag equation [Eq. (2–25)] are added, and the combined drag is equated to the gravitational force [Eq. (2–19)], or

$$F_v + F_s = F_g \tag{2–31}$$

and

$$3\,\pi\mu V_s D + (\rho_w V_s^2)\,(\pi D^2/4) = (\pi/6)\,(D^3 g)\,(\rho_s - \rho_w) \tag{2–32}$$

In this form, the settling velocity (V_s) of the grain is substituted for water velocity (U) in the drag equation because only the grain is moving against still water. Solving for velocity of the grain gives the general equation

$$V_s = \{[\tfrac{2}{3}D^3\rho_w\,(\rho_s - \rho_w)\,g + 36\,\mu^2]^{1/2} - 6\mu\}/D\rho_w \tag{2–33}$$

Settling velocities for natural sand grains (Figure 2–4) are in good agreement with Eq. (2–33). Grains of silt size follow the viscous drag equation, whereas grains coarser than sand follow the inertial drag equation. Grains of sand size show the effects of both viscous and surface drag.

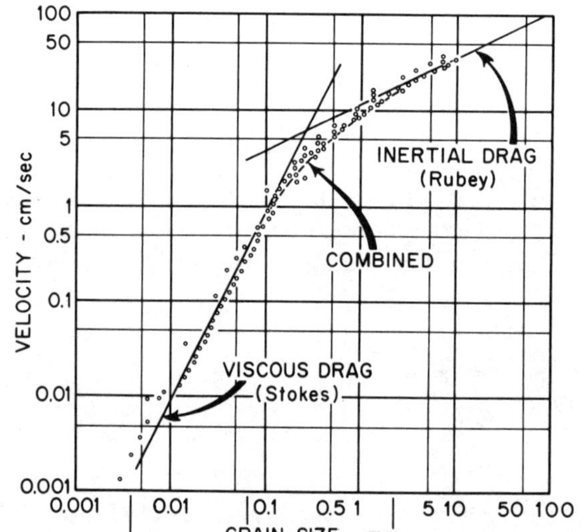

Figure 2-4 Settling velocities for natural grains (dots) compared with the general Eq. (2–33) that is the combined viscous-surface drag equations of Rubey (1933). Measurements of settling velocities from Lane (1938).

Transport Velocity

The derivation of settling velocity [Eq. (2–33)] is important here for several reasons. First, the interaction of drag forces and grain size is clearly shown. Second, the division between the fields of transport and deposition in the Hjulström diagram (Figure 2-2) approximates the settling velocity as given by Eq. (2–33). The difference between the two functions is less significant when it is remembered that there was a good deal of uncertainty in the observed function of Hjulström. More importantly, greater insight is afforded into the meaning of the transport field. Erosion of sand requires relatively high velocities of 15 cm/sec or more. Once the sand is eroded, then significantly smaller velocities are required to maintain movement because the grains are elevated above the stream bed and are fully exposed to flow. At this point, it can be imagined that the settling velocity of the grain becomes the minimum velocity for the transport of loose grains. In the case of transport, the water is moving, but the grain has no movement relative to the water and moves with the flow. No more precise definition of transport is needed here, and the emphasis in the following discussion can be placed on erosion.

Erosion Velocity

It is possible to calculate the approximate velocities for the erosion of sand if some simplifying assumptions are made; that is, if the grains are uniform spheres or spheroids in systematic packings. Then the effects of water flow on the grains can be simulated. The results show that grain size and shape, as well as adhesion tension, can be important variables in sedimentary processes.

The drag equations [Eqs. (2–22), (2–25), and (2–30)] derived above are expressed in terms of forces that act on grain surfaces or their projected areas. If the grains and water are in relative motion, the drag forces may be thought of as the impelling forces that may cause a grain to move. The gravitational force, as given in Eq. (2–19), is then the resisting force. At the threshold of grain movement, the two moments, drag and gravitational, are in balance [Figure 2-3(B)]. Any increase in the drag force, then, will cause the grain to move, and this increase in drag may be due solely to the increase in velocity if the other fluid properties remain constant. It is possible, therefore, to equate the proper expressions and then to solve for water velocity (U). This velocity will be the threshold velocity for grain erosion under different conditions of flow.

Experiments on settling velocities have shown that the combined viscous drag and surface drag equations apply to sand-size particles over a wide range of velocities (Figure 2–4). In fact, surface drag is effective at velocities of about 10 cm/sec or greater. Therefore, it may be possible to use surface drag alone to approximate erosion velocities, especially for the larger sand sizes. Again, it may be assumed that the grains are spherical as in the calculation of settling velocity, but because the grains rest on the bottom, the moments of the drag force and gravitational force must be considered as in the relation shown in Eq. (2–18) earlier; that is,

$$F_d = F_g \tan \alpha$$

Furthermore, grains at rest on the bed beneath the flowing water expose only part of their surface to the drag force. Thus, it is necessary first to consider how the grains are oriented in relation to water flow.

The conditions necessary for grain movement can be represented by packings of spherical grains (Figure 2–5). Well-sorted sands have porosities on the order of 40 percent and may be visualized as having orthorhombic packings (Figure 1–4). Some grains at the top of this packing may rest on underlying grains such that the angle of repose (α) is 30° [Figure 2–5(A)], and most of the projected surface area is exposed to water flow (White 1940). This orientation may apply only to a few grains at rest on the bottom. Another configuration would be that in which the orthorhombic packing is rotated so that the angle of repose (α) is 60° [Figure 2–5(B)]. In this orientation, a grain at the top of the packing has only one-half its surface exposed to flow, and movement of the grain would require higher velocities. Hence, this orientation might represent the condition that would apply to a large number of grains within a mass of bottom sediment.

Another possibility is that the grains may not be spherical but may have other shapes that can be represented as prolate spheroids [Figure 2–5(C) and (D)]. The spheroids have a long axis (A) and intermediate axes ($B = C$), so that the ratio (B/A) is equal to 0.73. In this case, the angle of repose (α) is 45° for a grain resting at the top of the packing, and the angle of repose is 70° for a grain within the packing. A smaller projected area is exposed if the grain is oriented with its long

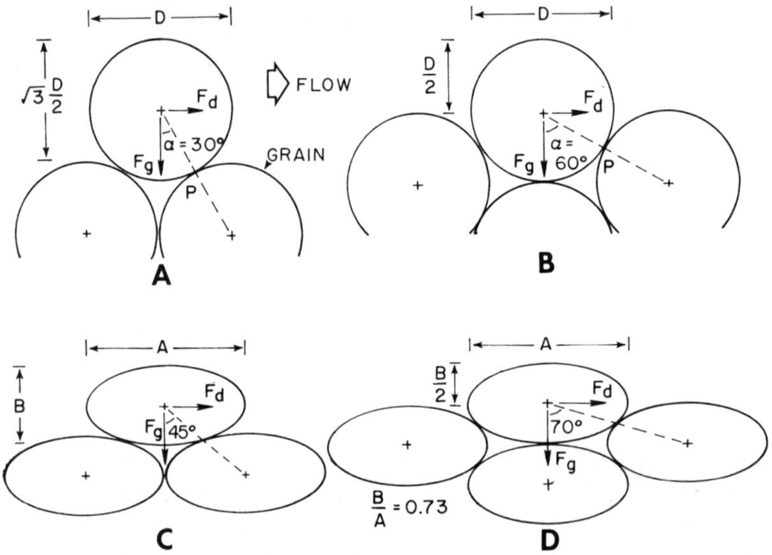

Figure 2-5 Cross section of grains in orthorhombic packings showing change in repose angle (α) with change in shape from spheres [(A) and (B)] to prolate spheroids [(C) and (D)] where the axial ratio $B/A = 0.73$.

axis parallel to the direction of flow. The prolate spheroid may be appropriate for representing finer sand sizes for which shape departs from spherical as a function of mean size.

For grains in these shapes and packings, the threshold equations must be modified for the different surface areas, or projected areas, exposed to flow. In order to calculate threshold velocities, other values must be chosen for the equations. Water density may be taken as 1 g/cm^3, grain density as 2.65 g/cm^3, and water viscosity as 0.01 poise. These values then allow calculation of approximate velocities under different conditions of flow.

Erosion velocities may now be estimated for differing grain orientations. Only surface drag will be considered and will be assumed to apply over the range of sand sizes, and grain shape will be assumed to be spherical. Thus, the only variables will be the angle of repose (α), the surface area exposed to drag, and grain size.

First consider the threshold velocity when the surface drag force [Eq. (2–25)] is balanced by the gravitational force [Eq. (2–19)] as the grain rests on the bottom, or

$$\rho_w U^2\,(\pi D^2/4) = (\pi/6)D^3 g(\rho_s - \rho_w)\tan\alpha \qquad (2\text{–}34)$$

and solving for velocity,

$$U = \left[\tfrac{2}{3}\,(Dg)\,(\rho_s - \rho_w)\tan\alpha/\rho_w\right]^{1/2} \qquad (2\text{–}35)$$

When the angle of repose (α) is 30°, nearly the entire surface of the grain is exposed to surface drag [Figure 2–5(a)]. Thus, inserting appropriate values, the threshold velocity becomes

$$U = 8D^{1/2} \qquad (2\text{–}36)$$

where D is grain diameter in millimeters. When $D = 1$ mm, then $U = 8$ cm/sec. This value may be taken as the threshold velocity for the transport of sand grains (Figure 2–6), and the value agrees well with that for the settling velocity for grains of the same size.

When the grains are more tightly packed, only half the surface area of the grain is exposed to surface drag [Figure 2–5(B)], and the balance of forces is

$$\rho_w U^2\,(\pi D^2/8) = (\pi/6)\,D^3 g\,(\rho_s - \rho_w)\tan\alpha \qquad (2\text{–}37)$$

and solving for velocity,

$$U = \left[\tfrac{4}{3}Dg\,(\rho_s - \rho_w)\tan\alpha/\rho_w\right]^{1/2} \qquad (2\text{–}38)$$

When the angle of repose (α) is 60°, and inserting other appropriate values, the threshold velocity is

$$U = 19D^{1/2} \qquad (2\text{–}39)$$

where D is grain diameter in millimeters. When $D = 1$ mm, $U = 19$ cm/sec. This value may be taken as the threshold velocity for erosion of sand grains of coarser

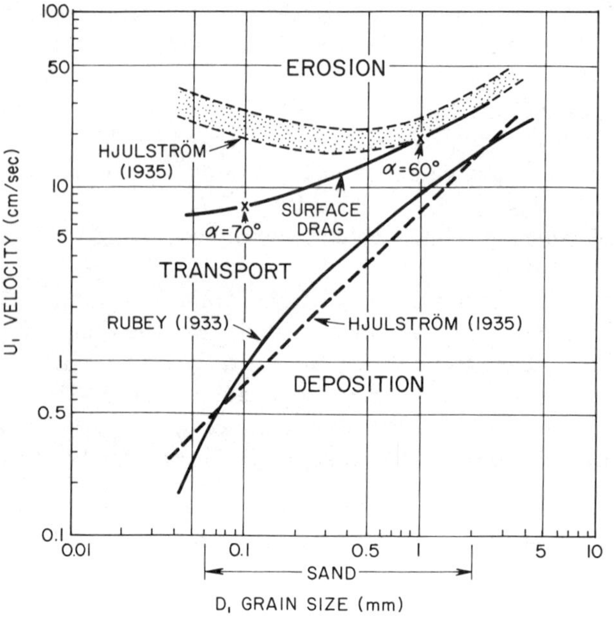

Figure 2–6 Velocities for transport and erosion of sand according to Hjulström (Figure 2–2) and calculated from the surface-drag equation and settling velocity equation (Rubey 1933).

size (Figure 2–6). However, for finer sizes, the angle of repose may increase ($\alpha = 70°$), and then inserting appropriate values, the threshold velocity is

$$U = 24D^{1/2} \tag{2–40}$$

where D is grain diameter in millimeters. When $D = 0.1$ mm, $U = 8$ cm/sec. This value may be taken as the threshold velocity for erosion of sand grains of finer size (Figure 2–6).

The calculated values for transport and erosion velocities are only approximations because the effects of viscous drag forces have been neglected. It may be expected that the viscous drag is a dominant force in the finer grain sizes rather than the surface drag. Nevertheless, the functions agree well with the measured velocities compiled by Hjulström (Figure 2–2), except for the erosion velocities of finer sand sizes. An increase in the angle of repose does not appreciably increase the required erosion velocity, nor in fact do other calculations of velocity based on viscous drag forces that act on spheroidal grains. Rather, some other factor such as cohesion may play a role in the erosion of fine sand and silt.

Effect of Cohesion

Sediments that have been partly dried in air may retain small amounts of water at points of grain contact. The adhering water tends to bind the grains together and to increase their resistance to the force of moving water. The adhesion of water gives the sediment mass a greater stability, or *cohesion,* under the flow of water. The force of adhesion depends on the amount of water present, its configuration, and the surface tension between air and water. When a dried sediment is suddenly submerged in flowing water, air may be trapped in the pores, and surface tension plays a major role in cohesion. If the air is subsequently driven out and the pores become completely filled with water, the surface tension disappears, and the sediment is affected only by the drag force of the water. Thus, the force of adhesion is a temporary one and does not act in a fully water-saturated sediment.

Water at points of grain contact has a surface form that resembles a short cylinder with curved sides (Figure 2–7). The curvature may be assumed to be spherical, or a surface of least energy. The adhesion tension (A_t) that holds water to the grain surface is given as

$$A_t = \sigma \cos \theta \tag{2–41}$$

where σ is the surface tension between air and water, and θ is the angle between the grain surface and fluid contact as measured through the more dense fluid, or water.

The adhesion tension acts along the circumference of water contact. The circumference of contact (C') is a function of the amount of water and its configuration as given by the angle β and diameter (D') of the cylinder of contact, so that

$$C' = \pi D' = \pi D \sin \beta \tag{2–42}$$

There is one circumference of water contact that binds the upper grain so that the total adhesion tension is the product of adhesion tension and circumference, that is,

$$A_t = (\sigma \cos \theta)(\pi D \sin \beta) \tag{2–43}$$

The total adhesion tension is a function of the amount of water at the grain contact. If the amount of water is large, then β is large, and the adhesion tension is near a maximum value. If the amount is small, then β is small, sin β approaches

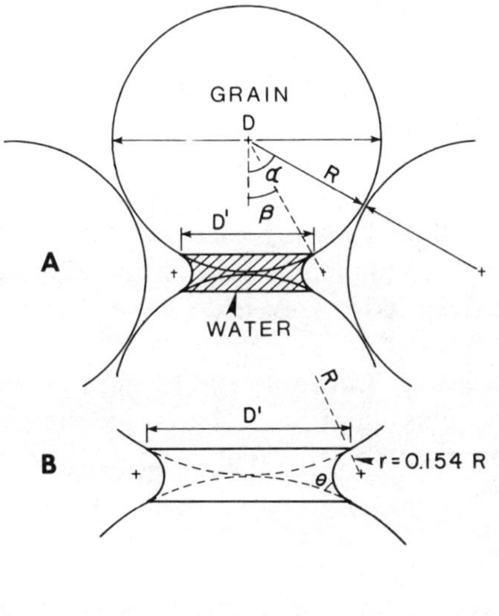

Figure 2–7 Configuration for water adhering to spherical grains at a point of contact: (A) dimensions of adhering water and grains; (B) enlargement showing contact angle (θ).

zero, and the adhesion tension is a minimum value. The fluid angle (θ) may be assumed to be constant and equal to zero since water tends to wet the grain surface.

The effect of adhesion tension alone may be evaluated by considering the balance of moments at the threshold of grain movement as given by Eq. (2–18). Then, the drag force (F_d) is related to the total adhesion tension (A_t) as

$$F_d = A_t \tan \alpha \tag{2-44}$$

Substituting equations for form drag [Eq. (2–30)] and adhesion tension [Eq. (2–43)] gives

$$(\rho U^2)(\pi D^2/8) = (\sigma \cos \theta)(\pi D \sin \beta) \tan \alpha \tag{2-45}$$

and solving for water velocity,

$$U = (8\sigma \cos \theta D^{-1} \sin \beta \tan \alpha)^{1/2} \tag{2-46}$$

Appropriate values may be chosen for a grain in tight packing where α is 60°, the air-water surface tension is 72 dynes/cm, and $\theta = 0$, so that

$$U = (9976 D^{-1} \sin \beta)^{1/2} \tag{2-47}$$

where the grain diameter (D) is in millimeters. When a small amount of water is present, then the angle (β) is small, as for example on the order of 5°, and sin 5° = 0.09, so that the velocity calculated by Eq. (2–47) is approximately

$$U = (100 D^{-1})^{1/2} \tag{2-48}$$

If the grain diameter is 1 mm, then $U = 10$ cm/sec, and this velocity is the additional flow rate required to initiate grain movement until the pores become completely water-filled.

The effect of adhesion tension is significantly greater for small grains, however. If the grain diameter is 0.1 mm, then $U = 32$ cm/sec, and this velocity is much greater than the erosion velocity of about 8 cm/sec estimated for a spheroid of the same size by drag force alone [Eq. (2–40)]. The total velocity required for movement is the sum of the two velocities and would be on the order of 40 cm/sec for a grain of 0.1 mm diameter. In addition, the effect can be expected to persist for finer sediment because a smaller pore size would prevent immediate saturation by water, and the influence of surface tension between air and water would be prolonged.

Thus, it is reasonable to imagine that adhesion tension is an effective force that helps a finer-grained sediment resist the erosive action of flowing water. The ultimate effect on partly dried sediment would be increasing resistance with finer grain size, exactly as that recorded by reversal of the erosion-velocity function of Hjulström (Figure 2–2).

The calculated velocities for transport and erosion are not rigorously derived and are based on simplifying assumptions. Therefore, the values may not have a high degree of accuracy, but they do show that grain size and shape exert a primary control in sedimentary processes.

BED FORM

Flowing water above a bed of cohesionless sand causes the formation of irregularities in the bed when velocities are sufficient for grain movement. These irregularities are called *bed forms* and can be defined in general terms as deviations from a flat bed (Simons, Richardson, and Nordin 1965). At low flow velocities, the bed develops small asymmetric wavelike structures, called *ripples,* in which the crests are transverse to flow direction. At higher velocities, the ripples grow to larger dimensions, called *dunes*. The formation of ripples and dunes from moving grains produces a characteristic internal structure of inclined laminae, the familiar *crossbedding* of sandy, current-transported sediment.

Ripples develop first as erosive features by the scouring of troughs, leaving low residual crests. These irregularities disturb the pattern of flow near the bottom, and there develops a zone of reverse flow in the troughs (Figure 2–8). As more sand is transported, ripples are constructed by the intermittent movement and deposition of grains.

Grains move along the upstream, or *stoss,* side of the ripple in a thin cloud near the bed, the larger grains pushed along by currents and smaller grains in partial suspension (Jopling 1967). Beyond the ripple crests there is a zone of separation in which the larger grains drop into the trough. At the same time, somewhat larger grains reach the crest and cascade down the lee slope. The finer grains continue in a cloud of suspended sediment above the trough to the next ripple crest.

Thus, the asymmetric ripples develop their characteristic form with coarser

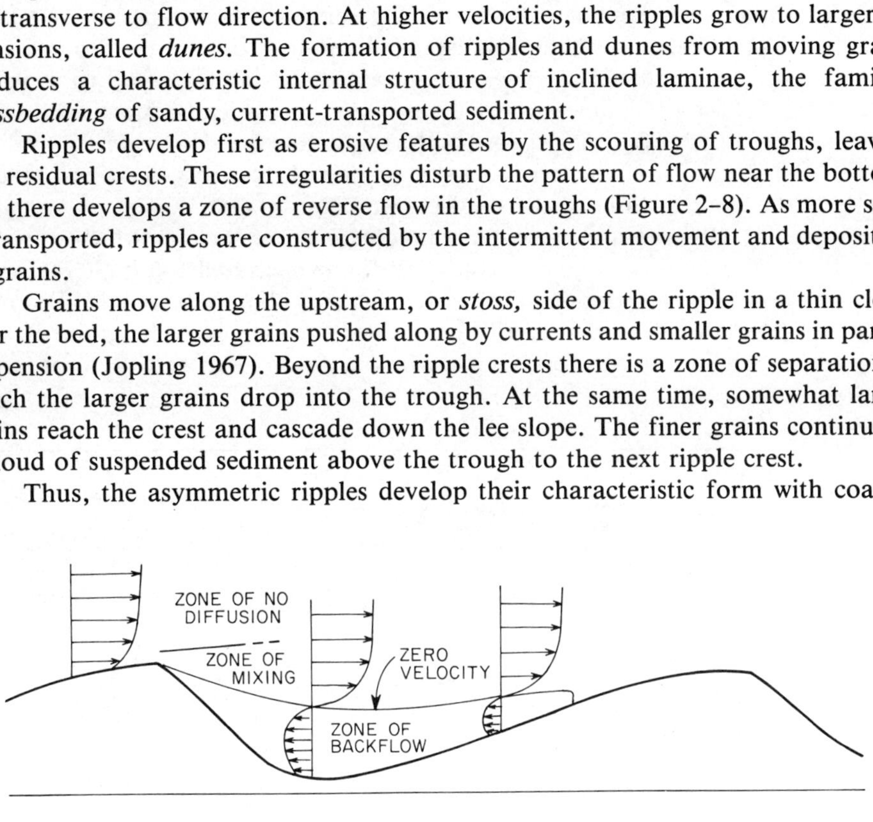

Figure 2–8 Flow patterns and deposition across a ripplied bed configuration. [After Jopling 1967.]

grains in steep laminae on the downstream side while the coarsest grains accumulate in troughs. The remaining sand becomes better sorted as the fine particles are carried downstream.

Flow Regime

Early flume studies helped to explain sand transport in channels but did little to explain the sedimentary structures observed in sandstones. About 1960, a series of experiments showed the relation of stream-bed configuration to water flow and grain size of bed material. These important observations helped to explain, finally, the sequence of sedimentary structures seen in vertical sections of fluvial sandstones. The bed configurations, or bed forms, define the *flow regime,* or velocity and grain size functions. The following explanation of these relationships is taken largely from the work of Simons, Richardson, and Nordin (1965).

Observations were made in a large flume through which water flowed over a bed of uniformly graded sand. The flume measured 100 ft (30 m) long and 8 ft (2.4 m) wide, and contained water depths up to about 1 ft (0.3 m). Bed material ranged from about 0.1 mm (very fine grained) to 1 mm (coarse grained). During flow, a succession of bed forms, or deviations from a flat bed, were observed (Figure 2-9).

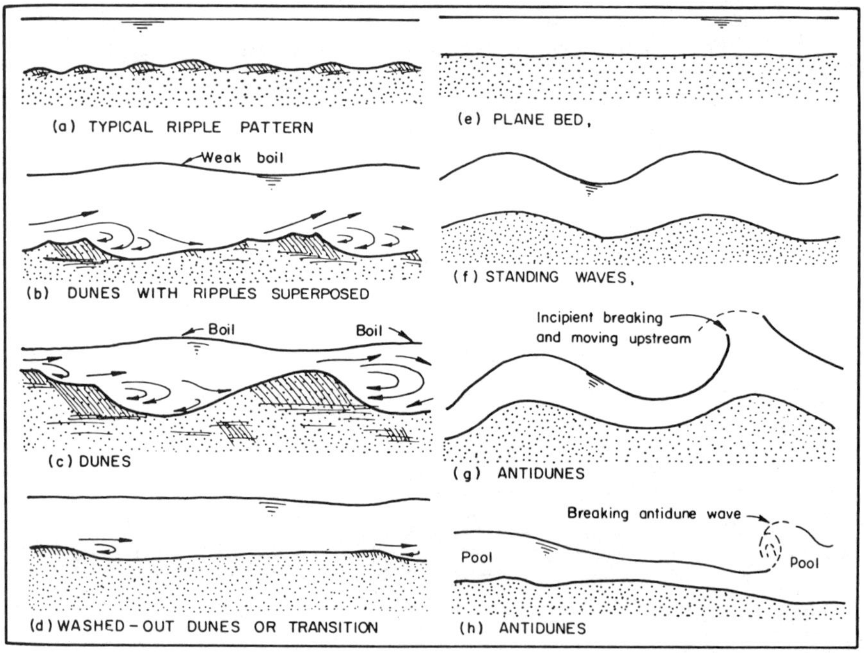

Figure 2-9 Bed forms generated by water flow in shallow flumes. [From Simons, Richardson, and Nordin 1965.]

The first bed form at low flow velocities was a small low-relief sand wave called a *ripple.* In the formation of ripples, the sand grains were carried largely by traction, and ripple height ranged up to about 0.2 ft (0.06 m). Increased velocities produced a larger wave form called a *dune,* during the formation of which the sand grains moved largely by saltation. At still higher velocities the dune forms were washed out, and the sand grains moved as a thin sheet several millimeters in thickness and formed a *plane-bed phase.* The highest velocities produced a wave form called an *antidune,* during the formation of which the bed form moved in an upstream direction. During the change in bed configuration, it was observed that the water surface was out of phase with the lower-velocity bed forms of dunes and ripples, whereas the water surface was in phase during the formation of antidunes. The antidune phase appeared to be unstable, and the surface waves as well as the bed form built to significant height, broke and collapsed, and then were rebuilt.

The authors [Simons, Richardson, and Nordin (1965)] summarized the results of a series of observations by a graph in which bed form was plotted as a function of stream power and grain size (Figure 2–10). Stream power ($\tau_0 v$) was defined as bed shear (τ_0) times velocity of flow (v). In other words, stream power is the rate of work done by the flowing water. Expressed as an equation, stream power is

$$\tau_0 v = (\rho g R S) \tag{2–49}$$

where ρ is density of the water, g is acceleration due to gravity, R is hydraulic radius, and S is slope.

The range of flow conditions is identified by bed form. The ripple and dune phases of bed-material transport are called *lower-flow regime,* and the plane bed and antidune phases are termed *upper-flow regime.* The resulting bed forms, therefore, signify the change in stream power or velocity. For grain sizes of bed materials

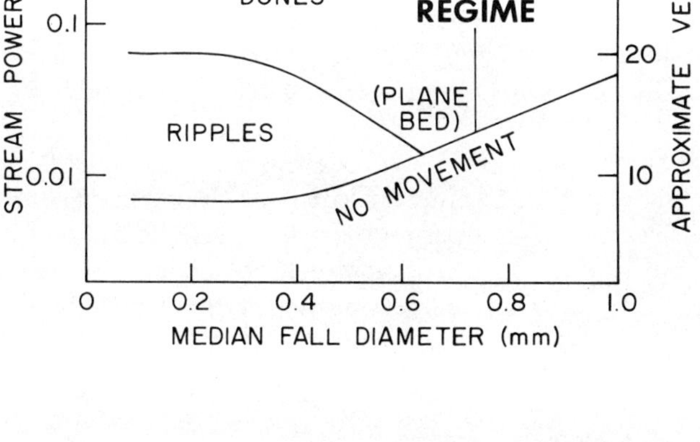

Figure 2–10 Relation of stream power, median grain size, and bed form. [From Simons, Richardson, and Nordin 1965.] Stream velocities shown at the right are approximate values for a grain size of 0.45 mm and extrapolated to 1-ft flow depth.

coarser than about 0.6 mm, there was no ripple formation, and with increasing velocity the bed material formed dunes of higher relief.

Sedimentary Sequence

The significance of the flow-regime diagram (Figure 2–10) is that it demonstrates the sedimentary structures that are generated during a decrease in flow velocity. It was observed that as velocity decreased, the bed form of the previous higher-velocity flow was preserved and the bed form of the lower-velocity flow developed on the previous bed form. This relationship means that during deposition of sand, the sedimentary structures represent bed forms of decreasing flow regime from the antidune and plane-bed phase, to dunes, and then to ripples. At the same time, there should be a change in grain size corresponding to the bed form for natural sands that are not well sorted. The upper-flow regime structures should be composed of coarser grains, whereas the lower-flow regime structures should be composed of successively finer grains.

The depositional sequence for decreasing channel flow is thus explained by the decrease in flow regime and velocity. The lowermost channel deposits should be coarse grained and should display the antidune or plane-bed phase. The succeeding finer sands should show the dune phase overlain by a ripple phase. Finally, the uppermost beds should represent very fine grained suspended materials that are deposited during the period of no sand movement. This total sequence is that which is commonly observed in stream-channel deposits. The lowermost beds are coarse-grained and range from massive to indistinctly laminated, probably representing highest flow regime. The succeeding beds are finer grained and crossbedded, representing the dune phase of transport. The uppermost sands are ripple-laminated and are overlain by clays.

The data from the original flow-regime experiments have been reinterpreted, but the general relationships (Figure 2–10) have been only slightly modified. For example, it is likely that the transitional bed form from lower- to upper-flow regime is a function of water depth (Southard 1971). In fact, in the original experiments, it was observed that the height of bed forms was related to water depth and that increasing depth resulted in an increase of dune height. It has also been suggested that for coarse bed materials, there is a plane-bed phase that is developed during the initial movement and before the dune phase of transport, essentially equivalent to the rippled bed form of finer-grained material. Despite the suggested modifications, the development of sedimentary structures in fluvial deposits can be explained in terms of flow regime.

Finally, another point can be made. Bed forms developed by channel flow are transient features and require a period of time for their development. Therefore, an abrupt decrease in flow velocity may not permit the orderly growth of bed forms. It can be expected, then, that rapidly deposited sand will not display the same sequence of bed forms expected from slowly decreasing flow velocities. This conclusion may help to explain why upper-flow regime structures, which are unstable bed forms, are not commonly identified in fluvial sandstones.

DENSITY FLOW

An important mechanism for generating flow is a difference in water density. More dense water may flow beneath a less dense body of water with velocities that can erode and transport sediment. Greater density may be achieved by increased salinity, decreased temperature, or particles in suspension. A type of density current, *a turbidity flow,* owes its greater density to suspended sediment. The turbidity flow can transport sediment for long distances, and it is an important mechanism by which sediment may be carried into deep ocean basins.

Experiments on density flows have been performed in laboratory flumes with both saline and sediment-bearing, or turbid, water. Early experiments were conducted with saline water, colored by dye to make it visible, and held behind a gate at the head of a flume. When the gate was lifted, the saline water flowed out and beneath the ambient water as a coherent mass of regular shape. In cross section, the flow had a leading head or snout with a slightly thicker body (Figure 2–11). The snout appears to override a thin layer of ambient water, and in wide flumes, the underlying layer is seen to be channeled below the overriding flow (Simpson 1969). Thus, the density flow seems to ride on a thin film of ambient fluid.

The primary motive force for density currents is the hydraulic head of a dense fluid that is approximated by the average height of the flow times the difference in specific weight between the fluids. The velocity of flow can then be derived by considering the energy relationships between the moving flow and the ambient water. The hydraulic head is the potential energy (E_p), and the flow has a kinetic energy (E_k). As the flow moves, there is an energy loss (ΔE) due to the frictional forces.

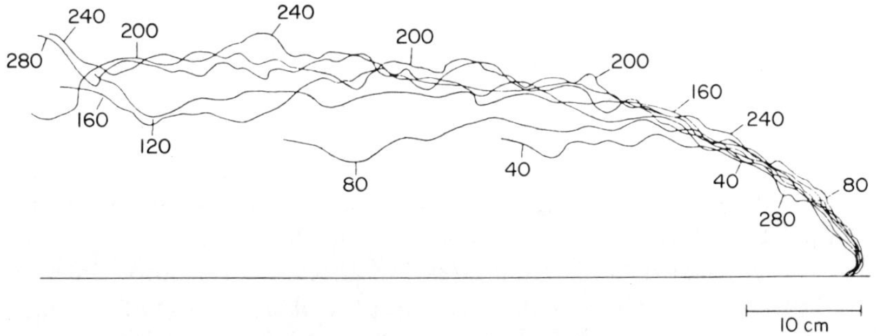

Figure 2–11 Shape of saline density flow with flume tilted 3° in the direction of flow. Numbers refer to distance of travel in centimeters. [From Kersey and Hsu 1976.]

When these energies are balanced, the flow has a uniform velocity on a slope of critical angle (θ_c) (Kersey and Hsu 1976). The energy balance is

$$E_p = E_k + \Delta E \tag{2–50}$$

The kinetic energy is

$$E_k = \tfrac{1}{2}(\rho_f U^2) \tag{2–51}$$

where ρ_f is flow density and U is velocity. The potential energy of the flow is equal to the average hydraulic head, or

$$E_p = \Delta\rho g(d/2) \tag{2–52}$$

where d is the thickness of the flow and $\Delta\rho$ is the difference in densities between the flow and the ambient water. The half thickness ($d/2$) in Eq. (2–53) results in an "average" potential energy for the flow. Substituting these expressions for energies in Eq. (2–50) and rearranging the terms gives

$$\tfrac{1}{2}(\rho_f U^2) = \Delta\rho g(d/2) - \Delta E \tag{2–53}$$

For the moment it may be assumed that the frictional forces are small, so that the term (ΔE) can be neglected. Then solving for velocity gives

$$U = (\Delta\rho g d/\rho_f)^{1/2} \tag{2–54}$$

This expression for velocity is an approximation because the frictional forces have been neglected. A velocity calculated by Eq. (2–54) can be expected to be slightly greater than the actual velocity for a flow of given density and thickness. Therefore, a true equation for velocity should have the form of

$$U = K\,(\Delta\rho d/\rho_f)^{1/2} \tag{2–55}$$

where K has a value of one or less.

Experiments with saline flows by Keulegan (1946*a*, 1946*b*, 1951, 1957, 1958) have shown that flow velocities are those given by Eq. (2–55) where K has a value of about 1 at Reynolds numbers less than 1200, and a value of 0.7 at Reynolds numbers greater than 1200. Thus, for the conditions of laboratory experiments, Eq. (2–55) seems to be a valid, although approximate, expression for velocity. Because K is smaller at higher Reynolds numbers, it may be assumed that the energy loss (ΔE) is due largely to turbulence, that is, to internal friction generated by turbulent flow. Furthermore, internal friction may be due largely to viscous forces, and thus the value of K is primarily dependent on flow viscosity.

A somewhat different equation applies to turbidity flows. Experiments have shown that the Chezy equation [Eq. (2–11)] can be used for density flows that contain suspended solids (Middleton 1966*a*, 1966*b*). The expression for body velocity of the flow (U_b) is

$$U_b = (8\Delta\rho g d S/\rho_f)^{1/2} \tag{2–56}$$

where d is flow thickness and S is slope, usually expressed as the tangent of the angle of inclination of the flume.

It seems reasonable that the Chezy equation [Eq. (2–11)] applies to turbidity flows in flumes because it also expresses velocity of flows in channels. However, Eq. (2–56) for turbidity flows appears to have a form similar to Eq. (2–55) for density flows in which the value of K involves the factor $(8S)^{1/2}$.

The approximate density-flow equations [Eqs. (2–55) and (2–56)] give some insight into the conditions that cause deposition of sediment. A decrease in velocity can be the result of a decrease in one or more variables such that the flow is no longer able to transport sediments. A decrease in flow thickness, as may be caused by overbank flow or by spreading flow at the mouth of a channel, may cause deposition. A decrease in flow density, as may be caused by mixing of the flow with the ambient water, may also cause deposition. Finally, a decrease in slope will also reduce velocity and cause deposition.

WAVE MOTION

Waves in the open ocean generally affect only the surface layers at shallow depths. During storms, however, large waves are generated by high winds, and these waves produce oscillatory water motion at greater depths. The motion may affect bottom sediment on the inner continental shelf, moving sediment to and fro without significant transport. Only when the waves reach shallow water near the shoreline is there lateral translation of the water mass and the generation of currents that are capable of moving large amounts of sand. Therefore, some knowledge of wave motion is helpful to understanding sand distribution on the marine shelf.

Waves are generated by winds. The prevailing winds that blow over uninterrupted reaches of open water set up a primary progressive wave or swell. Superimposed on the swell are smaller waves induced by more variable local winds until the sea surface becomes a confusing complex of wave forms. It is the primary waves that have the potential to affect the bottom sediments, especially when those waves are generated by major storms.

Wave Velocity

A good approximation of wave form is a sinusoidal function, although other mathematical expressions may be used. The wave has a crest and trough, a wave length (L) and a height (H) [Figure 2–12(A)]. Wave period is the time (T) of passage for one wave length. The wave propagates on the surface of water that has a depth (D). The wave moves at a phase speed, or celerity (c), which is given by the general equation

$$c = [(gL/2\pi) \tanh (2\pi D/L)]^{1/2} \tag{2–57}$$

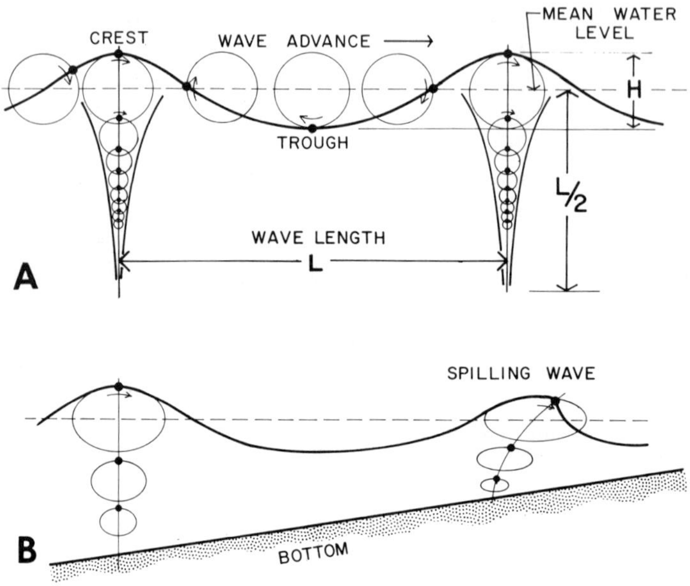

Figure 2–12 Wave diagrams; (A) deep-water waves and orbital motions of water particles; (B) shallow-water waves, particle orbits, and spilling or breaking wave.

This equation assumes that amplitude is negligibly small as compared with water depth. The relation between water depth (D) and wave length (L) is also significant. When depth is large as compared to length, then the function tanh $(2\pi D/L)$ approaches one, so that Eq. (2–57) simplifies to

$$c = (gL/2\pi)^{1/2} \tag{2–58}$$

This equation expresses velocities for waves in deep water. When depth is small compared with length, then the function tanh $(2\pi D/L)$ approaches $(2\pi D/L)$, and Eq. (2–58) simplifies to

$$c = (gD)^{1/2} \tag{2–59}$$

This equation expresses velocities for waves in shallow water.

Deep-water waves are considered to be those that have a depth-to-length ratio of greater than $\frac{1}{2}$. *Shallow-water waves* are those that have a depth-to-length ratio of less than 1/20. *Intermediate waves* have ratios between these limits, and velocities must be calculated by Eq. (2–57).

The equations for velocity are approximations because it is assumed that all the variables have constant relationships when waves pass from deep to shallow water. However, as waves move toward the shore, the wave length and height will change somewhat with respect to water depth.

Bottom Velocity

As a wave advances, water particles move in nearly circular orbits so that there is no net translation of water mass. Rotation is clockwise if wave advance is toward the right [Figure 2–12(b)]. The velocity (u) of water particles at the surface is

$$u = \pi H/T \tag{2-60}$$

Thus, the kinetic energy of the water particles is a function of wave height and period. The orbital paths decrease in diameter with depth away from the water surface. The horizontal velocity tangent to the bottom (u_b) is given by the equation (Weggel 1972)

$$u_b = [\pi H/T \sinh (2\pi D/L)] \tag{2-61}$$

The hyperbolic function (sinh) designates the rate of decrease of orbital diameter with depth. At a depth of one-half wave length, the orbital velocity is negligible [Figure 2–12(A)]. Thus, in deep water, progressive waves will have no effect on bottom sediment. However, as the waves move into shallow water, the effect on bottom sediment can be significant.

Maximum bottom velocities, (u_b) max, can be estimated as a function of water depth and wave height (Figure 2–13), and velocities for the erosion of sand can be attained at relatively shallow depth. For example, modest wave heights of 10 ft (3 m) can initiate movement of bottom sand at depths of about 150 ft (46 m). During major storms, wave heights commonly reach 50 ft (15 m) or more, and these waves

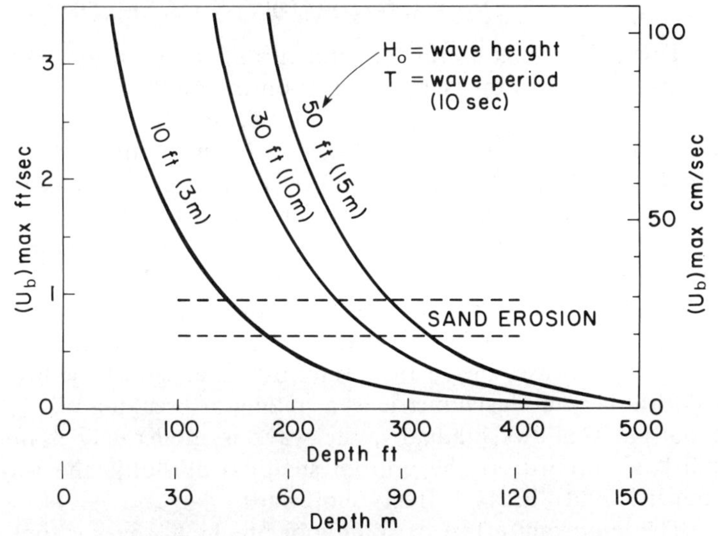

Figure 2–13 Maximum bottom velocities as a function of wave height and water depth for waves of period T = 10 sec. [After Weggel 1972, p. 14).]

can erode sand at depths of about 300 ft (91 m). The movement of sand at these depths can have two effects: (1) If silt and clay are also present, they will remain in suspension when wave motion declines, and the redeposited sand will be better sorted. (2) When the sand itself is raised from the bottom, it is subject to transport if currents are also present.

Sand Erosion

When progressive waves reach shallow water, the effect of the bottom becomes significant. The waves steepen, and the orbital paths become elongated horizontally in ellipsoidal form [Figure 2–12(B)]. Note that Eq. (2–61) for bottom velocity can be simplified for shallow-water waves (Allen 1970, p. 155). When depth is small compared with wave length, the function sinh $(2\pi D/L)$ approaches the value of $(2\pi D/L)$, and Eq. (2–61) becomes

$$u_b = \pi H/T(2\pi D/L) = HL/2DT \qquad (2\text{–}62)$$

The velocity for a shallow-water wave is

$$c = (gD)^{1/2} = L/T \qquad (2\text{–}63)$$

and rearranging terms,

$$(1/T) = (gD)^{1/2}/L \qquad (2\text{–}64)$$

Substituting this relation in Eq. (2–62) gives

$$u_b = (H/2D)\,(gD)^{1/2} = (H/2)\,(gD)^{1/2} \qquad (2\text{–}65)$$

This equation is useful for estimates of sand erosion. For example, consider the beach at Galveston Island, Texas, on the Gulf of Mexico, where the average sand size is 0.2 mm. During normal periods of fair weather, wave heights are only about 2 ft (0.66 m). Normal wave action, then, produces a maximum bottom velocity of 33 cm/sec at a depth of 30 ft (10 m). This velocity is sufficient to erode the entire range of sand sizes. Therefore, from this depth shoreward, the sand is kept in constant motion by normal wave action and results in a clean, quartz-rich sand due to the removal of clays and the abrasion of less resistant grains.

When water depth becomes still shallower, the effect of bottom friction retards the deeper water, whereas the surface water retains its kinetic energy and forward motion. In this case, the surface water moves ahead of the bottom water, and the wave form becomes asymmetric as a spilling or breaking wave [Figure 2–12(B)]. If the bottom shallows gradually, the wave is moderately asymmetric, and water "spills" shoreward. If the bottom shallows abruptly, the wave becomes highly asymmetric and "breaks" freely shoreward.

The important effect of spilling or breaking waves is that a mass of water is translated toward the shore and falls with considerable impact on the beach. The displaced water then returns seaward and flows as a rip current between succeeding waves. When waves impinge at an angle to the shoreline, the successive translation and return of water sets up a current that has a net direction parallel to the shore. These longshore currents are responsible for sand transport in large amounts.

Bed Forms

Oscillatory water motion produces bed forms in bottom sediments that are similar to low-flow regime bed forms produced by current flow (Figure 2–10). Sand movement begins at about 10 cm/sec for grain sizes of 0.1 to 0.2 mm (Figure 2–14). Velocities for movement increase with grain size, and velocities required for initial movement of grains are somewhat less than those for current flow (Figure 2–2) because the bottom sediment is unconsolidated and the degree of cohesion is low. With increasing velocities, symmetrical wave ripples are formed. At velocities greater than about 50 cm/sec, a plane-bed form is developed in which lamination is prominent. The three-dimensional aspect of the plane-bed form was not described, but it is possible that the bedding is that of the "hummocky" type in which the laminae are broadly convex upward [Figure 1–5(D)] (see Chap. 1). This type of bedding has been attributed to storm waves on the open shelf. The wave-induced plane bed might be similar to a "hummocky" dune of low amplitude and long wave length. Thus, the two bed forms, ripples and plane beds, are comparable to the lower flow-regime bed forms produced by channel flow.

By means of wave action, then, characteristic bed forms are developed. In deep waters only a small amount of transport results. The sand grains are moved about, and fine materials are winnowed and remain in suspension. In shallow water, sands are subject to erosion, and once in suspension, the grains are available for transport by currents. At any depth, the bottom sediment put into suspension by wave motion has the potential to form density currents that can result in long-distance transport of sediment in a seaward direction.

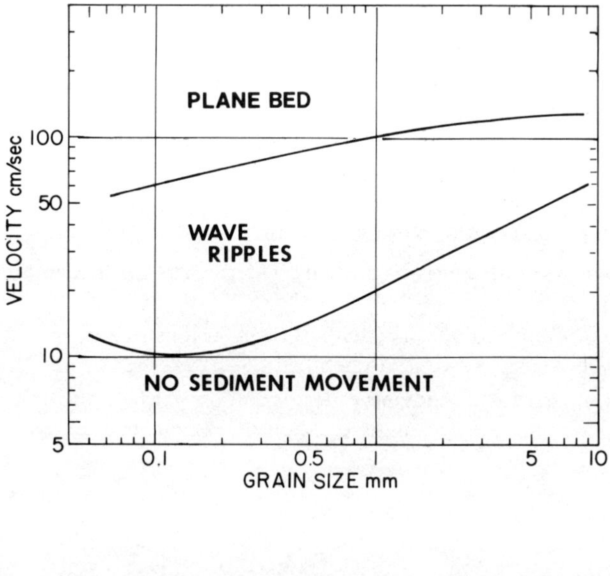

Figure 2–14 Bed forms as a function of near-bottom water velocity induced by waves and diameter of bottom sediment. [After Allen 1970, p. 170, from data by Manohar 1955 and Inman 1957.]

SUMMARY

Erosion, transport, and deposition of sediment by channel flow are selective processes that result in the segregation of well-sorted sands in the form of discrete bodies. These processes are complex but can best be understood by considering the forces that act on sand grains. These forces are the result of the interaction between the mass of fluids and grains when they are in relative motion. Minimum transport velocity is best explained by deriving the settling velocities for grains in still water. Erosion and velocity are explained by considering the force moments by which grains are rotated so that they expose their surface areas to the flowing water. It is seen that packing and orientation of sand grains may have an important effect on erosion velocities, and adhesion tension may be significant in finer-grained partly-dried sediments.

The formation of primary sedimentary structures during deposition has been explained by flume experiments in which bed forms in sand, or flow regimes, are a function of mean grain size and stream power or velocity.

The velocity for density flows can be approximated by considering the kinetic and potential energies of fluid that has a greater density than the surrounding fluid. The hydraulic potential, or "head," of a more-dense fluid causes a kinetic energy and movement as an underflow that provides a means for transport of sand into deep water.

All of the derived equations are idealizations that are helpful for visualizing the effects of current flow. They aid in understanding the final product, the sediments in vertical sequence, and in the interpretation of depositional environments and reservoir morphology.

REFERENCES CITED

ALLEN, J. R. L., 1970, *Physical processes of sedimentation:* New York, American Elsevier, 248 p.

COSTELLO, W. R., and J. B. SOUTHARD, 1981, Flume experiments on lower-flow-regime bed forms in coarse sands: Journal of Sedimentary Petrology, v. 51, p. 849–864.

FORTIER, SAMUEL, and F. C. SCOBEY, 1925, Permissible canal velocities: American Society of Civil Engineers Proceedings, Paper 1588, p. 941–956.

GILBERT, G. K., 1914, Transportation of debris by running water: U.S. Geological Survey Professional Paper 86, 263 p.

HJULSTRÖM, FILIP, 1935, Studies of morphological activity of rivers as illustrated by the River Fyris: Upsala Mineralogisk-Geologiska Institute Bulletin, v. 25, p. 222–527.

———, 1939, Transportation of detritus by moving water, *in* Parker Trask, editor, *Recent marine sediments:* Tulsa, Oklahoma, American Association of Petroleum Geologists, p. 5–51.

INMAN, D. L., 1957, Wave-generated ripples in nearshore sands: U.S. Army Corps of Engineers Beach Erosion Board Technical Memorandum 100, 65 p.

JOPLING, A. V., 1967, Origin of laminae deposited by the movement of ripples along a streambed: a laboratory study: Journal of Geology, v. 75, p. 287–305.

KERSEY, D. G., and K. J. HSU, 1976, Energy relations of density-current flows: an experimental investigation: Sedimentology, v. 23, p. 761–789.

KEULEGAN, G. H., 1946a, First progress report on Project 48: Model laws for density currents: Washington, D.C., National Bureau Standards, p. 1–17.

———, 1946b, Second progress report on Project 48: Model laws for density currents: Washington, D.C., National Bureau Standards, p. 1–17.

———, 1951, Fifth progress report on model laws for density currents. Distorted models in density current phenomena: Washington, D.C., National Bureau Standards Report 1188, p. 1–21.

———, 1957, Thirteenth progress report on model laws for density currents. An experimental study of the motion of saline water from locks into fresh water channels: Washington, D.C., National Bureau Standards Report 5168, p. 1–28.

———, 1958, Twelfth progress report on model laws for density currents. The motion of saline fronts in still water: Washington, D.C., National Bureau Standards Report 5831, p. 1–18.

LANE, E. W., 1938, Notes on the formation of sand: American Geophysical Union Transactions, 19th Annual Meeting, p. 505–508.

MANOHAR, M., 1955, Mechanics of bottom sediment movement due to wave action: U.S. Army Corps of Engineers Beach Erosion Board Technical Memorandum, v. 75, 121 p.

MIDDLETON, G. V., 1966a, Experiments on density and turbidity currents, I. Motion of the head: Canadian Journal of Earth Science, v. 3, p. 523–546.

———, 1966b, Experiments on density and turbidity currents, II. Uniform flow of density currents: Canadian Journal of Earth Science, v. 3, p. 627–637.

RAUDKIVI, A. J., 1976, *Loose boundary hydraulics:* London, Pergamon Press, 331 p.

RICHARDS, R. H., 1908, Velocity of galena and quartz falling in water: American Institute of Mining Engineers Transactions, v. 38, p. 210–235.

RUBEY, W. W., 1933, Settling velocities of gravel, sand and silt particles: American Journal of Science, v. 25, p. 325–338.

———, 1938, The force required to move particles on the bed of a stream: U.S. Geological Survey Professional paper 189 E., p. 120–141.

SIMONS, D. B., and F. V. RICHARDSON, 1960, Resistance to flow in alluvial channels: American Society of Civil Engineers Proceedings 86, no. HY5 (May).

———, E. V. RICHARDSON, and C. F. NORDIN, JR., Sedimentary structures generated by flow in alluvial channels in G. F. Middleton, editor, Primary structures and their hydrodynamic interpretation: Society of Economic Paleontologists and Mineralogists Special Publication 12, p. 34–52.

SIMPSON, J. E., 1969, A comparison between laboratory and atmospheric density currents: Quarterly Journal of the Royal Meteorological Society: v. 95, p. 758–765.

SOUTHARD, J. B., 1971, Representation of bed configurations in depth-velocity-size diagrams: Journal of Sedimentary Petrology, v. 41, no. 4, p. 903–915.

STOKES, G. G., 1850, On the effect of the internal friction on the motion of pendulums: Cambridge Philosphical Transactions, v. 9, pt. 2, p. 8–106.

SUNDBORG, AKE, 1956, The River Klaralven—a study of fluvial processes: Geografiska Annaler, v. 38, p. 127–316.

WEGGEL, J. R., 1972, An introduction to oceanic water motions and their relation to sediment transport, *in* D. J. P. Swift, D. B. Duane, and O. H. Pilkey, editors, *Shelf sediment transport: process and pattern:* Stroudsburg, Pennsylvania, Dowden, Hutchinson & Ross, Inc., p. 1–20.

WHITE, C. M., 1940, The equilibrium of grains on the bed of a stream: Royal Society of London Proceedings, v. 174, no. A958, p. 322–338.

3

Interpretation of Reservoir Morphology

Exploration for and development of sandstone reservoirs requires, above all, a reasonable prediction of sandstone occurrence and morphology. The morphology, in turn, is largely determined by the environment of deposition, and the environment is interpreted from the observed sequence of primary rock properties as displayed in cores from the subsurface.

The best interpretations are based on detailed knowledge of composition, texture, and sedimentary structures. Of these properties, the sedimentary structures are of greatest importance because they indicate most directly the nature of the processes that last acted upon the sediment and were largely responsible for its local distribution. The sequence of textural change is of next importance, and finally the composition may help to confirm the interpretation of process as well as to indicate the controls on porosity and permeability. The interpretation may be supported by other knowledge such as the regional stratigraphic setting, the nature of adjacent sediments, especially shales, the types of associated fossils, and observation of the lateral variations of the sandstone unit. The environmental interpretations from cores can be correlated with well logs for the purpose of mapping and predicting reservoir morphology and continuity.

DEPOSITIONAL ENVIRONMENTS

The interpretation of environment is the determination of the physical conditions at the time of sediment deposition. From a practical standpoint, the objective is to predict the size and shape of a reservoir sandstone from a single vertical sequence such as that exposed in a core. From the observed succession of composition, texture, and sedimentary structures, a judgment can be made of the character of the transporting medium, the conditions of flow at the time of deposition, and the nature of the depositional site. Then, an estimate of sandstone morphology may be deduced. The interpretation of environment implies certain biological and chemical conditions in addition to the physical processes, but it is the physical interpretation that leads most directly to the prediction of morphology.

The interpretations are best derived from a knowledge of modern environments of deposition. The deposits of streams, deltas, and beaches are well known and thus can be most easily extrapolated to the interpretation of sandstone sections. Interpretations by analogy become more hazardous in the lesser known environments of the marine shelf and deeper oceans.

Depositional environments are broadly classified as (1) continental, (2) transitional, and (3) marine. The *continental environment* includes all deposits of freshwater streams and lakes as well as windblown sediment, but the most common reservoir sandstones are those that originated as the fill of alluvial valleys (Figure 3–1). The *transitional environment* is that of the shore, and ranges from the fresh- to brackish-water deposits of the coastal plain to the shallowest marine. Common reservoir sandstones of the transitional zone are found in deltas at the mouths of major rivers and as barrier islands in the coastal interdeltaic zone. The normal *marine environment* is that of the shelf, slope, and deep ocean basins. Common reservoir sandstones of marine origin range from wave-washed and current-transported sands on the shelf to deeper-water submarine fans.

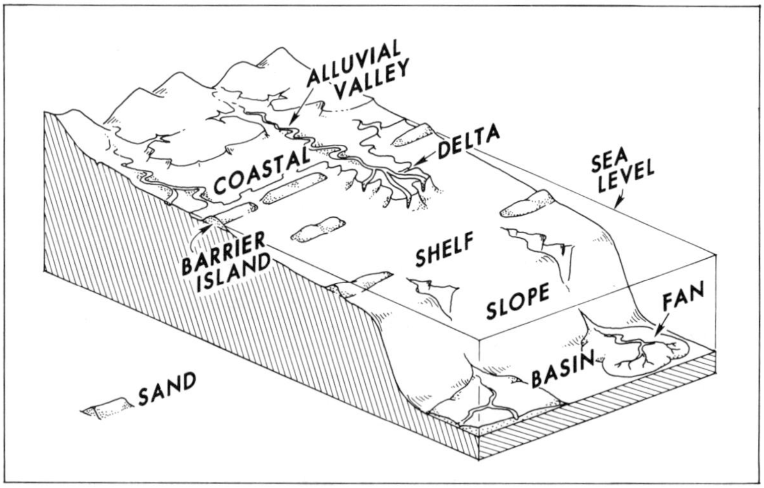

Figure 3–1 Diagram of major depositional environments for sandstones.

Modern Environments

Descriptions of modern environments are abundant in the literature. Some of the most useful volumes present models depicting the expected sequence and distribution of sand (Reading 1978; Walker 1979; and Davis 1983), whereas others illustrate details of sedimentary structures (Reineck and Singh 1975). Less commonly, the descriptions of environments are supplemented by illustrations of reservoir sandstones (Scholle and Spearing 1982).

Modern depositional environments in the Gulf of Mexico Province are a useful reference for many ancient sands (Table 3–1). The classification in Table 3–1 lists the types of significant sand accumulations that may become reservoirs, as well as the deposits of adjacent environments, which are generally those of nonreservoir sections. The continental environments are divided into eolian plain and alluvial-valley fill. The eolian plain may contain accumulations of sand that originated as windblown dunes. The alluvial valley contains deposits of sands of two general types—the deposits of the meandering, high-sinuosity stream channel and those of the braided, low-sinuosity channel. The relatively narrow channel deposits are bordered by finer-grained sediments of the fluvial flood plain.

The transitional environment consists of two categories: (1) the delta and (2) the coastal interdeltaic plain. Deltas represent a complex of sediment deposited by fluvial processes as streams enter a standing body of water such as the ocean. Consequently, the deltaic plain, or topset beds of the delta, commonly contain stream deposits of the same general type as those of the alluvial flood plain. The delta front, or foreset beds, is the mass of sediment deposited in a subaqueous environment at the river mouth. These deposits are generally gradational from the underlying marine environment upward through the fluvial sediments, and for this reason delta sediments have been termed *fluviomarine.*

In coastal areas between deltas, the shoreline consists of sands deposited in narrow linear beaches. These sands are chenier ridges or coastal barriers that are bordered on the landward side by continental deposits and on the seaward side by normal marine sediments. Both transitional environments, the deltas and coastal, have been major sites of sand accumulation throughout geologic time.

The normal marine environment extends from the shoreline to the deep basins of the ocean. The shelf environment is one of relatively shallow water that extends to the edge of the continental shelf. Many reservoir sandstones are known to have been deposited as shelf sands, although in the geologic column, shelf deposits are generally dominated by fine-grained sediments. Despite the general scarcity of sand, modern oceanographic studies are yielding valuable information concerning the transport of sand into the shelf environment. Beyond the shelf, the continental slopes and deep ocean basins contain significant amounts of sand, although, again, finer-grained sediments may be dominant.

Continental slopes are the sites of unstable sediment accumulation and may be subject to mass movement of bottom material. However, important sandstone

TABLE 3–1. CLASSIFICATION OF DEPOSITIONAL ENVIRONMENTS IN THE GULF OF MEXICO PROVINCE

- Continental
  - *Eolian plain*
    - *Sand dunes
      - Blow-out basin, intermittent lake
  - *Alluvial or fluvial flood plain*
    - Meandering stream flood plain
      - Meander belt
        - *Point bar (ridge and swales)
        - Natural levee
        - Abandoned channel (oxbow)
      - Flood basin
        - Semiarid or nonvegetated
        - Swamp, marsh, lake
        - Gathering stream
    - Braided stream
      - Channel (*dispersal, flood-plain gathering channels)
      - Interchannel flat, vegetated and nonvegetated
- Transitional
  - *Delta*
    - Deltaic plain
      - Abandoned straight distributary channel and natural levee
      - Meander belt (divisions same as alluvial flood plain)
      - Flood basin
        - Swamp (back swamp), marsh, lake, bay (transitional with interdistributary bay)
    - Delta front (subaqueous)
      - Delta fringe (shallow subaqueous; also called *proximal delta front* or *platform*)
      - *Distributary mouth bar (bar finger)
      - Distal delta front (distal bar)
        - Interdistributary bay, mud flat
        - Barrier island and spit
        - Chenier or beach
        - Tidal flat, delta and channel
      - Prodelta
        - Subdelta neritic muds
  - *Coastal interdeltaic plain*
    - (includes coastal fresh to hypersaline bays or lakes and littoral and/or very shallow neritic zone)
    - Chenier plain
      - Chenier (beach ridge)
        - Transgressive (onlap)
        - Regressive (offlap)
      - Mud flat (major regressive or offlap deposits)
      - Marsh
      - Lake and/or bay
      - Tidal pass, channel
    - Barrier island plain
      - *Coastal barrier, barrier island
        - Beach ridge
        - Sand dune
        - Shore face
        - Tidal channel
      - Bay and/or lagoon
        - Tidal channel
        - Mud flat
- Normal Marine
  - *Shelf* (transitional with subaqueous parts of deltaic and coastal interdeltaic)
    - *Inner (less than 120-ft water depths; includes subaqueous shoals, flats, bars, ridges, and delta-destructional bars)
    - Middle (120–300-ft water depths; includes mounds or rises and canyons)
    - Outer (300–600-ft water depths; includes mounds or rises and canyons)
  - *Slope*
    - Upper slope; includes mounds, canyons, silled basins
    - Lower slope or continental rise; includes scarp
  - *Basin*
    - Plain
    - *Fan or cone (turbidite)

SOURCE: Modified from Bernard and LeBlanc (1965).

*Environments of significant sand thickness.

reservoirs have been found in similar paleogeographic settings. The deep ocean basins are also sites of sand accumulation, primarily as a result of transport by turbidity currents, or bottom-following currents of suspended sediment in water. Knowledge of transport mechanisms in modern ocean basins remains uncertain despite the fact that turbidity currents are well known from laboratory experiments. Ancient deposits of the deep marine environments could be of increasing importance in the exploration for sandstone reservoirs.

The classification of modern environments (Table 3–1) is not all-inclusive because it is given for only a single geomorphic province. There are also other sites of significant sand accumulation. For example, glacial deposits are composed largely of coarse clastic materials, and glacial sediments may be delivered to the marine environment by ice flow. Alluvial fans are also composed of coarse clastic sediment, and they occur in modern areas of mountainous relief. Similarly, the processes active in alluvial fans may deliver large amounts of sand to the marine environment in areas of narrow, high-relief coastal plains. Both glacial-marine and alluvial-marine sediments, called *fan deltas,* are believed to form reservoir sandstones, but these deposits are relatively rare in the geologic record. Similarly, other environments of sand deposition may be imagined or observed, but these may be restricted in lateral or temporal extent and are considered unimportant as reservoir sandstones. Thus, the sandy sediments of the Gulf Coast Province offer a wide range of modern environments to serve as analogues for economically important sandstones.

Ancient Environments

The interpretation of depositional environment for single sandstone units permits the estimation of the depositional processes in a local area. For example, stream-channel deposits may be found in the continental environment of the alluvial valley, or they may also be present in the transitional environment of the delta plain. The local section, as exposed in a core, helps to define the mechanisms of sediment transport and deposition. However precise such an interpretation may be, other evidence is commonly needed to place the fluvial sandstone within its larger setting of meandering or braided stream, either within the alluvial valley or delta plain.

In a similar way, the interpretation of a turbidity-current deposit does not denote, a priori, a deposit of the deep ocean. The process of sand transport by turbidity currents is known to occur in shallow lakes and on marine shelves as well as in deep basins. Again, other evidence is needed to place the individual sandstone within its regional setting.

Evidence for the regional setting may come from a knowledge of paleogeography or paleontology. When such supporting knowledge is lacking or uncertain, important evidence for regional environment may come from the shales adjacent to the sandstones. For this reason, the bounding shales should be examined carefully and in no case should they be discarded.

For sandstones of marine origin, the major regional environments of deposition are *coastal, shelf, slope,* and *basin* (Figure 3–2). These terms may be used without definition of water depth and can refer merely to relative depth and distance from the shoreline at the time of deposition. The interdeltaic coastal environments are those that include brackish water to normal marine environments near the shoreline. The shelf environment is one of relatively shallow water in which sediments are deposited on a broad platform that extends out from the shoreline for variable distances in an offshore direction. The shelf area has a gentle inclination, commonly less than 1°. The slope environment has rapidly deepening waters, and the slope of the bottom sediments may be on the order of 2° or more. The basin environment has relatively deep marine waters, which, in many cases, have persisted over wide areas for significant periods of time. For ancient sediments, the shelf, slope, and basinal environments must be determined by regional mapping of stratigraphic units, or these environments may be inferred simply by their distance from a known shoreline position. No water depths can be estimated from the physical evidence alone, but some estimates may be made from the fossil assemblages in fine-grained rocks enclosing the sandstones.

In some sections, rather accurate estimates of water depth may be inferred from the associated faunas, particularly those of microfossils. Based on the study of modern microfaunas, the following environments can be recognized (Figure 3–2). The *littoral zone* is that of the shoreline itself within the tidal range. The *inner neritic environment* is that from mean sea level to a depth of 150 ft (50 m). The *middle neritic environment* extends from 150 ft (50 m) to 300 ft (100 m), and the *outer neritic* extends from 300 ft (100 m) to 600 ft (200 m). The *bathyal environment* extends from 600 ft (200 m) to 3000 ft (1000 m), below which depth is the *abyssal environment.* These environments can be identified with some confidence in younger Tertiary sediments but with less assurance in older rocks, as based on a comparison of the microfaunas with those of the modern.

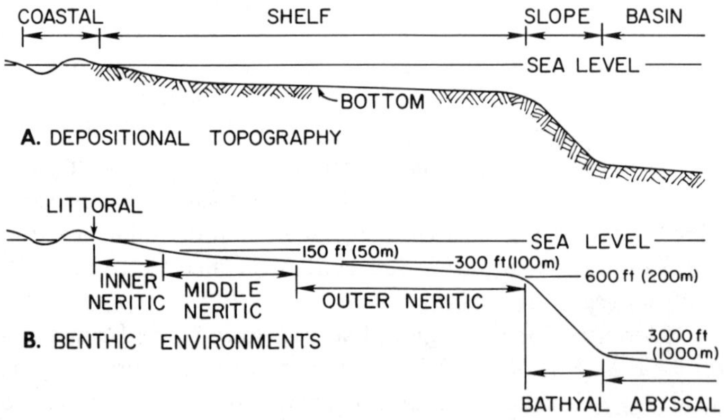

Figure 3–2 Definition of marine environments: (A) environments based on depositional topography and distance from shore without exact depth limits; (B) ecologic environments based on assemblages of benthic fauna.

The terms *neritic, bathyal,* and *abyssal* refer to the benthic or bottom environment as controlled by water depth; and in the modern oceans, these environments correspond to the shelf, slope, and basin environments, respectively. In most cases, however, the identification of bottom configuration and benthic environments is less certain for ancient rocks. Therefore, the general terms *shelf, slope,* and *basin* are used without depth connotation unless the benthic fauna in associated sediments can provide this evidence.

INTERPRETATION OF SEQUENCE

Sandstones commonly occur in regular successions that are characterized by vertical changes in sedimentary structures, texture, and composition (Visher 1965). These ordered sequences reflect the processes that transported sand and the conditions under which the sand was deposited. When observed in vertical sequence, as in a core, the succession of rock properties permits the interpretation of sedimentary processes and may allow a prediction of the size and shape of the sandstone body. Furthermore, the interpretation of sedimentary processes may permit an estimate of reservoir properties throughout the local extent of the sandstone.

Common Sequences

The most common, and therefore most important, sandstone successions are probably those transported and deposited in fresh or marine waters. The characteristics of these sequences can be summarized in a general manner (Figure 3–3). Individual sections may vary considerably from the ideal, depending on the local conditions of deposition, the nature of the source sediment, and other variables of sedimentary processes. In general, however, the idealized sections will form a guide for the interpretation of depositional environments in most reservoir sandstones, especially those that are likely to form stratigraphic traps.

Deposits in fluvial channels commonly show a sequence of sedimentary structures that range from highest flow regime at the base to lowest flow regime at the top. The dune phase of transport is prominent and is followed upward by ripples. Grain size tends to decrease upward, generally corresponding to the change in sedimentary structures. Composition is characterized by modest amounts of quartz and a smaller but significant component of rock fragments and feldspars. Matrix tends to increase upward, and again, the increase corresponds to the decrease in grain size near the top of the sequence.

The delta-front sequence may be characterized by two types of sections. The prograding-delta front sequence shows a gradation upward from bioturbated shallow-marine clays at the base and continues through a section that shows sedimentary structures of increasing flow regime upward. This section commonly contains abundant ripple laminations; but increasing numbers of higher-angle cross-beds, representing the dune phase of transport, are common near the top of the section. At

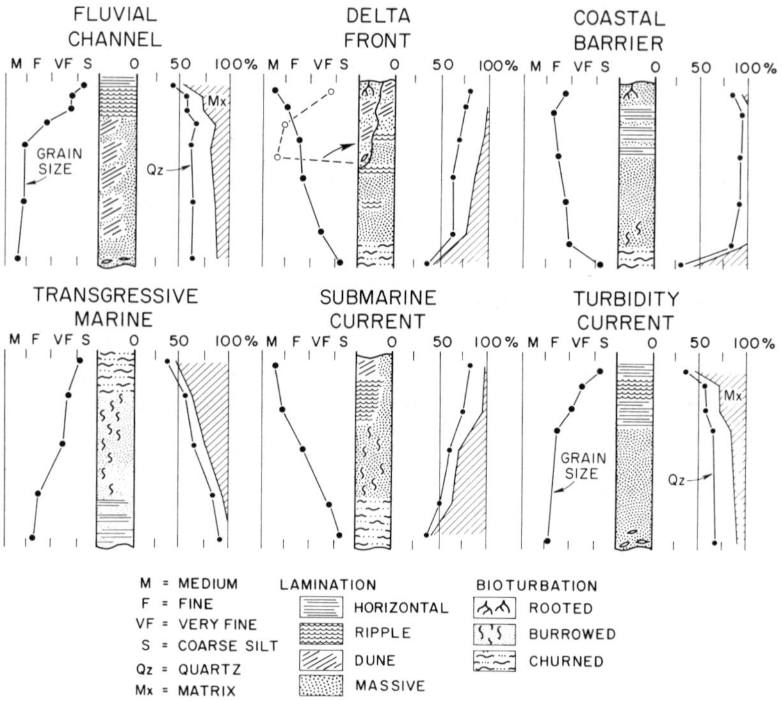

Figure 3–3 Common sequences of sedimentary structures, texture, and composition observed in reservoir sandstones of different origins. Diagrams have no vertical scale because thickness is not a criterion. [From Berg 1979, pp. 7–11.]

the same time, grain size increases upward in the section with structures of increasing flow regime. Composition is characterized by a "fluvial" aspect, with modest amounts of quartz and significant rock fragments and feldspars, but with decreasing matrix upward. This sequence represents the uniform, prograding, delta-front deposits. It is the only sequence that grades from marine (or other prodeltaic sediment) below to dominantly current-transported sediments above. Upward gradation from marine to fluvial is the principal indicator of delta-front deposition.

A major modification of the prograding delta is found where fluvial channels of the deltaic plain have cut deeply into the previously deposited delta-front section. In this case, the upward gradation is interrupted by a typical fluvial sequence of decreasing scale of sedimentary structures upward, a decreasing grain size, and a decreasing grain content upward and an increasing matrix. The top of the section may be bioturbated by rooting or may be overlain by coals indicating the presence of a delta-plain marsh or swamp.

Shoreline sandstones are represented by the typical sequence of the coastal barrier-island section. In this sequence, the normal-marine bioturbated muds of the inner neritic environment are succeeded by massive sandstones, also bioturbated, and near the top of the section, by laminated sandstones in which the laminae are inclined at low angles. The upper laminated zone represents the bedding produced by wave action at the beach face. This section may be overlain by eolian dune sandstones, which document the emergent part of the barrier-island section. The eolian sands may be bioturbated by rooting or may contain other evidence of subaerial exposure. Grain size tends to increase upward in the section, but the sands are generally well sorted, and textural gradation is slight. In some cases, the composition is characterized by a high quartz content and the lack of clay matrix. Persistent wave action tends to increase the content of more resistant grains and often reduces the more easily abraded components such as feldspars and rock fragments.

A transgressive marine sandstone commonly shows a sequence that is the inverse of the shoreline sands. Wave action along the encroaching shoreline first produced laminated sands at the base of the section, which were succeeded by massive bioturbated sands, and finally, at the top of the section, by the normal-marine bioturbated muds. Thus, the total sequence represents increasing water depths through time. Grain size decreases upward in a regular manner. Composition is generally highest in quartz content at the base and has increasing amounts of matrix upward.

Marine-shelf sandstones are of several types, but only one type deposited by submarine currents is illustrated (Figure 3–3). This sequence is characterized by an upward gradation from interbedded sandstone and shale below to dominant sandstone above. Structures range from small ripples below to increasing ripple size above, and in some cases, to a high-velocity dune phase of transport at the top. Bioturbation is abundant in the lower part and commonly is present throughout the section but in decreasing amounts upward. Grain size and quartz content generally increase upward, and the sandstones may have relatively coarse grain size. The morphology of marine-shelf sandstones ranges from large ovate bodies to narrow linear units of considerable length, and both types are generally elongate parallel to the distant shoreline.

In many aspects, the shelf sandstones resemble the vertical sequence of deltas, but they are distinctive in their more abundant bioturbation and a lack of evidence of subaerial exposure such as delta-plain rooting, coals, and fluvial channels. Furthermore, the shelf sandstones are commonly found as thin units within thick sections of dominant marine shales.

Deposition in deep marine basins includes the deposits of turbidity currents, or *turbidites*. In complete sequence, sedimentary structures range from a basal massive unit through laminated sandstones, overlain by rippled sands and finer-grained suspension deposits. The sequence is similar to that of fluvial channels in that it indicates decreasing flow regime, but it lacks the common dune phase of transport. Grain size decreases upward within the bed, corresponding to the change in sedimentary structures. In composition, turbidity current deposits are variable, and the quartz content may range from only 25 percent to much greater amounts, depending primarily on the composition of source materials. Matrix increases upward, corresponding to the decrease in grain size. Many turbidite sandstones lack the complete, or idealized, sequence of sedimentary structures, and in this respect the individual sandstones are variable and quite unlike other marine sandstones.

Evaluation of Criteria

Sedimentary structures. In practice, the interpretation of depositional environment depends largely on the interpretation of sedimentary structures in vertical sequence. From the primary structures, the processes of sediment transport that immediately preceded deposition may be deduced. The other properties of textural variation and composition may add significant support to the interpretation of processes, but composition and texture alone, without structures, are usually not diagnostic. Without knowledge of sedimentary structures, interpretation remains possible but not well documented.

Texture. Textural change in vertical sequence is the next most important criterion for interpretation. In fact, grain size is commonly a function of flow regime and, therefore, tends to change with sedimentary structures, as shown in the idealized sections (Figure 3–3). This relation means that textural changes can be useful when only small samples are available and sedimentary structures cannot be observed. Texture and composition are mutually dependent also, and the decrease in mean grain size is generally accompanied by a decrease in grain content and an increase in matrix. Texture is inherited from source materials, so that absolute size is not a criterion of environment. For example, many shoreline sands are fine grained and well sorted, but coarse-grained to pebbly sands may also be present.

Composition. Sandstone compositions show great variation because they are determined largely by the nature of the source material. Therefore, composition alone may be of limited value in the interpretation of environment for individual sandstones. On the other hand, if the composition of the source material is known, different depositional processes may have caused a change in composition. Hence, the use of composition may be important when there is a background of data from a single depositional system. In this case, compositional data may provide important clues to depositional environment.

The influence of source on composition is well illustrated by sandstones from the Oligocene Frio Formation in the Texas Gulf Coast (Figure 3–4). Sandstones from the upper Texas Coast tend to be high in quartz content, whereas sandstones from the middle and lower Texas Coast have increasing amounts of feldspar and lithic fragments. In South Texas, volcanic rock fragments form an important part of the composition. This variation probably reflects the contribution of volcanic rock fragments from the ancient Rio Grande River to the South Texas Coast. Farther north, fine-grained volcanics are less common, but glass fragments from airborne ash persist in small amounts.

Modification of original composition takes place by abrasion during sediment transport. Long-distance transport by rivers tends to reduce the number of feldspar grains and rock fragments, increasing the relative amount of more resistant quartz grains. When such sediment is delivered to the shoreline, reworking of the sediment by the waves generally results in a further increase in quartz content. However,

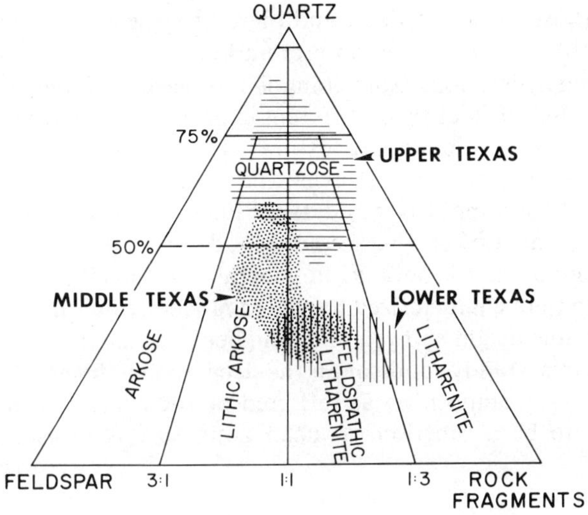

Figure 3-4 Composition of Oligocene Frio sandstones from the upper, middle, and lower Texas Gulf Coast. [After Loucks, Bebout, and Galloway 1977.]

short-distance transport by rivers results in little change in composition; and subsequent reworking by waves, although increasing quartz content, may produce a sand of only moderate quartz content. Therefore, two shoreline sands may be greatly different in quartz content because of the difference in the materials supplied to the beach.

Interpretation of composition. An early use of composition was for the interpretation of tectonic environment (Krynine 1948). Quartz-rich sandstones, called *orthoquartzites,* were known to be commonly associated with Paleozoic sediments of the stable craton. Clayey litharenites, or *graywackes,* were known to be geosynclinal sediments. Feldspathic sands, or *arkoses,* were thought to be common in basins adjacent to active orogenic uplifts. Essentially the same approach has been used more recently to relate sandstone composition to the site of deposition in a plate-tectonics framework (Dickinson and Suczek 1979). As might be expected, interpretations of tectonic environment from compositional data alone can be hazardous because the nature of local source materials determines the composition of the resulting sandstones.

An early use of composition for environmental interpretation was in the Oligocene Frio sandstones of the Seeligson field, South Texas (Nanz 1954). Frio reservoir sandstones showed fining-upward textures and a pattern of narrow, dip-trending, low-sinuosity bodies within the field area. A fluvial origin for these reservoirs was confirmed by compositional analysis. The reservoir sandstones contain about 50 percent quartz with abundant igneous, sedimentary, and volcanic rock fragments. This composition is similar to modern fluvial sands of the Texas Gulf Coast, whereas the modern beach sands show an enrichment in quartz to greater than 80 percent. It was assumed that the distribution of modern depositional environments was similar to those of the Tertiary, and therefore, the composition of Frio sandstones was compatible with a fluvial interpretation.

Other environmental interpretations from composition have been moderately successful against a background of data from different depositional environments. Statistical analysis of seven compositional and textural elements showed that environmental differences could be significant (Davies and Ethridge 1975). Samples were taken from the modern Mississippi River and delta, Eocene Wilcox sandstones from the Texas Gulf Coast, and Lower Cretaceous Muddy sandstones from the Powder River basin in Wyoming and Montana. In general, the fluvial and deltaic sediments of this varied suite were characterized by low to moderate quartz content, whereas sandstones of shoreline origin showed much higher quartz content.

For the Lower Cretaceous Muddy sandstones, the depositional differences are locally important (Figure 3–5). Composition as reflected by the content of monocrystalline quartz is shown to be a function of mean grain size as measured by

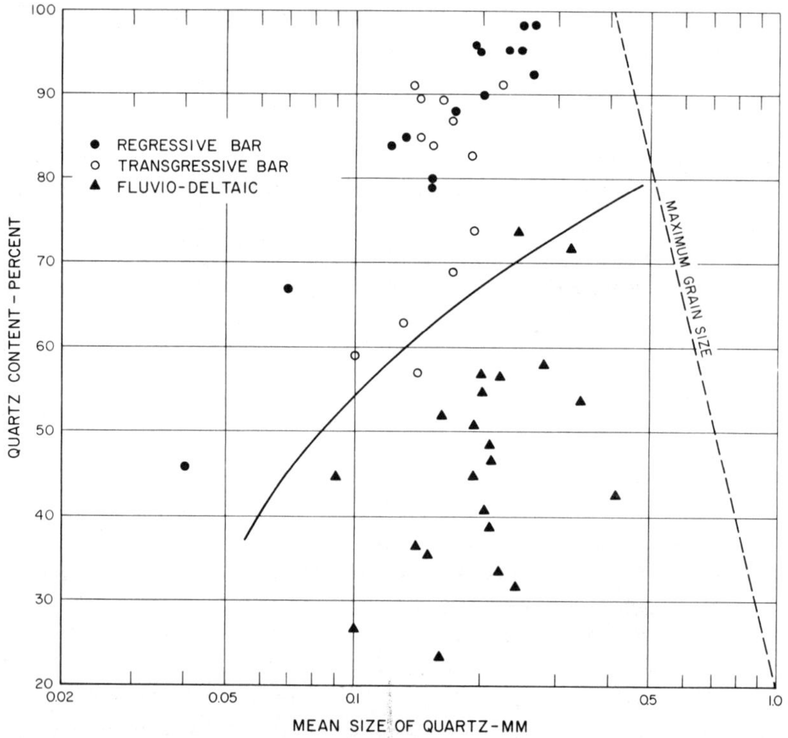

Figure 3–5 Quartz composition and texture for selected samples of the Lower Cretaceous Muddy sandstones, Powder River basin in Wyoming and Montana. Monocrystalline quartz content is a function of mean grain size of quartz for selected samples of fluvial-deltaic, barrier-island (regressive), and transgressive-marine sandstones. [From Berg 1970.]

monocrystalline quartz content in the range of 40 to 60 percent, with relatively large maximum grain size. This sediment was the source material supplied to the ancient Muddy shoreline.

Two types of marine sandstones can be distinguished. The first type consists of the barrier-island, or regressive, sands, which show an enrichment in quartz to greater than 90 percent, and the content of feldspars, rock fragments, and matrix was greatly reduced by wave action. In associated lagoonal sediments, quartz content decreases, largely due to the increase in clay matrix. The second type, the transgressive-marine sands, also show an increase in quartz content because of the reworking of fluvial-deltaic sands by wave action. Transgressive sands can be distinguished from barrier-island sands by the increase in matrix content upward with an attendant decrease in mean grain size. Both the barrier-island and the transgressive sands have greatly reduced maximum grain sizes as compared to the fluvial-deltaic source sediment.

The functional relation of quartz content and grain size (Figure 3–5) is consistent for more than 300 samples from 40 different cores, but the distinction in sand type could be made on the basis of only three different sections. Thus, a background of compositional and textural data can be significant to further environmental interpretations.

Once the distinctions have been established, environmental interpretations can be made with some confidence from composition and texture alone. Thus, as the amount of background data increases in any depositional system, the usefulness of composition becomes apparent. It has been proposed that similar reliable interpretations can be made from compositional and textural data from small samples such as sidewall cores (Ethridge, Gopinath, and Davies 1975). Again, the success of this method depends largely on the amount of background data available for comparison. In other words, the nature of the immediate source sediment must be known.

Despite the strong influence of source on composition, several generalizations can be made regarding composition and depositional environment:

1. Alluvial valley sands commonly have low to moderate amounts of monocrystalline quartz in the range of 40 to 70 percent, and the balance of grain composition is largely feldspar or rock fragments.
2. Delta-plain and delta-front sands have the same composition as fluvial sediment, and, generally, a low to moderate quartz content. However, these coastal plain sediments may be enriched in quartz content by long-distance transport and by reworking of previously transported coastal-plain sediment. Thus, the quartz content may be significantly increased to about 80 percent.
3. Shoreline sands, especially those subjected to prolonged wave action, may be greatly enriched in resistant mineral grains, and the content of monocrystalline quartz may reach values of 80 percent or more.
4. Marine-shelf and basin sands are generally not characterized by any single composition but owe their detrital elements to the immediate source sediment.

It is obvious that composition can be equivocal for interpretation, but in some cases, composition may be of supporting value for the interpretation of both tectonic and depositional environments. These data are best used with other information on rock properties.

INTERPRETATION FROM WELL LOGS

Interpretation of a cored section can be correlated to the response of the same section on well logs. With a knowledge of depositional environment for the cored section, differences in well-log response can generally be attributed to lateral changes in rock character, which, in turn, can be identified in relation to the section interpreted. For example, a thick fluvial-channel sandstone can be expected to change laterally to more thinly bedded sandstones and shales of an overbank section. And the response of the overbank section on well logs can be expected to be quite different from that of the thick fluvial sandstone.

Indentification of facies can then be made for well logs in an exploration area. At this point, when well control is widely spaced, the interpretation of the morphology of possible reservoirs cannot be mapped precisely. However, interpretation of a cored section can set certain limits to the morphology so that exploration efforts may then be concentrated in favorable areas.

With increased density of well control, such as in a developed field, the thickness and extent of reservoir sandstones may be mapped with increased confidence. The identification of lateral changes in well logs can be made with some assurance, and the final interpretation of morphology should confirm the interpretation of environment as determined by the cored section, or the interpretation must be revised. Mapping within field areas has the advantage that lateral changes can be identified readily, and the knowledge gained in field mapping can be applied with increased confidence to areas of exploration and sparse control.

Well logs provide information about the physical properties of the vertical sequence in sandstones, so that an interpretation can be made of the depositional environment. This use of logs is most reliable with some direct knowledge of the rock as from cuttings through the section or from cores in nearby wells. The interpretations become less reliable, even highly subjective, as the amount of direct knowledge decreases or as interpretations are carried to greater distances from known sections. Nevertheless, interpretations must be made for the purpose of stratigraphic mapping because well logs are always available as a record of the drilled section, whereas cores and even sample logs are not common.

SP and Gamma-Ray Logs

The most useful logs for interpretations of sequence are the SP and gamma-ray logs (see Table 1-6 in Chapter 1). Both logs reflect the relative amounts of sandstone and shale in vertical sequence. The SP response is greatest in sandstones that contain saline waters, and it is reduced when formation fluids are more highly resistive. The SP response is also affected by bed thickness and is greatly reduced in thin beds of 10 ft or less in thickness. The gamma-ray log is not affected by the formation fluids, and its response is due largely to the potassium isotope (^{40}K), which is common in clay minerals. Therefore, the natural gamma-ray log directly reflects the amount of clay or shale present. The gamma-ray log is also affected by bed thickness, but it responds to thin beds on the order of only 3 ft (1.0 m) in thickness.

Both the SP and gamma-ray logs respond to increasing sandiness by deflection to the left, increasingly negative for the SP curve and with decreased radiation for the gamma-ray log. This deflection is referred to as *coarsening* in sandstones for the reason that an increase in grain size is commonly accompanied by a decrease in the matrix, or shale, content. Thus, the logs are interpreted in terms of the interdependency of grain size and composition. However, the SP response may be reduced by the high resistivity of the formation so that a large amount of cement in an otherwise matrix-free sandstone may appear to be "shaly."

Despite the sometimes variable nature of log response, an experienced interpreter can use the logs to deduce the general nature of the sequence and thus the most likely depositional environment. Other logs, such as those that reflect porosity, are a useful supplement to the SP and gamma-log interpretations. The following examples illustrate the interpretation of a single type of sequence, that of a delta front, by two different combinations of logs. The examples show the principles of interpretation, principles that may be applied to many other sections of sandstones and interbedded shales.

Interpretation from SP log. The prograding sequence of delta-front sediments is easily identified in electric logs. An example is from the Lower Cretaceous Fall River Sandstone in the northeastern Powder River basin (Figure 3-6). The SP response is increasingly negative (to the left) upward from 5844 ft, reflecting a decrease in clay matrix and an increase in grain size upward. These changes were

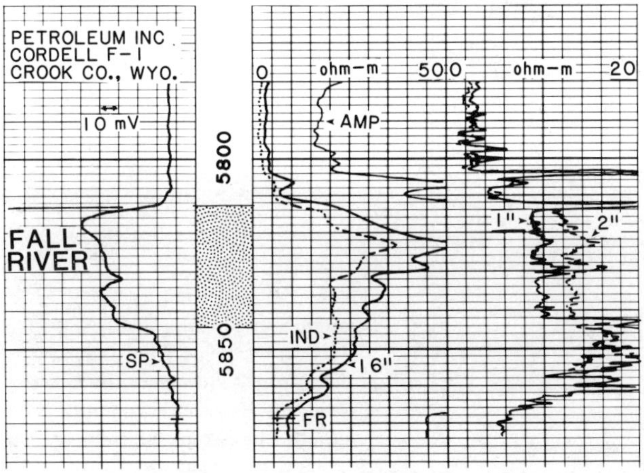

Figure 3-6 Spontaneous potential (SP), induction (IND) resistivity, and microresistivity logs (far right) through the Lower Cretaceous Fall River sandstone, northeastern Powder River basin, Wyoming. Sandstone section is stippled.

confirmed in a core of the Fall River section. At the same time, resistivity increases upward for the same reasons, but the sandstone contains gas, which also contributes to high resistivity. There is a distinct decrease in resistivity in the uppermost 10 ft (3 m) of the section, which probably reflects an increase in water salinity in a thin transgressive marine sandstone at the top of the delta-front section. The log was run in a well drilled in the same area as the core that illustrates the deltaic sequence of the Fall River in the Coyote Creek area (see Chapter 5).

The micro-resistivity log (Figure 3–6) indicates porous sandstone through the section where there is separation between the 1-in. and 2-in. curves. This porous sandstone represents the distal and distributary bars of the delta front that overlie a prodelta section of thinly interbedded sandstone and shale.

Interpretation from Gamma-Ray log. Gamma-ray logs also have a distinctive response in delta-front sections as shown by a log through a sandstone and shale sequence in the Lower Cretaceous Hosston Formation (Figure 3–7). The gamma-ray response indicates thinly interbedded sandstone and shale at a depth of about 14,650 ft, which represents the prodelta section. The gamma-ray response decreases upward, indicating increasing amounts of sandstone and decreasing amounts of thinly interbedded shales to the top of the deltaic section at 14,604 ft. The presence of a prograding delta-front section is confirmed in this well by the recovery of a core through the interval. A typical prograding sequence was observed in the core. Grain size and grain content increase upward as matrix decreases. The upper part of the section is characterized by inclined laminae that represent a dune

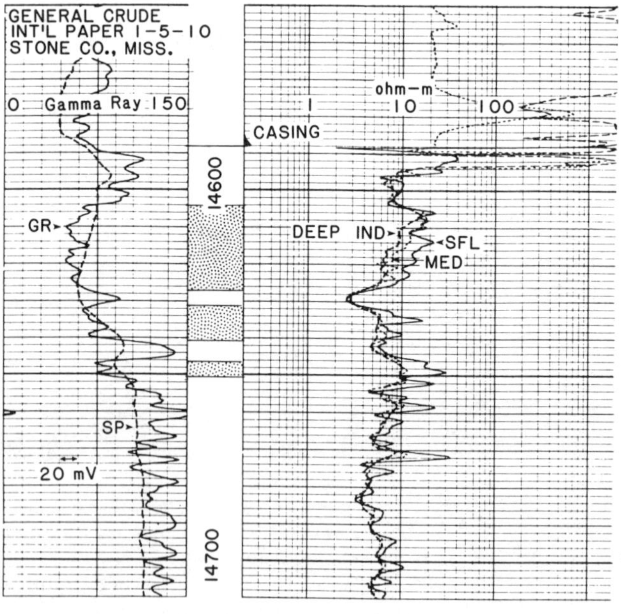

Figure 3–7 Gamma-ray (GR), spontaneous potential (SP), and induction resistivity (right) logs through a delta-front sequence in the Lower Cretaceous Hosston Formation, southern Mississippi. Sections of dominant sandstone are stippled.

phase of transport. Therefore, the gamma-ray log is correctly indicating the change in texture and composition.

Facies Change

The response of the SP or gamma-ray logs to the lateral as well as vertical change in rock type is illustrated by examples of logs taken through the cored sections described in this text (see Chapters 5-9). Fluvial sandstones commonly show an abrupt increase in grain size at the base where sandstones overlie eroded shales beneath the fluvial channel [Figure 3-8(A) and (B)]. The sandstone sections may show a rather uniform sequence with little significant variation, or the SP curve may show thinly interbedded shales by a more variable response. At the top of the sandstone, the SP log records an increasing amount of shale and decreasing grain size as the bed form changes to fine-grained, ripple-laminated sandstone or siltstone. Thus,

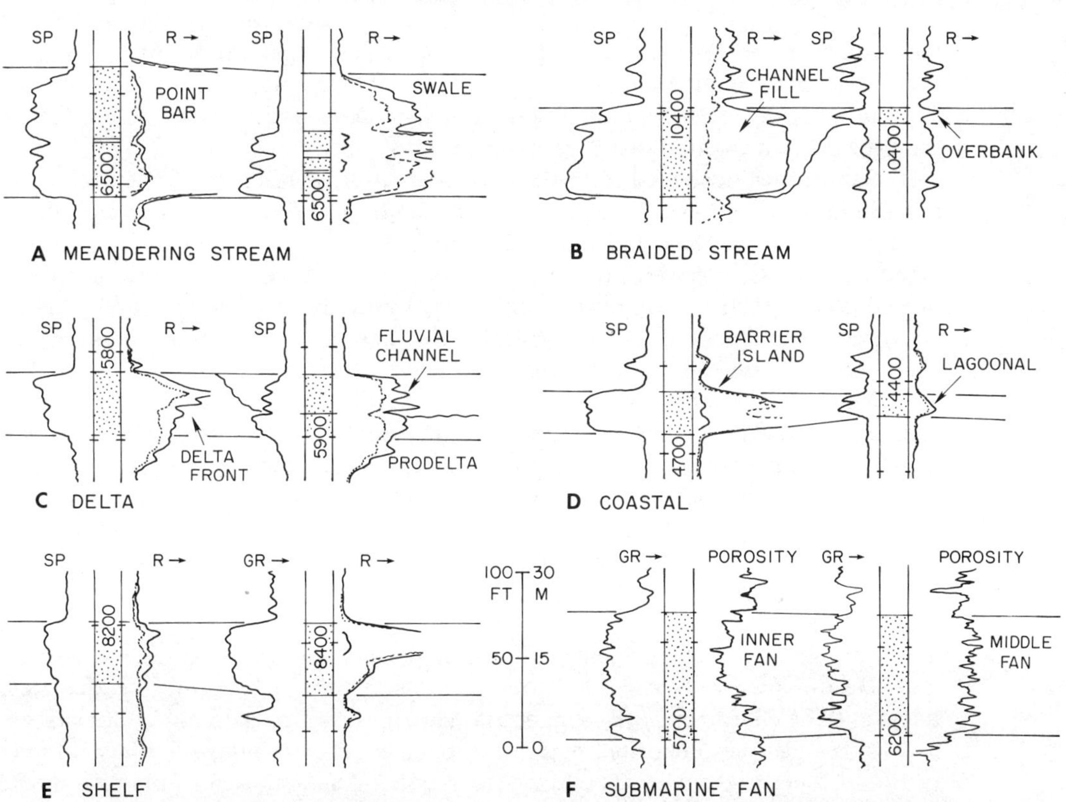

Figure 3-8 Typical response of spontaneous potential (SP) and gamma-ray (GR) logs in sandstone sequences. Log patterns are similar to textural changes (Figure 3-3). Dotted patterns indicate dominant sandstone as interpreted from the logs. Resistivity (R) or porosity logs are also shown.

the log records a grain-size change similar to the typical fluvial sequence (Figure 3–3).

Lateral changes in fluvial sections also can be identified on well logs. Thick fluvial sandstones that were deposited as point bars in a widely meandering stream may change abruptly to thinly interbedded sandstones and shales that were deposited in low areas, or swales, between the point bars [Figure 3–8(A)]. In this case, the lower part of the sandstone unit is continuous between wells, whereas the sandstone is replaced by shale in the upper part of the unit. In contrast, sandstones deposited as channel-fill in narrow braided streams show an opposite change [Figure 3–8(B)]. The upper part of the channel-fill section is continuous with thinly interbedded sandstones and shales of overbank deposits, whereas the lower part of the unit is replaced laterally by shales into which the channel was eroded. Thus, the lateral change indicated by well logs often suggests a morphology for the sandstone, whether highly sinuous and meandering, or narrow and braided.

Deltaic sandstones show an increasing sand content and grain size upward in the section that reflects the vertical gradation from marine prodelta shales below to the delta-front sandstones above [Figure 3–8(C)]. Typical variations in the delta-front sequence are also identified on logs as, for example, where fluvial channel-fill sandstones of the prograding delta plain are deposited in channels eroded into the previously deposited delta-front sequence.

Sandstones deposited in the coastal interdeltaic zones are represented by barrier-island and lagoonal sequences [Figure 3–8(D)]. The barrier-island or littoral sandstones are typically thin units of dominant sandstone bounded by marine or lagoon shales. In a seaward direction the barrier-island sandstones are replaced by marine shales, and in a landward direction, by thinly interbedded sandstone and siltstone of the lagoonal facies. Typically, the lagoonal section has a depressed, or shaly, response on the SP curve and is somewhat thinner.

Marine-shelf sandstones formed by submarine currents are typically found as isolated units within thick marine shales [Figure 3–8(E)]. These sandstones commonly show an increase in sand content and grain size upward, similar to the delta-front section, but both the SP and gamma-ray logs suggest a large amount of interbedded shale. The SP response may be greatly reduced in shelf sandstones that are highly bioturbated and in which the sand and shale are completely mixed by burrowing [Figure 3–8(E) left]. In other cases, the log may indicate a section of clean, relatively shale-free sandstone at the top [Figure 3–8(E) right]. Both clean and shaly sands may be present within a single shelf-sandstone unit.

Deep-marine or basin sandstones are characterized by thinly bedded turbidites, and the character of the sequence is clearly shown by gamma-ray logs [Figure 3–8(F)]. Sandstones deposited within a submarine fan are units of thinly interbedded sandstones and shales. The inner-fan sequence is composed of massive stacked channel sandstones that represent the fill of slope channels that supplied sand to the deeper submarine fan. The gamma-log character shows the dominant sandstone of the inner fan, but the irregularity of the gamma-ray response suggests the thinly bedded nature of the sequence. The middle part of a submarine fan contains more interbedded shale, which is reflected in the highly variable response of the gamma-ray log. The outer part of the submarine fan is increasingly dominated by shale until only marine shales are present beyond the limits of the fan. Unlike the other sandstone sequences, the log response does not define single thin sandstone beds, but the overall character of the sequence is readily identified by the logs alone.

In many cases, the general nature of the sandstone section can be interpreted from the well logs themselves, and a prediction can be made for the areal extent of reservoirs. However, increased confidence of interpretation and prediction of morphology can be achieved by correlation of well-log response with cores. In the following reservoir descriptions (Chapters 4–9), each cored section is accompanied by some combination of well logs that shows a typical response to the sequence. It must be remembered, however, that each log reflects a particular combination of rock properties as well as fluids, so that the response is essentially unique to that section.

SUPPLEMENTAL DATA

Interpretation of sandstones can be supported by a knowledge of the local stratigraphic succession and of the nature of adjoining beds, especially shales. In many basins, even a rudimentary knowledge of the regional geology can tentatively assign a given sandstone to a continental, transitional, or marine environment. And for basins in which the stratigraphy is well known, the trends of depositional "dip" and "strike" are even more clearly established and assist in determining the local trends of sandstone bodies.

Shales

The nature of the adjacent shales can also provide an important clue to the character of the depositional site. For example, a high degree of bioturbation in shales denotes a shallow-marine environment of well aerated waters that supported a diverse benthic fauna. On the other hand, lack of bioturbation indicates a generally hostile environment for bottom-dwelling organisms that could be either freshwater, a restricted anaerobic setting, or excessively deep marine water. In no case, however, does the character of bioturbation alone establish depositional sites.

In the marine environment, abundant bioturbation is commonly associated with sand. This relationship suggests that submarine currents not only transported sand but also supplied nutrient-rich waters for maintaining a robust benthic community of organisms. Some types of burrows, as well as other trace fossils, do suggest evidence for water depth. These distinctive forms may have a wide range of occurrence and are not indicative of specific depths, but their association with other forms in an assemblage of trace fossils is often suggestive of environment.

Microfossils

Fossils in adjacent shales, especially microfossils, can provide important indicators of environment and even an estimate of water depth. The use of microfossils for paleobathymetry is well established in many areas, especially in the Tertiary section of the Gulf Coast Province and other areas. During the 1940s, studies of foraminiferal assemblages from bottom sediments in the modern Gulf of Mexico provided the basic knowledge for distribution of microfossils according to depth. Foraminifera are one-celled animallike organisms that secrete shells that may be preserved in ancient sediments. The shells are small, normally less than 1 mm in diameter, and they survive intact in drill cuttings. Hence, the use of Foraminifera became widespread in the early history of Gulf Coast drilling as reliable stratigraphic markers. Their use in paleobathymetry became established soon thereafter.

Benthic Foraminifera are those that live on or near the bottom, and their shells may be either calcareous or arenaceous. Calcareous shells are either glassy in appearance (hyaline), or they are translucent (porcellaneous) in appearance. Arenaceous shells consist of fine-grained clastic materials that are held together by organic substances secreted by the organisms.

Planktonic Foraminifera are those that live in the water column above the bottom and have a floating habit. They may be attached to other organisms such as algae. The shells are generally calcareous and globular in shape.

Modern microfaunas. Studies of foraminiferal distributions in the modern oceans have shown that some benthic species are adapted to relatively narrow depth ranges. Distribution of benthic species is influenced largely by temperature and pressure. On the other hand, the planktonic species are independent of bottom conditions, and their distributions are influenced primarily by the temperature of the surface waters. The principles of paleobathymetric interpretation are based on these different habits. Assemblages of species, or faunas, are characterized by certain benthic forms, some of which may have restricted depth ranges. The benthic faunas are particularly abundant in shallow-marine waters but are less common in the deeper, more hostile environments. Consequently, planktonic species become increasingly abundant in the deeper-marine environment because on the death of the organisms, the shells sink to the bottom and become a dominant part of the faunal assemblages.

Early studies in the Gulf of Mexico showed the relative abundance of foraminiferal species according to water type and depth (Lowman 1949). These conclusions are summarized in a traverse from the deltaic plain of the Mississippi River, extending across the shoreline and shelf to the abyssal depths in the Gulf of Mexico (Figure 3–9). The traverse was made where the continental shelf was narrow. This section, therefore, is somewhat compressed as compared with a normal horizontal scale for the Gulf, but it provides a wide range of depth conditions.

The fresh to brackish-water environment of the delta plain is characterized by arenaceous Foraminifera of the species *Difflugia* in fresh water, and *Ammobaculites*

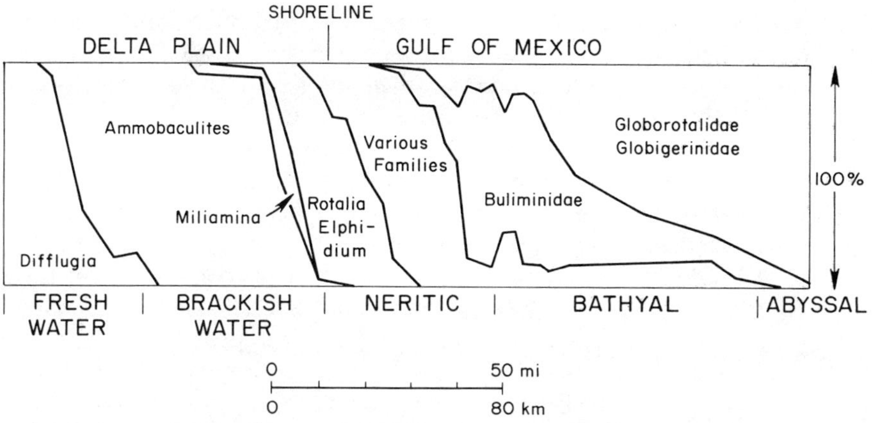

Figure 3–9 Distribution of Foraminifera in bottom sediments from the Mississippi delta and Gulf of Mexico shelf and slope. [From Lowman 1949, p. 1952.]

in the fresh to brackish transition. Another finely arenaceous form, *Miliamina*, occurs at the outer margin of the transition zone. The strongly brackish to normal marine environments of the shoreline are characterized by the genera *Rotalia* and *Elphidium*. These are robust calcareous forms that often survive even the strong wave action at the shore face.

The inner and middle neritic zones contain a diverse fauna of both calcareous and arenaceous species that represent different families. These zones of the normal marine environment are the most favorable for many types of organisms as well as the Foraminifera, and hence the number of individuals as well as species tends to be high.

The outer part of the neritic zone, in water depths greater than about 300 ft (91 m), is characterized by a decrease in diversity of microfauna and the dominance of several genera of the family Buliminidae. Also in the outer neritic zone, a significant number of planktonic Foraminifera begin to appear in the bottom assemblage.

The bathyal environment of the continental slope shows a steady increase in planktonic forms and a decrease in benthic forms as water depth increases. In the abyssal zone of water depths greater than 6000 ft (2000 m), the generally hostile bottom conditions result in a near disappearance of benthic genera, and planktonic species of the families Globigerinidae and Globorotalidae are dominant. Bottom temperatures at these depths are commonly on the order of 1° to 2°C. Only a few arenaceous benthic species survive under these rigorous conditions. The dominance of planktonic Foraminifera in progressively deeper water means that the ratio of planktonic to benthic species is a useful environmental indicator in the outer neritic, bathyal, and abyssal zones.

Bathymetric ranges for modern bottom assemblages have been defined in many areas, and the distributions are similar to those illustrated (Figure 3–9). It is now possible to estimate the depth ranges for fossil microfaunas in the younger rocks of the Tertiary. The fossil assemblages can be assigned depth ranges with considerable confidence because many fossil species are the same as, or similar to, living species. In the older Tertiary section, many of the fossil species are not now living, and in this case, related species are assumed to have had similar depth ranges to those of living species. For still older microfaunas, it is not possible to assign depth ranges with precision because most of the fossil species are now extinct, and even genera and larger groups are missing. Still, it may be assumed that similar morphological types had depth ranges on the same order as those of living species, and approximate depths may be assigned to older fossil assemblages, largely on the basis of gross morphology or phylogenetic relationships.

Ancient microfaunas. Application of these principles to the Tertiary faunas of the Gulf Coast became well established during the 1960s, and paleoecological interpretations have been summarized for many sections (Albers 1966; Scull et al. 1966; Skinner 1966; and Tipsword, Setzer, and Smith 1966). Today paleoecological zonation according to depth ranges is commonly used in oil and gas exploration.

Paleobathymetric intrepretations based on microfaunas must be used with some caution. Two factors compound the interpretation problem, and these are (1) the displacement of benthic species by currents, and (2) the destruction of calcareous forms by abrasion or solution. Older microfaunas, particularly those of the larger and more resistant types, can be eroded and transported with clastic materials. An experienced micropaleontologist will readily recognize the transported shells by their abraded character and stratigraphic displacement. More difficult is the recognition of contemporaneous microfossils that may have been displaced from shallow water into deeper water by bottom currents. Therefore, use of microfaunal assemblages for paleobathymetry may result in serious errors. In these cases, only the deepest ranging species can be used for accurate depth determinations (Figure 3–10). The hazards of interpreting displaced faunas has long been recognized in many areas, especially in the deep-water Tertiary sections of California (Bandy and Arnal 1960; and Natland 1968).

Another problem for intrepretation is the physical conditions of the environment, which may determine whether the bottom fauna is preserved. For example,

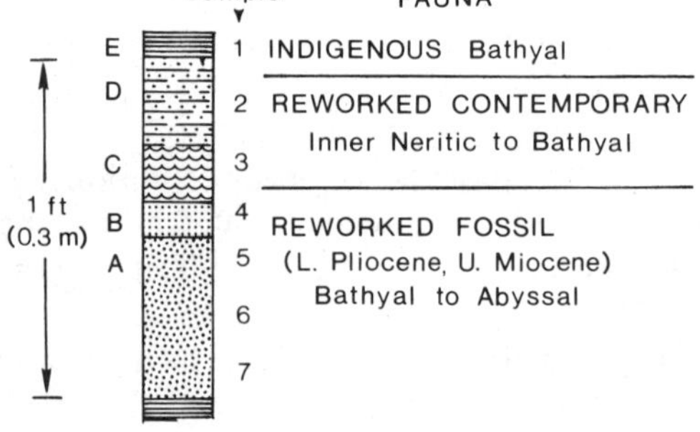

Figure 3–10 Microfauna in a deep-water turbidite sequence. [From Natland 1968.] Only the upper pelagic shale has a fauna that represents true water depth.

at great depths in the ocean, about 4000 m, the bottom waters are undersaturated with respect to calcium carbonate, and the rate of calcium-carbonate supply is approximately equal to the solution rate. This depth is called the *calcite compensation depth* (CCD), and it occurs at variable depths in modern oceans depending on the temperature, which is controlled in turn by oceanic circulation patterns. Below the CCD, the shells of calcareous species are dissolved, and only arenaceous Foraminifera survive in the bottom sediments. Consequently, a deep-water fauna may be depleted in planktonic forms, and the occurrence of rare arenaceous individuals might suggest a quite different environment such as that of brackish water. Commonly, stratigraphic position and location within a basinal sequence can aid in establishing whether calcareous forms are absent because of excessive water depths.

In summary, the use of Foraminifera in paleobathymetric interpretations can be briefly stated as follows. An abundance of single arenaceous species but low faunal diversity indicates fresh to brackish water. Robust, calcareous Foraminifera indicate the shoreline and inner neritic environments. A highly diverse, calcareous fauna indicates an inner to middle neritic environment. Finally, increasing planktonic Foraminifera characterize the outermost neritic and deeper-water environments.

Stratigraphic Sequence

The nature of the stratigraphic sequence is also helpful to interpretation. The succession of beds that includes the reservoir sandstone may be termed *transgressive* if sandstones are dominant in the lower part of the section and grade upward to marine shales in the upper part. The succession is *regressive* if marine shales in the lower part grade upward to dominant sandstones. The terms *transgressive* and *regressive* imply the advance or retreat of the shoreline, respectively, but the successions may have been the result of a decreasing or increasing supply of coarse sediment in a continuously downwarped basin of deposition.

The nature of the succession is commonly obvious in well logs, and the interpretation of individual sandstones may be aided in the following manner. In a regressive sequence, the first thick sandstones encountered above marine shales may be of shoreline origin, either barrier-island or delta-front, and above the transition zone, sandstones may be of fluvial origin in a coastal plain setting. In a transgressive sequence, the reverse succession is expected, perhaps with modifications of the shoreline sands by succeeding processes of wave action and submarine-current erosion and transport. In other words, sandstones in the regressive sequence may retain their original morphologies, whereas those in the transgressive sequence can be altered greatly by reworking.

Core Analysis

Extrapolation of morphology in exploration and production areas can be supported by indirect evidence of rock properties. For example, measurements from core analysis may be available long after the core itself has been discarded or lost. In this case, the record of permeability is particularly useful. Permeability is partly a function of grain size, and a vertical change in permeability may suggest a similar change in texture, which may aid in the environmental interpretation in areas where a knowledge of rock type is lacking. Supplemental data such as the interpretation of drill cuttings and microfossils may prove to be of great importance when projecting interpretations far beyond the location of cored sections. The most reliable interpretations, however, must still be based on a detailed knowledge of the rock section, its composition, its texture, and the sedimentary structures in vertical sequence.

Morphology

Each environment of deposition has a characteristic distribution of sediment (LeBlanc 1972), and the morphologies of individual sandstones are determined primarily by the sedimentary processes that last affected the sediment. However, sands that were transported by channelized flows may have the same general morphology whether they were carried by fluvial or by submarine currents. Thus, narrow sinuous patterns of sandstone distribution may represent either the alluvial-valley, deltaic plain, or submarine-fan environment (Table 3–1). It is obvious that the interpretation of environment from morphology alone can be hazardous. The distinction between the several types of channel deposits on the basis of cores or other data may be important in the prediction of reservoir properties.

METHODS OF STUDY

Direct knowledge of rock properties in the subsurface must come from *cores,* or cylinders of rock, recovered during the drilling of wells. Cores are commonly taken for the purpose of measuring the secondary properties of porosity, permeability, and fluid saturations in a potential reservoir rock. Therefore, cores may be taken after the discovery of an oil field when more detailed knowledge of the reservoir is required for economic evaluation. Less commonly, cores are recovered in wildcat wells.

Full-diameter Cores

The method of obtaining a full-diameter core is quite simple in principle (Figure 3–11). A special pipe, called a *core barrel,* is attached at the lower end of the drilling pipe. This core barrel is equipped with a bit that may consist of either rotary or stationary cutting heads, or for hard rock, the bit face is studded with small industrial diamonds. The core barrel is lowered to the bottom of the hole, and rotation of the drilling string allows the bit to penetrate the rock, cutting a thin cylindrical "hole." The inner barrel contains the cylinder or core of the rock. During coring, drilling fluid or mud is pumped downward through the drilling pipe, and a valve at the top of the barrel allows the core to pass upward, displacing mud from the inner barrel while mud flows downward around the outside of the core barrel to lubricate the bit.

When a desired length of core has been cut, the drilling string is raised slightly and rapidly rotated. This causes the rock cylinder within the core barrel to separate at the bottom. The drilling string is then withdrawn from the hole. During withdrawal, the core is prevented from leaving the core barrel by spring clips and toggles mounted in a "core catcher" at the bottom of the barrel.

Cores recovered by conventional methods may be 20 to 60 ft (6 to 18 m) in length and from about 2.5 to nearly 4 in. (6.2 to 10 cm) in diameter. At the surface, the core is removed from the barrel, the depths are marked with chalk, and the cores are boxed for shipment to the laboratory. There, small cylindrical "plugs" about 1 in. (2.5 cm) in diameter are taken from the core for the purposes of measuring porosity, permeability, and fluid saturation.

The coring operation is expensive because it requires removing the drill pipe from the hole, attaching the core barrel, entering the hole for coring, and recovery of the core barrel by again removing the pipe from the hole. Therefore, two "round trips" with the drill string are required before normal drilling can proceed. Furthermore, coring requires treatment of drilling mud to maintain optimum properties, which is another added expense.

Sidewall Cores

Full-diameter cores can be recovered only in compact consolidated rocks because friable sands are lost through the core catcher. Small samples in unconsolidated formations may be obtained by means of a sidewall sampler. This device is run in a borehold on logging cable. Small metal cylinders are driven laterally into the walls

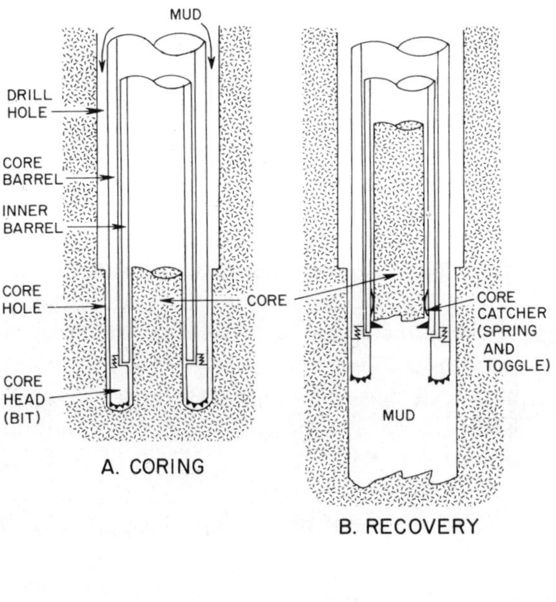

Figure 3–11 Schematic diagram of a core barrel and diamond bit showing (A) the coring operation, and (B) the recovery of core from the hole.

of the hole by powder charges that are ignited electrically. The metal cylinder penetrates the formation and is filled by a small plug of the sediment. When the device is withdrawn, a heavy wire attached to the metal cylinder brings it to the surface. By this method, as many as 40 shots may be made from a single run. The sidewall samples are about 1 in. (2.5 cm) in diameter and up to about 2 in. (5 cm) in length.

The properties of the sidewall samples may be altered by the impact of the core cylinder and by contamination from the drilling mud. Nevertheless, sidewall cores provide a sample of the rock at levels of interest within the borehole. Therefore, sidewall cores are usually taken after the well has reached its total depth and the borehole logs have been run. The levels at which the sidewall cores should be taken can then be determined by interpretation of the borehole logs. Despite their altered nature, some primary properties of the rock, particularly composition and texture, may be determined from the sidewall samples.

Obviously, the full-diameter core is best for geologic interpretation. The core may be cut to provide a vertical slab for the examination of sedimentary structures, and samples may be taken at selected horizons for petrographic analysis. Thus, the rock properties may be observed in vertical sequence, providing that the core has been carefully handled and properly preserved.

Core Study

Because of the added expense of coring, any core samples recovered from the subsurface should be studied in detail and should be preserved for future reference. The vertical sequence observed in slabbed full-diameter cores presents an exposure of a section that has a maximum width of 4 in. (10 cm). Despite the narrowness of this section, great detail of the sequence can be observed because the cores represent an unweathered, and relatively undisturbed, sequence of rock properties. The primary purpose of core recovery may be for economic evaluation of the reservoir, that is, for the measurement of porosity, permeability, and fluid saturations. Nevertheless, the geologic information that can be obtained from the core slab is invaluable for the practical purpose of prediction of reservoir morphology and of the behavior of fluids during production. Furthermore, core samples are necessary for studying the effects of composition and texture on reservoir properties.

In the following studies of reservoir sandstones (Chapters 4–9), full-diameter cores were slabbed, and detailed descriptions were prepared of the sequence of rock types and sedimentary structures. For permanent reference, a thin slab about 0.5 in. (1.2 cm) in thickness was cut, and the continuous slabs were mounted on 3-ft (1 m) plywood strips by means of epoxy resin. Representative samples were then taken from selected levels of the sandstones for petrographic study of texture and composition by standard techniques described in Chapter 1. Samples were selected so that they were representative of bedsets, and in sufficient number that the vertical change of composition and texture could be observed through the entire sequence. Additional samples were taken to establish variations in composition and their effect on porosity and permeability. The interpretation of depositional environment as deduced from vertical sequence was then confirmed by mapping the reservoir morphology by subsurface control.

The fields studied in these chapters represent significant oil accumulations, although few of the "major" category of 100 million barrels are represented, and some are quite small. All of the fields are stratigraphic traps or those in which stratigraphic variations within the reservoir section provide a significant part of the trap. The fields are specific reservoirs that have certain individual characteristics of lateral and vertical sequence. Nevertheless, these reservoirs also display more general characteristics that are common to many other reservoir sandstones.

SUMMARY

Deposition of sand produces a regular sequence of composition, texture, and sedimentary structures. From observation of these properties in vertical sequence, as in a core, the morphology of the sandstone can be predicted. The sum of the physical conditions under which a sand was deposited is termed its *environment of deposition*. Current flow is responsible for long-distance transport of sand, and other processes, such as wave action, may act on sands to produce other distinctive sequences. Thus, the result of these processes in a single core sequence allows an estimate of the size and shape of the reservoir.

For the best interpretations, all the primary properties must be observed. When full-diameter cores are not available, texture and composition may be measured in small samples as in sidewall cores. Other information helpful to the interpretation is a general knowledge of the stratigraphic sequence, location of the section within a depositional system, and fossils from the adjacent shales. A knowledge of the secondary properties, porosity and permeability, may be useful when the cores are no longer available.

The interpretation of environment from one or more cores may be correlated to the response of borehole logs. Then the distribution of the reservoir can be mapped by means of logs from other wells in the local area. Thus, the morphology may be determined with some confidence, and it allows predictions of lateral change and reservoir continuity.

REFERENCES CITED

Albers, C. C., 1966, Foraminiferal ecologic zones of the Gulf Coast: Gulf Coast Association of Geological Societies Transactions, v. 16, p. 345–348.

Bandy, O. L., and R. E. Arnal, 1960, Concepts of foraminiferal paleoecology: American Association of Petroleum Geologists Bulletin, v. 44, p. 1921–1932.

Berg, R. R., 1970, Identification of sedimentary environments in reservoir sandstones: Gulf Coast Association of Geological Societies Transactions, v. 20, p. 137–143.

———, 1979, Exploration for sandstone stratigraphic traps: American Association of Petroleum Geologists Continuing Education Course Notes Series 3 (revised edition), 48 p.

Bernard, H. A., and R. J. LeBlanc, 1965, Resume of the Quaternary geology of the northwestern Gulf of Mexico Province, *in* H. E. Wright Jr., and D. G. Frey, editors, *The Quaternary of the United States:* Review volume, VII Congress International Association of Quaternary Research, Princeton, N.J., Princeton University Press, p. 137–185.

Davies, D. K., and F. G. Ethridge, 1975, Sandstone composition and depositional environment: American Association of Petroleum Geologists Bulletin, v. 59, p. 239–264.

Davis, R. A., Jr., 1983, *Depositional systems, a genetic approach to sedimentary geology:* Englewood Cliffs, N.J., Prentice-Hall, Inc., 669 p.

Dickinson, W. R., and C. A. Suczek, 1979, Plate tectonics and sandstone compositions: American Association of Petroleum Geologists Bulletin, v. 63, no. 12, p. 2164–2182.

Ethridge, F. G., T. R. Gopinath, and D. K. Davies, 1975, Recognition of deltaic environments from small samples, *in* M. L. Broussard, editor, *Deltas, models for exploration:* Houston, Texas, Houston Geological Society, p. 151–164.

Krynine, P. D., 1948, The megascopic study and field classification of sedimentary rocks: Journal of Geology, v. 56, p. 130–165.

LeBlanc, R. J., 1972, Geometry of sandstone reservoir bodies, *in* T. D. Cook, editor, *Underground waste management and environmental implications:* American Association of Petroleum Geologists Memoir 18, p. 133–189.

Loucks, R. G., D. G. Bebout, and W. E. Galloway, 1977, Relationship of porosity formation and preservation to sandstone consolidation history—Gulf Coast Lower Tertiary Frio Formation: Gulf Coast Association of Geological Societies Transactions, v. 27, p. 109–120.

Lowman, S. W., 1949, Sedimentary facies in Gulf Coast: American Association of Petroleum Geologists Bulletin, v. 33, p. 1939–1997.

Nanz, R. H., Jr., 1954, Genesis of Oligocene sandstone reservoir, Seeligson field, Jim Wells and Kleberg Counties, Texas: American Association of Petroleum Geologists Bulletin, v. 38, p. 96–117.

Natland, M. L., 1968, Presidential address: Paleontology and turbidites: Journal of Paleontology, v. 37, p. 946–951.

Reading, H. G., editor, 1978, *Sedimentary environments and facies:* New York, Elsevier, 557 p.

Reineck, H. E., and J. B. Singh, 1975, *Depositional sedimentary environments:* New York, Springer-Verlag, 439 p.

Scholle, P. A., and Darwin Spearing, 1982, Sandstone depositional environments: American Association of Petroleum Geologists Memoir 31, 410 p.

Scull, B. J., C. J. Felix, S. B. McCaleb, and W. G. Shaw, 1966, The inter-discipline approach to paleoenvironmental interpretation: Gulf Coast Association of Geological Societies Transactions, v. 16, p. 81–117.

Skinner, H. C., 1966, Modern paleoecological techniques, an evaluation of the role of paleoecology in Gulf Coast exploration: Gulf Coast Association of Geological Societies Transactions, v. 16, p. 59–79.

TIPSWORD, H. L., F. M. SETZER, and F. L. SMITH, JR., 1966, Interpretation of depositional environment in Gulf Coast petroleum exploration from paleoecology and related stratigraphy: Gulf Coast Association of Geological Societies Transactions, v. 16, p. 119–130.

VISHER, G. S., 1965, Use of vertical profile in environmental reconstruction: American Association of Petroleum Geologists Bulletin, v. 49, p. 46–61.

WALKER, R. G., 1979, Facies models: Geological Association of Canada, Geoscience Canada Reprint Series 1, 211 p.

4

Eolian Sandstones

Windblown sands may form extensive deposits when conditions for their formation are optimum. These eolian-dune sands are the sediments most likely to deserve the description of "blanket" deposits. However, their distribution is finite, and the natural limits of eolian dunes can result in abrupt termination of the sediment body.

Eolian-dune sands occur in arid lands regardless of temperature. They are commonly found in vast continental areas ranging from the equatorial regions to the Arctic and Antarctic. Many places of the world have thick and extensive deposits such as the Sand Mountains of the Rub' al Khali, or Empty Quarter, of Saudi Arabia and the vast reaches of the Sahara in North Africa. The climatic requirement for dune formation is little or no rainfall so that the sand grains lack cohesion and are readily moved by the winds. In some desert areas, periodic moisture and the resulting vegetation tend to prevent the transport of sand and to control the form of dunes. In most deserts, however, stabilization of migrating sand is only a temporary phenomenon, and the sands are generally thick and widespread.

Although the conditions for the development of extensive dune fields may be unique, such dune fields are known to occur over wide areas, both in modern land areas as well as in ancient ones. The process of wind transport is also distinguished by the production of relatively well-sorted grains, and thus the extensive sandstones of eolian origin may have uniform reservoir properties. The gross morphology as well as the internal character provide ideal reservoir sandstones. In addition, the process of wind transport can act on any subaerial sediment, and so the effects of the wind may be seen in sediments associated with fluvial-channel, flood-plain, or coastal environments of deposition. In these associations, the eolian-dune facies may be quite limited in extent, adding to but not dominating the sediment sequence.

The eolian process, therefore, can be an important modifier in various depositional settings.

WIND TRANSPORT

The principles of erosion and transport of sand grains by the wind are the same as those for water flow (Chapter 2). An important difference, however, is the low density of air. Therefore, greater velocities are required for the threshold of grain movement by air than are needed for water transport. Much work on wind transport, both theoretical and experimental, has been conducted, and especially important are the studies by Bagnold (1941), Chepil (1945), and Sundborg (1956), all of whom contributed greatly to the knowledge of grain and air interactions.

The movement of sand by the wind takes place largely near the surface and perhaps within a meter or so of the ground (Bagnold 1941, p. xviii). The common mode of sand movement is by *saltation,* in which the grains rise steeply, are carried downwind a short distance in suspension, and then fall to strike the ground at a low angle, again rising upward after impact. Wind-tunnel observations have shown that about 1 ft (33 cm) downwind from the point at which the grains first move, they obtain sufficient speed to begin saltating (Bagnold 1941, p. 33). When transport is fully developed, a "sand cloud" of saltating grains is maintained.

Beneath the "sand cloud," larger grains are moved by *surface creep* in which movement is induced by saltating grains that strike the surface. Thus, the grains in surface creep are largely unaffected by the wind above. Only the smallest grains are transported by *suspension.* These small grains remain airborne by turbulence, which allows them to be transported long distances before returning to the ground.

Observations of natural soils have demonstrated the wind velocities that are necessary for grain movement in both the field and laboratory (Chepil 1945). Most easily moved are grains in the range of 0.1 to 0.15 mm, and higher velocities are required to move both larger and smaller grains (Figure 4–1). The effect is similar to that in water erosion except that the minimum value for wind transport is displaced from medium grained to the very fine-grained size. The experimental data compare well with empirical functions derived for the threshold of movement in unstable and stable "stratification" (Sundborg 1956).

The grains most commonly moved, that is, those from about 0.05 to 0.5 mm, are the grains that tend to move by saltation. Coarser grains from 0.5 to 1 mm are those that move by surface creep under normal conditions with moderate wind velocities. Finer grains less than 0.05 mm tend to be carried in suspension at velocities high enough to erode such grains.

The differences between water and wind velocities required to move sand are due to the different physical properties of fluids. The density for hot dry air may be taken as 1.2×10^{-3}g/cm^3, which is on the order of 0.1 percent of that for water. The ratio of fluid densities suggests that wind velocities must be about 1000 times greater than water velocities for the movement of sand grains. The difference of

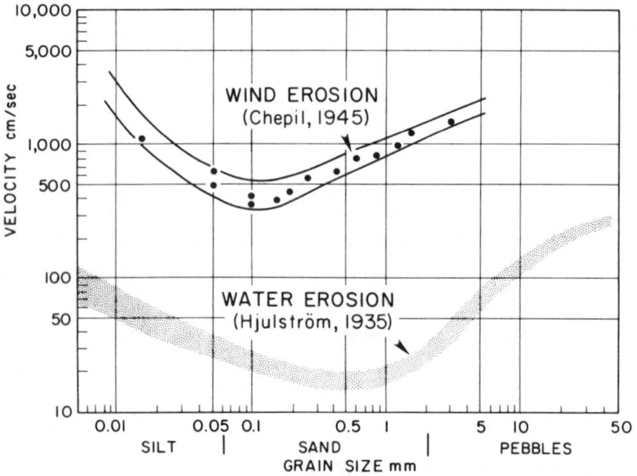

Figure 4–1 Wind velocities required to move soil grains. [From Chepil 1945.] Velocities were measured 6 in. (15 cm) above the soil surface. Shown for comparison is the water velocity required for sand erosion. [Based on Hjulström 1935.]

100 times shown in the diagram (Figure 4–1) is probably due primarily to the differences in height of measurement above the grain surface, being only 6 in. (15 cm) for air and 1 m for water.

The movement of sands by the wind at low velocities creates a characteristic ripple form of long wave length. The ripples are asymmetric with steep sides downwind, similar to those formed by water flow. A major difference is that the wind ripple may have a minimum height of only a few millimeters and a maximum height on the order of 2 cm, whereas water ripples most commonly have heights of 2 to 6 cm. Grains ranging from about 0.1 to 0.4 mm have an increased tendency to become rippled at velocities above the thresholds for erosion (Chepil 1945, p. 403). This is the grain size that moves largely by saltation rather than by creep.

Another characteristic of wind ripples is that the crests of the ripples tend to become coarser grained than the sand below. The crests form small ridges from which the finer grains are removed by direct impact of the wind or by saltation. The result is that an individual ripple may form a thin bed or laminae in which the grain size appears to increase upward (Bagnold 1941, p. 155). The opposite condition is true for water ripples in which the coarser grains tend to be concentrated in the troughs.

Large sand ridges may also form under certain conditions. Ridges with heights of 60 cm and wave lengths of 20 m are common in some deserts (Bagnold 1941, p. 155). These ridges begin to form when increasing wind velocities are sufficient to move the dominant grain size by saltation but leave coarser grains behind. When wind velocities increase still more, the large grains move by saltation and form ridges of coarse sand.

The formation of very large waves, or dunes, apparently depends on surface irregularities such as existing mounds or patches of sand. The patch disturbs the wind flow, causing dispersion of wind energy and resulting in deposition. In wind-tunnel experiments, sand accretes rapidly on the windward side of the obstruction

(Bagnold 1941, pp. 169–70). This process appears to be similar to that of antidune formation in the case of high-velocity water flow. Presumably, by means of flow disruption, sand continues to accumulate in prominent landforms as long as the sand supply is abundant.

MODERN DUNE SANDS

Eolian dunes are hills or ridges of sand that are constructed by the wind. They are formed in a great variety of shapes depending on the constancy of the winds and sand supply. The term *dune* refers to the geomorphic expression of the aggradational feature, and thus descriptive terms do not necessarily signify the relative extent or importance of this feature. Nevertheless, the classification and description of dune forms can serve to explain the great variety of these desert features.

Classification of Dunes

A recent classification has been adapted especially for interpretation of earth-satellite imagery by McKee (1979*a*). The most extensive eolian sands are called *sheets* and *stringers* (Table 4–1). Strictly speaking, these are not dune forms because they display broad, nearly flat surfaces. Viewed on a more regional scale, however, they are certainly constructional features and, in the case of sheet sands, these eolian deposits may form impressive mountainlike masses on the earth's surface.

The smaller, more familiar dune forms are of several types (Figure 4–2). These are characterized by a general asymmetry and one or more *slipfaces,* that is, steeper surfaces that in most cases are inclined in the direction of dominant wind direction. The *barchan dune* is a small form of crescent shape, thicker in the middle and with thin ends pointing in the direction of the dominant winds [Figure 4–2(A)]. Barchans are commonly small and are the result of sparse sediment supply. The *barchanoid ridge* forms at right angles to the dominant wind direction and consists of a row of connected crescents so that the ridge is cuspate on its slipface [Figure 4–2(B)]. Barchanoid ridges form under conditions of unidirectional winds when the supply of sand is greater than in the case of single barchan dunes. The *transverse ridge* forms at right angles to the dominant wind direction when the sand supply is abundant. These ridges are asymmetrical and may be slightly irregular or cuspate on their lee sides [Figure 4–2(C)].

A *parabolic dune* is U-shaped, and the broadly curved ridge points in the direction of the prevailing winds. The form of this dune may be partly controlled by vegetation (McKee 1979*a*, p. 11.). A *linear (seif) dune* is a long symmetrical ridge that is parallel to the direction of the prevailing winds. These ridges are unusually straight and regular and are composed of two slipfaces, one on each side. Linear dunes are believed to be formed by two dominant wind directions.

A *reversing dune* is an asymmetrical ridge, cuspate but irregular in form, which is constructed by opposing bidirectional winds. A *star dune* has a central peak and

TABLE 4–1. CLASSIFICATION OF MODERN DUNES

Name	Morphology	Number of slipfaces	Wind direction	Sand supply (relative)	Relative size
Sheet	Sheetlike with broad flat surface	None	Multidirectional	Abundant	Large, thick
Stringer	Thin elongate strip	None	Unidirectional	Sparse	Extensive, thin
Dome	Circular or elliptical mound	None	Bidirectional	Sparse	Small, thin
Barchan	Crescent	1	Unidirectional	Sparse	Small, thin
Barchanoid ridge	Row of connected crescents	1	Unidirectional	Abundant	Large, thick
Transverse ridge	Asymmetrical ridge	1	Unidirectional	Very abundant	Large, thick
Parabolic	U-shape	1 or more	Unidirectional	Abundant	Large, thick
Linear (Seif)	Symmetrical ridge	2	Bidirectional	Sparse	Extensive, thin
Reversing	Asymmetrical ridge	2	Bidirectional		Small, thin
Star	Central peak with 3 or more arms	3 or more	Multidirectional	Sparse	Small, thin
Blowout	Circular rim of depression	1 or more	Multidirectional	Variable	Small

SOURCE: Adapted from McKee 1979*a*.

three or more arms and is formed by shifting multidirectional winds. A *blowout dune* is a circular ridge that surrounds a depression or area of deflation.

Many dune fields are characterized by a single dune type, one that can be recognized over extensive areas. However, other dune fields are of the compound type and are composed of two or more dune forms in adjacent areas, or they may have one form superimposed on another. A variety of other types can be distinguished, and there is no unified explanation to account for this variety of observed forms. Nevertheless, the constancy of the eolian process is such that internal characteristics of dunes should be uniform despite the great variety of external shapes.

Reservoir sandstones of eolian origin are usually a composite of many single dunes, and they form generally thick and extensive bodies. It is not expected that the type of individual dunes can be discerned in such a composite section, but the variety of modern dune types illustrates the complexity of forms that might compose a composite body.

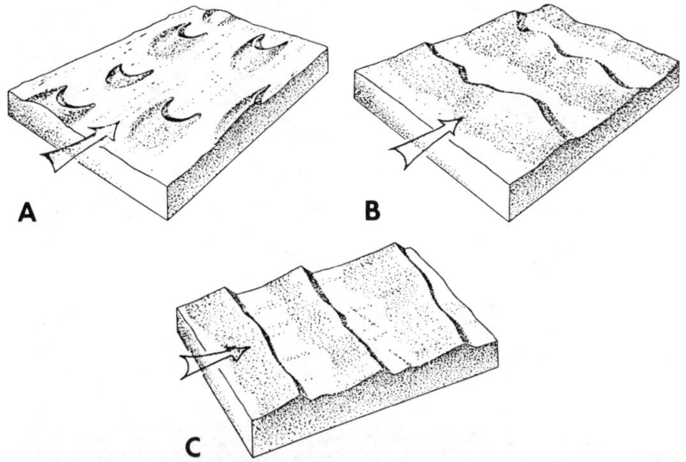

Figure 4–2 Diagrams of eolian dunes; arrows show dominant wind directions: (A) barchan dunes; (B) barchanoid ridges; (C) transverse dunes. [From McKee 1979*a*.]

Sedimentary Structures

The internal structures of modern dunes have been studied by extensive trenching in many parts of the world. Bedding is characterized by cross-stratification in which the cross-strata have angles that range up to 30° or 35°, the angle of repose of dry sand. Cross sections of modern dunes display these structures (Figure 4–3). Barchans show sets of cross-strata that have a maximum thickness on the order of 5 to 10 ft (2 to 3 m). Cross-strata are truncated by the bounding surfaces between sets, and dips are variable and range from horizontal to 33°. Barchanoid ridges show the same variations, but some individual sets appear to be much thicker and may extend up to 25 ft (8 m). Linear dunes show equally high dips, and in general the dips of cross-strata within sets tend to parallel the two opposing slipfaces.

All of the dune cross sections shown in Figure 4–3 illustrate the maximum dip for sets and cross-strata. Sections that are normal to those illustrated would display dips that are generally lower and commonly in the range of 10° or less. Therefore, any vertical section might display a wide range of dips ranging from horizontal to a maximum of 35°.

As viewed on a somewhat smaller scale, cross-strata can be characterized as of two types (Figure 4–4). These types illustrate those found in sandstones believed to be of eolian origin, but they serve equally well for modern dune sands. The most abundant type is *planar,* in which the cross-strata are truncated at the boundaries of sets. The planar sets may be either *tabular* or *wedge-shaped.* The internal cross-strata tend to be regular planar surfaces, but they may be broadly curved and concave downward. The *trough type* of cross-stratification (Figure 1–5) is extremely

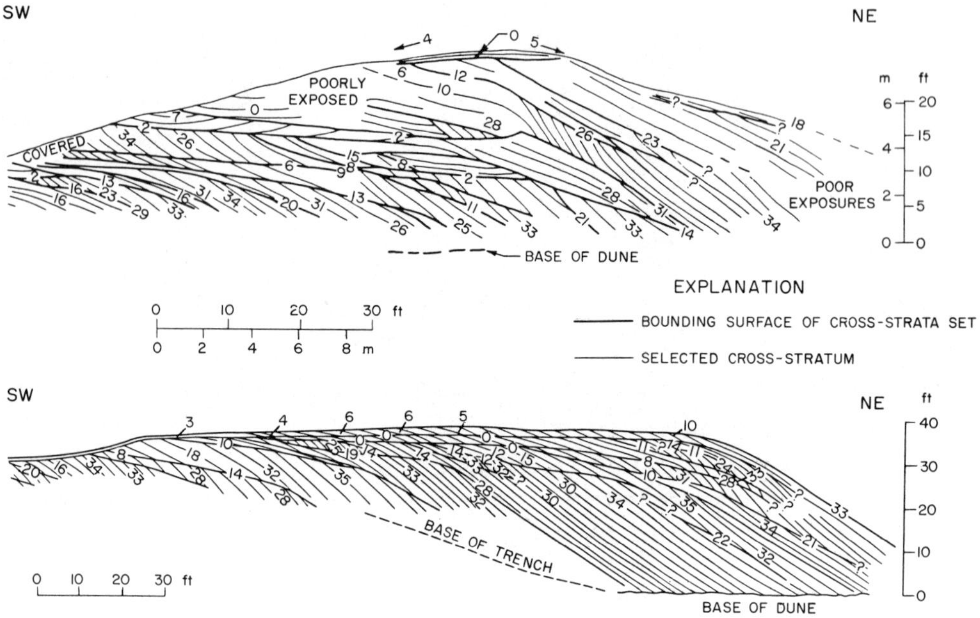

Figure 4–3 Cross sections of dunes exposed in trenches: (A) barchanoid ridge dune with section parallel to dominant wind direction; (B) transverse dune with section parallel to dominant wind direction. Small numbers refer to apparent dip of cross-strata in degrees. [Diagrams from McKee 1979*b*.]

rare and is not commonly recorded for modern dunes, although giant troughs, called *festoon structures,* have been described from ancient eolian deposits.

Stratification originates by several mechanisms. The most common features on the surface of dunes are various types of ripple structures. *Wind ripples* have ridges that are normal to the prevailing wind directions on the stoss sides of dunes but may be oriented vertically on the lee sides because of eddy currents. Characteristically, these wind ripples have long wave lengths and low amplitudes so that the length/amplitude ratio, or "ripple index," is commonly greater than 15 and, in most cases, is on the order of 30 or more. The long stoss sides of the ripples tend

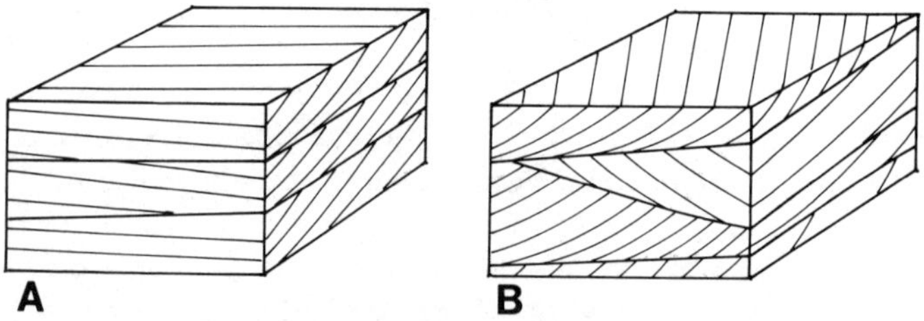

Figure 4–4 Common forms of cross-strata recognized in eolian sandstones: (A) tabular-planar; (B) wedge-planar. [From McKee 1979*b*.]

to have a concentration of the coarser grains as a small-scale lag deposit. A sequence of migrating ripples may form a succession of surfaces along which the coarser grains are concentrated, thereby producing a fine-scale lamination in which the texture coarsens upward (Figure 4–5). This type of lamination is different from that produced by water flow in which the coarser grains are concentrated in the troughs between ripple crests, and the succession of ripples produces laminations in which grain size decreases upward (Figure 2–8).

Another type of structure is called *adhesion ripples*. These ripples are formed on the surface of water-saturated sand. Because of the cohesion between grains, the stoss side of the ripples are steeper than the lee sides, and the stoss side is the result of erosion.

Whatever their origin, ripples of long wave length are characteristic of the eolian process and result in parallel, even laminae that are essentially parallel to the surface exposed to the wind. The thickness of the individual laminae is equal to the height of the ripple and commonly on the order of a few millimeters to about 1 cm. When viewed on a small scale, the parallel, even, and continuous laminae are distinctive.

Other processes produce well-defined laminations on the lee sides of dunes. The slipfaces are characterized by *grain fall* from suspension, and deposition can result in fine-scale laminations that decrease upward in grain size. Lee sides of dunes attain the greatest slopes, essentially that of the angle of repose for dry sand, but this slope is unstable. Consequently, thin sheets of sand tend to move down the lee slope as small-scale slides. During sliding, heavy minerals, if present, appear to move downward and concentrate at the "shear" surface (Inman, Ewing, and Corliss 1966). Thus, a series of slides may produce *avalanche stratification* with widely spaced laminae due to the internal variation of composition and texture. The thickness of the laminae is on the order of 2 cm.

Sliding of large sand masses on the lee slopes can produce *contorted bedding*. Sliding tends to obliterate previously formed lamination, and mechanical adjustments between grains during sliding may cause destruction of the laminae, resulting in a massive or structureless appearance. If the sand is slightly moist, internal cohesion may preserve the original stratification, and this results in laminations that are

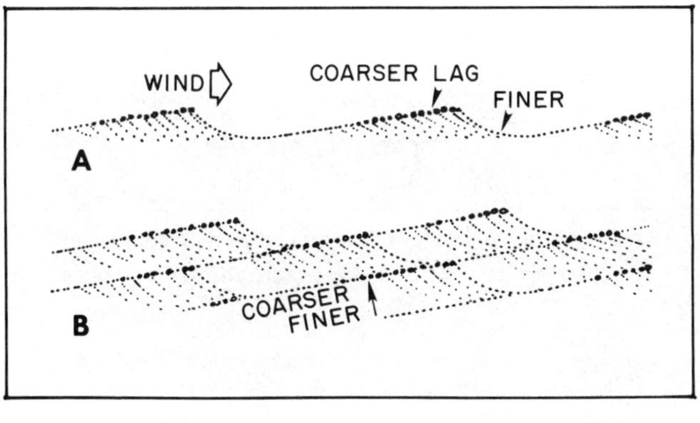

Figure 4–5 Diagram of long wave-length wind ripples showing development of laminae in which grain size coarsens upward. [Adapted from Kocurek and Dott 1981.]

folded, or even overturned. Such contorted bedding has been observed in modern dunes (Bigarella 1972 and Steidtmann 1973).

Other postdepositional structures are found on modern dunes and are commonly preserved in ancient dunes: These include *raindrop imprints, animal tracks,* and *bioturbation by rooting.* Bioturbation, especially, is an important process in partly vegetated dunes. Here the fine roots of grasses and small shrubs extend deeply into the stabilized dune sands. However, bioturbation by rooting tends to destroy the primary stratification, resulting in a massive or structureless appearance. Because of the well-oxidized environment, no trace of organic matter may remain.

Composition and Texture

The composition and texture of dune sands are determined by source sediment (Table 4–2). For example, the dunes of Guerrero Negro in Baja California are derived from adjacent beaches and contain a relatively high content of quartz (Inman, Ewing, and Corliss 1966). These sands are unusual in their high content of heavy minerals—hornblende, apatite, epidote, and magnetite. The heavy grains are significantly smaller than the quartz and feldspar and are concentrated along the laminations. As another example, the Algodones dunes of the Salton Basin, California, occur on a flood plain, and their source sediment is derived from the Colorado River (Van de Kamp 1973). Here the dune sands are somewhat coarser grained than the fluvial sands, but maximum grain size is reduced from that of the fluvial sediment. There is no difference, however, in sorting as expressed by the Trask coefficient (Table 4–2). Again, the dune sands of the modern barrier island at Galveston, Texas, are somewhat coarser grained than the beach and upper shoreface sands from which they are derived. These dune sands are better sorted and slightly enriched in their content of monocrystalline quartz. Based on these examples, it is

TABLE 4–2. TEXTURE AND COMPOSITION OF SELECTED DUNE AND SOURCE SANDS

		Grain size[a]			Detrital composition[b]				
Location	Type	Average (mm)	Max (mm)	Sorting	Qz (%)	F (%)	Rx (%)	Oth (%)	Mx (%)
Guerrero Negro[c]	Eolian	0.15	—	0.24 ϕ	78	3	—	19	—
Algodones[d]	Eolian	0.24	0.30	1.1	68	14	13	4	—
Algodones[d]	Fluvial	0.20	0.60	1.1	70	13	8	8	tr
Galveston[e]	Eolian	0.16	—	0.42 ϕ	80	7	12	1	—
Galveston[e]	Beach	0.14	—	0.44 ϕ	78	4	16	2	—

[a]Average = median or mean size; sorting = phi ϕ deviation or minimum Trask coefficient.
[b]Qz = quartz, F = feldspars, Rx = rock fragments, Oth = other minerals, Mx = matrix.
[c]Inman, Ewing, and Corliss (1966, p. 799): Oth = hornblende, apatite epidote, magnetite.
[d]Van de Kamp (1973): Oth = mostly carbonate grains.
[e]Ethridge (1970): Rx = rock fragments including polycrystalline quartz.

doubtful that dune sands could be distinguished by composition and texture, especially if the source sediment is not known. When large numbers of samples are available, however, discriminant function analysis has been successful in distinguishing dune sands from their source sands (Moiola and Spencer 1979).

Sediments associated with extensive dune sands (Ahlbrandt and Fryberger 1981) are those of other desert environments: These include *wadi* (fluvial), *playa* (lacustrine), and *sabkha* (arid tidal flat). Wadi sediments are generally coarse grained and represent lag deposits in the channels of ephemeral streams. Playa sediments are fine grained and represent suspension deposits in temporary intradune lakes. Sabkha deposits are also fine grained and may contain interlaminated evaporite minerals, dense dolomites, and algal structures. Less extensive dune sands may be associated with alluvial valley, lacustrine, lagoonal, or beach environments.

ANCIENT EOLIAN SEDIMENTS

Sandstones interpreted to be eolian in origin have been described from many parts of the world (McKee 1979*b*). These sandstones range in age from Late Paleozoic to Middle Mesozoic, and many are reservoir sandstones. A well-known reservoir of Permian age is the Rotliegendes Sandstone of the North Sea basin, a thick section of finely laminated sandstone of dune and interdune origin (Glennie 1972). Other reservoirs are Permo-Pennsylvanian sandstones in the Rocky Mountain area, the Tensleep Sandstone of Wyoming (Fox, Lambert, and Mast 1975 and Mankiewicz and Steidtmann 1979), the equivalent Weber Sandstone of northwestern Colorado (Fryberger 1979), the Casper Sandstone of southeastern Wyoming (Steidtmann 1974), and the Lyons Sandstone of northeastern Colorado (Walker and Harms 1972). Another prolific reservoir of eolian origin is the Jurassic Nugget Sandstone of the Overthrust belt in western Wyoming (Lindquist 1983 and Picard 1975).

Most of these sandstones were formed during periods of widespread aridity as indicated by associated evaporites or red beds. In the Gulf Coast basin, the Jurassic Norphlet Sandstone is an eolian reservoir that was deposited as coastal dunes following the evaporites of the Louann Salt.

NORPHLET SANDSTONE, HATTERS POND FIELD, ALABAMA

Jurassic formations of the Gulf Coast province have long been known as reservoirs for oil and gas. The Smackover Limestone and Cotton Valley Group sandstones were well established as reservoirs in the 1930s in southern Arkansas and northern Louisiana. The underlying Norphlet was known in that area as a thin nonproductive unit between the Smackover and the underlying Louann Salt. The Norphlet thickens and becomes a major sandstone unit southeastward in Mississippi and Alabama, and oil production was finally established in 1967 near Jackson, Mississippi. The

Norphlet Sandstone then became a target for deep drilling in the area and is now an important reservoir in several large fields.

Geologic Setting

The Mississippi salt basin extends across southern Mississippi and into southwestern Alabama as an area of deep-seated salt structures (Figure 4–6). Regional dip is toward the southwest, and an area of no salt domes separates the Mississippi salt basin from the Louisiana salt-dome province on the south. The lack of salt structures coincides with a prominent east-trending structure called the Wiggins arch. This deep structure may have produced local thinning in the underlying salt with a resultant lack of domal structures.

The Norphlet Sandstone reaches a maximum thickness of more than 700 ft (213 m) in the deeper part of the area. The underlying Louann Salt (Table 4–3) is present, presumably, throughout much of the area, but its thickness is unknown. The overlying Smackover Limestone has a maximum thickness of 500 ft (152 m), but it thins northward and is truncated at the updip margin of the area where it is overlapped by the Cotton Valley Group. The Smackover also thins downdip and has a thickness of only 100 ft (30 m) in the Hatters Pond area. The lower Smackover contains laminated mudstone and wackestone, and the upper part contains oolitic or oncolitic grainstones (Mancini and Benson 1980). In the downdip areas, such as Hatters Pond, the Smackover section is largely dolomitized. The succeeding Buckner Formation contains anhydrite and some interbedded dense carbonates.

The general character of the Jurassic sequence, then, appears to denote marginal-marine to evaporitic environments. The sandstones of the Norphlet represent

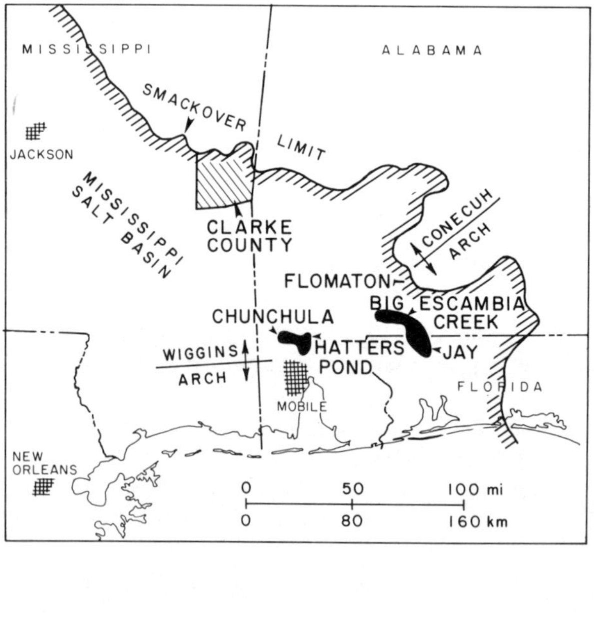

Figure 4–6 Map of southeastern Mississippi and southwestern Alabama showing the updip limit of the Jurassic Smackover Limestone and the location of major field areas. [Adapted from Oxley, Minihan, and Ridgeway 1967 and Wilson 1975.]

TABLE 4–3. GENERALIZED STRATIGRAPHIC SECTION OF THE JURASSIC SYSTEM IN MISSISSIPPI AND ALABAMA

Series	Group	Formation
Upper Jurassic	Cotton Valley	Schuler/Bossier
	Louark	Haynesville-Buckner Smackover Norphlet
Middle Jurassic		Louann
Lower Jurassic		Werner

SOURCE: Modified from Dinkins 1968.

a continental facies that was dominant after deposition of the Louann Salt. Carbonate mudstones and grainstones in the Smackover probably represent beach and shallow-marine deposits that were present at a maximum stage in the late Jurassic transgression. The overlying Buckner represents a return to marginal-marine or lagoonal evaporites.

Drilling History

Smackover reservoirs are found on small anticlinal closures in the Mississippi salt basin, and these structures are related to low-relief domes or "swells" that resulted from flowage of the underlying Louann Salt (Hughes 1968). Some of these structures also have smaller areas of production in the Norphlet. First production of Norphlet oil was at the Pelahatchi field (Hartman 1968), and this discovery was followed by others such as those in Clarke County, Mississippi (Figure 4–6). An early Norphlet discovery in southwestern Alabama, Flomaton field, was later eclipsed by a major producing area in the Smackover as the Jay field in Florida was extended northward into the Big Escambia Creek field. Finally, a significant Smackover discovery was made at Hatters Pond in 1975, 12 mi (19 km) north of Mobile, Alabama. The structure included a thick section of oil-productive Norphlet Sandstone, and this discovery established the Norphlet as a significant reservoir.

The structure at Hatters Pond shows a local, relatively small closure along the upthrown side of a major normal fault (Figure 4–7). Dry holes along the fault have encountered piercement salt, which has cut out the producing section. The oil column has a vertical height of about 350 ft (107 m), and the field is separated from the Smackover producing area in the Chunchula field by a shallow syncline.

Cores of the Norphlet were examined from two wells in the field, the Getty Klein 1 and Getty Creola Minerals 2 (Figure 4–8). The top of the Norphlet Sandstone is marked on the gamma log by an increase of about 30° API units. The Norphlet section extends from 18,120 ft (5523 m) to a total depth of 18,511 ft (5642 m) in the Klein 1 discovery well. A nearly complete section of slabbed core was available for study from the Creola Minerals 2 well.

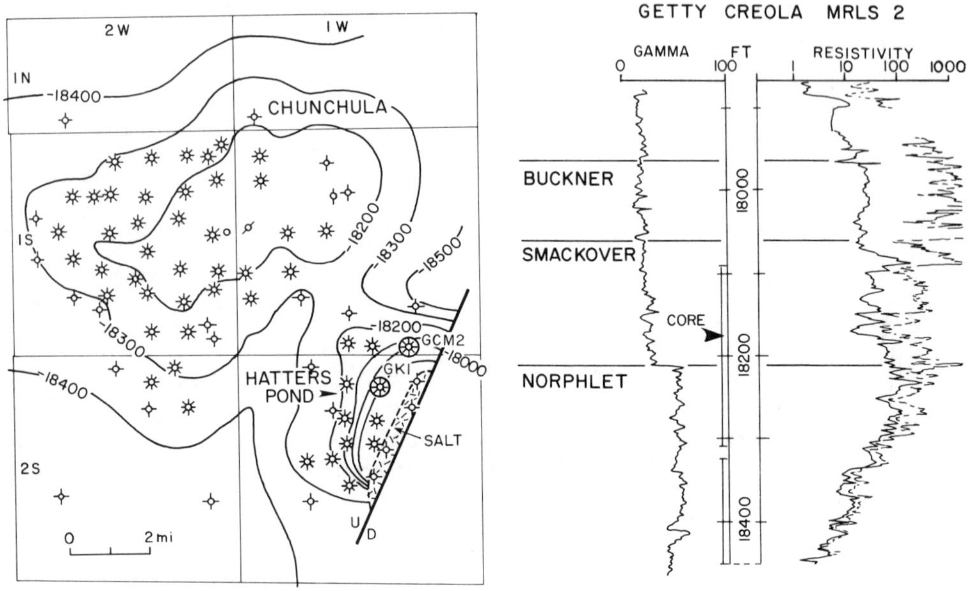

Figure 4–7 Structure on top of the Smackover Formation (left) in the Hatters Pond and Chunchula fields, Mobile County, Alabama. Contour interval 100 ft (30 m). Cores were examined from the circled wells. [Adapted from Mancini and Benson 1980, Figure 13.] Gamma-ray and induction logs (right) of cored well in the Hatters Pond field. Location of well is shown on map (GCM2).

Description of Sandstone

Sedimentary structures. The Norphlet section is characterized by inclined lamination. The laminae are planar, even, continuous, and parallel. Dip angles range from nearly horizontal to 30°, and dip is uniform within relatively thick bed sets that range from a few feet to more than 30 ft (9 m) in thickness.

The well-laminated aspect of the sandstone is shown by slabbed cores in nearly continuous sections (Figure 4–8). Many of the laminations are darkly colored by secondary alteration of minerals along bedding planes, and this alteration results in a distinctive banded appearance. Even where laminae are not dark, the parallel, even lamination is still observed. Spacing between laminae is commonly on the order of 0.5 to 1 cm, although some sets are more closely or widely spaced. Widely spaced laminae, up to 2 cm, appear to be more common in steeply dipping sections. Most of the lamination represents the low amplitudes and long wave lengths that are typical of the ripple structure of windblown sands. It is likely that some of the more widely spaced laminae with steeper dips represent avalanche lamination of the dune face. In the narrow cored section it is not possible to determine the geometric relations along the laminae.

It is also difficult to establish the exact thickness of bed sets in the section

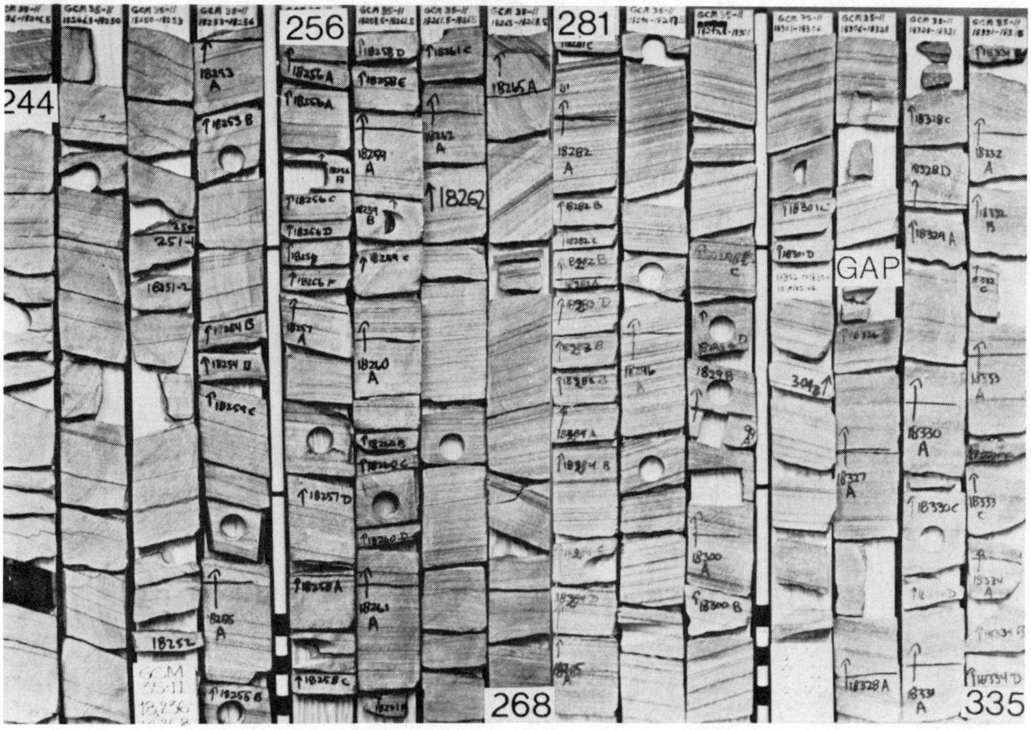

Figure 4–8 Slabbed core of Norphlet Sandstone, Getty Creola Minerals 2, Hatters Pond field. Section is from depths of 18,244–18,335 ft (5561–5588 m); note gaps in section. Scale bars are divided into feet and inches.

illustrated (Figure 4–8). The reason for this is probably the result of previous handling. The core had been broken into short segments for sampling, and the slab may not have been cut in the plane of maximum dip. Nevertheless, it is apparent that the dip is uniform through relatively thick intervals to 10 ft (3 m) or more. Dip changes, when they do occur, are rather abrupt, as the change from about 15° to nearly horizontal that is seen at a depth of 18,256 ft (5564 m). Despite the condition of the core, the gross character of the eolian section is clearly shown.

More detailed observation of the sandstone reveals the nature of the lamination and the presence of physical contortion in parts of the section (Figure 4–9). The planar, even, and parallel laminae are evident through most of the section, but selected laminae are enhanced by dark coloration, which represents diagenetic alteration on preferred bedding planes. Along these thin zones, pressure solution has taken place to produce a distinct suturing or small-scale "stylolitic" structure (Figure 4–9). During compaction, quartz was dissolved at the points of grain contact, moved laterally, and then redeposited. The resulting dark layer is, in part, a residuum of insoluble materials. Apparently, pressure solution took place along only those laminae that had a small but significant amount of matrix because the pres-

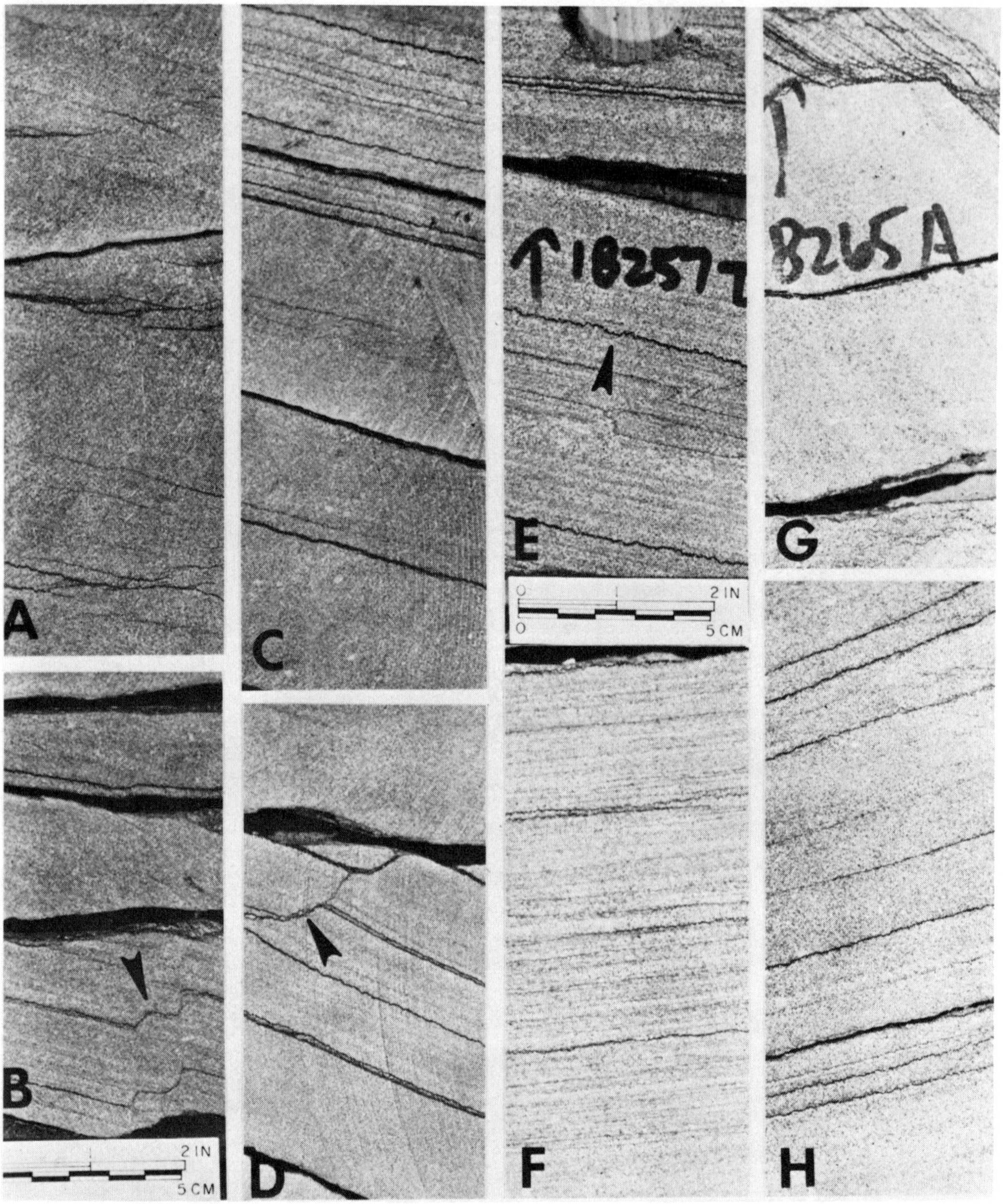

Figure 4–9 Sedimentary structures in the Norphlet Sandstone, Getty Creola Minerals 2, Hatters Pond field: (A) massive to indistinctly laminated upper unit—18,233 ft (5557 m); (B) small faults (arrow) in laminated sandstone—18,246 ft (5561 m); (C) closely to widely spaced laminae—18,248 ft (5562 m); (D) small faults in laminated sandstone—18,249 ft (5562 m); (E) and (F) finely laminated sandstone—note styolitic laminae in (E) (arrow)—18,257 ft and 18,262 ft (5565 m and 5566 m); (G) small fault overlying massive or indistinctly laminated sandstone with nearly vertical dip—18,265 ft (5567 m); (H) widely spaced laminae; note slightly convergent laminae in upper part—18,267 ft (5568 m).

ence of clay is believed to have promoted pressure solution (Weyl 1959). Stylolitic zones within the Norphlet are common and have a mean amplitude of 0.4 mm and, in some places, a maximum amplitude of 1.5 mm (Calgero 1979).

Physical deformation of the section is also conspicuous. Laminae may be offset by small faults (Figure 4–9). At other levels, the sandstone is highly disturbed with strongly tilted laminations and more massive sandstone apparently homogenized by contortion or folding (Figure 4–9).

The uppermost 20 ft (6 m) of the Norphlet has a somewhat different character (Figure 4–9). Some indistinct, parallel, even laminae are present, but more obvious are a few gently curved laminae. At one place, distinct ripple lamination is present. The upper generally massive section may represent a disturbance, perhaps reworking, at the top of the Norphlet during transgression, which resulted in deposition of the overlying Smackover Limestone.

Petrography. Samples were selected from the Klein 1 core for thin-section study (Table 4–4). The sandstones are relatively coarse grained; the average mean size is 0.29 mm (medium grained) and the maximum size is 0.63 mm (coarse grained). However, sorting is good as reflected by an average standard deviation of 0.10 mm (0.5 ϕ). There is no apparent textural gradation through the section, but the number of samples studied was relatively small as compared with the total thickness.

Total quartz content averages 65 percent (Table 4–4). The upper massive unit is more highly quartzose, which is consistent with the interpretation of the unit as having been reworked during transgression. For the total section, monocrystalline quartz is dominant, and polycrystalline quartz averages only 10 percent of the total detrital composition. Feldspar grains average 14 percent, and potassium feldspar is more abundant than plagioclase. Rock fragments average 8 percent and include fine-grained volcanic and metamorphic rocks as well as chert. The volcanic fragments appear to be highly altered. Matrix is common and averages 12 percent; a significant amount of the matrix may be authigenic. A part of the matrix includes rims of chlorite, which were observed on quartz grains.

Cements are both silica and carbonate. Silica is found as overgrowths on quartz grains, and carbonate appears to be mostly dolomite, which occurs as small rhombic crystals, obviously of authigenic origin. A minor amount of iron oxide is present.

Composition of the Norphlet at Hatters Pond is different from that reported in an area farther north in Clarke County, Mississippi (Badon 1975, p. 384). In that area the upper Norphlet is more highly quartzose with an average of 90 percent of total quartz (Table 4–4). This composition suggests a different source for detritus, perhaps to the north of Clarke County, whereas the Norphlet at Hatters Pond may have received contributions of sand from the northeast in the vicinity of the Conecuh arch (Figure 4–6).

Halite is an unusual cement that has been reported from the Norphlet at the Pelahatchie field (Hartman 1968). The interstitial halite comprised about 20 percent of the bulk volume and caused reduced permeability and porosity of about 4 md and 8 percent, respectively. The halite may have formed early from solution of the

TABLE 4–4. AVERAGE PROPERTIES OF THE NORPHLET SANDSTONE AT HATTERS POND FIELD, ALABAMA, AND IN CLARKE COUNTY, MISSISSIPPI

			Quartz size[a]			Detrital composition[b]					Cement[c]			
Area	Unit	Number of samples	Mean (mm)	Max (mm)	σ (mm)	Qz (%)	F (%)	Rx (%)	Oth (%)	Mx (%)	Sil (% of total)	oth (% of total)	Permea-bility (md)	Porosity (%)
Hatters Pond	Massive	1	0.41	1.10	0.14	83	9	3	1	4	—	4	3.0	11.9
Hatters Pond	Laminated	14	0.29	0.63	0.10	65	14	8	1	12	5	3	8.2	13.5
Clarke County	Massive	10	0.28	—	—	89	7	2	—	2	—	3	—	14.0
Clarke County	Laminated	10	0.34	—	—	90	7	1	—	2	—	1	—	17.9

SOURCE: Data in Clarke County, Mississippi, from Badon 1975.

[a]Long-axis measurements; σ = standard deviation.

[b]Qz = total quartz, F = feldspars, Rx = rock fragments, Mx = matrix, Oth = other minerals.

[c]Sil = silica as grain overgrowths, Oth = dolomite and anhydrite.

Louann Salt by ground water and movement upward through the Norphlet sands by capillary rise and evaporation. If halite were an early pore-filling cement, it may have prevented tight compaction of the framework grains. From a depth of 17,000 ft (5182 m), sandstone samples with salt removed by dissolution had porosites of 27 percent, an unusually high value.

Porosity and permeability. Analysis of the Creola Minerals 2 core (Figure 4-10) shows an average permeability of 1.2 md and an average porosity of 8.8 percent. Both permeability and porosity increase downward and reach maximum av-

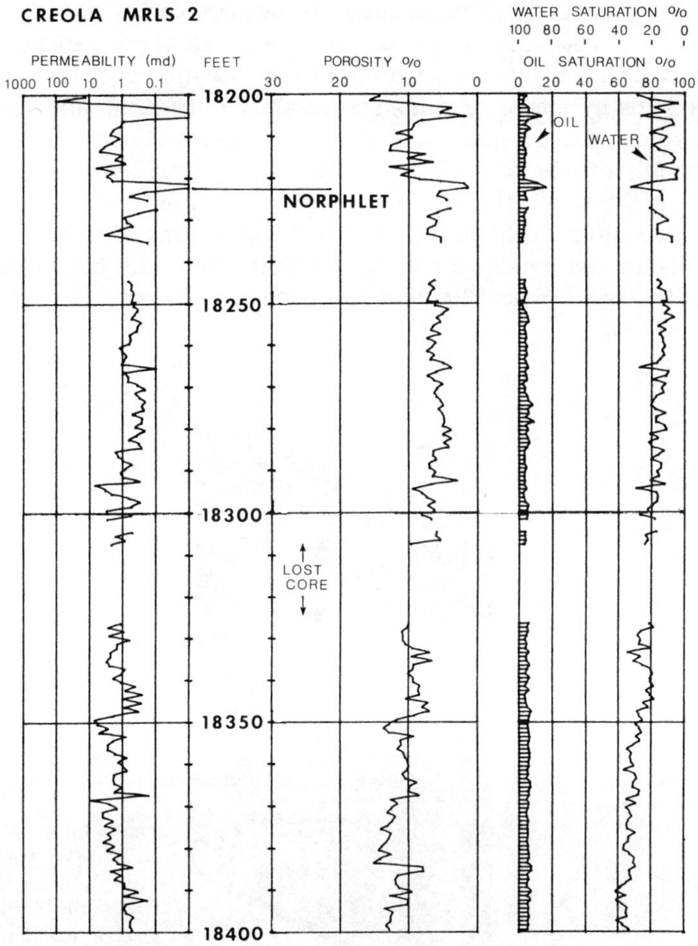

Figure 4-10 Permeability, porosity, and fluid saturations in the upper part of the Norphlet Sandstone, Getty Creola Minerals 2, Hatters Pond field.

erage values of 2.7 md and 12.8 percent in the section extending from 18,368 to 18,385 ft (5598 to 5604 m). In the lower Norphlet, porosities remain relatively high at an average 12 percent, but permeability is somewhat decreased. The average porosity of about 9 percent is that expected by compaction and diagenesis below a depth of 18,000 ft (5486 m). Based on the limited number of samples studied, lower porosites tend to be associated with higher matrix content. Analysis for the Klein 2 core indicates higher average values of 11 md and 13.3 percent. It may be assumed that some diagenetic change, perhaps the development of secondary porosity, may be responsible for somewhat higher porosities laterally in the Norphlet.

Eolian sandstones are commonly anisotropic with regard to the flow of fluids because of their pronounced lamination. An excellent example of permeability variations in eolian sandstones comes from the Permian Rotliegendes Formation (Figure 4–11). The Rotliegendes sandstone is a well-known eolian reservoir in the North Sea (Glennie 1972). Permeabilities were measured at closely spaced intervals along thin beds by means of a mini-permeameter. Permeabilities are uniform along the beds but differ by ten times or more between adjacent beds. The differences are probably caused by variations in both grain size and composition, especially cements. Consequently, production of fluid can be expected to be controlled by the dip and direction of the inclined bed laminations. Detailed permeability measurements are not available for the Norphlet sandstone, but its character is similar to the Rotliegendes, and flow properties also can be expected to be strongly controlled by bedding.

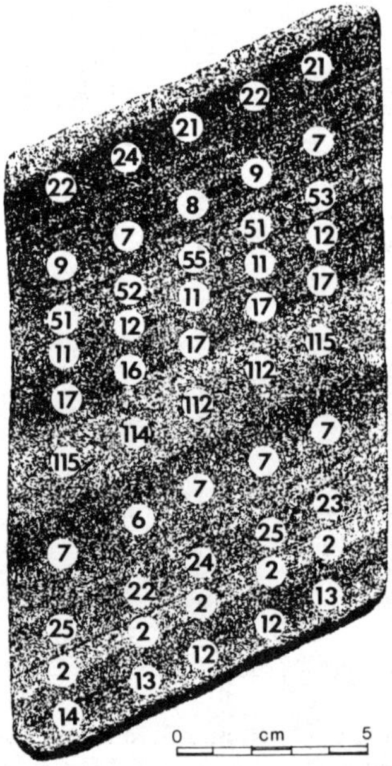

Figure 4–11 Detailed permeability measurements along thin beds in the Permian Rotliengendes sandstone from a field in the North Sea. Permeabilities are given in millidarcys. [Measurements provided by Koenraad J. Weber, Shell Oil Company.]

Dip Log Interpretation

The regular bedding in eolian sandstones produces a distinctive response on dip logs (Table 1-6), logs that show, in general, units of uniform higher dips in dune sands and irregular lower dips in interdune deposits (Lupe and Ahlbrandt 1979). A dip log through the Norphlet section in the Getty Klein 1 well illustrates this character (Figure 4-12). Within certain intervals, the dip magnitudes are relatively constant, and the dip azimuths are essentially unidirectional. For example, in the units designated 1, 2, 3, and 4, the average dips and directions are, respectively, 19 ° at 280°, 23° at 180°, 9° at 185°, and 18° at 245°. The units range in thickness from 15 to

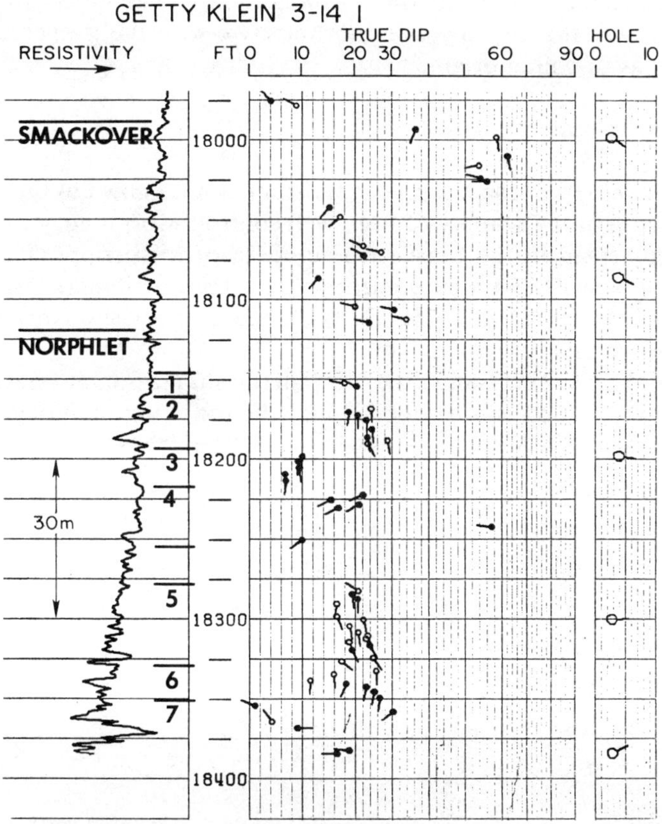

Figure 4-12 Interpreted dip log through the Norphlet Sandstone, Getty Klein 1, Hatters Pond field. Position of dots indicates amount of dip; short lines indicate dip direction relative to north (vertical). The interpreted dips have been corrected for hole deviation and are called "true" dips.

35 ft (4.5 to 11 m) and probably represent single bed sets that have internally consistent dips.

Other units, such as 5 and 6, show a similar uniformity in both dip magnitude and direction. Only one unit (7) shows significant scatter and may indicate much thinner sets. In some intervals, no dips are recorded, probably because of the lack of correlation among the resistivity curves. This situation may be caused by the highly resistive nature of the section, either because of oil or gas saturation, or because the section is compact and well cemented. Highly contorted zones may also yield poor dip interpretation.

The interpreted dips (Figure 4–12) are called "true" dips because the hole deviation has been removed. Structural dip is still present, however, and must be considered before any interpretation can be made of the significance of stratigraphic dips. Across the Klein well, structural dip is about 3° to the northwest (Figure 4–7), and this low dip does not interfere with the interpretation of the sandstone dips as of eolian origin.

Morphology

Distribution of the Norphlet Sandstone is not known in the Hatters Pond field area because of the scarcity of deep well control. Morphology of the formation is better documented in the Clarke County area of Mississippi (Figure 4–6). In this area, there is sufficient well control so that the distribution of the Norphlet could be established (Badon 1975). Furthermore a composite Norphlet sequence is known from cores.

In Clarke County, the Norphlet has a maximum thickness of 500 ft (152 m), and the section can be divided into the following lithologic units:

Unit	*Thickness ft (m)*
Upper Member	
1. Massive quartzose sandstone.	30 (9 m)
2. Laminated quartzose sandstone.	300 (91 m)
3. Feldspathic graywacke.	
Lower Member	
4. Red siltstone, anhydritic.	150 (46 m)
5. Black shale.	30 (9 m)

This section has been interpreted as follows (Badon 1975, pp. 386–90). After deposition of the Louann Salt, a local basin developed in which the thin black shale unit (5) was deposited under stagnant conditions. The red siltstone unit (4) represents a supratidal mud flat, probably deposited under arid conditions. The feldspathic graywacke unit (3) represents a low-tidal sand flat succeeded by a supratidal mud flat with intertidal sands; local scouring and cross-strata suggest that part of the section may be fluvial in origin. The laminated quartzose sandstone unit (2) contains plane, even, and parallel lamination with dips of 10 to 30°; this section represents eolian coastal dunes. The uppermost unit (1) represents a thin transgressive section that signaled the beginning of marine conditions that resulted in the succeeding Smackover Limestone.

Thickness of the upper member of the Norphlet reflects the distribution of the sandstone facies, both the feldspathic graywacke and quartzose sandstone (Figure 4–13). A major part of this section, however, represents the sandstones of eolian origin, and therefore the distribution of coastal dune sand is approximated. The Norphlet is absent north of a line that represents the updip limit of the Louann Salt. South of this line the upper member attains a thickness of more than 300 ft (91 m). The sandstone is extensive through a large part of the area, but it thins abruptly within a distance of about 2 mi from the Nancy to the West Nancy fields.

The upper member is missing across Pachuta Creek field, a salt structure that is known to have influenced the deposition of Jurassic sediments. Early uplift of salt probably controlled the formation of oolite bars in the Smackover, which form the principal reservoir in the field (Weber 1980). In the same area, a thin section of the lower member of the Norphlet remains. Therefore, it is possible that upper-

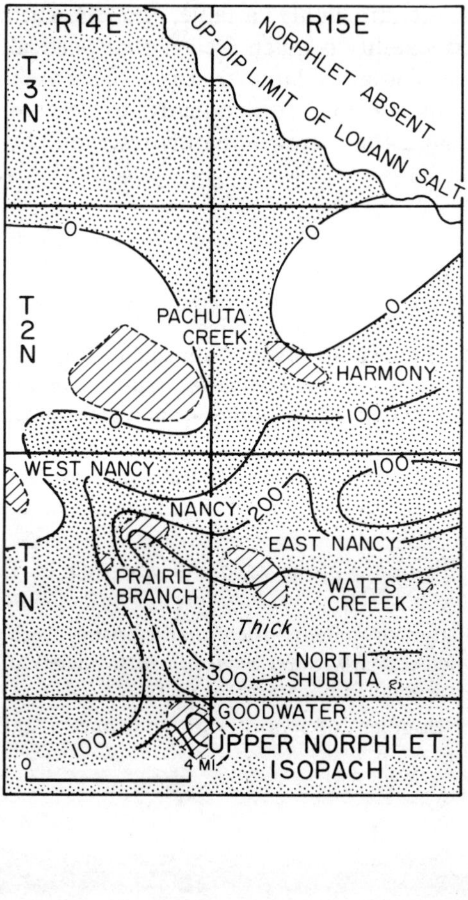

Figure 4–13 Thickness of the upper member of the Norphlet Formation in part of Clarke County, Mississippi. Oil fields shown by ruled pattern. Contour interval 100 ft (30 m). [Adapted from Badon 1975, p. 383.]

member sandstones were deposited across the entire area but were removed by erosion over local salt structures before Smackover deposition.

SUMMARY

Sandstones of eolian origin are characterized by even, parallel, continuous laminae that are inclined at angles that range up to 30°. Bed sets are generally thick, and within sets the laminae have uniform dip. Beds are bounded by erosional surfaces that commonly separate cosets that have different dips. Physical deformation of bedding by slumping takes the form of small-scale normal faults, or there may be evidence of greater deformation by folding, even overturning of some thin beds, presumably while the sand was moist and partly cohesive.

Composition and texture are more or less uniform throughout great thicknesses of section. Quartz content may be high, but dune sands, like others, owe their composition to the nature of the source sediment. Feldspar and rock fragments can be a significant part of the detrital composition. Matrix is largely of authigenic origin.

Porosity and permeability are highly variable, and lamination produces strong anisotropy. Horizontal permeability is much greater than vertical, but permeability can vary by 100 times between adjacent laminae. Dune sands are the sediments most likely to attain wide areal extent and, consequently, to deserve the term *blanket* deposits. Therefore, these sandstones are commonly reservoirs by structural, rather than stratigraphic, closure.

REFERENCES CITED

AHLBRANDT, T. S., and S. G. FRYBERGER, 1981, Sedimentary features and significance of interdune deposits: *in Recent and ancient nonmarine depositional environments: Models for exploration,* Society of Economic Paleontologists and Mineralogists Special Publication 31, p. 293–314.

BADON, C. L., 1975, Stratigraphy and petrology of Jurassic Norphlet Formation, Clarke County, Mississippi: American Association of Petroleum Geologists Bulletin, v. 59, p. 377–392.

BAGNOLD, R. A., 1941, *The physics of blown sand and desert dunes:* London, Methuen & Company, Ltd., 265 p.

BIGARELLA, J. J., 1972, Eolian environments: their characteristics, recognition, and importance, *in Recognition of ancient sedimentary environments:* Society of Economic Paleontologists and Mineralogists Special Publication 16, p. 12–62.

CALOGERO, FRANK, 1979, Development of stylotites in sandstones of the Gulf Coast: Texas A&M University M.S. thesis, 83 p.

CHEPIL, W. S., 1945, Dynamics of wind erosion: II. Initiation of soil movement: Soil Science, v. 60, p. 397–411.

DINKINS, T. H., JR., 1968, Jurassic stratigraphy of Central and Southern Mississippi: Mississippi Geological Survey Bulletin 109, p. 9–37.

ETHRIDGE, F. G., 1970, Quantitative petrographic criteria for recognition of environments of deposition: Texas A&M University Ph.D. dissertation, 169 p.

FOX, J. E., P. W. LAMBERT, and R. F. MAST, 1975, Porosity orientation in the Tensleep and its equivalent, the Weber Sandstone, western Wyoming, *in* D. W. Bolyard, editor, *Deep drilling frontiers of the central Rocky Mountains:* Rocky Mountain Association of Geologists Guidebook, p. 185–216.

FRYBERGER, S. G., 1979, Eolian-fluvial (continental) origin of ancient stratigraphic trap for petroleum in Weber Sandstone, Rangely oil field, Colorado: Mountain Geologist, v. 16, p. 1–29.

GLENNIE, K. W., 1972, Permian Rotliegendes of northwest Europe interpreted in light of modern desert sedimentation studies: American Association of Petroleum Geologists Bulletin, v. 56, p. 1048–1071.

HARTMAN, J. A., 1968, The Norphlet sandstone, Pelahatchie field, Rankin County, Mississippi: Gulf Coast Association of Geological Societies Transactions, v. 18, p. 2–11.

HJULSTROM, FILIP, 1935, Studies of morphological activity of rivers as illustrated by the River Fyris: Upsala Mineralogisk-Geologiska Institut Bulletin, v. 25, p. 222–527.

HUGHES, DUDLEY, 1968, Salt tectonics as related to several Smackover fields along the northeast rim of the Gulf of Mexico basin: Gulf Coast Association of Geological Societies Transactions, v. 18, p. 320–330.

INMAN, D. L., EWING, G. C., and CORLISS, J. B., 1966, Coastal sand dunes of Geurrero Negro, Baja California, Mexico: Geological Society of America Bulletin, v. 77, p. 787–802.

KOCUREK, GARY, and R. H. DOTT, JR., 1981, Distinctions and uses of stratification types in the interpretation of eolian sand: Journal of Sedimentary Petrology, v. 51, p. 579–595.

LINDQUIST, S. J., 1983, Nugget Formation reservoir characteristics affecting production in the Overthrust belt of southwestern Wyoming: Journal of Petroleum Technology (July), p. 1355–1365.

LUPE, ROBERT, and T. S. AHLBRANDT, 1979, Sediments of the ancient Eolian environment-reservoir inhomogeneity, *in* E. D. McKee, editor, A study of global sand seas: U.S. Geological Survey Professional paper 1052, p. 241–252.

MANCINI, E. A., and D. J. BENSON, 1980, Regional stratigraphy of the Upper Jurassic Smackover carbonates in southwest Alabama: Gulf Coast Association of Geological Societies Transactions, v. 30, p. 151–165.

MANKIEWICZ, DAVID, and J. R. STEIDTMANN, 1979, Depositional environments and diagenesis of the Tensleep sandstone, eastern Big Horn basin, Wyoming: Society of Economic Paleontologists and Mineralogists Special Publication 26, p. 319–336.

MCKEE, E. D., editor, 1979*a*, A study of global sand seas: U.S. Geological Survey Professional Paper 1052, 429 p.

——, 1979*b*, Ancient sandstones considered to be eolian: U.S. Geological Survey Professional paper 1052, p. 187–251.

MOIOLA, R. J., and A. B. SPENCER, 1979, Differentiation of eolian deposits by discriminant analysis: U.S. Geological Survey Professional Paper 1052, p. 53–58.

OXLEY, M. L., EDWARD MINIHAN, and J. M. RIDGEWAY, 1967, A study of the Jurassic sediments and portions of Mississippi and Alabama: Gulf Coast Association of Geological Societies Transactions, v. 16, p. 24–48.

PICARD, M. D., 1975, Facies, petrography, and petroleum potential of Nugget Sandstone (Jurassic), southwestern Wyoming and northeastern Utah, *in Deep Drilling Frontiers of the Central Rocky Mountains,* Rocky Mountain Association of Geologists Guidebook, p. 109–127.

STEIDTMANN, J. R., 1973, Structures in the moist, cold-climate sand dunes of southwestern Wyoming: *in Greater Green River Basin,* Wyoming Geological Association Guidebook, p. 209–213.

——, 1974, Evidence for eolian origin of cross stratification in sandstone of the Casper Formation, southernmost Laramie Basin, Wyoming: Geological Society of America Bulletin, v. 85, p. 1835–1842.

SUNDBORG, AKE, 1956, The River Klaralven—a study of fluvial processes: Geografiska Annaler, v. 38, p. 127–316.

VAN DE KAMP, P. C., 1973, Holocene continental sedimentation in the Salton basin, California: a reconnaissance: Geological Society of America Bulletin, v. 84, p. 827–848.

WALKER, T. R., and J. C. HARMS, 1972, Eolian origin of Flagstone beds, Lyons Sandstone (Permian), type area, Boulder County, Colorado: Mountain Geologist, v. 9, p. 279–288.

WEBER, A. J., 1980, Pachuta Creek (Smackover) field, Mississippi: Gulf Coast Association of Geological Societies Transactions, v. 30, p. 233–242.

WEYL, P. K., 1959, Pressure solution and the force of crystallization—a phenomenological theory: Journal of Geophysical Research, v. 64, p. 2001–2025.

WILSON, G. B., 1975, Early differential subsidence and configuration of the northern Gulf Coast basin in Southwest Alabama and Northwest Florida: Gulf Coast Association of Geological Societies Transactions, v. 25, p. 196–206.

5

Fluvial Sandstones

River systems and their sediments are among the best known of all depositional environments. They have been studied extensively for many years, and current transport has also been investigated thoroughly in laboratory flumes.

Largely because of these studies, ancient fluvial deposits are readily recognized. Still, there has been no systematic study of fluvial sandstones that will permit us to predict their morphologies with confidence. Ancient fluvial deposits are the result of processes that have acted through time to produce a complex of intermixed sediments. For this reason, it is difficult to interpret a limited vertical section of sandstones in terms of its lateral extent, whether locally or regionally. Nevertheless, it may be assumed that ancient fluvial processes were governed by the same physical principles that operate in the modern environment, and so patterns of ancient fluvial systems may be interpreted in terms of the patterns observed in modern systems. Thus, it may well be possible to predict local morphologies for fluvial sandstones as well as give some estimate of their regional extent.

STREAM TRANSPORT AND DEPOSITION

Sediment is transported in three ways. Coarse grains are moved by traction and roll or slide along the bed in continuous contact with the bed. Finer grains are moved by saltation in which the grains bounce along the bottom in alternating periods of movement and rest. The finest grains are carried by suspension in the water column with no contact with the bed for long periods.

The total sediment moved by a stream is called the *load*. The *bed load* is the sediment that moves by traction or saltation. The *suspended load* is the sediment carried in the water column above the bed and is supported by turbulence. During any period of flow, the character of the bed load and suspended load depends on flow velocity and grain-size distribution of the total load. In general, the bed load is composed of coarser sand, whereas the suspended load consists of finer sand, silt, and clay.

Distribution of load is illustrated by sediment concentrations in the Mississippi River at flood stage (Figure 5–1). Sand grains larger than 0.125 mm (fine grained) have the highest concentrations near the base of the water column, whereas finer grain sizes tend to be more evenly distributed throughout the column. All of the sizes shown probably represent the suspended load above the lowest point of measurement, about 1 ft (0.3 m) from the river bed. The bed load consists of the coarser sediment below that point. It can be expected that with decreasing velocity, a part of the suspended sand fraction will become the bed load, while the previous bed load will become stationary, that is, it will be deposited. It is apparent, then, that the size of distribution in bed load or suspended load changes continuously with the velocity of flow.

The nature of water flow in channels determines the sequence of sedimentary structures and the textures in stream deposits. Therefore, a knowledge of channel flow is essential to understanding fluvial sequences.

In straight channels the distribution of velocities tends to be symmetrical. Highest velocities are concentrated at and just below the water surface in the center

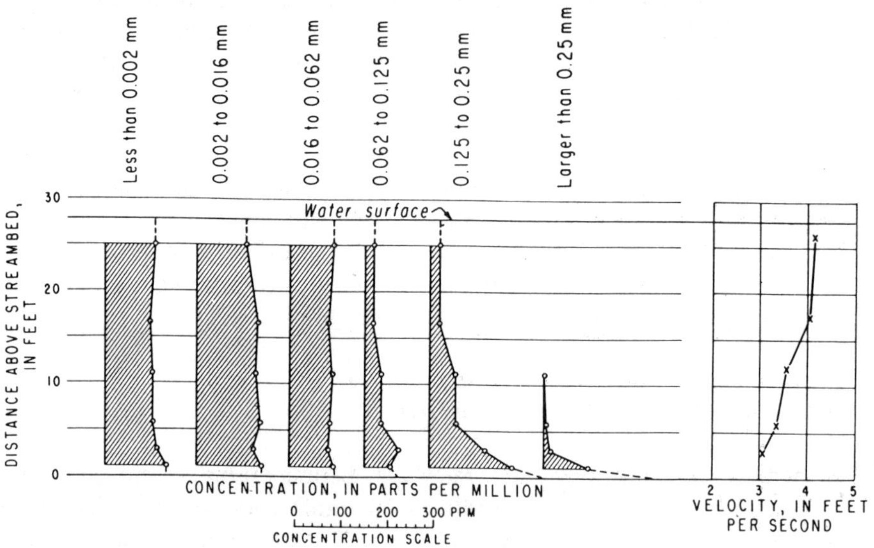

Figure 5–1 Distribution of sediment classes as a function of distance above stream bed in the Mississippi River at St. Louis, Missouri, April 24, 1956. [From Colby 1963, p. A25.]

of the channel. Lowest velocities are around the perimeter of the channel because of the frictional drag along the bottom and channel margins. This symmetrical pattern of flow is altered in curved channels (Figure 5–2). The highest velocities are displaced toward the outer side of the channel, and water piles toward the outer edge and exerts an outward-directed force against the channel margin. Outward displacement of velocities also results in scouring of the channel bottom toward the channel margin. Conversely, lower velocities are present on the inside of the bend, and deposition of sediment may take place.

Because of the asymmetric velocity distribution in curved channels, an asymmetric channel is formed and maintained. Lateral erosion and channel scouring at the outside of the curve cause lateral migration of the channel and deposition on the inside of the channelway. Therefore, the asymmetric channel profile is not only developed by curvature, but it is maintained by flow and is accompanied by erosion and deposition.

The line of maximum velocity is not only diverted toward the outside of the bend, but it also exerts a lateral force on the downstream side of the channel margin. This downstream displacement results in bank erosion and a tendency for the curve to migrate in a downstream direction as well as laterally.

Outward displacement of the high-velocity tube also causes helical flow within the stream channel. The displaced tube of water is itself replaced by water that flows upward and inward along the inside and sloping bottom of the asymmetric channel. This water is returned as top flow in a downstream direction. The helical flow pattern has been demonstrated in laboratory flumes as well as documented in natural streams (Allen 1971, p. 131).

The helical flow pattern in asymmetric channels has important consequences for deposition on the inside of the curved channel. The rising and inward-flowing

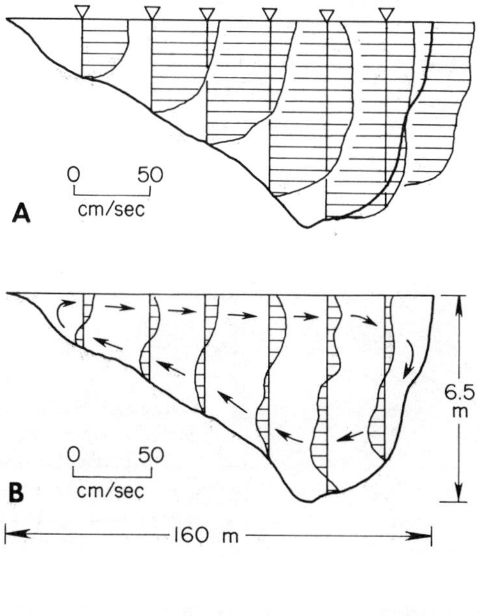

Figure 5–2 Velocity distribution in symmetric and asymmetric stream channels. [From Allen 1971, Figure 4.4.]

bottom water decreases in velocity as it climbs toward the inner margin of the channel. Velocity decrease means that coarser grains may be deposited in the deeper parts of the channel, whereas the smaller grains may be deposited on the shallow parts of the channel inside of the bend. As water velocity decreases, so does the flow regime, and the form of the bed changes with the deposition of grains. A dune phase forms with deposition of coarser grains, and the ripple phase forms with deposition of finer grains. With continued deposition on the inside of the bend, additional sediment accretes on the face of the developing deposit. Thus, a significant body of sediment develops on the inside of the bend.

With further channel migration, the resulting deposits build laterally, or *prograde*, to form a sediment body that is characterized by a regular sequence of sedimentary structures and grain-size change (Figure 5–3). Coarser sediments with inclined laminae represent the dune phase and are succeeded by finer sediments in a more thinly bedded ripple phase. These bed forms and textural changes are formed during times of flow. Their formation depends on continued flow, channel migration, and deposition. Thus the depositional sequence is the result of the dynamic interaction of the channel flow processes during flood.

The coarsest particles, granules, pebbles, or even boulders, may be left within the deepest part of the channelway where movement of large particles is intermittent. The depositional sequence may build out over the lag sediments, and therefore the coarsest particles may be incorporated at the base of the depositional sequence. Also, a rapid decrease in flow velocity may result in deposition of coarse grains within the channel itself in the upper flow-regime of the antidune phase or plane-bed phase of deposition. If deposition takes place from the high flow-regime, a coarser, poorly laminated sediment may be incorporated at the base of the depositional sequence.

Another important result of helical flow is that the dune phase of transport may result in large sand waves in the deeper parts of the inside channel margin. Because helical bottom flow is at an angle to the direction of maximum velocity, the crests of sand waves are at an angle to the main direction of sediment transport.

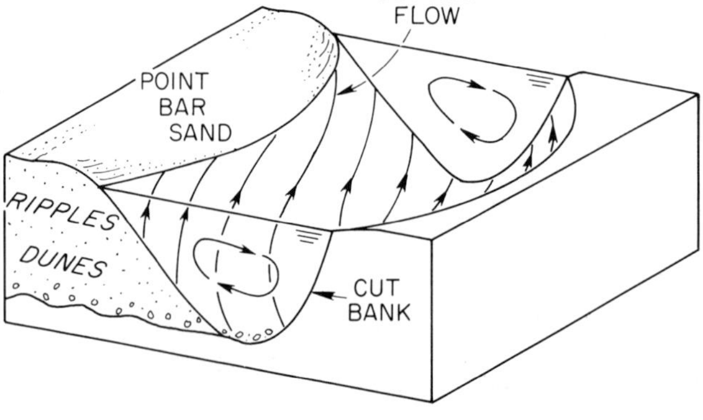

Figure 5–3 Helical flow in asymmetric, curved channels that results in a characteristic depositional sequence from a coarse-grained dune phase below to a fine-grained ripple phase above. [From Allen 1971, Figure 4.5.]

The sand waves thus formed may show crests that diverge about 60° from the direction of water flow with the crests pointing in a generally downstream direction.

When stream flow decreases to low velocities, the water level declines and coarse sediment transport ceases. Very fine sediment may be deposited from suspension to form a clay "drape" over the previously deposited coarser sediment. The channelway itself is filled with deposits of decreasing grain size upward and decreasing flow regime from dune to ripple phase. Finally, very fine-grained sediment from suspension is deposited within the restricted channelway. At the low stage of flow, sediment transport is at a minimum. Abandonment of the channelway may result in final filling of the channel with fine-grained sediment from suspension to form an arcuate clay plug.

The sequence of fluvial deposition described above may take place within channels that are only slightly to highly curved. When the sinuosity of channels is low, the resulting deposits, although curved, are elongated in a downstream direction and are termed *longitudinal bars*. Around such features, stream flow generally separates during the declining water level and results in isolated islands of bar sands separated by branching and joining channels. The resulting stream pattern is termed *braided*, and the longitudinal bars have been called *braid bars*. On the other hand, the term *point bars* generally is reserved for the arcuate deposits within the bends of highly sinuous streams. The distinction, therefore, between braid bars and point bars rests largely with the curvature of the stream channels themselves.

The depositional sequence within fluvial channels remains the same regardless of the amount of channel curvature. The sequence itself is the result of decreasing velocity and decreasing flow regime.

MODERN FLUVIAL SEDIMENTS

Stream Patterns

The patterns formed by modern streams are highly variable and range from essentially straight to highly curved and winding. In order to classify stream patterns, the degree of curvature, or sinuosity, is used. *Sinuosity* is defined as the ratio of actual channel length (L_a) to the direct down-valley length (L_d). Straight channels have essentially no curvature, and therefore have a sinuosity of unity. Braided channels have low sinuosity, whereas meander channels have high sinuosity. For purposes of classification, the braided channels are defined as those that have a sinuosity of less than 1.5, and meander channels are those that have a sinuosity equal to or greater than 1.5 (Leopold et al. 1964, p. 281).

The use of the term *braided* is not entirely satisfactory to distingush single-channel patterns because the word suggests interweaving, or the branching and joining of multiple channels. Therefore the term *anastomosing* has been used as a synonym for braiding. In other countries, however, this term refers to a more broadly branched and joined system of channels rather than to an intricate interweaving of single channels (Schumm 1968, p. 1580). The latter use of the term is perhaps a good one, for it serves to denote a pattern for a large-scale, multiple-channel system.

In order to visualize the meaning of sinuosity, several idealized curves may be compared (Figure 5–4). Natural channels do not follow regular patterns such as mathematical functions, but for purposes of comparison, circular segments are arranged in series to depict sinuosity. Low sinuosity of 1.1 is shown by arcuate segments subtended by an angle (θ) of 90° [Figure 5–4(A)]. Higher sinuosity of 1.5 is approximated by semicircular segments subtended by an angle of 180° [Figure 5–4(B)]. High sinuosity of 2.0 is shown by segments subtended by an angle greater than 200° [Figure 5–4(C)]. These contrasting sinuosities can be recognized rather easily by noting the approximate value of the angle (θ).

Another useful term to describe the patterns of multiple channels is *distributary*, which designates a branching and splaying pattern in a downstream direction in which the individual channels do not rejoin. Thus the word *distributary* has an opposite meaning from that of *anastomosing*.

In general, *braiding* is a pattern commonly attributed to the upper parts of alluvial valleys, whereas *meandering* more commonly characterizes the lower parts of alluvial valleys. However, natural streams are not consistent in local configuration, and patterns may change from meandering to braided to straight within relatively short distances.

Channel patterns appear to be controlled by local gradient and discharge (Figure 5–5). On the basis of many measurements, it has been shown that low gradient or low discharge results in meander patterns, whereas high gradient or high discharge results in braided patterns (Leopold and Wolman 1957). The diagonal line

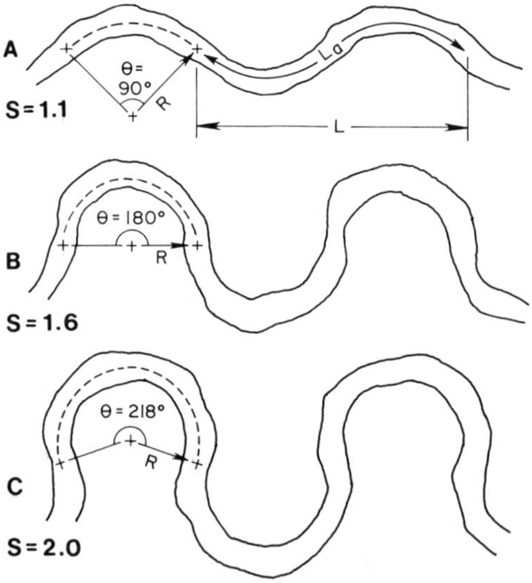

Figure 5–4 Depiction of stream sinuosities by circular segments in series: (A) Low sinuosity, S = 1.1 and θ = 90°; (B) higher sinuosity, S = 1.6 and θ = 180°; (C) high sinuosity, S = 2.0 and θ = 218°. Note the increase in angle (θ) with increasing sinuosities.

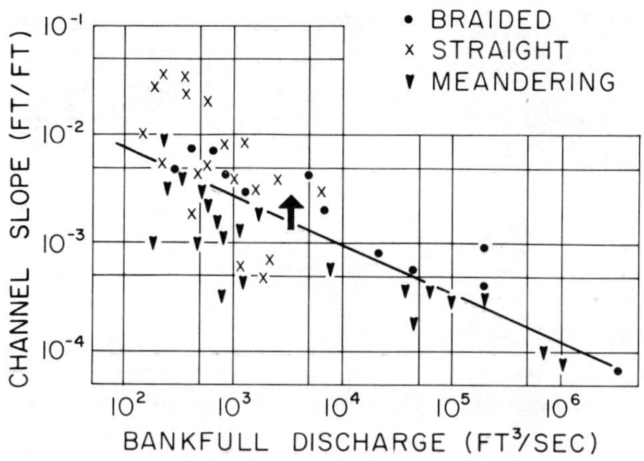

Figure 5–5 Channel patterns as functions of slope and discharge in modern streams. [From Leopold and Wolman 1957.]

across the diagram (Figure 5–5) separates most of the braided and meander patterns, but for the same set of measurements, straight channels showed no regular relationship to gradient or discharge.

In other words, local changes in gradient or discharge directly influence flow patterns of braided and meandering streams, whereas straight reaches may be largely the result of bank resistance or local geological alignments such as joints or faults. It may be possible to extrapolate these relationships to large river systems that change through time. For example, climatic change may cause an increase in discharge with time and thereby produce a change in channel pattern from meandering to braided. In the same way, regional tectonic movement may cause an increase in gradient and thereby produce a similar change from meandering to braided, even if the discharge remains constant (Figure 5–5, arrow).

Stream Sediment

The fluvial sequence is well documented from a variety of modern streams. This is the sequence deposited by decreasing flow velocity in channels (Figure 3–2). Bed forms are preserved in succession from the high flow-regime of the antidune phase or plane-bed phase to the lower flow-regime dune phase and ripples (Figure 2–9). At the same time, coarser sand is deposited at higher velocity and finer grains at lower velocity (Figure 2–2). Thus, the total sequence shows decreasing upward scales of sedimentary structures and decreasing grain size. This well-known sequence, however, can be variable in thickness of bedding units, grain size, and composition. Differences in texture and composition, of course, depend on the type of available sediment as well as on the competence of stream flow. The bedding sequence may also show differences depending on the position of deposits within the channel. A number of modern stream-channel deposits are selected to illustrate these variations (Figure 5–6).

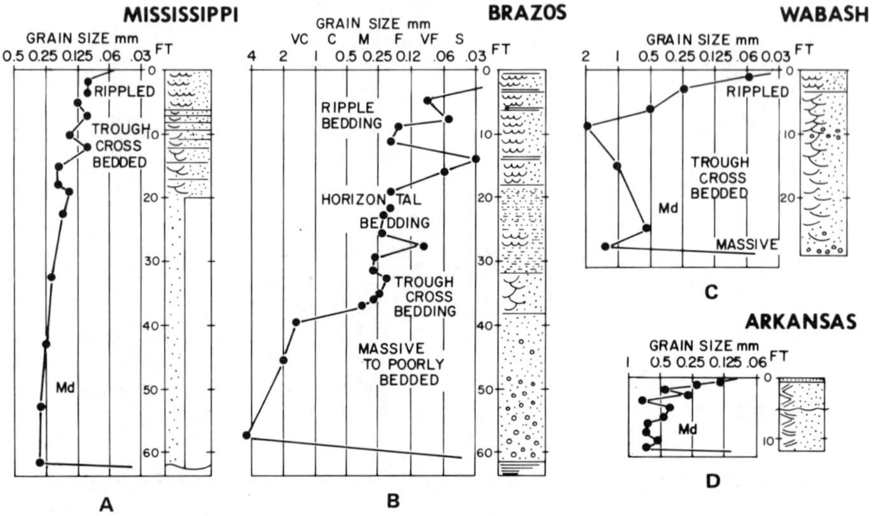

Figure 5–6 Variations in bedding sequence in selected modern fluvial deposits: (A) Mississippi River; (B) Brazos River at Richmond, Texas; (C) Wabash River; (D) Arkansas River. [(A) from Pryor 1973; (B) from Bernard et al. 1970; (C) from Jackson 1976; (D) from Steinmetz 1972.]

Thickness. *Point bars* are represented by thick sequences. In the lower Mississippi River, point bars are at least 60 ft (18 m) thick (Kolb and Van Lopik 1968), although details of sedimentary structures are known for only the upper parts (Pryor 1973). Trough crossbeds are present in sets that are 3-ft (1-m) thick and decrease upward to 1 ft (0.3 m). These sets are succeeded by ripple laminations in the uppermost 6 ft (2 m). The sands are medium grained (0.28 mm) in the lower part, and grain size decreases to fine grained (0.14 mm) at the top of the section.

A more detailed section is known from the lower Brazos River [Figure 5–6(B)]. Maximum thickness of the total sequence is 60 ft (18 m), and the section consists of a lower pebbly and massive unit, succeeded by a trough crossbedded unit, a parallel and ripple-laminated unit, and a rippled section at the top (Bernard et al. 1970). Grain size decreases upward from coarse grained at the base to very fine grained at the top.

A thinner point-bar section is well known from the Wabash River [Figure 5–6(C)]. This section has a maximum thickness of 30 ft (9 m), and the total sequence was inferred from the lateral succession of exposed beds and those in shallow trenches (Jackson 1976). The section is coarse grained and gravelly at the base and contains a dominant trough crossbedded section and a thin ripple-laminated unit at the top. The trough crossbedded unit is coarse to very coarse grained in the lower part, and mean size decreases upward to fine grained at the top. Thickness of trough sets apparently decreases upward, and some thin planar crossbed sets may be included near the top.

A *braided-stream sequence* is that of the Arkansas River [Figure 5–6(D)]. The section is 12-ft (3.6-m) thick and was deposited during a single flood of the river (Steinmetz 1972). The sequence consists of thin trough sets succeeded by planar sets. Fine-grained horizontally laminated sands and muds were deposited as a thin overlying bed.

Each of the sequences described in Figure 5–6 forms a composite set and has in common the succession of larger-scale trough sets in the lower part, decreasing thickness of sets upward, and small-scale planar and ripple sets at the top. The most notable difference is the greater thickness of point-bar deposits in contrast to those of the braided stream. The total thickness of sequence is determined largely by the depth of stream flow, laterally to the deposit in the case of point bars and essentially along the channel in the case of the braided streams. It might be concluded that meandering streams are characterized by thick composite sets and braided streams by thin composite sets.

Such a conclusion might be valid only in the most general cases. The larger rivers such as the Mississippi, Brazos, and even the Wabash are meandering flow systems characterized by lateral accretion of deposits within broad meander loops. Thus, the areal extent of point-bar sands within the meander loops can be expected to be relatively large. On the other hand, the braided streams represent the cut and fill of single channels during flood and ebb flow. Therefore, it can be expected that large braided channels could cut more deeply (Figure 5–6), thereby producing a sequence that might resemble that of a point-bar deposit. In the same way, a small meandering stream might produce relatively thin point bars; but in this case, the lateral extent of such a sequence could also be expected to be limited because of the small stream size.

Therefore, the conclusion might be restated for a more general case that the thick fluvial sequence signifies a deposit of greater lateral extent, whereas the thin fluvial sequence represents a more limited areal extent. Then, the thickness of these composite sets may yield an approximation of the areal distribution of a sand deposit. Based only on the modern sequences, the more limited extent may be indicated by sets on the order of 20 ft (6 m) or less.

Texture. Grain size decreases with long distance of transport in fluvial systems. A well-known example is the lower Mississippi River through its length of nearly 1100 mi (660 km) from Cairo, Illinois, to the Gulf of Mexico (U.S. Army Corps of Engineers 1935). Mean grain size is generally coarse to very coarse sand along most of this length, and maximum size is variable due to the input from tributaries (Figure 5–7). However, there is a pronounced decrease from the vicinity of Vicksburg, Mississippi, at 600 mi (360 km) below Cairo, to the delta mouth. If only the sand fraction is considered, the mean size decreases from about 0.6 mm at Cairo to 0.2 mm at the mouth, and the overall change is nearly linear. Furthermore, there is a striking change from the alluvial valley to the delta plain, located about one-half the distance between Baton Rouge and New Orleans, Louisiana. At

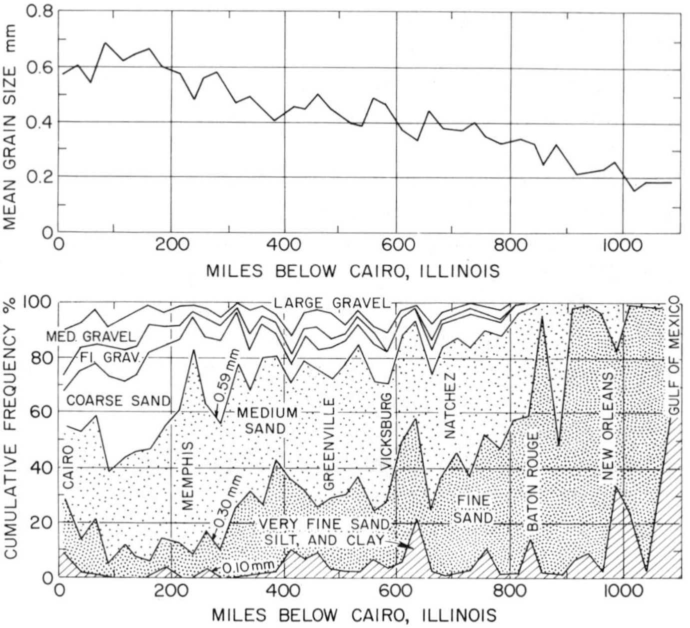

Figure 5-7 Change in grain size of bed load in the Mississippi River from Cairo, Illinois, to the mouth. [From U.S. Army Corps of Engineers 1935.]

this point the mean size of the sand fraction remains fairly constant at 0.2 mm until the delta mouth is reached.

These changes in texture reflect grain-size reduction by abrasion as well as the influence of local source. Tributaries to the alluvial valley cause local variations in mean size, whereas the total sand fraction is gradually reduced in size. When the delta plain is reached, a new source of sediment is present in the form of previously deposited finer sediment that is subject to erosion during flood. Only the finest fractions reach the delta mouth.

Composition. Fluvial sediments owe their grain composition to available source materials. In general, fluvial sands are characterized by a modest amount of quartz, whereas feldspars, rock fragments, and accessory minerals make up a significant part of the composition. If the source area consists of igneous and metamorphic rocks, the fluvial sediment can be expected to be low in quartz content, say on the order of 25 to 40 percent. On the other hand, a source area composed of sedimentary rocks may yield a high quartz content of 70 percent or more, and sedimentary rock fragments can be expected to be present.

It was long believed that stream transport could greatly modify composition by abrasion and chemical weathering of selected grains, but this hypothesis does not seem to hold for Mississippi River sands (Russell 1937). Samples of a single size class (0.15 to 0.21 mm) were found to change only slightly in feldspar content, from

25 to 20 percent, through a distance of 1100 mi (1760 km) of transport from Cairo, Illinois, to the delta mouth. A later study on Mississippi River point-bar sands (Ethridge 1970) indicates a somewhat greater change in the ratio of quartz, feldspar, and rock fragments, from 61:13:26 upstream to 74:7:19 downstream over the same distance, but this study used only a small number of samples and locations. It may be concluded that abrasion and weathering during transport do not induce large compositional changes in the Mississippi.

Some mineral varieties may be susceptible to change in transport. An example is polycrystalline quartz, which is believed to disaggregate during stream transport into small grains of the monocrystalline variety (Blatt 1967). An example of this possible change is found in sands of the Neuse River of North Carolina (Cleary and Conolly 1971).

The Neuse River flows largely through the Piedmont province where soils are derived from granite, gneiss, and schists. These soils may contain 50 to 75 percent quartz, both monocrystalline and polycrystalline varieties. During 200 mi (32 km) of transport, the sediment is enriched in monocrystalline quartz from a minimum of 20 percent to a maximum of 90 percent at the river's mouth (Figure 5–8). Poly-

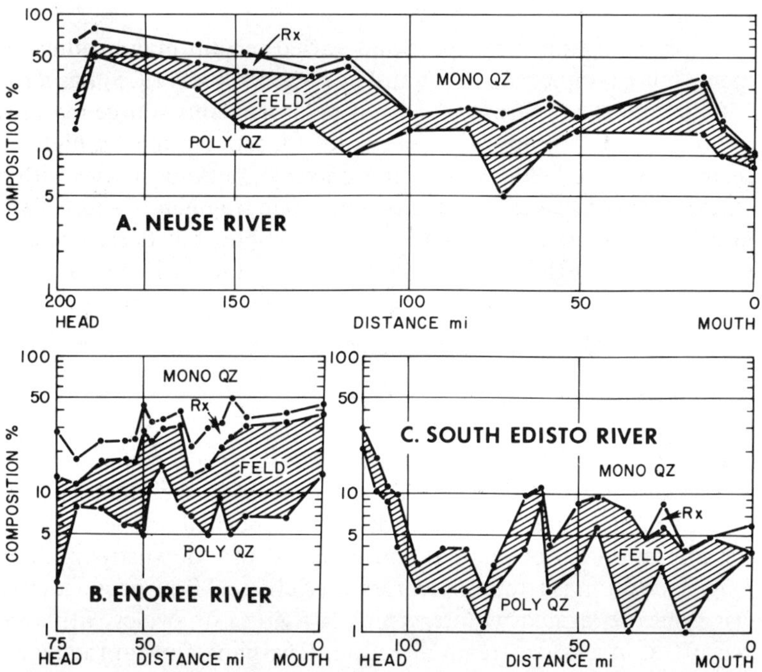

Figure 5–8 Contrasts in composition of fluvial sediment in the Piedmont streams—the Neuse and Enoree rivers, and in a coastal-plain stream—the South Edisto River. Poly qz = polycrystalline quartz, Feld = feldspars, Rx = rock fragments, and Mono qz = monocrystalline quartz. [Adapted from Cleary and Conolly 1971, p. 2763.]

crystalline quartz is more susceptible to decrease than monocrystalline quartz, which is derived from the disaggregation of the polycrystalline variety.

Another stream, the Enoree River of South Carolina, flows entirely within the Piedmont province and shows the effects of tributaries (Figure 5–8). Monocrystalline quartz is reduced from about 70 to 55 percent along the distance of transport, and this change probably is due to the local contribution of polycrystalline quartz and feldspars.

Fluvial sediments may also change abruptly in composition due to another type of change in immediate source sediments as, for example, when a river passes from a terrain of weathered igneous and metamorphic rocks to a coastal plain composed of previously eroded and transported sediment. An example is the South Edisto River, which is confined to the coastal plain of South Carolina (Figure 5–8). At its head, the sediment contains about 70 percent of monocrystalline quartz with polycrystalline quartz and feldspars, a composition inherited from the Piedmont streams. Within the coastal plain, however, the fluvial sediment is immediately enriched in quartz to a total of 90 to 95 percent, nearly all of which is monocrystalline quartz. Quartz enrichment is believed to be the result of ''dilution'' by local erosion of quartzose sands of Cretaceous and Tertiary age in the coastal plain (Cleary and Conolly 1971).

Short-distance transport and varied terrain may produce a quartz-poor sediment. For example, the Rio Nautla of Mexico (Self 1975) has a total length of about 150 km and flows through areas of heterogeneous source materials from volcanics, granites, and limestones to alluvium. Quartz content is highest in the headwater segment, which includes the granitic source, but it decreases with transport distance, whereas volcanic rock fragments, limestone fragments, and feldspars increase. Mean grain size ranges from coarse sand to cobbles but decreases abruptly to medium-grained sand in the coastal segment. Thus, the sediment delivered to the shoreline by fluvial processes can be highly variable, and composition depends largely on source materials.

Sediment Morphology

Sediment distribution within a meandering stream system is well known (Figure 5–9). As discussed earlier, during times of flood the channel bank is eroded at the outside of meander loops, and sands are deposited on the inside as point bars. Channel migration is both lateral and diagonal in a downstream direction. In the same way, deposition is lateral. Recurrence of channel flow produces a succession of point-bar ridges separated by intervening low areas or swales with each succeeding point bar offset in a downstream direction. Overbank flow at the high-water stage results in levee deposits that border the outside of the channel. Within the meander loop, the swales may also be occupied by flowing water.

Abandonment of the channel may take place when the river cuts through the narrow neck between two meander loops to produce a *neck cutoff*. Alternatively, a new channel may be established through a low swale to produce a *chute cutoff*.

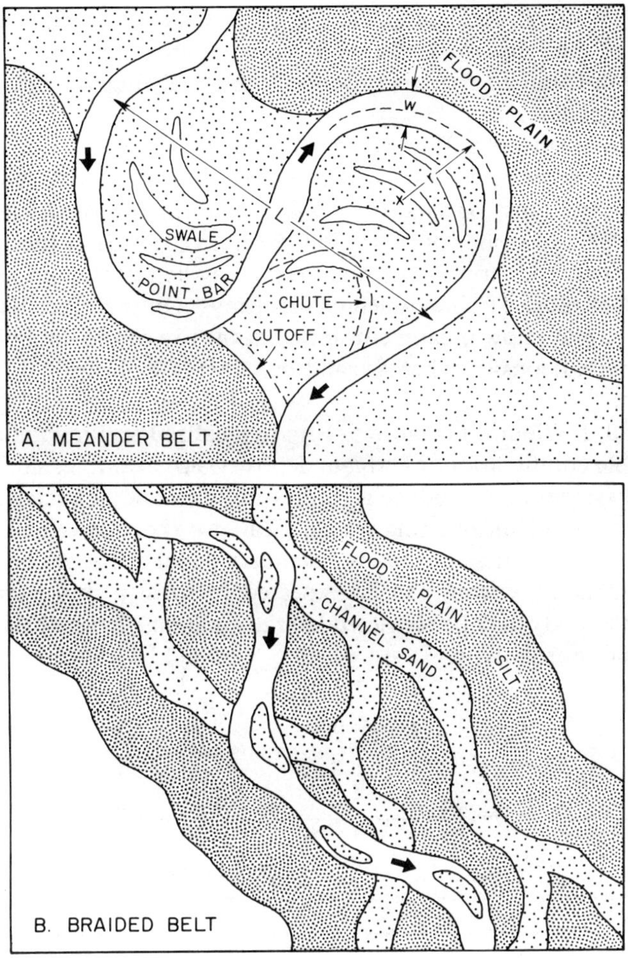

Figure 5–9 Plan view of sediment morphology in a meander system (A) and in a braided system (B).

The abandoned channel remains as an *oxbow lake* to be filled by fine-grained sediment.

The pattern of sediment distribution within a braided system may be distinctly different from meandering because of the relative narrowness of the low-sinuosity channel. During times of flood, the channel may be scoured into previously deposited sediment, cutting downward with only minor lateral translation of the channelway. When flow decreases, sand is deposited vertically, and the channel is filled by sand. Overbank flow may result in levees that border the channel on both sides as a thin apron of fine-grained sediment.

Abandonment of the braided channel takes place during decreased flow when the channelways become choked with sediment. Recurrence of flood may then produce a new channel path that is largely independent of the previous channelway.

These contrasting patterns of flow and deposition may result in contrasting sand distributions (Figure 5–10). Lateral deposition of point bars within the meander loop produces a body of sand that is relatively wide as compared to its thickness. Vertical deposition within the braided channel produces a relatively narrow body of sand as compared to its thickness.

Another consequence of different flow patterns is the continuity of channel-sand deposits. Point bars seem to be essentially continuous through a wide lateral extent, and variations appear as loss of sand downward from the top of the body as a consequence of the ridge and swale configuration, whereas the base of the sand body may be nearly continuous. Continuity of point bars may be interrupted completely by narrow mud-filled plugs of abandoned channelways. On the other hand, the braided-channel deposit may show greater continuity at the top of the sand body from channel-fill to overbank deposits, whereas sand may increase abruptly downward in channel locations.

The diagrams of sand distribution (Figure 5–10) are idealized and may not be representative of all possible combinations of stream deposits. For example, shifting channelways of a braided system may produce a composite sand body of relatively wide extent and identical to a small meandering system. Still, it is the relative amount of sand that is important. The thickness of sand is essentially equal to the maximum flow depth, and a small shallow stream can be expected to form a thin narrow deposit. Thus, the composite body may be restricted laterally and may have a braided aspect whatever its internal morphology.

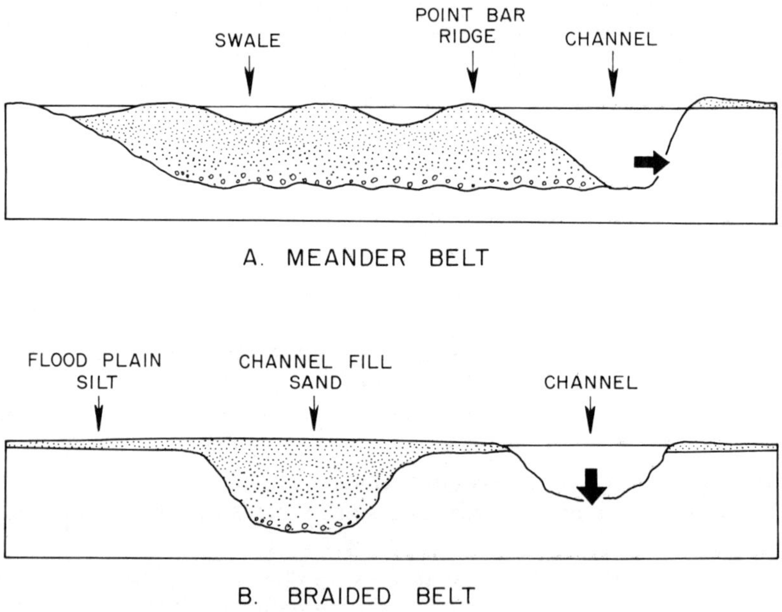

Figure 5–10 Diagrammatic cross sections of sand distributions within a meander belt (A) and in a braided belt (B).

In meandering streams, there are consistent relationships among certain channel parameters. Measurements were made of meander length (L), radius of curvature (r), and bank-full channel width (w) for small streams on glacial ice, medium to large continental rivers, and giant meanders of the Gulf Stream in the western Atlantic (Leopold and Wolman 1960). Plots of these measurements (Figure 5-11) show the following average relationships:

$$L = 10.9w^{1.1} = 4.7r^{0.9} \qquad 10.9$$

These relationships appear to fit reasonably well with most streams. Although the reasons for these consistencies in meander form are not known, it is possible that meandering flow represents a stable configuration of least energy dissipation (Leopold and Wolman 1960). The meander functions may have predictive value when only one measurement is known, and such predictive use would assume a stable condition of flow.

The meander relationships in Figure 5-11 are those for streams with sinuosities of 1.5 or more, as previously defined. Braided streams of lower sinuosities cannot be expected to show the same fit. For example, the radius of curvature for braided streams should be larger for a given wave length, and braided channels should plot below the average function of L and r for meandering channels.

Other channel relationships have been suggested. For example, it may be assumed that stream discharge, both water and sediment load, exerts a control on channel width, depth, sinuosity, and other channel measurements. Therefore, the

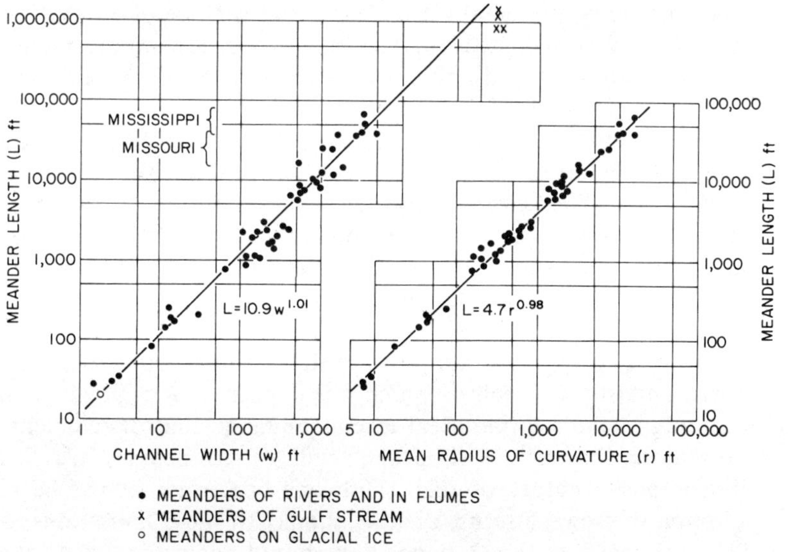

Figure 5-11 Meander length (L) as a function of bank-full channel width (w) and radius of curvature (r) for a wide range of natural streams. Measurements defined in Figure 5-9. [From Leopold and Wolman 1960.]

type of sediment in banks and within the channel may also be related to these parameters (Schumm 1968, 1972). If this were true, the composition of fluvial sediment as determined in a single vertical sequence might allow the prediction of local channel morphology. However, the relationships observed in modern streams have been applied to ancient sediments with only limited success, largely because of the lack of data for ancient stream morphologies (Ethridge and Schumm 1978).

ANCIENT FLUVIAL SEDIMENTS

There is little difficulty in recognizing ancient stream deposits when the vertical sequence of composition, texture, and sedimentary structures is known (Allen 1965). Sandstones of fluvial origin have commonly been identified as reservoirs in both stratigraphic and structural traps. A more difficult problem is the prediction of morphology in exploration and in field development, and accompanying this problem is the prediction of reservoir heterogeneity for the most effective fluid recovery.

Reservoir sandstones of fluvial origin are composite bodies. Like all other ancient sediments, they represent a succession of deposits formed through a period of time and under conditions that may have varied with time. Nevertheless, individual sandstone units may be treated as if they were the products of essentially uniform conditions and by processes similar to those that operate at the present. Morphologies can then be predicted by observation of bed thickness in cores to predict narrow as opposed to widespread deposits. The patterns of the composite bodies may be interpreted as braided or meandering, and the local extent of meander deposits may be predictable by comparison with modern-stream relationships (Figure 5–11). From such attempts to relate properties and morphologies of sandstones may come an improved ability to anticipate reservoir occurrence.

Cores from the subsurface are not always available for observation of bed thickness. Still, the designations of two distinct patterns of flow—*meandering* and *braided*—will be useful in the subsurface to describe the form of channel as mapped in plan view. These terms may not always describe the actual form of individual channels, however, because in mapping ancient sandstones, it is the total body of porous sediment that is commonly mappable. Such sandstone bodies may be, in fact, a composite of individual channel sandstones each of which may have different forms.

For example, it is conceivable that a single sandstone unit could consist of several sandstones, each of which was deposited in braided channels, but the total unit might have a broad areal extent similar to that of sediments deposited within meandering channels. Without the details of sedimentary structures in several wells, the composite nature of such sandstone bodies might not be suspected. The description of flow patterns, then, is subject to these limitations, and extended conclusions regarding fluvial deposition should not be drawn on the basis of channel pattern alone. Other properties of the rock, particularly those of sedimentary structures in cores, should also be considered for a reliable interpretation of channel type.

Reservoir sandstones of fluvial origin are described in the following sections of this chapter. The examples were chosen to illustrate, as far as possible, all properties of the rock: composition, texture, bedding sequence, and morphology. The first example is the Middle Pennsylvanian Robinson Sandstone of Illinois, a small composite body that was deposited in a braided channel. Next is the Middle Pennsylvanian Bartlesville (Bluejacket) Sandstone of southeastern Kansas, which is composed of thin clayey sandstones in narrow overlapping point-bar deposits. The Upper Cretaceous Tuscaloosa Sandstone, Mississippi, shows a succession of older meander-belt deposits succeeded by braided-belt deposits, a sequence that is best explained by transgression as a result of basin downwarp and increased stream gradients with time. Finally, the Lower Cretaceous Fall River ("Dakota") Sandstone of the eastern Powder River basin, Wyoming, forms large arcuate point bars in a meandering channel that was cut through previously deposited deltaic sediments. These examples illustrate a variety of compositions and morphologies, as well as reservoir properties, a range in properties that might be encountered in other sandstones of fluvial origin.

ROBINSON SANDSTONE, FRY AREA, ILLINOIS

Sandstones of Pennsylvanian and Late Mississippian age were, in earlier years, the major oil reservoirs in the Illinois basin. Many of these reservoirs were developed during the period of 1904 to 1913 along the LaSalle anticline in Clark, Crawford, and Lawrence counties, southeastern Illinois. Production from these sandstones reached a maximum of about 30 million bbls/yr (4.8 million m^3/yr), but production declined rapidly after about 1915 (Miller 1968, pp. 33–36). Beginning in 1944, production was renewed from many of these reservoirs by water flooding in order to extract a remaining and significant amount of oil from the sandstones. Other secondary and tertiary recovery projects were soon initiated.

In the Fry area, an *in situ* conbustion project was attempted in the Pennsylvanian Robinson Sandstone. For such projects the detailed characteristics of the reservoir sandstone must be known, and core drilling was begun in a small 3.3-acre (1.3-ha) area for a pilot study.

The cores revealed the bedding sequence of a small sandstone body that probably represents the deposit of a single fluvial channel. Although the reservoir is a narrow isolated body, its characteristics may be typical of the many thin sandstones that are found in the Pennsylvanian of the Illinois basin.

Geologic Setting

The Fry area is located on the crest of the LaSalle anticline in southwestern Crawford County, Illinois (Figure 5–12). The anticline trends south-southeastward and is strongly asymmetric with very gentle dips to the east and dips of 3° to 6° to the southwest. Pennsylvanian sandstones produced oil from small structural closures along the crest and gentle east flank of the LaSalle anticline.

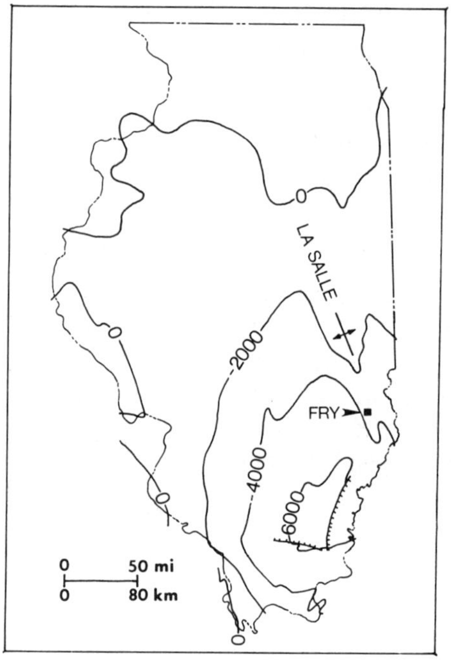

Figure 5-12 Structure map of southern Illinois showing contours at base of Mississippian rocks and location of Fry Unit on the LaSalle anticline. Contour interval 2000 ft (610 m). [Adapted from Illinois and Indiana-Kentucky Geological Societies 1968.]

Deposition in southern Illinois during the Late Mississippian (Chesteran) resulted in a dominant shale section that contains limestone sheets and elongate sandstone bodies. The maximum thickness of the Chesteran Series is 1400 ft (427 m) in the southern part of the Illinois basin. Cyclic deposition characterized the epoch, with marine limestones alternating with prograding deltaic sandstones that reached southwestward across the shallow seas. Near the end of the Mississippian, the region was uplifted, subjected to erosion by southwest-flowing streams, and cut by valleys as much as 200-ft (61-m) deep.

Pennsylvanian deposition was dominantly nonmarine, and the section has a maximum average thickness of 2000 ft (610 m). Multiple sandstones are present throughout the section (Table 5-1), and these form elongate discontinuous bodies that may be relatively straight or highly meandering. Many of the sandstones are deeply channeled in underlying beds, and correlations are difficult. A major seaway developed to the east, and the direction of stream flow was generally in that direction. Increasing numbers of coal beds appeared in the section during early Desmoinesian time, and the numerous thin marine shales and fossiliferous limestones found in the section permit the definition of many cyclothems, each characterized by a systematic sequence of sandstone, shale, and limestone. In addition to the sandstones as oil reservoirs, the coals have been of great economic importance.

The numerous sandstones of the Pennsylvanian have been studied in some detail by Potter (1963). The Robinson sandstone is not separately identified in the stratigraphic sequence because the name is an informal one and applies to oil res-

TABLE 5-1. STRATIGRAPHIC NOMENCLATURE FOR PENNSYLVANIAN SANDSTONES IN THE ILLINOIS BASIN

Series	Group	Formation	Sandstone members
Missourian		Mattoon	
	McLeansboro	Bond	Mt. Carmel
————		Modesto	Tivoli
			Gimlet
Desmoinesian		Carbondale	Anvil Rock
	Kewanee		Vermilionville
		Spoon	Palzo
————		Abbott	Murray Bluff
Atokan			Finnie
	McCormick		Grindstaff
————			
		Caseyville	Pounds
Morrowan			Battery

SOURCE: Adapted from Potter 1963, p. 16, and Bond, et al. 1971, p. 1172.

ervoir sandstones in Crawford County. However, the Robinson sandstone is probably at the level of the Palzo Sandstone Member of the Spoon Formation (Table 5-1).

Drilling History

Oil was discovered in the Fry area in 1906, but the area soon became part of a larger field known as the Main Consolidated field. Early production rates were high, but decline in production was rapid. Water flooding was initiated in the Fry area in July 1952 on a 160-acre tract, but it was unsuccessful. The Fry unit was then selected as a research site for an *in situ* combustion test, and operations began in the spring of 1960. Field operations and performance of the test were described by Clark et al. (1965*a*, 1965*b*).

A preliminary air-injection test began in 1960, and the four wells nearest the air-injection well responded by increased oil production from 5 to 30 bbls/day and water production from 42 to 91 bbls/day. New wells were drilled and cored, and the reservoir was ignited in 1961. The combustion front first reached the nearest well to the west, located about 275 ft (84 m) from the injection well. It next reached the two wells located at the same distance to the north and south of the injection well.

After two years of operation the combustion test was considered successful.

Initial oil in place was estimated to be 171,600 bbl in the 3.3-acre area, and recovery during the test period was 101,586 bbl or about 59 percent of oil in place. The oil had a specific gravity of 28.7° API and a viscosity of 40 cp at a reservoir temperature of 65°F.

Both preliminary air-injection tests and the combustion test indicated significant directional control by the reservoir on fluid movement. The combustion front moved rapidly to the west, and then later affected performance in wells to the northeast. Furthermore, combustion tended to follow the middle part of the reservoir sandstone and particularly along the base of this zone. Apparently, the inhomogeneous nature of the reservoir controlled the movement of combustion as well as the production of fluids.

The characteristics of the reservoir are clearly shown by a core from Ohio Wampler 14, a production well drilled to the west of the air-injection well. Composition, texture, and the sequence of sedimentary structures illustrate the control of primary rock properties on reservoir performance as well as the details of rock properties of a typical fluvial sandstone.

Description of Sandstone

Sedimentary structures. The Robinson Sandstone displays a sequence of sedimentary structures that form three major units, which have been termed the A, B, and C zones (Hewitt and Morgan 1965). The characteristics of these units are as follows, in descending order:

Unit	*Thickness ft (m)*
A. Sandstone, very fine grained, clayey, ripple-bedded.	5 (1.5)
B. Sandstone, medium grained, cross-laminated; 2-ft (0.6-m) bed with abundant shale pebbles at base.	24 (7.3)
C. Sandstone, medium grained, cross-laminated and ripple-laminated in thin sets, with 1-ft (0.3-m) bed of coarse-grained sandstone at base.	22 (6.7)

The lowest unit (C) is distinctive because of the alternating character of the bedding. This unit is composed of composite sets that appear to be about 1 to 3 ft (0.3 to 1 m) in thickness. Each composite set can be further subdivided into a lower cross-laminated sandstone and an upper ripple-laminated sandstone with interbedded shale [Figure 5–13 (G) and (H)]. The cross-laminae have relatively steep dips of 15° to 20° and form sets that are 1 to 2 ft (0.3 to 0.6 m) in thickness. The ripple-laminated sets have thin interbeds of brownish gray shale that are on the order of 0.5 cm in thickness. The shale laminae outline ripple lenses of sand that have maximum thicknesses of 2 or 3 cm and internal cross-laminae that dip about 10°. The ripple lenses

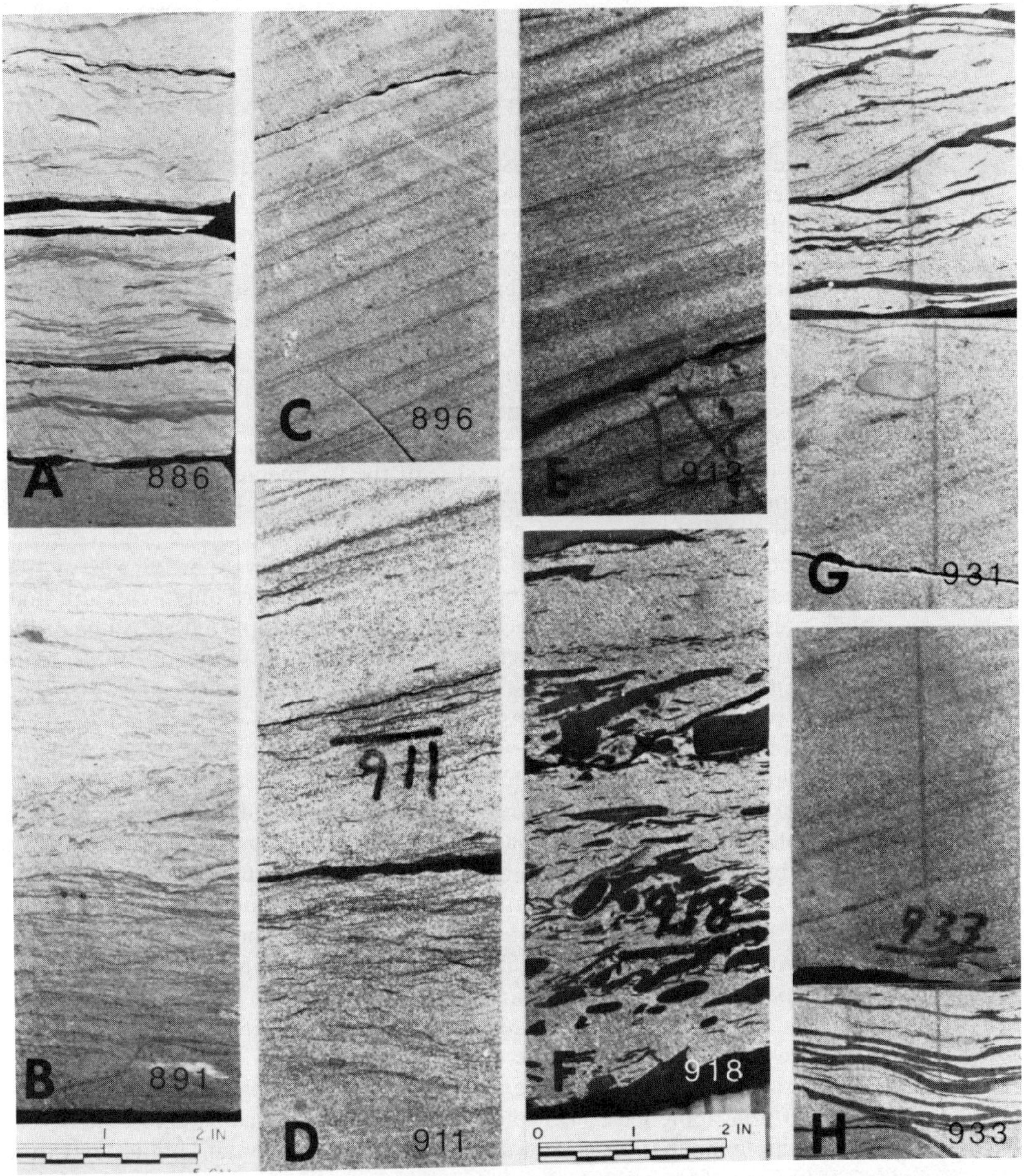

Figure 5–13 Sedimentary structures in the Robinson Sandstone, Ohio Wampler 14, Fry area, Illinois. Depths in feet given in lower right corner of photographs: (A) very fine-grained ripple-laminated sandstone and interbedded clay—886 ft (270 m); (B) fine-grained ripple-laminated sandstone—891 ft (272 m); (C) cross-laminated sandstone showing dip of 20°—896 ft (273 m); (D) finely ripple-laminated sandstone overlain by cross-laminated sandstone—911 ft (278 m); (E) cross-laminated sandstone showing dip of 15°—912 ft (278 m); (F) shale-pebble conglomerate—918 ft (280 m); (G) and (H) composite set of cross-laminated sandstone overlain by ripple-laminated sandstone and interbedded shale—931–933 ft (284 m).

have wave lengths that appear to be about 10 to 15 cm. Thus, the composite unit of the C zone exhibits a distinctive cyclic character. The cross-laminated beds represent a dune phase in which the individual dune heights were on the order of 0.5 to 1.0 ft. The overlying ripple-laminated beds represent a lowest flow-regime bed form that alternated with periods of no flow, during which significant thicknesses of clay were deposited from suspension.

The next overlying unit (B) has a basal conglomerate composed of rounded shale pebbles in a medium-grained sand matrix [Figure 5-13(F)]. The pebbles are generally discoid, well-rounded, and average about 2 cm in long diameter. The shale of pebbles appears to be identical in character to that of the underlying C unit.

The basal pebble zone is succeeded by cross-laminated sandstone in which the laminae have an average dip of 20° and range from 15° to 25°. The cross-laminated sandstones appear to be in sets that are about 1 to 6 ft (0.3 to 2 m) in thickness, although this is difficult to ascertain because of the fragmented nature of the core. Each set is composed of a dominant crossbedded unit in which the laminae dip 15° to 20° [Figure 5-13(C), (D), and (E)], but at the top is a thin bed of rippled sandstone only about 5 to 10 cm thick [Figure 5-13(D)]. The ripple bedding is of fine scale, and the amplitude of ripples appears to be on the order of only 1 cm, although no shale is present to define more precisely the individual ripple lenses. Some of the ripples appear to be of the "climbing" type denoting rapid sediment fallout from low flow-regime currents.

At 902 to 907 ft, the laminae are strongly curved and suggest that the beds were deformed by slumping soon after deposition. Small rounded pebbles of shale, identical to those of the basal sandstone, are present at several levels through the B unit. The pebbles are scattered through the sandstone and do not appear to form discrete pebble zones. In the upper part of the B unit, the laminae dip at relatively high angles, but there appears to be a decrease in the thickness of individual sets to about 1 ft (0.3 m). The overall uniformity of bedding in the B unit suggests that the dune phase of sediment transport was dominant throughout its period of deposition but with decreased dune height near the top of the unit.

The uppermost (A) unit shows an abrupt decrease in the scale of primary sedimentary structures. The lower half of the sandstone is distinctly ripple-bedded in relatively thin lenses that are on the order of 1 to 2 cm in thickness. Increasing amounts of clay upward form a fine matrix, and in some places, thin laminae of gray shale outline ripple lenses of sand. The uppermost part of the A unit is very clayey, and ripple lamination gives way to a thin and horizontal lamination in the upper 2 ft (0.6 m) of the unit. This gradational sequence of primary structures suggests decreasing flow regime and, finally, deposition of the suspended load by settling.

Petrography. The sandstone has an overall mean grain size of 0.18 mm (fine grained). In the B and C units there is little textural variation within bed sets (Figure 5-14). Mean size ranges from only about 0.21 mm in cross-laminated sandstones to about 0.19 mm in ripple-laminated sandstones. In the A unit, mean size decreases

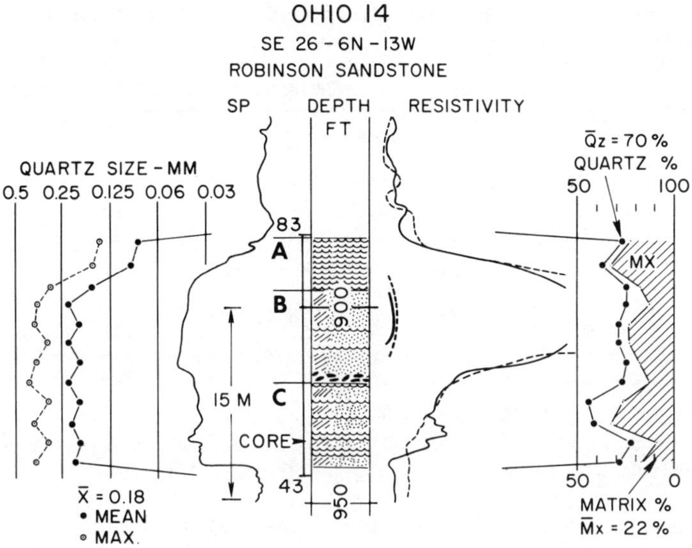

Figure 5–14 Texture, composition, and electric logs of the Robinson Sandstone in Ohio Wampler 14, Fry area. Letters A, B, and C refer to bedding units described in text.

abruptly to 0.11 mm (very fine grained) with an increase in clay matrix and fine-scale lamination.

Detrital composition is dominated by a moderately high quartz content of 68 percent and clay matrix of 22 percent (Table 5–2). Other components are minor in amount, with feldspar 5 percent, rock fragments 4 percent, and other minerals only 1 percent. Cement averages 3 percent of bulk composition.

Quartz content is that of the monocrystalline type. Feldspar is chiefly orthoclase with a trace of plagioclase in some samples. Rock fragments are about half polycrystalline quartz grains, and the remainder is composed of chert, phyllite, schist, and a few finely crystalline igneous fragments. Other minerals are mica and opaque grains. The major cement appears to be sparry calcite, but part may be siderite with a trace of iron oxide.

The matrix is composed largely of clay minerals. The composition of clay pebbles in the conglomerate zone (917–919 ft) is unusual in that the clay has been partly replaced by siderite (Hewitt and Morgan 1965, p. 338). Small pieces of carbonaceous material are present along some laminae, and a few large pieces of carbonized wood were found in the conglomerate. Generally, the amount of carbonaceous material in the section is negligible.

Porosity and permeability. The reservoir properties of the sandstone vary with the character of the bedding units. There is a pronounced decrease upward through the section in porosity and permeability. The C zone has the highest average horizontal permeability of 387 md and porosity of 19.9 percent. The A zone shows

TABLE 5-2. PROPERTIES OF ROBINSON SANDSTONE IN OHIO WAMPLER 14, CRAWFORD COUNTY, ILLINOIS

		Quartz size[a]			Detrital composition[b]					Cement		
Depth (ft)	Unit	Mean (mm)	Max (mm)	σ (mm)	Qz (%)	F (%)	Rx (%)	Oth (%)	Mx (%)	(% of total)	Permea-bility (md)	Poros-ity (%)
883–896	A	0.11	0.28	0.03	70	1	2	1	26	1	61	18.9
896–819	B	0.20	0.39	0.06	71	4	4	tr	20	2	306	19.9
919–941	C	0.20	0.36	0.06	65	8	4	1	22	5	387	19.9
Means		0.18	0.30	0.06	68	5	4	1	22	3	294	19.7

SOURCE: Petrography by F. G. Ethridge 1971.

[a]Long-axis measurements; σ = standard deviation.

[b]Qx = monocrystalline quartz, F = feldspars, Rx = rock fragments including polycrystalline quartz, Mx = matrix, Oth = other minerals, tr = trace.

a decrease to an average permeability of only 60 md and porosity of 18.9 percent (Table 5-2, Figure 5-15).

The C zone also shows the greatest permeability variation in individual beds with a maximum of about 750 md in cross-laminated sandstone and a minimum of less than 100 md in zones of interlaminated shale and rippled sandstone. Other cores of the Robinson Sandstone show somewhat greater variation in horizontal permeability with a maximum in a few samples of more than 1000 md (Hewitt and Morgan, 1965, Figure 7).

Permeability is significantly affected by bedding characteristics, as shown by measurements on core plugs and whole cores (Hewitt and Morgan 1965, pp. 341–42). Vertical permeability is very low in the A zone because of the large amount of clay. Vertical permeability is commonly 75 percent to 95 percent of horizontal

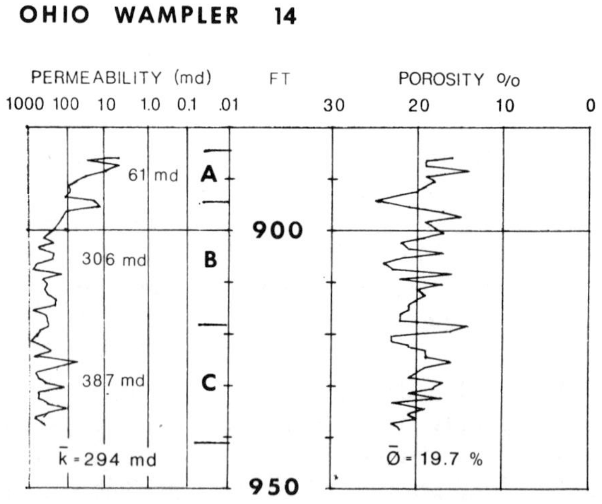

Figure 5-15 Porosity and permeability of the Robinson sandstone in Ohio Wampler 14.

permeability in the B zone. Furthermore, maximum permeability in the B zone is in the dip direction of cross-laminae. Thus, the direction of maximum permeability within the reservoir is probably parallel to the long axis of the sandstone body and in a southwesterly direction. Shale laminae at the top of cosets in the C zone probably form effective barriers to vertical flow, but the cross-laminated beds of the C zone showed no consistent variation in direction for measurements of horizontal permeability.

Morphology

The Robinson Sandstone displays the characteristics of a fluvial channel deposit. The cored section shows sedimentary structures typical of current flow, and these range from cross-laminae that dip at an average of 20° to small-scale ripple laminae. The cross-laminated and rippled sets form composite bed sets that were deposited during decreasing flow. The composite sets are very thin, however, ranging from only 1 to 6 ft (0.3 to 2 m), and the thin beds suggest deposition in shallow channels.

An isopach map of the Robinson sandstone (Figure 5–16) shows a narrow elongate body that trends southwestward. Maximum thickness is 40 ft (12 m), and average thickness is about 30 ft (9 m). Total width is only 2000 ft (610 m), so that lateral termination of the sandstone is relatively abrupt. According to core control, all of the sandstone units (A, B, and C) are present throughout the section except for the distal margins. Although the central axis is somewhat irregular, it can be estimated that the average sinuosity is only 1.1 for the composite section. Thus, the pattern is that of braided-stream deposits, which may have been predicted from the thinness of depositional units observed in the core.

A well-log section through the composite body (Shelton 1973) suggests that the entire unit is enclosed in a dominantly shaly section. The margins of the body appear to be "feathered," or thinly interbedded, with shale. In other words, there seems to be no evidence for an erosional channel that was subsequently filled by sand. Rather, the composite body appears to have been built by the stacking of numerous small channel deposits.

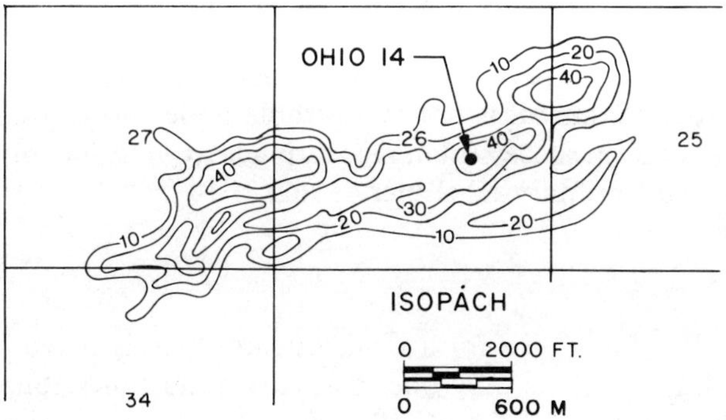

Figure 5–16 Thickness of Robinson Sandstone in the Fry area, Crawford County, Illinois, showing the nature of the composite channel body. Contour interval 10 ft (3 m).

An oriented core in the vicinity of Ohio Wampler 14 shows that the average direction of cross-laminae is westerly at angles of 15° to 30° about the general reservoir trend of S55°W. The dips represent laminae of the dune phase of sand transport in which the crests of dunes were oriented essentially normal to the average direction of stream flow. In this case, the dip directions are relatively good indicators of the channel trend.

Conclusions

Sedimentary structures revealed in the Robinson core indicate the association of thin composite bed sets with a narrow low-sinuosity pattern of a fluvial channel. This association is not uncommon. Although maximum thickness is significant at 40 ft. (12 m), the presence of thin composite sets could be used to predict a braided morphology for the total sandstone body.

BARTLESVILLE SANDSTONE, CHETOPA AREA, SOUTHEASTERN KANSAS

The Middle Pennsylvanian Bartlesville Sandstone has been one of the largest producers of oil in Oklahoma. It is estimated that approximately 1.5 billion barrels of oil has been produced from this sandstone. The earliest production was in southeastern Kansas from a discovery well drilled in 1892 (Weirich 1953, p. 70). Major producing areas were soon established in Oklahoma, however, after completion of the first commercial oil well in the state within the townsite of Bartlesville during 1897. Other discoveries soon followed and resulted in the development of at least six major fields, each of which produced more than 100 million barrels of oil. The largest, Cushing field, has produced more than 420 million barrels since its discovery in 1912. Most of the Bartlesville reservoirs were found at relatively shallow depths of from 1000 to 3000 ft (305 to 915 m).

Close to the outcrop, the Bartlesville Sandstone is commonly saturated with heavy oil, and early studies suggested that the amount of this resource might be as great as 1.6 billion barrels of oil (Enright 1964). Recent estimates, however, give a more modest figure of about 200 million barrels of oil in place in southeastern Kansas (Ebanks, James, and Livingston 1977). The reduced size of the resource is based mainly on the discontinuous nature of the sandstone bodies as well as on their thin and shaly character. Most of this oil is not recoverable under present conditions, but tertiary methods have been successful in recovering heavy residual oil that remains after production of light primary oil (Emery 1962).

Geologic Setting

Bartlesville is an informal subsurface name for a sandstone unit that is called Bluejacket at the outcrop. The Bartlesville is only one of several sandstones found in the lower part of the Cherokee Group of the Middle Pennsylvanian Desmoinesian

Series. Although individual sandstones are small lenticular bodies, the Bartlesville sandstones are extensively present throughout eastern Kansas from the vicinity of Kansas City southward into northeastern Oklahoma (Figure 5–17). These sandstones are all part of a major alluvial valley-deltaic system.

Only the lower part of the Cherokee Group is present in the Chetopa area (Table 5–3). The Krebs Formation of the lower Cherokee is composed largely of shale, but sandstones are developed at two levels, the Bartlesville in the upper section and the Warner sandstone in the lower section. The Krebs Formation rests unconformably on Mississippian limestone, and the older Pennsylvanian Morrow and Atoka are missing. The overlying Cabaniss Formation consists largely of shale, but it also contains several thin sandstones and beds of coal or highly carbonaceous shale. One of these thin beds, the Weir-Pittsburgh coal, forms a persistent marker in the Chetopa area.

The Cherokee Group is more than 2000-ft (610-m) thick in the McAlester basin in east-central Oklahoma, but it thins rapidly northward. In southeastern Kansas,

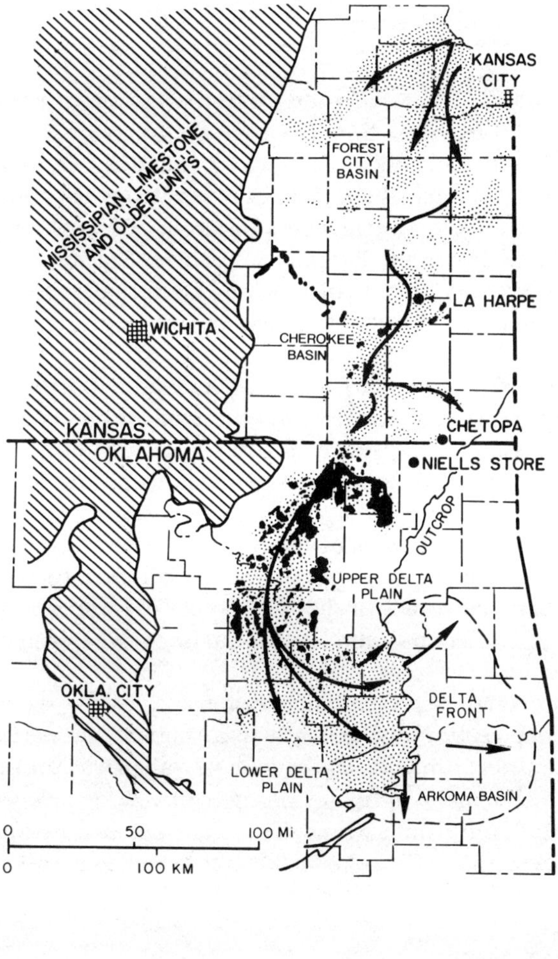

Figure 5–17 Distribution of the Bartlesville (Bluejacket) Sandstone in eastern Kansas and northeastern Oklahoma showing inferred transport direction in a major alluvial valley-deltaic plain system. Oil fields (black) are largely stratigraphic traps in fluvial-channel sandstones (stippled). [Map adapted from Weirich 1953 and Visher 1968.]

TABLE 5–3. STRATIGRAPHIC SECTION OF THE LOWER CHEROKEE GROUP, CHETOPA AREA, LABETTE COUNTY, SOUTHEASTERN KANSAS

System Series	Formation	Member	Depth ft (m)	Thickness ft (m)
	Cabaniss (part)		Surface	38 (11.6)
		Weir-Pittsburg Coal	38 (11.6)	2 (0.6)
		Unnamed shale	40 (12.2)	45 (13.7)
Pennsylvanian Desmoinesian	Krebs	Bluejacket (Bartlesville) Sandstone	85 (25.9)	70 (21.4)
		Unnamed shale	155 (47.3)	160 (48.8)
		Warner Sandstone	315 (96.1)	15 (4.6)
		Unnamed shale	330 (100.7)	45 (13.7)
Mississippian		Limestone	375 (114.4)	

its total thickness is about 400 ft (122 m). To the south the section contains marine shales, and the Bartlesville (Bluejacket) Sandstone at the outcrop shows various aspects of prodelta, delta-front, and fluvial-channel sections (Visher 1968). In the subsurface of northeastern Oklahoma, the Bartlesville Sandstone is found as elongate lenslike bodies that have an average thickness of about 100 ft (30 m), but the sandstones are highly variable and grade laterally into shale. Subsurface mapping indicates that these sandstones may be part of a deltaic plain that has a wide areal extent northwest from the delta-front section of the outcrop (Visher, Saitta, and Phares 1968).

Bartlesville sandstones are more narrowly distributed in northeasternmost Oklahoma and, presumably, represent more restricted channel deposits of the lower alluvial valley. Thinner sandstones are present in southeastern Kansas where individual sandstones commonly have thicknesses on the order of 20 to 50 ft (6 to 15 m), and both coal and limestone are thin or absent, which suggests a dominantly nonmarine origin rather than deltaic. Cores of the Bartlesville Sandstone show that it is of fluvial origin in the northern part of its extent. In addition to the Chetopa area (Figure 5–17), point-bar sandstones in the Bartlesville have been identified at Niells Store, Craig County, Oklahoma, and in the LaHarpe area, Allen County, Kansas (Schumacher 1976).

The Cherokee Group, then, represents a major transgressive sequence that grades from a thicker marine section in the south to a thinner nonmarine section in the north. The major transport direction in the alluvial valley was south, parallel to the rising landmass of the Nemaha anticline, along which older Paleozoic rocks were exposed during earliest Pennsylvanian time. The Cherokee Group thins toward the Nemaha anticline, but eventually, Cherokee sediments extended across the uplift.

Drilling History

In the Chetopa field, the Bartlesville Sandstone is found at a depth of only 100 ft (30 m) and about 45 ft (15 m) below the Weir-Pittsburg Coal (Table 5-3). Maximum thickness of the sandstone is about 60 ft (18 m), and it contains variable saturations of heavy viscous oil that range up to 50 percent of pore space. This oil presented an attractive resource if it could be extracted economically. Core drilling was begun

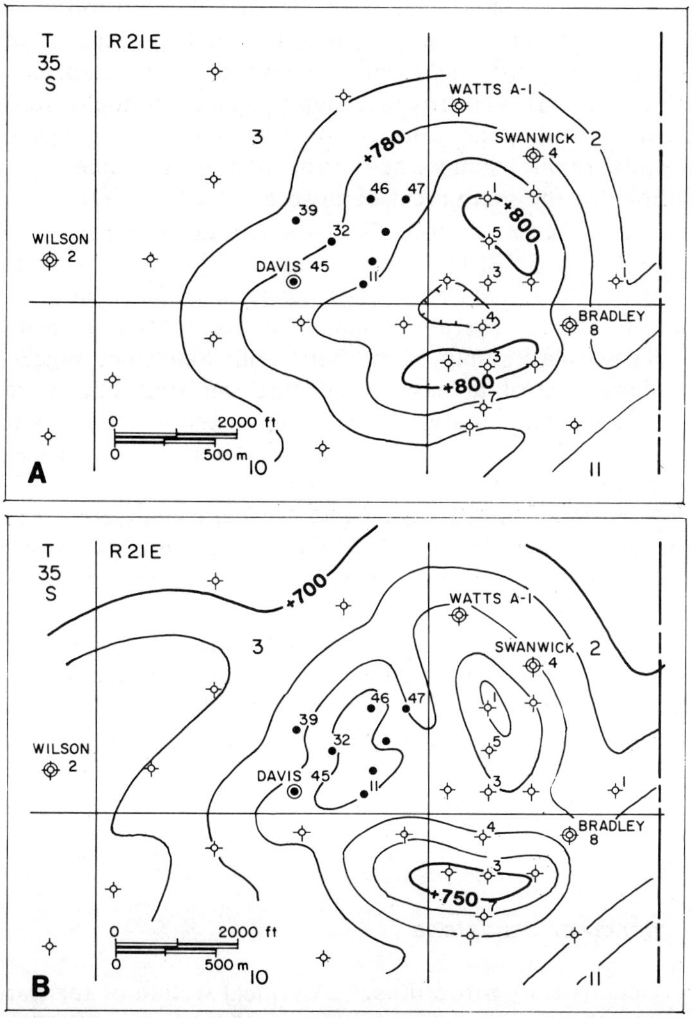

Figure 5-18 Structure maps of the Chetopa area, LaBette County, Kansas: (A) structure on top of the Weir-Pittsburgh coal; (B) structure on top of the Bartlesville Sandstone. Contour interval 10 ft (3 m). Cores of the Bartlesville Sandstone were examined from the circled wells.

in 1964 by Tenneco Oil, and a large number of holes were drilled in Secs. 2, 3, 4, and 9, 10, 11, T35S, R21E, LaBette County. Thicker sections of oil-saturated sandstone were found in SE Sec. 3, and the area was selected for steam injection. However, the project was not successful and was abandoned in 1968.

The regional dip across the Chetopa area is toward the northwest at the rate of about 20 ft/mi (4 m/km). Local structure shows considerable variation, largely because of stratigraphic change at the top of the Bartlesville Sandstone and also because of compaction of overlying sediments (Figure 5–18).

The structure at the top of the Bartlesville Sandstone shows an arcuate pattern of small structural closures [Figure 5–18(B)]. Three separate areas of structural closure can be mapped, and locally, reversal of gradient appears to be as much as 80 ft/mi (15 m/km). The structure map was made on top of the first prominent sandstone that could be identified in borehole logs, well cuttings, or in cores. It is likely that stratigraphic changes at the top of the sandstone, or sand buildup, are responsible for the small areas of closure.

A structure map on the Weir-Pittsburg coal shows two small arcuate areas of closure [Figure 5–18(A)]. Local reversal of dip is on the same order of that of the sandstone. The arcuate patterns on the persistent coal marker probably are the result of differential compaction of shale over the underlying sandstone.

The structure map of the Bartlesville Sandstone suggests that in the area of more dense control, at least three distinct sandstone bodies are present, all of which are broadly arcuate, curved in an updip direction. These patterns are suggestive of point-bar sandstones that were deposited on the inside of meander loops of an ancient fluvial channel.

Sandstone cores, borehole logs, and core analyses were available from 35 wells in the area. The cores from 5 wells were selected for study of the section (Johnson 1973). These cores are indicated by the circled and named wells on the structure map (Figure 5–18).

Logs from three of the cored wells illustrate the highly variable character of the Bartlesville section (Figure 5–19). Sandstones are found at three levels. Thin sandstones are present in the upper Bartlesville of the Swanwick 4 well, and a middle sandstone is present in the Bradley 8 well overlying a thick shale and a lower sandstone. A single prominent sandstone is present in the Davis 45 well at a level below the upper sandstone in Bradley 8 but somewhat above the lower sandstone. The character of these units is shown by the cored sections.

Description of Sandstone

Sedimentary structures. A typical section of the Bartlesville Sandstone is shown by the core from Tenneco Davis 45 located in SW SW SE Sec. 3 (Figure 5–18). Total thickness of the section is 61 ft (19 m), but the section can be divided into several distinct units as follows, in descending order:

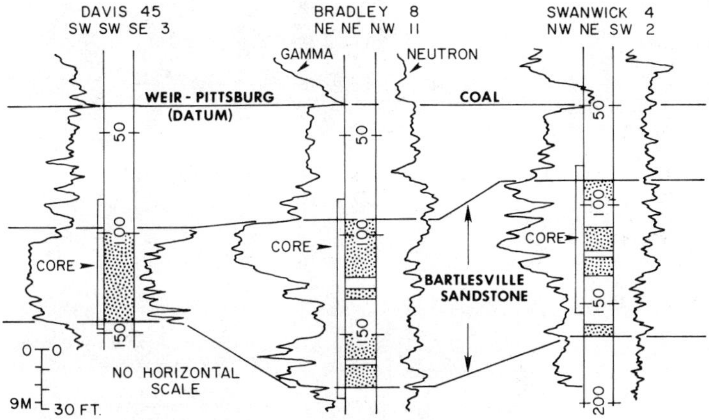

Figure 5–19 Comparison of borehole logs of the Bartlesville Sandstone, Chetopa area, LaBette County, Kansas. Location of wells shown in Figure 5–18. Thicker sandstones in the Davis 45 and Bradley 8 wells contrast with interbedded sandstones and shales of the Swanwick 4 well.

Unit	*Thickness ft (m)*
1. Sandstone, ripple-laminated.	6 (1.8)
2. Shale, gray, thinly interlaminated with very fine sandstone.	8 (2.4)
3. Sandstone, very fine grained, ripple-laminated with thinly interlaminated siltstone.	7 (2.1)
4. Sandstone, medium to fine grained, laminated to massive, with basal conglomerate.	40 (12.2)
5. Shale, gray, laminated to nodular.	4 (1.2)

The lowest unit (5) is a highly weathered shale. Near the base of the core, distinct laminations can be observed, with alternating light gray and red claystone. However, most of the section shows tubular to globular masses of calcareous material (Figure 5–20). These nodules represent caliche formation at a preexistent soil horizon before the deposition of the sandstone.

The next overlying unit (4) is the main sandstone section. At the base are abundant pebbles, some of angular shale clasts and other of silicified wood [Figure 5–20(J)]. Pebbles are imbedded in a matrix of medium-grained sandstone and are scattered through a thickness of 3 ft (0.9 m). The sandstone appears to be generally massive, but it is distinctly laminated at numerous levels [Figure 5–20(D)–(H)]. The laminae are parallel, even, and inclined at angles of 10° to 15° with a few laminations as high as 20°. Laminae are marked by finely divided carbonaceous material

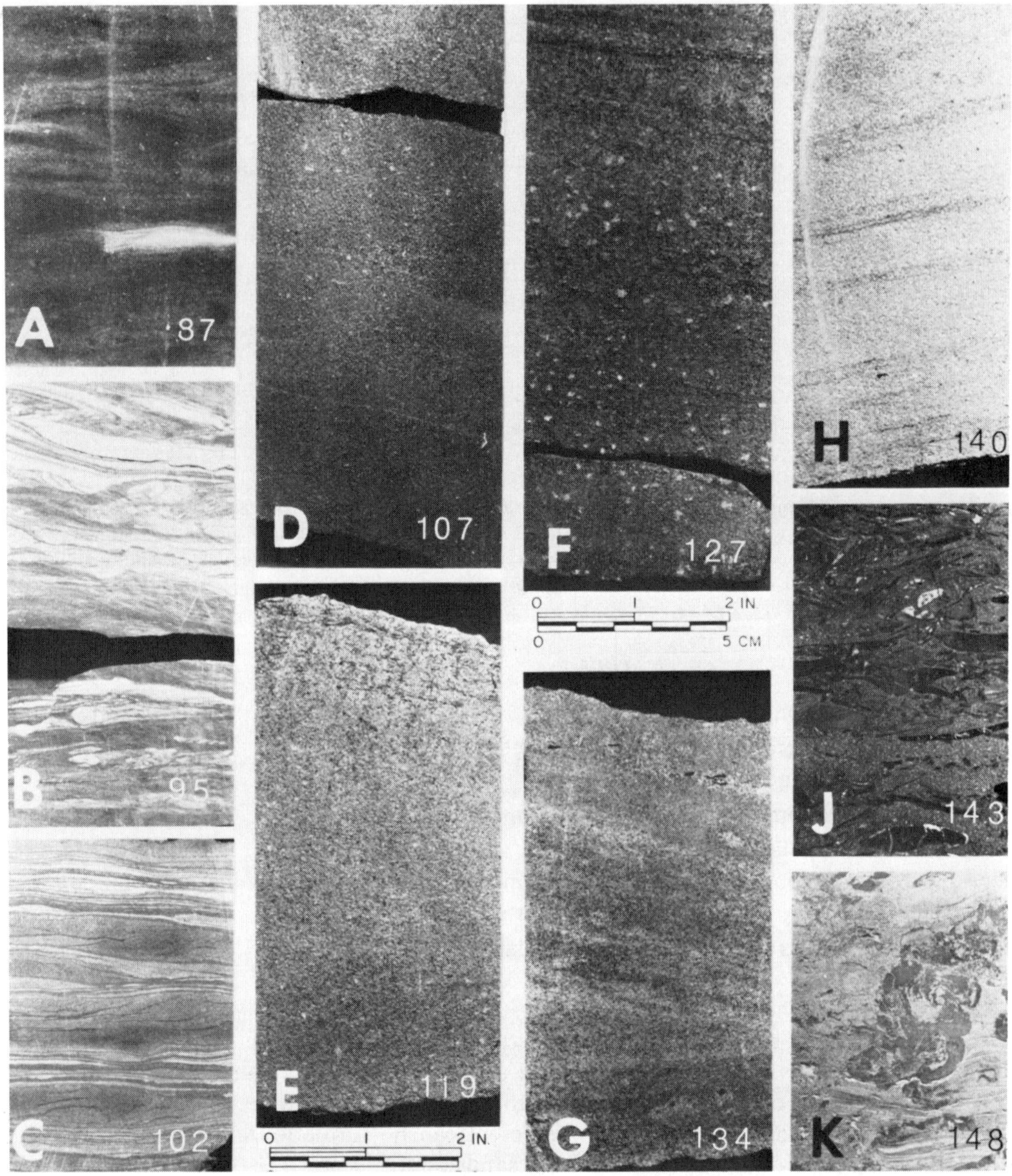

Figure 5-20 Sedimentary structures in Bartlesville Sandstone, Tenneco Davis 45, Chetopa area, Kansas: (A) fine-grained ripple-laminated sandstone at the top of the Bartlesville section—87 ft (26.5 m); (B) interlaminated clay and silt showing contorted bedding—95 ft (29 m); (C) very fine-grained rippled sandstone at the top of the main Bartlesville sandstone—102 ft (31 m); (D)-(H) cross-laminated sandstone, oil-stained above and water-saturated below—107 ft, 119 ft, 127 ft, 134 ft, and 140 ft (32.6-42.7 m); (J) conglomerate of silicified wood fragments, shale, and coal in medium-grained sand matrix—143 ft (43.6 m); (K) weathered nodular clay below main Bartlesville sandstones—148 ft (45.1 m).

[Figure 5–20(H)] or are visible only because of slight differences in color [Figure 5–20(F) and (G)]. The uppermost part of the sandstone is saturated with dark heavy oil, which tends to obscure bedding, but indistinct laminae are still discernible [Figure 5–20(B)].

The next overlying unit (3) is finely laminated with distinct curved nonparallel laminations [Figure 5–20(C)]. Thin shale laminae are also present and become increasingly abundant upward in the unit. The sandstone laminations represent ripple structure formed during the lowest flow-regime.

The overlying shale unit (2) contains numerous thin beds of very fine-grained sandstone [Figure 5–20(B)]. Many of the thin lenses, no more than 1-cm thick, are irregular to contorted, which suggests a significant amount of soft sediment deformation. This section probably represents flood-plain deposition from an adjacent channel. The shale was deposited rapidly after flooding and was contorted by slight movement during compaction.

Fine-grained sandstone is again present in the uppermost unit (1), which is uniformly and finely laminated [Figure 5–20(A)]. The laminae are indistinct but continuous, nonparallel, and slightly curved and form a lenslike structure. This lamination represents ripple structure formed during low flow-regime. The section also appears to be the result of overbank deposition because it is separated from the main body of the underlying sandstone.

The total Bartlesville section is typical of point-bar deposition in a fluvial channel. The lowermost pebble conglomerate represents the channel lag deposit. The main sandstone unit (4) represents the dune phase of sediment transport in the upper part of lower flow-regime. The dune form is indicated by the common laminations that have relatively high dips of 10° to 20°. These sands were deposited laterally to the main channel. The overlying ripple sandstone unit (3) represents the upper part of the point-bar section that was deposited in shallow water inside of the meander loop. This unit (3) represents the last deposition of the point-bar sequence. The overlying shale unit (2) was fine-grained overbank sediment that was deposited during a later flood stage.

The return of rippled sandstone in the uppermost unit (1) probably also represents overbank deposition, but in this case, of a somewhat coarser-grained texture but still lower flow-regime. Therefore, this sediment must represent an apron of natural levee deposits that narrowly border a stream channel developed at a later stage.

A contrasting facies to the point-bar section is shown in the core from the Tenneco Bradley 8 well located in NE NE NW Sec. 11 (Figure 5–18). In this section, the lateral equivalent of the point-bar section in Davis 45 consists of interbedded shale and sandstone. This section represents the fine- to coarse-grained fill of the main channel on the outside of the meander loop. The channel-fill section is overlain in turn by a second point-bar sandstone that represents the deposits of a later flood stage when the partially filled channel was reoccupied by flood waters. The sequence in the Bradley 8 core is as follows, in descending order:

Unit	*Thickness ft (m)*
1. Sandstone, very fine grained, ripple-laminated.	10 (3.0)
2. Sandstone, fine grained, inclined laminae, with basal pebble zone.	27 (8.2)
3. Shale and siltstone, brown to gray with some fine-grained rippled sandstone.	31 (9.5)
4. Sandstone, thin-bedded, with scattered to abundant clasts of siltstone and claystone.	25 (7.6)

The two lower units (3 and 4) represent the channel fill, whereas the upper units (1 and 2) represent the overlying point bar.

The lowermost sandstone unit (4) is thinly bedded with abundant interbedded siltstone and shale. The sandstones are generally massive, and laminations are only poorly developed. Clasts of siltstone and claystone commonly range from 2 to 10 cm in length and are discoid, angular to rounded, and some are contorted. These conglomeratic sandstones represent multiple sections of basal lag deposited after floods in the open channel.

The succeeding channel-fill units (3) consist largely of brown to dark gray clayey siltstone with parallel, continuous, and even laminae. Some ripple-laminated sandstone is also present in thin beds. This section represents fine-grained deposits in the open channel during an extended period of no stream flow. At the time of its deposition, the channel was probably cut off when the main stream changed course. Then the channel was left abandoned, probably as an oxbow lake, which filled later with fine-grained sediment deposited largely from suspension with only occasional low flow-regime currents.

The succeeding point-bar units (1 and 2) were deposited when the stream reoccupied the old channel. The sequence of a basal pebble conglomerate, inclined laminations in the sandstone, and overlying ripple-laminated sandstone is identical to the sequence found in the earlier point bar of Davis 45. The total section, however, is somewhat thinner and has only a 37-ft (11-m) thickness.

The contrasting sections of the Davis 45 and Bradley 8 wells show a succession of point bars within a single channel and illustrate the character of the channel-fill sediment. The section of the Swanwick 4 well again shows a channel-fill sequence (Figure 5–19). Thinner shaly sandstones are interbedded with gray shales. The sandstones are cross-laminated to rippled, and shale pebbles are found at several levels. This uppermost section represents the final deposition before abandonment of the local river channel.

Petrography. The sandstones have an overall mean grain size of 0.19 mm (fine grained), but mean grain size varies according to bedding type (Table 5–4). The cross-laminated sandstones are somewhat coarser grained and have a mean grain size of 0.25 mm. The ripple-laminated sandstones are the finest grained, and mean grain size is 0.11 mm (very fine grained). Consequently, the point-bar sand-

TABLE 5–4. AVERAGE PROPERTIES OF BARTLESVILLE SANDSTONE, CHETOPA AREA, KANSAS

			Quartz size[a]		Detrital composition[b]					
Facies	Bedding type	Number of samples	Mean (mm)	Max (mm)	Qz (%)	Oth (%)	Mx (%)	Cement[c] (% of total)	Permeability (md)	Porosity (%)
Upper point bar	Ripple-laminated	6	0.11	0.20	46	16	38	4	97	20.8
Lower point bar	Cross-laminated	13	0.25	0.45	40	21	39	4	383	26.3
Channel fill	Rippled, comglomeratic	9	0.15	0.28	45	19	36	3	28	16.9
Averages			0.19	0.34	43	19	38	4	208	22.1

SOURCE: Petrography by Johnson 1973.

[a]Long-axis measurements.

[b]Qz = monocrystalline quartz, Oth = feldspars and rock fragments including polycrystalline quartz, Mx = matrix.

[c]Cement = silica as grain overgrowths.

stones show a decrease in grain size upward, commonly from about 0.30 mm at the base to about 0.10 mm at the top (Figure 5–21).

The shale-pebble sandstones have a matrix that is intermediate in grain size and averages 0.15 mm (fine grained). Mean grain size is variable in these sandstones and ranges from 0.10 to 0.21 mm.

Composition is similar in all sandstones regardless of bedding. Average composition is 43 percent monocrystalline quartz, 19 percent feldspar and rock fragments, and 38 percent matrix. Feldspars comprise about 10 percent of composition and are both orthoclase and plagioclase. Rock fragments are largely polycrystalline quartz and chert grains. A small amount of detrital mica is present, especially in the finer-grained, ripple-laminated sandstones. The matrix appears to be composed largely of clay minerals that have been altered in part to fine-grained sericite. In

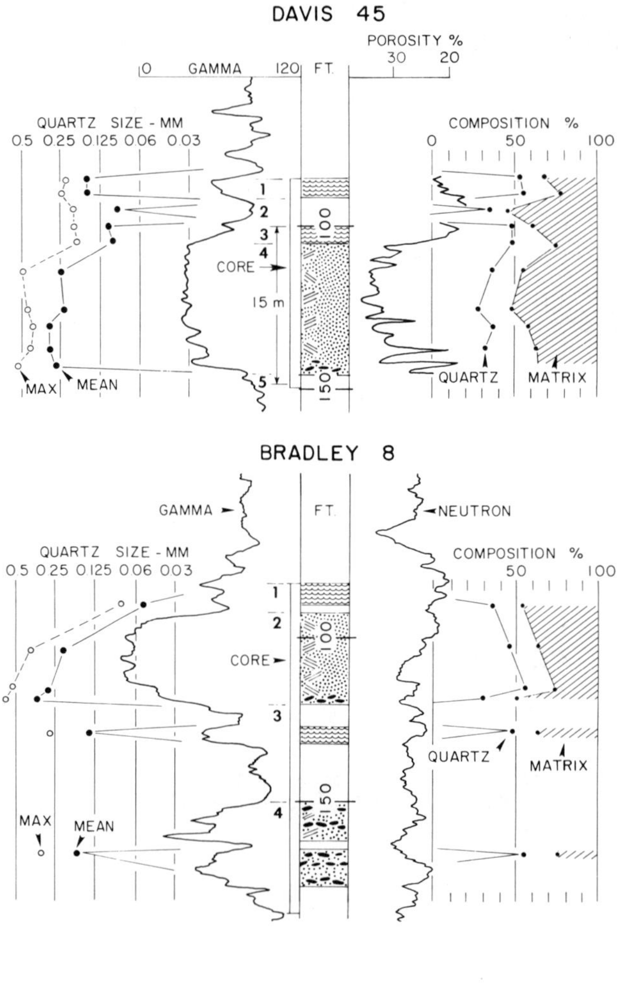

Figure 5–21 Texture, composition, and borehole logs of the Bartlesville Sandstone in the Tenneco Davis 45 (above) and in the Tenneco Bradley 8 (below) wells. Location of wells shown in Figure 5–18.

some places the matrix is intimately mixed with calcite cement. In the cross-laminated sandstones, a part of the matrix appears to have been small shale and chert fragments that were deformed during compaction of the sediment. Therefore, the matrix appears to be largely of detrital origin, derived in part from rock fragments in the coarser-grained sandstones and from altered clay minerals in the finer-grained sandstones.

Cements consist of both silica as overgrowths on quartz grains and as intergranular carbonate. Silica overgrowths are more common, and calcite cement is present in minor amounts only in the coarser-grained sandstones. Therefore, much of the carbonate cement may have been derived by recrystallization of carbonate rock fragments.

Both texture and composition tend to confirm the fluvial origin of the sandstones. The decrease upward in mean grain size corresponds to change in bedding type and further supports the interpretation of deposition in fluvial channels. Composition is characterized by modest amounts of monocrystalline quartz together with a significant amount of feldspar, rock fragments, and matrix, most of which appears to be detrital in origin.

Porosity and permeability. The reservoir properties correspond closely to bedding units. Crossbedded sandstones have the highest average porosities and permeabilities (Figure 5–22). In the Davis 45 core, the main body of crossbedded sandstone has an average permeability of 989 md and an average porosity of 32.4 percent. The lower conglomeratic part of the point bar, however, has a greatly reduced permeability of 2.5 md and a reduced porosity of 26.9 percent because of the large numbers of pebbles and some calcite cement. The finer-grained rippled sandstones in the upper part of the section have an average permeability of 22 md and an average porosity of 19.8 percent.

A similar relationship is illustrated in the Bradley 8 core. The main section of crossbedded sandstone has an average permeability of 290 md and an average porosity of 26.1 percent. In the basal conglomeratic section, average permeability and porosity are reduced to 8.4 md and 20.9 percent, respectively. Overlying ripple-laminated sandstones have an average permeability of 9.1 md and an average porosity of 18.1 percent.

Sandstones of the channel-fill units also have reduced permeabilities and porosities of 50 md and 20.4 percent. Values in the shale-pebble conglomerates are highly variable and range from less than 1 md to a maximum of 246 md.

Oil saturations are greatest in the upper parts of crossbedded sandstones. Maximum residual oil saturations are on the order of 50 percent and decrease to about 10 percent in the lower part of crossbedded sandstone. Increased oil saturation is present in the basal conglomeratic parts of the crossbedded sandstones as well as in the ripple-laminated sandstones above. This oil could not be moved by steam injection, probably because of its low content of volatile constituents and high viscosity.

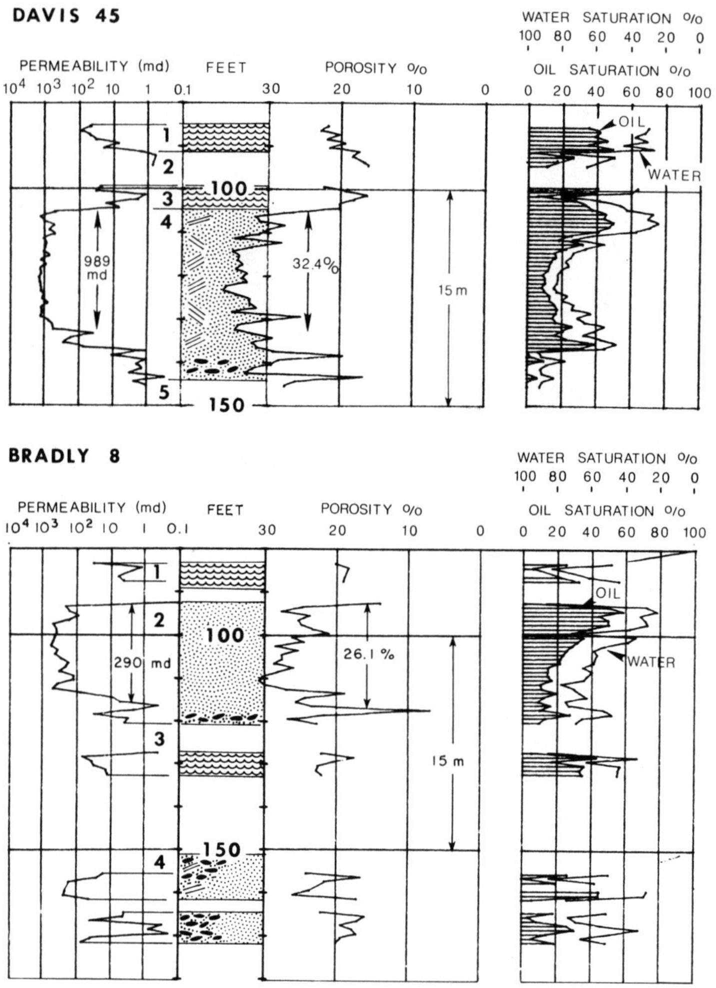

Figure 5–22 Permeability, porosity, and fluid saturations in the Bartlesville Sandstone, Tenneco Davis 45 (above) and Tenneco Bradley 8 (below). Location of wells shown in Figure 5–18.

Morphology

The two principal facies of the Bartlesville, as defined in the cored sections, can be identified by means of borehole logs and core analyses in adjacent wells. The point-bar sandstones are commonly thicker, vertically continuous sandstones and have higher values of permeability and porosity. The channel-fill sections contain interbedded sandstones and shale, and the sandstones are thinner, commonly contain shale pebbles, and have lower values of permeability and porosity. The distribution of these facies was determined by use of all data; where borehole logs were not run, gamma logs of cored sections, as well as core analysis, made facies identification possible.

Stratigraphic correlations can be established by cross sections through the area (Figure 5–23). The lower Bartlesville point bar is restricted largely to the area of Davis wells in SE Sec. 3 and is referred to as the Davis point bar. The upper Bartlesville point bar consists of two sandstones. One is restricted to the area of Swanwick wells in SW Sec. 2 and is referred to as the Swanwick point bar. The other is

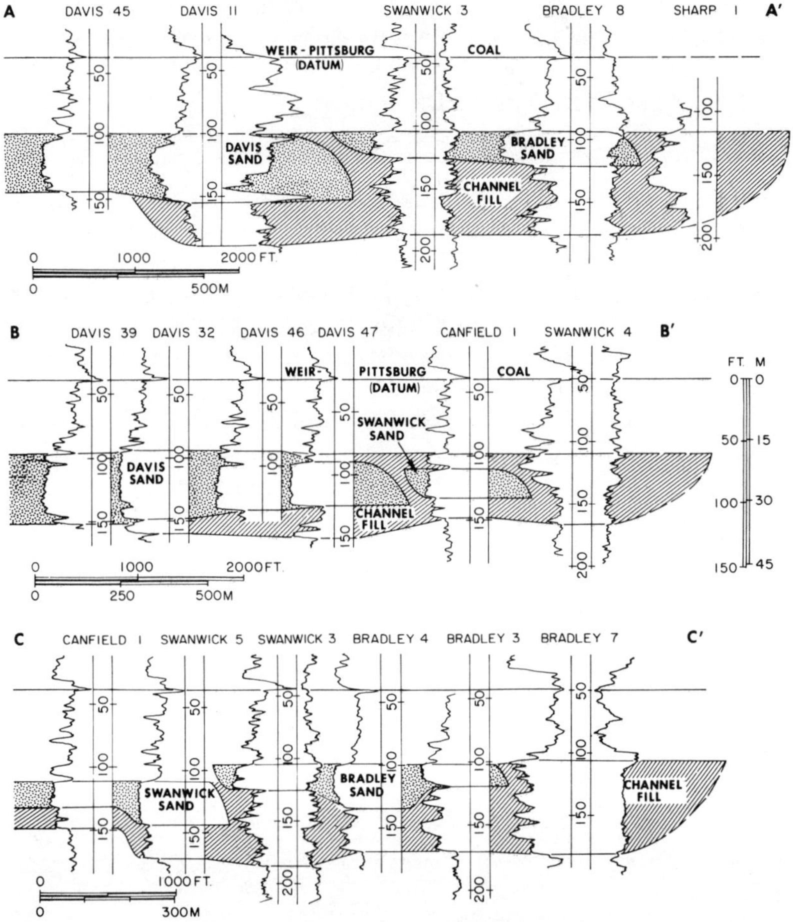

Figure 5–23 Correlation of Bartlesville sandstone in Chetopa area showing the relationship of upper and lower Bartlesville Sandstone with the channel-fill facies. Datum is base of Weir-Pittsburgh "coal" (carbonaceous shale). Location of profiles shown in Figure 5–24.

in the area of Bradley wells in NW Sec. 11 and is referred to as the Bradley point bar.

The Davis point bar has a thickness that ranges from about 30 to 55 ft (9 to 17 m) and is replaced laterally by the channel-fill facies shown in the lower part of the Bradley 8 core (Figure 5–21). The channel-fill facies has a maximum thickness of about 60 ft (18 m) beyond the area of the Davis point bar, and it extends into the Davis area where log correlations indicate a thickness of 34 ft in the Davis 11 well (Figure 5–23, A-A′). The channel fill is not present, however, below the point-bar sandstone in the Davis 45 core where weathered gray and red claystone immediately underlies the point-bar section.

The Swanwick point bar has a maximum thickness of 36 ft (11 m) in the Swanwick 5 well (Figure 5–23, C-C′). It occurs at a level slightly below the Bradley point bar, and laterally, its position in some wells is occupied by rippled sandstone similar to the upper unit (1) in the Davis 45 core.

The Bradley point bar has a maximum thickness of 34 ft (10 m) in the Bradley 4 well (Figure 5–23, C-C′). As seen on the correlation sections, the Bradley point bar is a thin lenticular body that overlies the older channel-fill section and is also replaced laterally by channel-fill sediments, as in the Swanwick 4 well (Figure 5–23, B-B′).

Distribution of point-bar sandstones is shown by an isopach of thicker continuous sandstone sections (Figure 5–24). All three point-bar sandstones are included on the map, and thin sandstones of the channel-fill facies have been omitted. The resulting patterns of point-bar sandstones show small arcuate bodies and suggest that each of the principal sandstones are themselves composite bodies. The

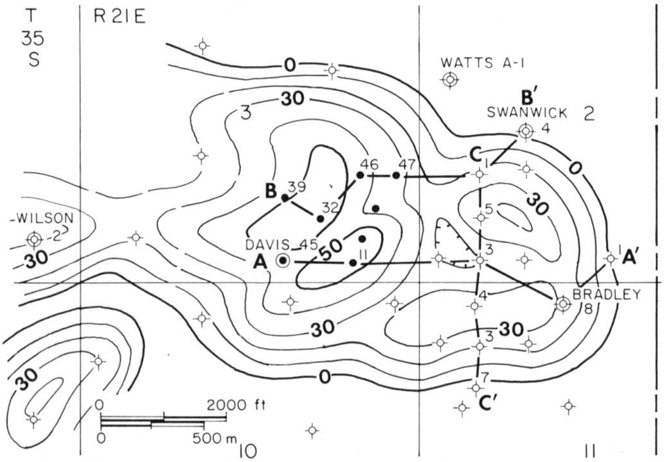

Figure 5–24 Thickness of point-bar sandstones in the Bartlesville section, Chetopa area. Contour interval 10 ft (3 m). Correlation profiles A-A′, B-B′ and C-C′ are shown in Figure 5–23.

Davis sandstone in SW Sec. 3 shows two parallel curved areas of thicknesses greater than 50 ft (15 m). However, correlation of borehole logs provides no clue for distinguishing the two sandstone bodies.

The Swanwick and Bradley sandstones in Sec. 2 and Sec. 11 are divided into two distinct bodies. Together they form an arcuate, nearly continuous trend. The Swanwick point bar is developed at a somewhat lower stratigraphic position and is overlapped to the south by the Bradley point bar. Because of this overlapping relationship, it may be inferred that the southern point bar is younger, and therefore stream flow was generally to the south in the direction of overlap.

A broad area of channel fill completely surrounds the area of point-bar sandstones. The extent of the channel-fill facies is not known precisely, but a single point of control east of the map area suggests that its width is at least 2000 ft (610 m). The extent of the sandstones to the west is also poorly known. A thick point-bar section is present in the Wilson 2 well (SE SE Sec. 4), but subsurface control is not sufficient to determine the lateral extent or trends in the western area.

Conclusions

The Bartlesville section contains sandstones of point-bar origin. These sandstones display a sequence of sedimentary structures that range from dominant cross-laminated in the lower part to ripple-laminated in the upper part. The entire sequence is relatively thick and on the order of 30 to 55 ft (9 to 17 m). The point-bar sequence is a single thick composite bed set that contrasts markedly with the thinner sets of braided-stream deposits as, for example, in the Robinson sandstone described previously.

Despite their significant thicknesses, the Bartlesville point bars are small, areally restricted, and partly overlapping bodies. Furthermore, these sandstones are isolated in what appears to be a greater thickness and extent of channel-fill deposits that are dominated by shales, shaly sandstones, and shale-pebble conglomerates. The total Bartlesville section is about 80 ft (24 m) in thickness, and on the average, less than half this thickness consists of point-bar sandstones. In areal extent, the point-bar sandstones are also somewhat diminished by the apparent wide extent of the shale-filled channel, which approaches, in maximum width, nearly the lateral extent of the individual point-bar sandstones.

Thus, the ancient Bartlesville River appears to have been dominated by a suspended load rather than a bed load. The source of sand must have been inadequate to provide for more extensive sandstone bodies.

The large suspended load of the Bartlesville River may be reflected in the average composition of point-bar sandstones, which have nearly 40 percent of fine-grained matrix. A part of this matrix, that of clay size, may be diagenetic in origin, perhaps by alteration in place of feldspars and rock fragments. Nevertheless, a detrital origin for much of the matrix is consistent with the shale-dominated nature of the adjacent channel-fill deposits.

The meander-belt deposits of the Chetopa area may not be typical of the main

Bartlesville transport system. It is obvious, however, that the "Chetopa bend" was formed by a highly meandering stream that may have been a major tributary in the Bartlesville River system.

LOWER TUSCALOOSA SANDSTONES, MALLALIEU FIELD, MISSISSIPPI

Upper Cretaceous sandstones of the Lower Tuscaloosa Formation produce oil and gas at more than 40 fields in the salt basin of south Mississippi. These reservoir sandstones are variable in thickness and limited in lateral extent within the producing area, and they have been interpretated as fluvial sandstones, primarily because of their meanderlike patterns in local distribution. This study of sandstone petrography and geometry at Mallalieu field supports the fluvial interpretation but also demonstrates that there are two distinct sandstone morphologies—a lower section of meander-belt deposits and an upper section of braided-belt deposits. These two types have characteristic local patterns, and their recognition may be important in future exploration and field development.

Lower Tuscaloosa oil was first discovered in southwest Mississippi (Figure 5-25) at the Cranfield and Brookhaven fields in 1943 and then at Mallalieu and Baxterville fields in 1944. These fields are large domal structures and have significant reserves. Further exploratory drilling has had only limited success in the years following 1944 (Karges 1962); but during the period of 1958–1961, increased drilling resulted in the discovery of other important fields such as Little Creek, McComb,

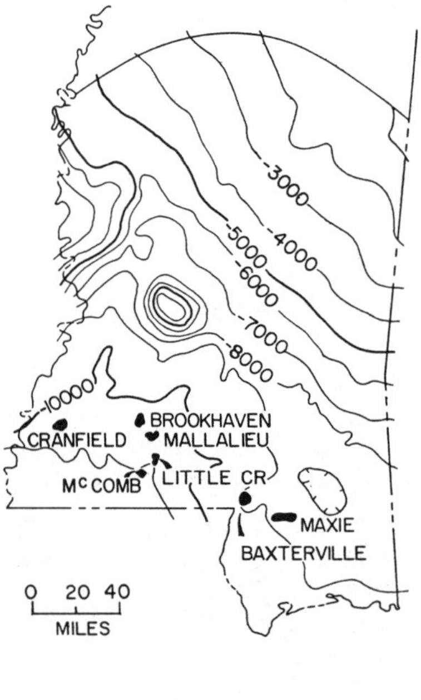

Figure 5-25 Index map of Mississippi showing structure at base of Upper Cretaceous and location of principal oil fields producing from Lower Tuscaloosa sandstones. [After Nunnally and Fowler 1954. Reprinted from Berg and Cook 1968 with the permission of the Gulf Coast Association of Geological Societies.]

and Smithdale. More recent exploratory activity has been moderate but persistent and has resulted in several new field discoveries of moderate size. Current exploration has shifted to the deeper Tuscaloosa section downdip.

Geologic Setting

The Lower Tuscaloosa Formation of the subsurface in southern Mississippi is the basal unit of the Gulfian Series. It lies unconformably on Comanchean rocks and overlaps successively older formations northward. The Lower Tuscaloosa reservoir sandstones are overlain conformably by marine shales of the Middle Tuscaloosa Formation. Thus, the overall sequence is transgressive in nature.

The Lower Tuscaloosa consists of interbedded shales, siltstones, and sandstones, and the formation has a maximum thickness of nearly 600 ft (183 m). Subsurface nomenclature (Mississippi Geological Society 1957) differentiates a lower ''massive member,'' which is dominantly sandstone up to 250-ft (76-m) thick, and an upper ''stringer sand member,'' which averages 300-ft (92-m) thick. The massive member is present only in a limited area of south central Mississippi and is not present at Mallalieu field.

Lower Tuscaloosa sandstones of the type that occur at Mallalieu have been attributed to various environments of deposition. Braunstein (1950, p. 15) called the producing sandstones marine because they are calcareous and ''glauconitic,'' but Karges (1962) believed that local sandstone distribution patterns indicated meandering stream channels in a deltaic environment. Scull et al. (1966, pp. 110–14) proposed that the ''stringer sands'' at Smithdale field represent point bars because of texture, bedding, and meanderlike patterns. These fluvial interpretations, however, were not supported by detailed lithologic descriptions, nor did they distinguish between the different sandstone patterns that are evident in published isopach maps.

Drilling History

The structure on the Lower Tuscaloosa at Mallalieu field (Figure 5-26) shows a large, nearly symmetrical dome, which probably occurs over a deep-seated salt structure. The structure has been drilled to a depth of 15,000 ft (4575 m) without encountering salt. Structural closure is greater than 100 ft (30 m) and is probably on the order of 150 ft (46 m). Reversal of dip to the north is on the order of only 50 ft/mi (0.5 m/km). The field was mapped by seismograph before drilling of the discovery well.

The producing area covers some 2900 acres, and oil columns are about 90 ft (28 m). The reservoir sandstones are relatively ''tight'' and have an average porosity of 27 percent and a permeability in the range of 5 to 25 md. The field has produced 35 million barrels of 38° API gravity oil and has now essentially reached the limit of primary production. A major part of this production has come from the west half of the field area and from the reservoir called herein the C sandstone.

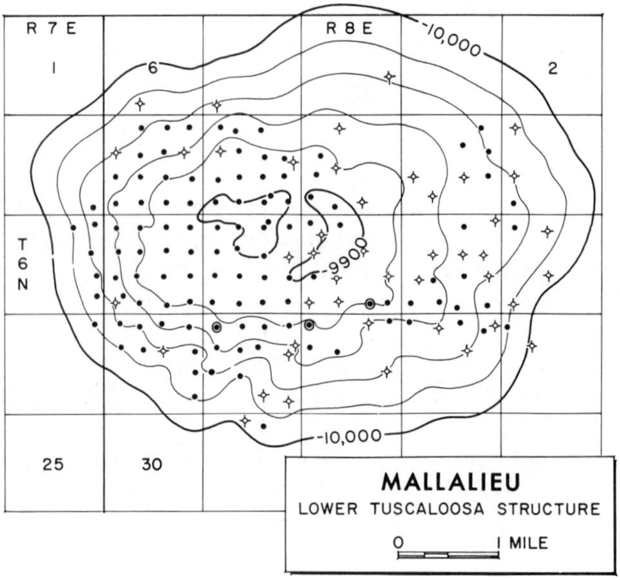

Figure 5–26 Structure map of the top of the Lower Tuscaloosa formation, Mallalieu field, Lincoln County, Mississippi. Contour interval 20 ft (6 m). [Modified from Karges 1962.] Core samples were obtained from the circled wells. [Reprinted from Berg and Cook 1968 and published with the permission of the Gulf Coast Association of Geological Societies.]

At Mallalieu field, the Lower Tuscaloosa has a thickness of about 300 ft (92 m) and consists of interbedded gray, green, or red mudstones and gray to white silty sandstones and siltstones. Porous sandstones are found at six different levels, but not all zones contain sandstone in any one well. The zones are persistent locally, however, and are designated by letter for convenience (Figure 5–27). The uppermost A zone is a thin siltstone or silty sandstone that has an average thickness of only 5 ft. The B zone has sandstone at three levels. The upper level (B-1) commonly consists of one or two thin sandstones, which reach a maximum thickness of about 15 ft (4.5 m) in only one well. The middle level (B-2) is a thin siltstone in a few wells and has a maximum sandstone thickness of 30 ft (9.2 m) in only one well; in most wells the middle level is a silty mudstone. The lower level (B-3) has a maximum sandstone thickness of 45 ft (14 m) and is the most widely developed part of the B zone.

The C sandstone is the main oil reservoir; it is present chiefly on the west half of the structure and has a maximum thickness of 55 ft (17 m). The D sandstone is also thick and apparently is present over a wide area, but many wells were not drilled into the D sandstones. Only a few deeper wells have penetrated the lowermost sandstones of the Lower Tuscaloosa, and the distribution of these sandstones is not established by existing control.

Description of Sandstones

The characteristics of Lower Tuscaloosa sandstone and adjacent rocks were established by examination of core samples from four wells (Figure 5–26). This study is based largely on the work of B. C. Cook (1968), the major part of which was pub-

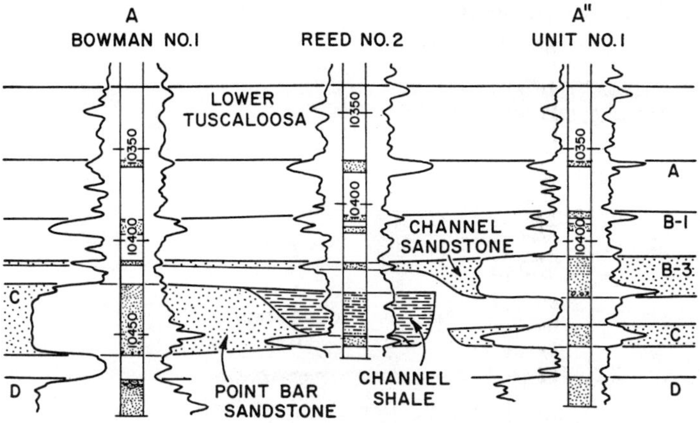

Figure 5–27 Electric logs of the Lower Tuscaloosa Sandstone, Mallalieu field, showing the variable nature of the several sandstones identified by letters A through D. Locations of wells shown in Figure 5–26. [From Berg and Cook 1968 and reprinted with the permission of the Gulf Coast Association of Geological Societies.]

lished previously (Berg and Cook 1968). The original study was based on petrography of small core samples and on sandstone mapping. Full-diameter cores had not been preserved, and sedimentary structures in vertical sequence could not be described in detail. Nevertheless, the data suggest a fluvial origin, and this interpretation is further supported by the distinctive morphologies of the reservoir bodies.

Petrography. A typical thick section of C sandstone is shown in the Bowman 20–1 core (Figure 5–28) and has a range in mean grain size from 0.47 mm at the base to 0.24 mm at the top. Average mean size is 0.35 mm (medium grained). Samples were too friable to make thin sections, and therefore mean size was measured in loose grain mounts. Composition could not be accurately determined, but the dominant grain type is quartz.

Composition of the C sandstone, however, is probably similar to the underlying D sandstone, which has a similar thickness and distribution. The D sandstone consists of 66 percent quartz, 32 percent matrix, 1 percent feldspar, and 1 percent mica and other grains (Table 5–5). Quartz includes both monocrystalline and polycrystalline varieties plus chert; only a small amount of chert is present. Calcite cement comprises an average 2 percent of bulk composition. Matrix of the D sandstone, and of the other sandstones, includes clay minerals, finely divided micaceous minerals, and larger grains of chlorite. The chlorite is believed to be of diagenetic origin and probably was derived by alteration of clay minerals. Therefore, chlorite is included with other matrix minerals because of its secondary origin from clays. Chlorite grains are visible under low magnification and apparently were called glauconite grains in older core descriptions. Mineral composition is that of an immature quartzose sandstone and is not unusual for sands of fluvial origin.

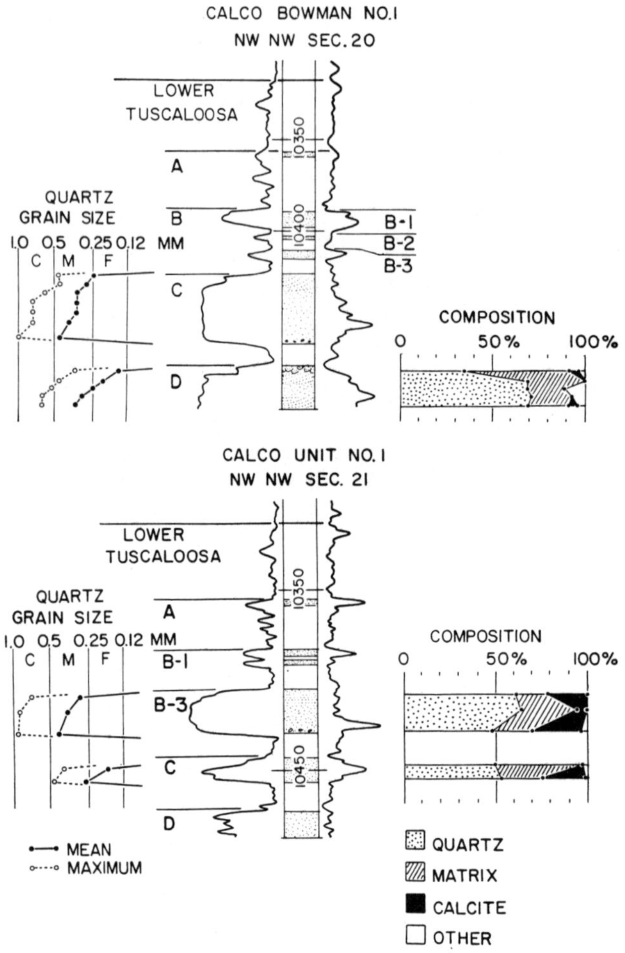

Figure 5-28 Characteristics of Lower Tuscaloosa sandstones showing grain size variation and mineral composition. Location of wells shown in Figure 5-26.

Some samples retain evidence of bedding. The C sandstone grades from massive in the lower part to laminated or cross-laminated in the upper part. This sequence of bedding, together with a vertical decrease in grain size, is typical of fluvial sandstones.

Thick sandstone of the B zone is shown in the Board of Supervisors 3-8 well (Figure 5-27). The sandstone has a range in mean grain size of quartz from 0.22 mm at the base to 0.12 mm at the top. Average mean size is 0.18 mm (fine grained). The dominant constituents are 64 percent quartz and 32 percent matrix with 2 percent feldspar and 2 percent muscovite (Table 5-5). Calcite cement averages 4 percent of bulk composition. This composition is similar to that of the D sandstone. Bedding, as seen in short core samples, grades from massive and cross-laminated in the lower part to ripple-laminated at the top. Rocks adjacent to the B-3 sandstone range

TABLE 5–5. AVERAGE PROPERTIES OF LOWER TUSCALOOSA SANDSTONES, MALLALIEU FIELD

				Quartz size[a]		Detrital composition[b]				Cement[c]		
Well	Sand	Depth (ft)	Number of samples	Mean (mm)	Max (mm)	Qz (%)	F (%)	Oth (%)	Mx (%)	(% of total)	Permea-bility (md)	Poros-ity (%)
Board of Supervi-sors 3-8	A	10,374	2	0.12	0.26	54	1	8	37	1	<0.01	20.4
Board of Supervi-sors 3-8	B	10,410	8	0.18	0.36	64	2	2	32	4	129.	25.9
Bowman 21-1	B	10,410	3	0.37	0.82	70	1	-	29	18	160.	30.4
Bowman 21-1	C	10,440	2	0.24	0.44	59	1	-	40	12	121.	23.4
Bowman 21-1	C	10,435	7	0.35	0.65	(not measured)					65.	25.0
Bowman 20-1	D	10,480	5	0.25	0.50	66	1	-	33	2	83.	21.2

SOURCE: Petrography from Cook 1968.
[a]Long-axis measurements
[b]Qz = monocrystalline and polycrystalline quartz, F = feldspar, Mx = matrix, Oth = other minerals.
[c]Cement = calcite.

from dark green mudstone to dark gray shale. Carbonaceous fragments are common along some laminations.

A thinner B sandstone in the Bowman 21-1 well is somewhat coarser grained with a mean size of 0.34 mm (medium grained). Quartz content is also greater at 70 percent, and the sandstone also has a large amount of calcite cement that averages 18 percent (Table 5–5).

The A sandstone is very fine grained and silty and has a mean grain size of 0.12 mm. It is thinly bedded with carbonaceous laminations and contains a large amount of muscovite. A fossil freshwater clam was found in mudstone immediately above the A sandstone.

The source area for Tuscaloosa sands may have been Paleozoic rocks that were exposed to the north on the Pascola arch, an eastward extension of the Ozark dome. The arch now lies below Cretaceous rocks of the Mississippi embayment. The Tuscaloosa Formation on the flank of the Pascola arch in western Tennessee and northern Mississippi consists of poorly sorted chert gravel that was derived largely from Devonian and Mississippian rocks (Marcher and Stearns 1962). Locally, an eastern facies is a very fine-grained, well-sorted quartzose sandstone that contains pebbles of vein quartz and quartzite. These sands may have been derived from weathering of Cambrian and Ordovician rocks, and this finer fraction was carried farther south by streams to the Tuscaloosa shoreline. Therefore, the high quartz content at Mallalieu, relative to other grain types, appears to be the result of high quartz in the source sediment.

Porosity and permeability. Reservoir properties of lower Tuscaloosa sandstones are variable. Average permeabilities range from a low of 14 md in thin sandstones to a high 160 md in thick sandstones (Figure 5–29), and maximum permeabilities are as high as 700 md in the basal sections of the B sandstones. In all of the sandstones, permeabilities decrease upward, probably reflecting a decrease in grain size upward. Average porosities are also variable and range from a low of 21 percent in the D sandstone to a high of 30 percent in the B-3 sandstone.

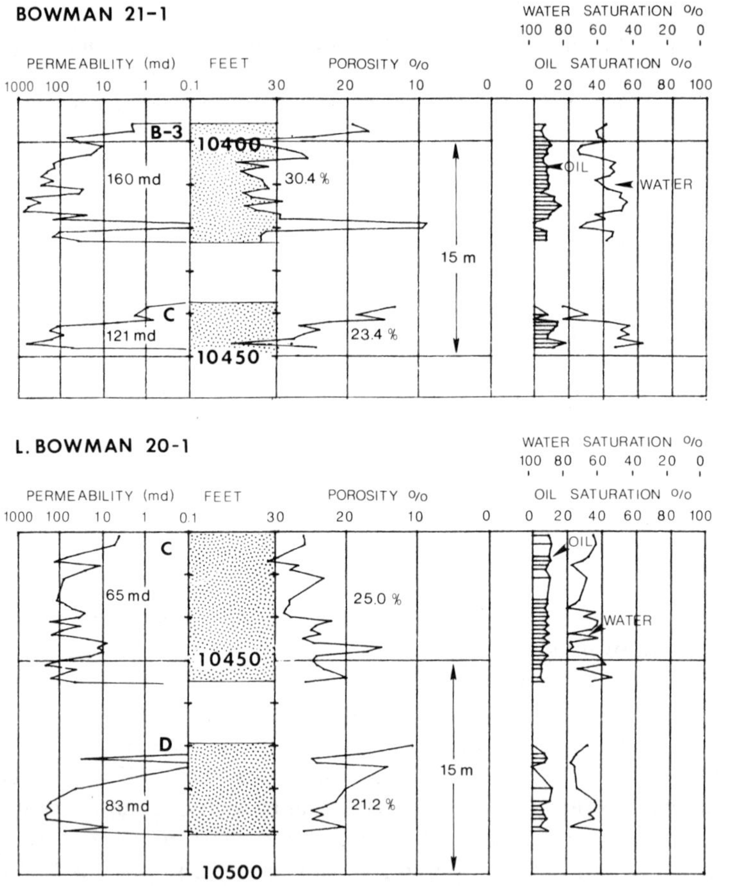

Figure 5–29 Permeability, porosity, and fluid saturations of Lower Tuscaloosa sandstones, Mallalieu field. Location of wells shown in Figure 5–26, and petrography illustrated in Figure 5–28.

There is no apparent relationship between average permeabilities and texture or composition (Table 5–5). In fact, higher values of average permeability are associated with larger amounts of calcite cement when averages for the total sections are considered. The explanation for this may be that some of the samples selected for petrographic study may have been less friable and more calcareous, and thus not representative of the total composition.

Fluid saturations are also somewhat variable. Residual oil saturation ranges from about 10 percent to 15 percent in oil-producing sandstones, whereas the residual water saturation is commonly greater than 60 percent. The high water-saturations may be the result of absorbed water in the abundant clay matrix.

Morphology

The C sandstone occurs principally on the west half of the Mallalieu structure (Figure 5–30). The average thickness is about 30 ft (9 m), commonly in one continuous section, and the maximum thickness is 55 ft (16.8 m). Several wells in the northwest part of the field have 10 ft (3 m) or less of the C sandstone at the base of the interval, with an upward gradation to interbedded sandstone and shale. The main body of the C sandstone is terminated abruptly along a broadly arcuate line that trends across the central and north parts of the field and encloses a semicircular area that is still open to the southwest beyond the field limits. Within the main sandstone area, arcuate trends of thicker sandstone parallel the edge of the body. Outside of the main sandstone area, the C zone consists of silty mudstone and shale. In at least one well (Figure 5–30), the C sandstone is largely replaced by dark gray shale that is quite different from the dark green mudstone that occurs in the C zone in most other wells.

Smaller areas of sandstone are present in the south and northeast parts of the field. These sandstones occur at slightly lower levels but still within the C zone. They attain a maximum thickness of 30 ft (9 m), and although of limited extent, they appear to form small arcuate patterns like that of the main body of C sandstone. These sandstones are somewhat finer grained than those of the main body.

The pattern of C sandstone distribution is similar to that found in a meander belt where point bars are deposited on the inside of meander loops, and the topography of the meander belt is one of alternating arcuate ridges and swales. The stream channel erodes laterally on the outside curve of the meander. If the meander is cut off during flooding, the abandoned channel forms an oxbow lake, which is filled subsequently with fine-grained sediment. The dark gray shale found in the south part of the field may represent such a filled channel, but the channel cannot be identified in other wells either because it is too narrow or because it is filled with dark green mudstone like the deposits of the eroded bank. Examination of additional well samples would be required to determine the extent and nature of channel fill.

The B-3 sandstones (Figure 5–30) reach a maximum thickness of 45 ft (13.7 m), and porous sandstone averages about 25 ft (8 m) along a narrow channel that

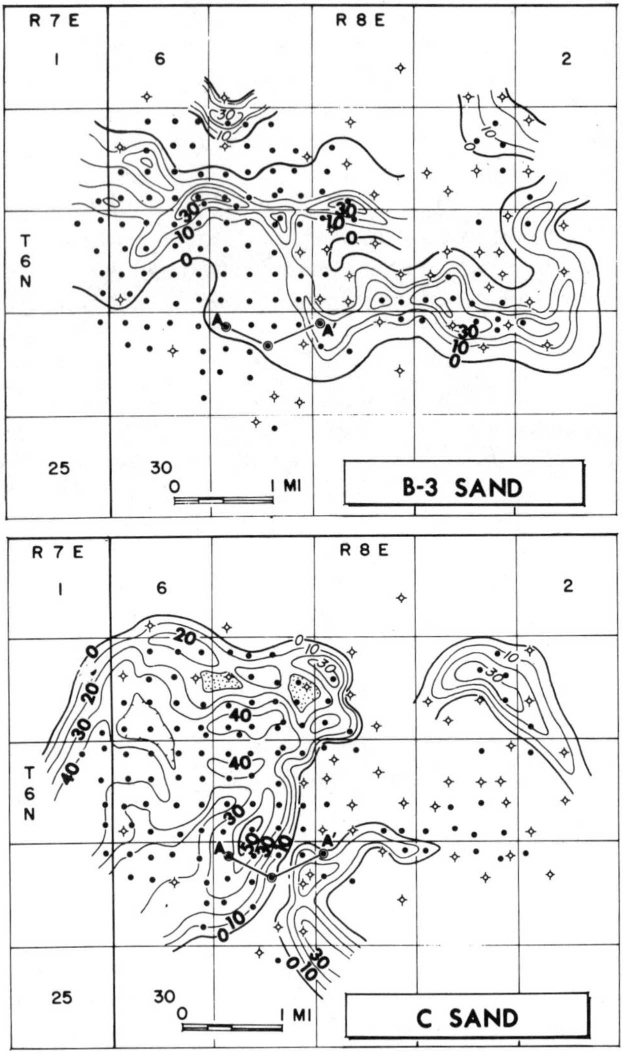

Figure 5–30 Thickness of Lower Tuscaloosa C and B-3 sandstones, Mallalieu field. Contour interval 10 ft (3 m). [Adapted from Berg and Cook 1968 and published with the permission of the Gulf Coast Association of Geological Societies.]

has a sinuous trend across the central and southeast parts of the field. Outside this channel, the B-3 zone is represented by a silty sandstone or siltstone that has a thickness of less than 10 ft (3 m). The siltstone occurs widely throughout the area, but even this thin unit is absent at the field margins at some distance from the channel.

The B-3 sandstone distribution is similar to that of a narrow stream channel that is eroded downward during flooding and fills with sand when flow decreases. The widespread siltstone at the same horizon may represent a flood-plain deposit that is adjacent to the main channel. This pattern lacks the more regular curves of meander loops and does not have an adjacent abandoned channel of the meander belt. The channel fill is the coarser detritus of the bed load itself. Therefore, the pattern is distinctly different, and is best described as *braided*. The channel-fill sandstone and its adjacent flood-plain siltstone may be called *braided-belt deposits*.

The overlying sandstones of the B zone show patterns similar to that of the B-3 sandstone, but their thickness rarely exceeds 10 ft (3 m) (Cook 1968). The associated shales and mudstones are dark gray to greenish gray and may be carbonaceous, but some brown mudstones are also present.

The A sandstone is widespread throughout the field area, and its thickness is commonly 10 ft (3 m) or less. The uniform thickness and distribution of the A sandstone suggest that it is a flood-plain deposit. Above the A zone, dark shales of the Lower Tuscaloosa become increasingly more uniform and less silty and less carbonaceous, and they grade upward into the marine shales of the overlying Middle Tuscaloosa section.

The difference in morphology of these deposits is also illustrated by comparing the vertical sequences shown by electric logs (Figure 5–27). The point-bar sandstones of the C zone have their base at essentially the same stratigraphic level throughout the field area. Loss of sandstones between wells occurs from the top of the section downward as shown by the change from maximum thickness in the Bowman 20–1 well to thinner sections in the Reed 2 and Bowman 21–1 wells. The same change is found within the large arcuate body of C sandstone in the west half of the field (Figure 5–28), where loss of sand in the swale areas also takes place from the top downward. An opposite change in sand thickness is found in the braided-channel sandstones of the B-3 zone. The top of the zone is everywhere at essentially the same stratigraphic level, both within the thicker channel-fill sections as well as in the flood-plain siltstones lateral to the main channels. Sandstone thickness increases from the top downward as shown by the change between the Reed 2 and Unit 1 wells (Figure 5–27).

These vertical changes in thickness seem to be consistent throughout the field area, and furthermore, they suggest that the morphology of the sandstones may be anticipated by the correlation of well logs when subsurface control is sparse. Thickness changes between wells, resulting in an abrupt increase of sandstone downward, suggest narrow braided-belt deposits. Conversely, thickness changes with a decrease of sandstone downward suggest widespread meander-belt deposits. Thus, when only a few well logs are available, the direction of thickness change may provide a clue to the morphology of the sandstone bodies.

The Lower Tuscaloosa sandstones were probably deposited in fluvial channels within a coastal-plain environment. The adjacent shales are generally nonmarine in character, and their colors range from gray to greenish gray and brown. Thin beds of carbonaceous material or lignite were observed in the cores. The overall sequence, however, is that of transgression. Regionally, the Lower Tuscaloosa sandstones rest unconformably on older Comanchean rocks, and they are replaced upward in the section by dark marine shales of the Middle Tuscaloosa. The change in fluvial channel patterns from meandering below to braided above seems anomalous for a transgressive sequence. Braided channels of modern streams are often associated with alluvial-valley environments rather than with coastal-plain environments.

The reason for change in fluvial morphology from meandering to braided probably lies in the change of channel slope or discharge through time. Leopold and Wolman (1957, p. 59) measured meandering, braided, and straight reaches of modern streams. They observed, generally, that braided reaches have higher channel slope and greater discharge than meandering reaches (Figure 5–5). From a plot of these measurements, it can be concluded that for a given discharge, the braided reach has the higher slope, and conversely, for a given slope, the braided reach has the greater discharge.

These relationships might be extrapolated to explain a change through time for local stream patterns of a given river system. For example, a regional tilt that increased channel slope might produce a change from meandering to braided if discharge remained relatively constant through time. For the Lower Tuscaloosa sandstones, a regional tilt or basin downwarp is indicated by the transgressive nature of the total sequence, and this tilt could have produced the observed changes in the stream morphology. Thus, in an overall transgressive sequence like the Lower Tuscaloosa, the more extensive point-bar sandstones can be expected to be lower in the section, whereas the less extensive sandstones of braided channels are located higher in the section.

Conclusions

The Lower Tuscaloosa sandstones at Mallalieu field are fluvial in origin. This conclusion is based largely on textural gradation from coarser grained below to finer grained above in the individual sandstones. The interpretation was not supported by observation of sedimentary structures because of the friable and incomplete nature of the cored sections. Sandstone distribution supports the fluvial origin, however, and shows two contrasting patterns of deposition—a meander-belt morphology for the older C sandstone and a braided-channel pattern for the overlying B sandstone. The upward change in channel morphology was probably the result of increasing stream gradients through time caused by downwarp that produced the overall transgressive sequence. These conclusions might be applied to the prediction of channel morphology in similar transgressive sequences.

FALL RIVER SANDSTONE, COYOTE CREEK AND MILLER CREEK FIELDS, NORTHEASTERN WYOMING

The Lower Cretaceous Fall River Formation, the "Dakota" sandstone of the subsurface, produces oil at eight fields on the northeast flank of the Powder River basin (Figure 5–31). Most of this oil has been found in stratigraphic traps. Because the Fall River is largely of marine and deltaic origin, the reservoir sandstone was be-

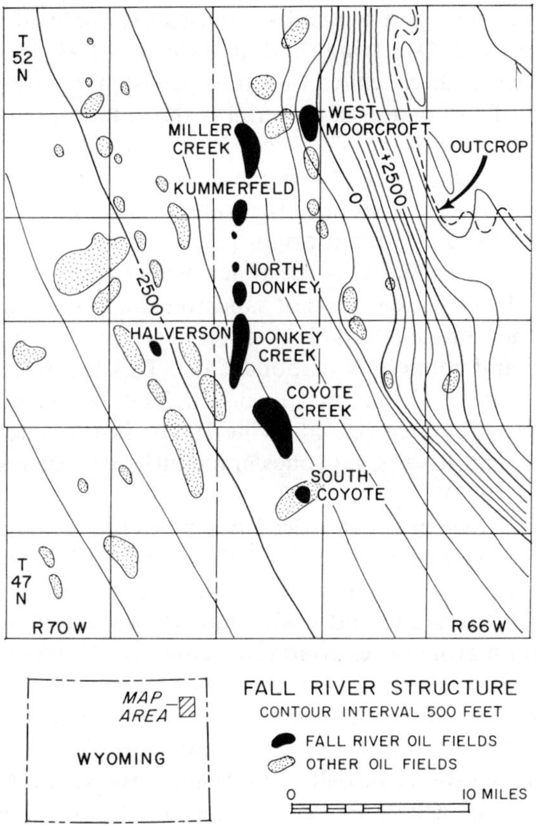

Figure 5–31 Index map of Fall River oil fields in the northeast Powder River basin, Wyoming. [Reprinted from Berg 1968 and published with the permission of the American Association of Petroleum Geologists.]

lieved to be the result of deposition in a littoral environment, perhaps as a series of barrier-bar sands flanked by lagoonal shales (Miller 1962). On the other hand, a fluvial origin for these sandstones was proposed at the West Moorcroft field (Mettler 1966) and at Coyote Creek field (Bolyard and McGregor 1966, p. 2236). A fluvial interpretation best explains the distribution of sandstone within the oil fields and can be widely applied in subsurface stratigraphic analysis.

Parts of the following discussion were published previously (Berg 1968), but a detailed description of Fall River cores has not been presented before. The sequence of sedimentary structures, texture, and composition confirm that the sands were deposited in fluvial channels.

Geologic Setting

The Fall River Sandstone is part of an overall transgressive sequence that begins with nonmarine shales of the Jurrasic Morrison Formation (see Chapter 6, Figure 6–9). This section is overlain by Lower Cretaceous sandstones and shales of the Lakota Formation. The Lakota has a thickness of about 150 ft (45 m) and consists of interbedded carbonaceous shale and lenticular sandstones, some of which are coarse grained and conglomeratic. Overlying the Lakota is a thin shale, 10- to 20-ft (3- to 6-m) thick. This shale, formerly called Fuson, is persistent over a wide area. The shale has a high content of carbonaceous material at the outcrop, but it appears to be of marine origin in the subsurface. The succeeding Fall River Sandstone is, in turn, overlain by the Skull Creek shale, a black platy mudstone of marine origin. Thus, the position of the Fall River Sandstone in the transgressive sequence suggests that it is of marginal-marine or coastal-plain origin.

At the outcrop along the west flank of the Black Hills uplift, the Fall River Sandstone consists of 100 to 150 ft (30 to 46 m) of sandstone and interbedded shale. The sandstones are of both the lenticular and blanket types, and associated siltstones and shales are dark-colored and commonly carbonaceous. This facies has been interpreted as mixed sediments of deltaic and marine origin deposited as a delta complex was built westward across the area (Dondanville 1963). In the outcrop at the north end of the Black Hills, Fall River sandstones apparently were deposited in an estuary (Campbell and Oaks 1973).

Farther west in the subsurface of the Powder River basin the formation grades into interbedded sandstones and dark gray shale. This facies has been interpreted to be composed largely of littoral marine deposits (Miller 1962), which grade westward into a thinly interbedded siltstone and shale sequence deposited as offshore marine sediment. The littoral marine facies commonly consists of two or three distinct sandstone beds separated by shale, and these units are persistent throughout a wide area (Stapp 1967). Hence, this facies has a uniform "regional" aspect, and it is broadly gradational to the adjacent facies to the east and west. The regional facies of the Fall River is now known to be deltaic in origin in the vicinity of Coyote Creek field. A description of this deltaic sequence is presented in Chapter 6.

In some local areas a more variable facies is found in which sandstones are not laterally persistent but may grade abruptly into shale and siltstone within a short distance. Local variation occurs in irregular areas up to about 15 sq mi (39 km^2). As might be expected, lithologic variation is common in the oil-producing trend and is responsible for oil accumulation. Lateral changes are well illustrated at Coyote Creek field, the largest of the Fall River fields.

Coyote Creek Field

Thick Fall River Sandstone pinches out abruptly updip at Coyote Creek, trapping an estimated 20 million bbl (3.2 million/mm^3) of oil. Regional dip is southwest at about 175 ft/mi (33 m/km), and locally the structure is uniform and not modified by any significant variation (Trotter 1962). Distribution of sandstone within the field is well known; the pattern of porous and permeable sandstone (Figure 5–32) shows linear and arcuate bands of thick sandstone that curve updip. Porous sand thickness is variable and ranges up to a maximum of 90 ft (27 m). The sandstone is composed of fine well-sorted grains, and the sharp change to less permeable shale and silt is clearly shown by electric logs (Figure 5–33). The maximum thickness of

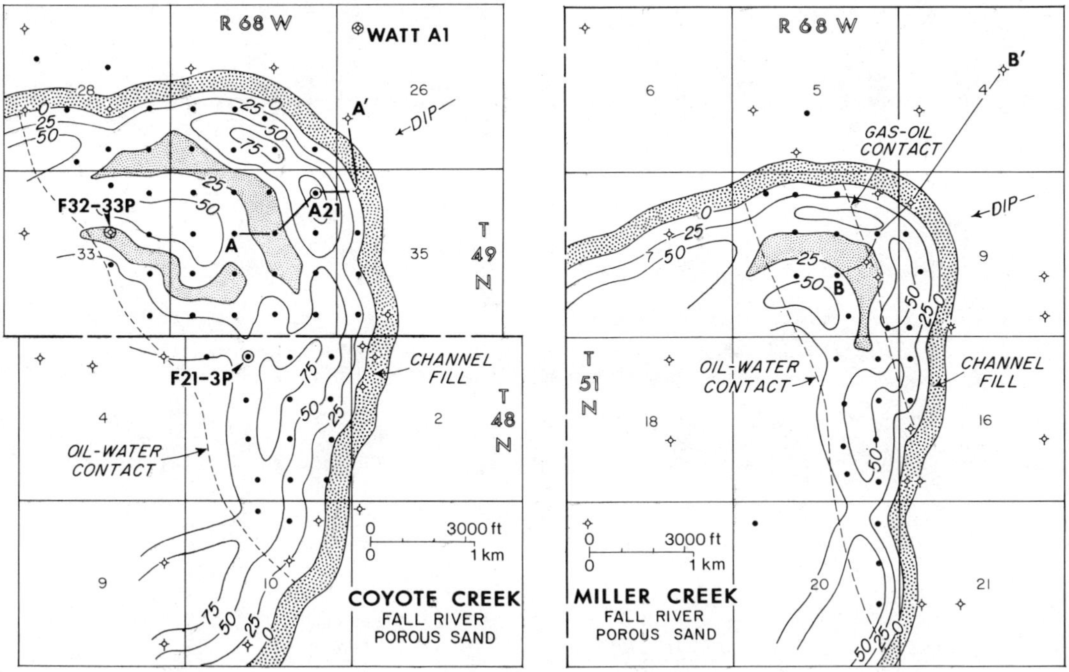

Figure 5–32 Isopach map of net porous sandstone at Coyote Creek field (left) and at Miller Creek field (right). Contour interval 25 ft (8 m). Cores were examined from circled wells in Coyote Creek field. Profiles A-A′ and B-B′ are shown in Figure 5–33. [Maps modified from Berg 1968 and published with the permission of the American Association of Petroleum Geologists.]

producing sandstones occurs at well A-21 in a continuous section. Immediately updip at well A-22, the Fall River interval is completely changed to thinly interbedded shale and siltstone. The less permeable zone has an average width of about 500 ft (152 m), a remarkably narrow band for such an abrupt lithologic change. On the northeast, the Fall River Sandstone again is a thick section that has variable porosity as in well B-2.

Downdip from the area of maximum thickness is an arcuate band of low permeability sandstone (Figure 5–32) in which the upper Fall River Formation is composed of tight sandstone, silt, and shale, but some permeable sandstone is found in the lower Fall River (Figure 5–33, well 1). Greater but variable thicknesses of permeable sandstone are present farther downdip. Well control is sparse below the oil-water contact, but continuous thick sandstone apparently is present for at least 2 mi (3.2 km) downdip.

The pattern of the Fall River sandstone at Coyote Creek (Figure 5–32) is similar to a meander-belt pattern. The narrow permeability barrier represents the deep abandoned channel, the adjacent high-permeability sandstones are the last-formed point bar, and the lower permeability band immediately downdip is a complex of

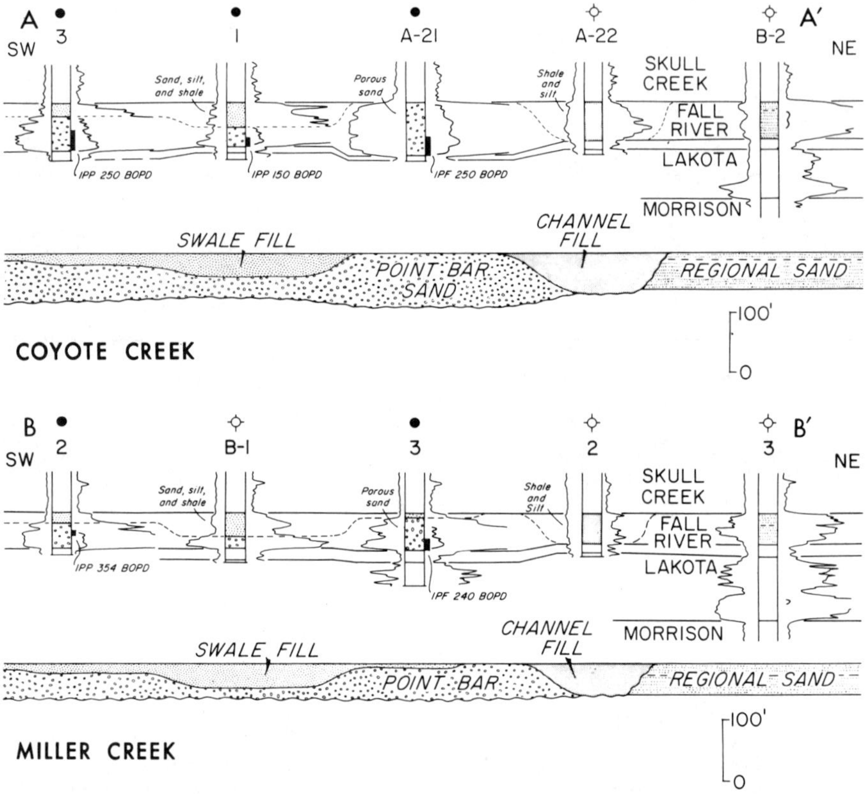

Figure 5–33 Electric-log correlation section at Coyote Creek field (A-A′) and at Miller Creek field (B-B′) showing interpretation of meander-belt facies. Location of sections shown in Figure 5–32. [Reprinted from Berg 1968 and published with the permission of the American Association of Petroleum Geologists.]

alternating point-bar and swale deposits. This interpretation not only explains the narrow updip barrier and the distribution of permeable sandstone, but it also explains the continuous arc of the permeability barrier, which provides an unusual amount of lateral closure and results in nearly 250 ft (76 m) of oil column.

Miller Creek Field

Distribution of porous Fall River Sandstone in the Miller Creek field (Truchot 1963) is similar to that in Coyote Creek. A prominent ridge of porous sandstone, arcuate updip, is bounded by a narrow filled channel (Figure 5–32). Maximum sandstone thickness of 52 ft (16 m) is present at the south end of the field, but thicknesses of 40 to 50 ft (12 to 15 m) are common near the channel-fill barrier. Just southwest of the maximum sandstone ridge is a relatively deep clay-filled swale. Two dry holes within the swale have less than 20 ft (6 m) of porous sandstone at the base of the

Fall River, and wells immediately downdip from the swale produce oil from a sandstone that is stratigraphically higher than the sandstone underlying the swale area. The distribution of meander-belt facies is clearly shown by electric logs of wells at the north end of the field (Figure 5–33).

A similar interpretation can be made for other Fall River oil fields (Figure 5–31). Sandstone isopachs have been published for Kummerfeld field (Miles 1963) and West Moorcroft (Mettler 1966), and these maps can also be contoured to show a pattern of sandstone ridges, arcuate and bounded by narrow filled channels. Kummerfeld and West Moorcroft are small fields and evidently are bounded by limited segments of abandoned channels. A similar situation exists in the South Coyote Creek field where a few wells in the center of the field produce oil from the Fall River Sandstone.

Description of Sandstone

Sedimentary structures. The character of Fall River sandstones is clearly shown by cores from three wells at Coyote Creek that show differences in both thickness and rock type (Figure 5–34). A thick section of stacked point bars is shown in well F21–3P; a nearby section of point-bar and swale-fill deposits is shown in well F32–33P; and a thick section at the updip edge of the reservoir, the "bar margin," is shown in well Watt A-21. These three facies have distinctive bed associations.

The point-bar section in well F21–3P shows nearly the entire thickness of the Fall River Sandstone. The cored section can be divided into the following units, in descending order:

Unit	*Thickness ft (m)*
1. Sandstone, mostly ripple-laminated in the upper part, cross-laminated in the lower part in bed sets about 1-ft thick.	14 (4.3)
2. Sandstone, ripple-laminated in the top 7 ft (2.1 m), remainder with indistinct to inclined laminae.	43 (13.1)

The lower unit (2) appears to form a single point-bar sequence. In general, sedimentary structures are difficult to observe because of the highly quartzose nature of the sandstone. At many levels, however, laminations are present and have dips that range from 10° to 20° [Figure 5–35(F) and (G)]. A thin interval in the lower part of the unit (2) contains abundant shale clasts in a very fine-grained sandstone [Figure 5–35(E)]. Clasts range from thin angular wafers, partly contorted by compaction, to large angular fragments that are more than 1-cm thick. Similar zones of shale clasts appear in other cored sections within the field. Convergent dips, representing truncation surfaces, are present at several levels [Figure 5–35(F)].

Near the top of the lower unit (2), the sandstone is distinctly ripple-laminated

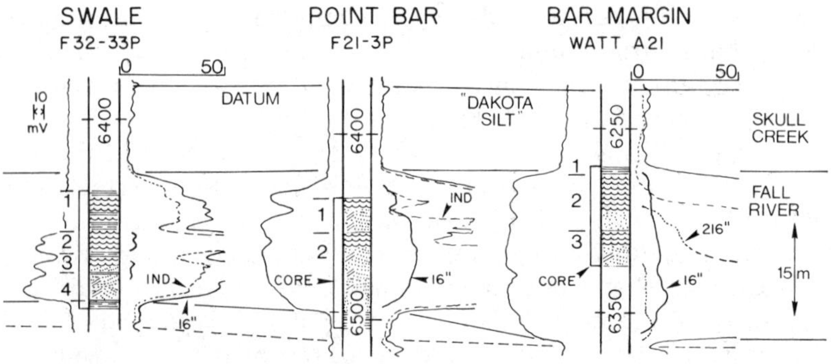

Figure 5–34 Comparison of contrasting sections of the Fall River Sandstone, Coyote Creek field. Location of wells shown in Figure 5–32, no horizontal scale. Numbers to the left of the depth columns refer to bedding units described in text.

[Figure 5–35(C)]. Ripple amplitudes range from 1 to 3 cm and decrease upward. Thus, the entire lower unit (2) represents a single fluvial-channel sequence that is steeply cross-laminated in the lower part and ripple-laminated at the top. The sequence is apparently a continuous composite set. Individual bed sets are not obvious in the cross-laminated sandstone, and this suggests that the bed sets are relatively thick.

A second, thinner point-bar sequence is represented by the overlying unit (1). Cross-laminated sandstone is found in thin sets only about 1 ft in thickness in the lower half of the unit [Figure 5–35(B)]. The overlying sandstone is dominantly ripple-laminated, and the amplitude of ripples appears to be relatively small and on the order of 1 cm [Figure 5–35(A)].

Thus, the section in F21–3P consists of two stacked point bars that represent nearly the entire thickness of the Fall River sandstone in this well. The uppermost 15-ft section of the Fall River section was not cored, but according to the well log, it appears to consist of a few feet of sandstone, probably rippled, and overlain by a shaly unit that may represent the final suspension deposits of the point-bar section.

A typical thin-bedded section representing largely swale deposits is shown in the core of the F32–33P well (Figure 5–34). The section consists of four distinct units that are in desending order as follows:

Unit	*Thickness ft (m)*
1. Shale, dark gray, some interbedded very fine-grained sandstone with small-scale ripple lamination.	23 (7.0 m)
2. Sandstone, ripple-laminated in upper half, some horizontal laminae in lower half, minor interbedded shale.	11 (11.4 m)

(*cont.*)

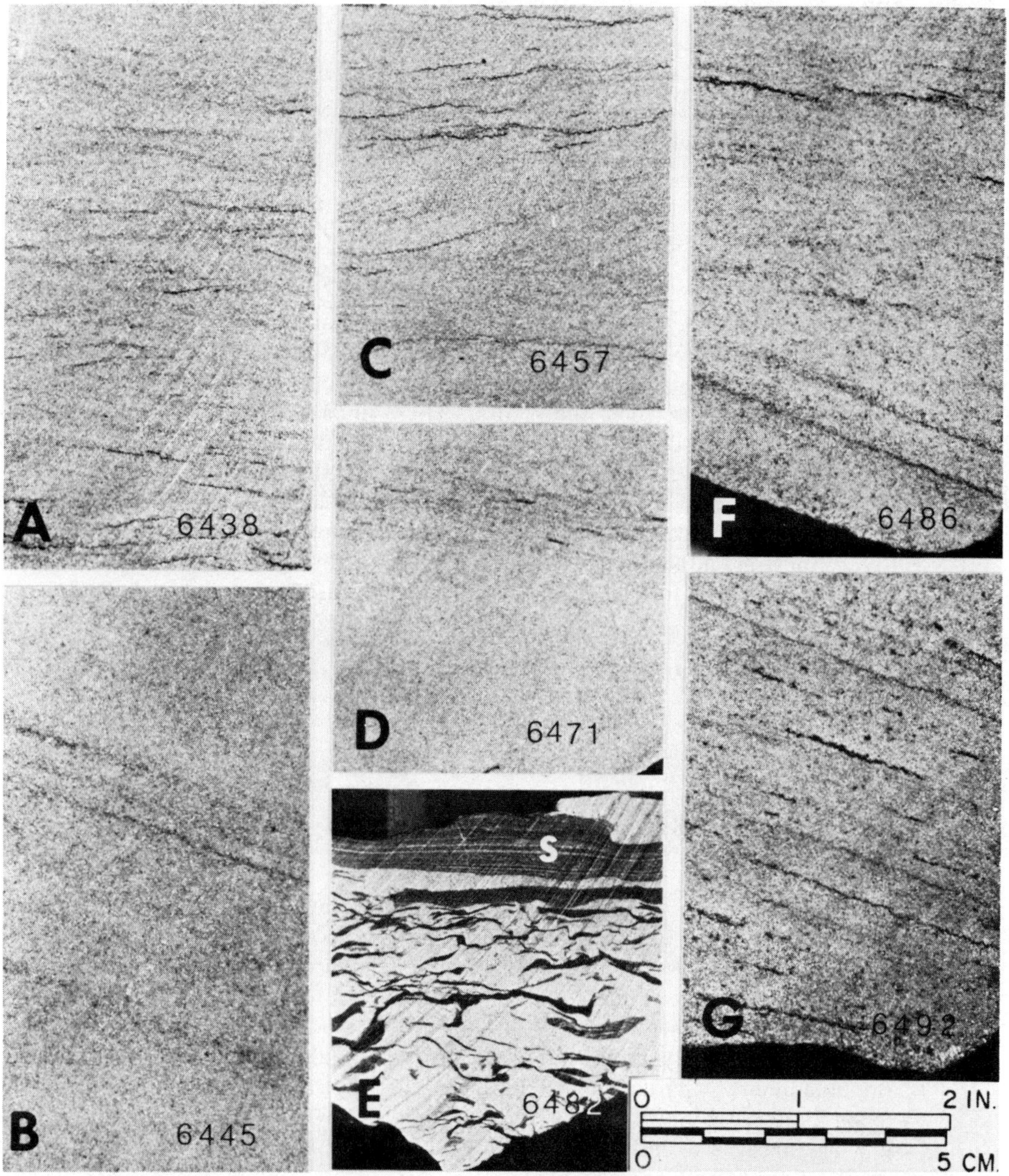

Figure 5–35 Sedimentary structures in the Fall River Sandstone, Mobil F21-3P, Coyote Creek field. Depths in feet given in lower right corner of photographs: (A) ripple-laminated sandstone—6438 ft (1964 m); (B) cross-laminated sandstone—6445 ft (1966 m); (C) ripple-laminated sandstone—6457 ft (1969 m); (D) indistinctly laminated to massive sandstone—6471 ft (1974 m); (E) large shale clast(s) and small deformed clasts in fine-grained sandstone—6482 ft (1977 m); (F) and (G) cross-laminated sandstone—6486 ft (1978 m) and 6492 ft (1980 m).

Unit	*Thickness ft (m)*
3. Sandstone, ripple-laminated and shaly in the upper half, indistinctly laminated to massive in lower half.	10 (3.0 m)
4. Sandstone, indistinct laminae with much interbedded shale in upper part, inclined lamination near base.	16 (4.9 m)

The lowermost unit (4) shows some inclined lamination with dips up to 15°, and this unit represents somewhat higher flow-regime, perhaps a thin point-bar sequence. The overlying units, however, become dominantly ripple-laminated and increasingly shaly upward in the section. These units (2 and 3) probably represent lowest flow-regime deposits within the swale areas during times of flood. The uppermost unit (1) is dominantly shale and represents the final fine-grained fill of the swale area.

The final cored section in well Watt A-21 shows the upper half of the Fall River sandstone adjacent to the abandoned channel (Figure 5–34). The section can be divided into the following units, in descending order:

Unit	*Thickness ft (m)*
1. Shale, dark gray, partly bioturbated at the top, with a few ripple lenses of sand.	5 (1.5 m)
2. Sandstone, mostly ripple-laminated, some thin massive beds.	27 (8.2 m)
3. Sandstone, ripple-laminated in top 9 ft (2.7 m), massive to indistinctly laminated below.	21 (6.4 m)

The lowermost unit (3) appears to represent a point-bar sequence. Lamination is not well developed, but some indistinct laminae dip about 15° in the lower part of the unit. The upper part is distinctly ripple-laminated. The lowermost Fall River section was not cored, but it may be assumed that a similar point-bar sequence is present and has a thickness of about 30 ft (9.2 m).

The next overlying unit (2) is seemingly anomalous because of its dominantly ripple-laminated character. The amplitude of ripples ranges from 1 to 2 cm and tends to decrease upward through the section. A few thin beds near the base of the unit are massive to indistinctly laminated, and there is little evidence for a dune phase of sediment transport.

The uppermost shale unit (1) appears to be bioturbated and largely of normal-marine origin at the top. Some of the shale, however, has a sooty or coaly appearance, and near the base of the unit are irregular sideritic masses that appear to represent fillings of root traces. Therefore, the top of the Fall River sandstone appears to have been vegetated, perhaps by a marsh, before deposition of the overlying Skull Creek shale.

The thick ripple-laminated section of unit (2) is not a normal point-bar sequence but is distinctive in its dominant low flow-regime character. The total Fall River section is also unusual in showing a maximum thickness of 86 ft (26.2 m). The location of this section adjacent to the shale-filled abandoned channel may be significant. At a late stage in its depositional history, the upper Fall River point bar adjacent to the inactive channel may have been subject to low-velocity tidal currents. There is little evidence in the core, however, for reversing tidal currents, primarily because the core slabs were not oriented and were cut at various angles to the bedding. Alternatively, the point-bar sands adjacent to the abandoned channel could have been subject to low-velocity flow during times of flood from an adjacent active channel. In either case, low flow-regime currents reworked the top of the point-bar sands as a final episode in meander-belt deposition.

Petrography. The point-bar sandstones of well F21–3P have an overall mean grain size of 0.19 mm (fine grained). Grain size tends to decrease upward from cross-laminated to rippled sandstones, but grain size is somewhat variable within the point-bar sequences (Figure 5–36). Maximum grain size also decreases upward from 0.40 mm at the base of the section to 0.24 mm at the top.

The quartz content is relatively high. Average composition is 84 percent monocrystalline quartz, 8 percent polycrystalline quartz, 1 percent feldspar, 3 percent rock fragments, 2 percent other grains, and 2 percent matrix. Rock fragments are largely chert, and when polycrystalline quartz is included, the total rock fragments comprise 11 percent of the detrital composition (Table 5–6). Other minerals include iron oxides, micas, and heavy minerals. Cement is largely silica overgrowths on quartz grains.

The swale section in well F32–33P is generally finer grained than the point-bar sandstones (Figure 5–36). The lowermost sandstone unit (4) of probable point-bar origin contains the coarsest sand. Grain size decreases from 0.30 mm (medium grained) at the base to 0.06 mm in the overlying section. At the same time, maximum grain size decreases from 0.45 mm to 0.14 mm. Sandstones of the overlying units show a decrease upward from a mean grain size of 0.15 mm in the massive sandstone unit (3) to 0.05 mm in the shale unit (1) at the top of the section. The sandstones are still relatively high in quartz content when the shaly sections are excluded (Table 5–6). The lowermost unit (4) has 78 percent monocrystalline quartz and 13 percent rock fragments, of which about half are composed of polycrystalline quartz. The overlying sandstone units still have an average 73 percent monocrystalline quartz but show an increase to 12 percent of other minerals, much of which is mica and organic material. Silica cement shows a striking increase to an average 14 percent as compared with lower values in the point-bar sandstones.

The bar-margin section of Watt A-21 shows a somewhat finer grain size than the point-bar section. The overall mean grain size is 0.13 mm, and mean grain size decreases upward from 0.16 mm at the base to 0.12 mm in the uppermost rippled sandstone. Composition is still highly quartzose with 84 percent monocrystalline

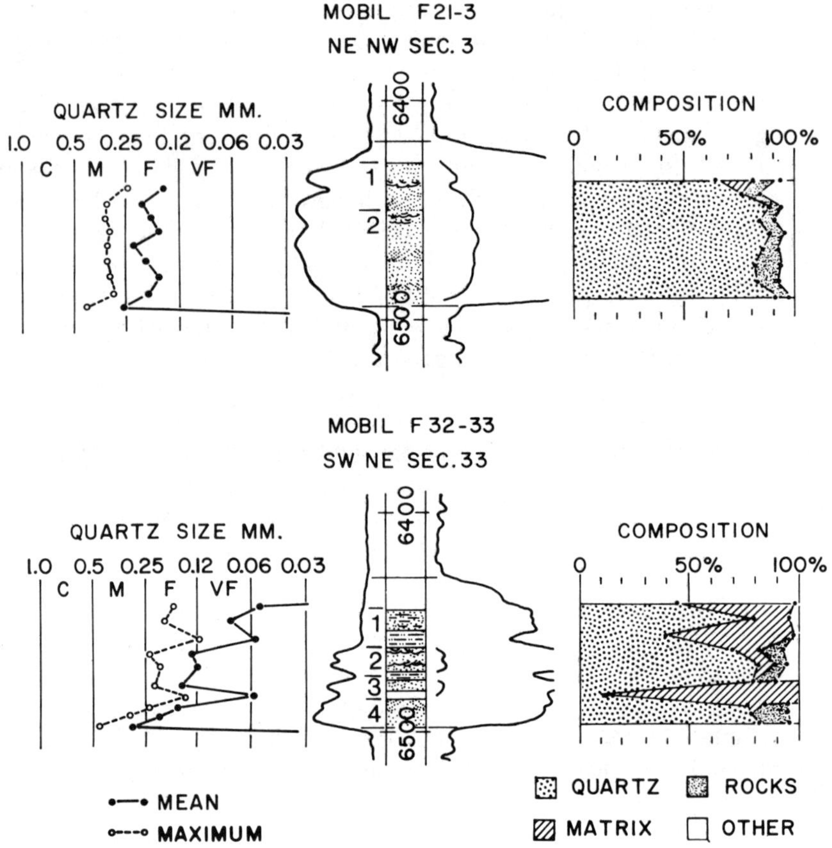

Figure 5-36 Texture, composition, and bedding units in Fall River Sandstone, Coyote Creek field. Numbers at left of center column refer to bedding units described in text.

quartz and 6 percent polycrystalline quartz and rock fragments. Matrix content is 9 percent and consists mainly of detrital clays. One sample, at 6319 ft, contains much authigenic clay in the form of chlorite, which has replaced the margins of quartz grains. Silica overgrowths on quartz grains average 16 percent of bulk composition.

All of the Fall River sandstones are similar in composition, and the only significant difference is in the amount of quartz overgrowths (Table 5-6). The point-bar sections have average silica overgrowths in the amount of 4 percent, whereas the swale and bar-margin sandstones have an average of 15 percent. Although composition was determined for only three cores, the difference in silica overgrowths apparently is related to facies and, perhaps, may be related to the amounts of interbedded shales or the presence of nearby shales. Silica is provided to the sandstones by water from compaction of shales, which are abundantly present in the

TABLE 5–6. PROPERTIES OF FALL RIVER RESERVOIR SANDSTONES, COYOTE CREEK FIELD

				Quartz size[a]			Detrital composition[b]					Cement[c]		
Well	Facies (unit)	Average depth (ft)	Number of samples	Mean (mm)	Max (mm)	σ (mm)	Qz (%)	F (%)	Rx (%)	Oth (%)	Mx (%)	(% of total)	Permeability (md)	Porosity (%)
F21-3	Upper point bar (1)	6445	3	0.18	0.29	0.05	81	2	14	1	2	3	137	15.5
F21-3	Lower point bar (2)	6470	6	0.20	0.32	0.06	84	1	11	2	2	4	320	17.2
F32-33	Swale (1,2,3)	6464	5	0.11	0.19	0.03	73	tr	9	12	6	15	14	12.7
F32-33	Point bar (4)	6492	3	0.22	0.33	0.05	78	1	13	4	4	13	130	16.0
Watt A-21	Bar margin (2,3)	6300	9	0.13	0.29	0.05	84	tr	6	1	9	16	25	12.6

[a]Long-axis measurements; σ = standard deviation.

[b]Qz = monocrystalline quartz, Mx = matrix, Oth = other minerals.

[c]Cement = silica as grain overgrowths.

swale section. Although not observed in the bar-margin section, shales are present in the adjacent abandoned-channel section. In contrast, the point-bar section is located in an area of dominant sandstone, relatively far from nearby swale areas (Figure 5–32). Thus, point-bar sands may have lacked a nearby supply of silica and remain relatively free of cement.

Porosity and permeability. The most permeable sandstones are found in the point-bar facies of well F21–3P (Figure 5–37). The cross-laminated sandstone of the main point-bar section has the highest average permeability of 320 md and a porosity of 17.2 percent. The overlying section, dominantly ripple-laminated, shows an average permeability of 137 md and a porosity of 15.5 percent.

The swale facies of well F32–33P shows higher permeability only in the lowermost unit (4), where average permeability is 130 md, and the porosity is 16 percent. The overlying units (2 and 3) show reduced porosity and permeability, largely because of the decrease in grain size.

The bar-margin facies of well Watt-21 shows a lower average permeability of 25 md and a porosity of 12.6 percent. The reduced values are probably the result of smaller grain size and larger amounts of silica overgrowths as compared with the point-bar section. Increasing amounts of silica cement above the basal sandstones are reflected in the decrease of permeability of the ripple-laminated sandstone of unit (2) and the upper part of unit (3).

Oil was produced from the point-bar section of well F21–3P and the bar-margin section of Watt A-21. In both sandstones, residual oil saturation averaged about 21 percent and residual water saturation averaged about 34 percent. On the other hand, the swale section of well F32–33P contained an average residual oil saturation of about 15 percent and an average residual water saturation of about 42 percent. The base of the most permeable unit (4) is at the oil-water contact, and the well was abandoned without significant oil production.

Morphology

Distribution of porous Fall River Sandstone at Coyote Creek is identical to that within meander belts of modern rivers (Figure 5–32). The ridge and swale pattern, the loss of sand upward in the section, and the narrow abandoned channel all correspond closely to modern deposits. Sandstone patterns that result from meander-belt deposition and the shape of adjacent channels should be comparable to meanders of modern rivers. Leopold and Wolman (1960, pp. 771–76) found that meander patterns appear to be the result of stream-flow dynamics and that there is a constant relationship between channel width, curvature, and meander length (wave length).

Measurements of these parameters may be estimated for Coyote Creek as well as for Miller Creek field assuming that their shale barriers represent channel fills (Figure 5–38). Measurements are only approximate because complete meander patterns were not outlined by well control or were not preserved. Radius of curvature (r) is the most easily determined, but meander length (L) must be estimated from

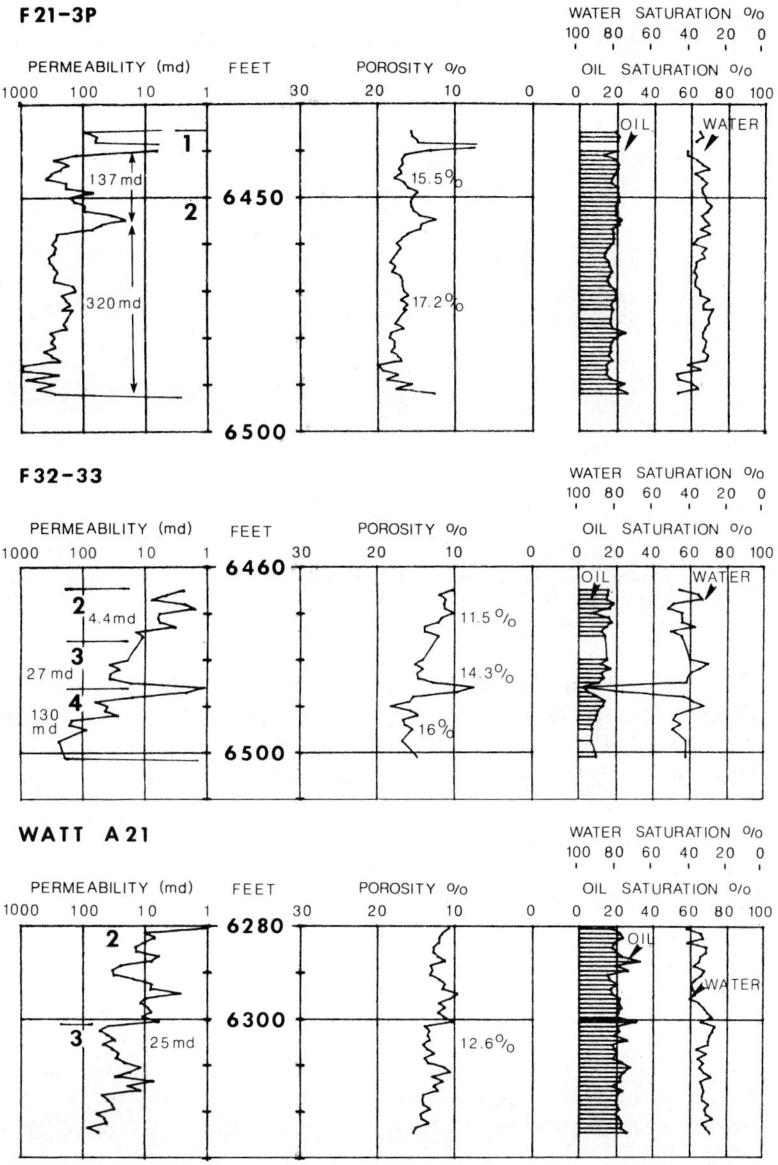

Figure 5–37 Permeability, porosity, and fluid saturations in Fall River Sandstone, Coyote Creek field. Numbers at left of center column refer to bedding units described in text.

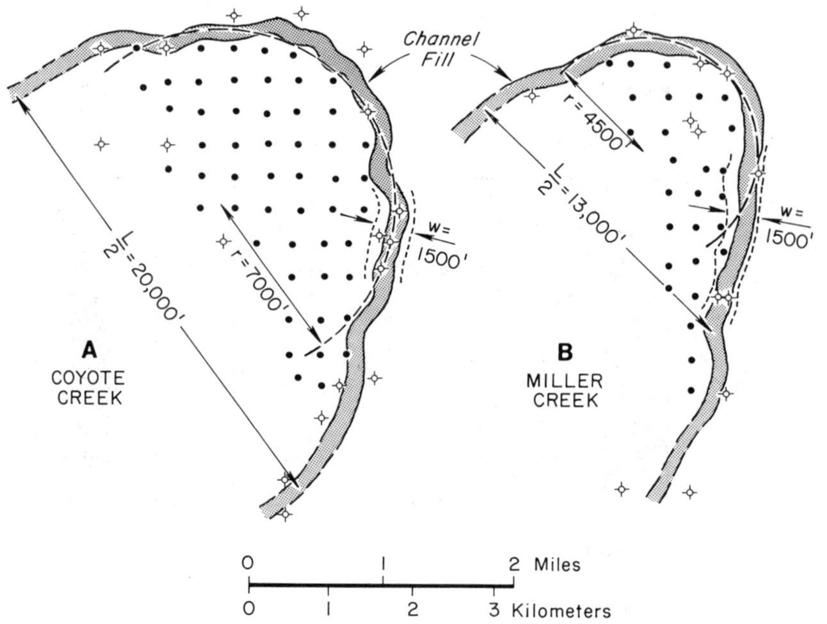

Figure 5–38 Estimates of one-half meander length ($L/2$), radius of curvature (r), and bank-full channel width (w) for Fall River reservoirs at Coyote Creek and Miller Creek fields.

half length ($L/2$). Channel width (w), or bank-full width for modern streams, is indicated by the maximum extent of the channel-fill shale. This width is considerably greater than the zero sand "channel" of the isopach map (Figure 5–32). Even with quarter-mile spacing of wells, maximum channel width is difficult to estimate.

Measurements of Fall River point-bar sandstones compare reasonably well with those of modern rivers (Figure 5–11) when meander length is plotted as a function of radius of curvature (Figure 5-39). The Fall River sandstones at Coyote and Miller Creek fields have values in the range of those for the lower Missouri and Mississippi rivers, and they suggest comparable size for the ancient Fall River. The Tuscaloosa C sandstone at Mallalieu field (Figure 5–30) appears to represent a river of similar size. The Fall River sandstones at Kummerfeld field (Miles 1963) were probably deposited by a smaller stream, as were the Bartlesville sandstones at Chetopa (Figure 5-24) and in the nearby La Harpe area (Schumacher 1976). The agreement of measurements between ancient and modern streams suggests that the morphology of point-bar sandstones might be predicted in exploratory or oil-field drilling.

Conclusions

Fall River sandstones were deposited within large meander loops and form a series of point bars and swales. The vertical sequence of sedimentary structures and the textural gradation support the interpretation that these sandstones are of fluvial

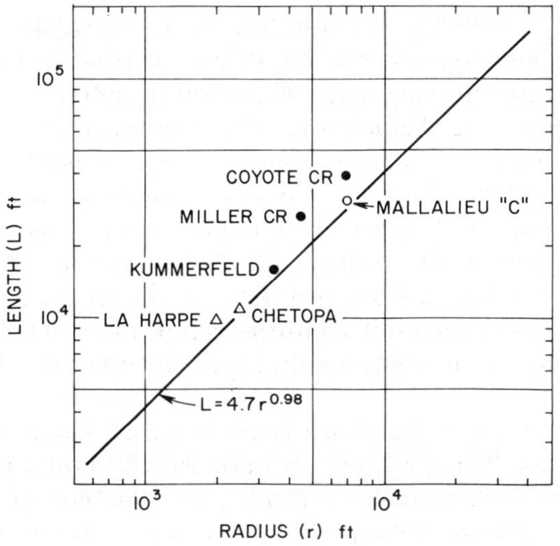

Figure 5–39 Measurements of meander length as a function of radius of channel curvature for point-bar sandstones compared with the average function for modern rivers (Figure 5–11). Black circles are Fall River sandstones; open circle is Tuscaloosa C sandstone; and triangles are Bartlesville sandstones.

origin. The sections range from thick sandstones composed of stacked point bars to thinner sections of basal sandstones succeeded by shales that were deposited in swale areas. A third distinct facies, called the *bar margin*, is present adjacent to the abandoned shale-filled channel and shows a dominance of ripple-laminated sandstones that were transported by low-velocity currents during a final stage of deposition.

The meander-belt sections have a characteristic variability in the field areas with rapid lateral change from thick point-bar sandstones to thin sandstones in adjacent swales. Typically, the electric logs reflect the abrupt basal change from underlying shale to fluvial sandstone, and the entire succeeding section may be composed of sandstone or may be gradational upward into swale deposits.

These fluvial sandstones have a high content of quartz, a composition that was determined largely by the immediate source materials lateral to the meander channels. The adjacent deposits, also quartz-rich, are gradational upward from shale below to sandstone above and represent deltaic sequences. Thus, the Fall River sandstones are the deposits of large meandering streams of a deltaic plain that were eroded into previously deposited deltaic sediments.

SUMMARY

Fluvial sandstones are readily recognized by their distinctive sequence of sedimentary structures from a dominant cross-laminated dune phase of transport in the lower part to a rippled phase above. Textural gradation follows the decrease in flow regime from coarser grained below to finer grained above. Composition of fluvial sandstones depends, as in many other deposits, on the nature of the source sediment, but generally quartz is modest in amount, whereas rock fragments and feldspars are relatively abundant. Exceptions are delta- or coastal-plain channel sandstones in which quartz content may reach 90 percent or more.

Reservoir morphology can be characterized as of two general types—braided or meandering. The braided channel-fill sandstones have low sinuosity and an average width on the order of 2000 ft (610 m), whereas meandering point-bar sandstones have high sinuosity and much wider distributions on the order of 10,000 ft (3050 m) or more. An exception is the meander fill of streams that carried a large suspended load so that the point bars are relatively small because of sand-poor sediment. Maximum thickness of reservoir sandstone can be essentially the same in both braided and meandering systems, whereas the lateral distributions of sandstone are quite different.

Prediction of local reservoir morphology possibly can be made on the basis of meander form as expressed by the ratio of meander length to radius of channel curvature. The reliability of such estimates depends on the amount of subsurface control, so the method may be more useful in field development than in exploration.

REFERENCES CITED

ALLEN, J. R. L., A review of the origin and characteristics of recent alluvial sediments: Sedimentology (special issue), v. 5, 191 p.

——, 1971, *Physical processes of sedimentation*: New York, Elsevier Publishing Company, 248 p.

BASS, N. W., 1936, Origin of the shoestring sands of Greenwood and Butler Counties, Kansas: Kansas Geological Survey Bulletin, 23, 135 p.

BERG, R. R., 1968, Point-bar origin of Fall River Sandstone reservoirs, northeastern Wyoming: American Association of Petroleum Geologists Bulletin, v. 52, no. 11, p. 2116–2122.

——, and B. C. COOK, 1968, Petrography and origin of Lower Tuscaloosa sandstones, Mallalieu field, Lincoln County, Mississippi: Gulf Coast Association of Geological Societies Transactions, v. 18, p. 242–255.

BERNARD, H. A., C. F. MAJOR, JR., B. S. PARROTT, and R. J. LEBLANC, SR., 1970, Recent sediments of southeast Texas: University of Texas Bureau of Economic Geology Guidebook 11, 16 p.

BLATT, HARVEY, 1967, Original characteristics of clastic quartz grains: Journal of Sedimentary Petrology, v. 37, no. 2, p. 401–424.

BOLYARD, D. W., and A. A. MCGREGOR, 1966, Stratigraphy and petroleum potential of Lower Cretaceous Inyan Kara Group in northeastern Wyoming, southeastern Montana, and western South Dakota: American Association of Petroleum Geologists Bulletin, v. 50, p. 2221–2244.

BOND, D. C. et al., 1971, Possible future petroleum potential of Region 9 - Illinois Basin, Cincinnati Arch, and northern Mississippi Embayment: American Association of Petroleum Geologists Memoir 15, v. 2, p. 1165–1218.

BRAUNSTEIN, JULES, 1950, Subsurface stratigraphy of the Upper Cretaceous in Mississippi: Mississippi Geological Society, 8th Annual Field Trip Guidebook, p. 13–24.

BROWNFIELD, R. L., G. L. CARPENTER, H. R. SCHWALB, and A. E. SMITH, 1968, List of Illinois basin oilfields with accumulative production of 5,000,000 bbls or more, *in Geology and Petroleum Production of the Illinois basin*: Evansville, Ind., Illinois and Indiana-Kentucky Geological Societies, p. 49–51.

CAMPBELL, V. C., and R. Q. OAKS, JR., 1973, Estuarine sandstone filling tidal scours, Lower Cretaceous Fall River Formation, Wyoming: Journal of Sedimentary Petrology, v. 43, p. 765–778.

CHARLES, H. H., 1941, Bush City Oil Field, Anderson County, Kansas, *in* A. I. Levorsen, editor, *Stratigraphic Type Oil Fields*: Tulsa, Oklahoma, American Association of Petroleum Geologists, p. 43–56.

CLARK, G. A., R. G. JONES, W. L. KINNEY, R. E. SCHILSON, H. SURKALO, and R. S. WILSON, 1965, The Fry *in situ* combustion testfield operations: Journal of Petroleum Technology (March), p. 343–347.

——, 1965*b,* The Fry *in situ* combustion test-performance: Journal Petroleum Technology (March), p. 348-353.

CLEARY, W. J., and J. R. CONOLLY, 1971, Distribution and genesis of quartz in a Piedmont-Coastal Plain environment: Geological Society of America Bulletin, v. 82, p. 2755–2766.

COLBY, B. R., 1963, Fluvial sediments — a summary of source, transortation, deposition, and measurement of sediment discharge: U.S. Geological Survey Bulletin 1181-A, 47 p.

COOK, B. C., 1968, Depositional environment of the "Stringer Sand" Member, Lower Tuscaloosa Formation (Cretaceous), Mallalieu field, Mississippi: Texas A&M University M.S. thesis, 97 p.

DILLARD, W. R., D. P. OAK, and N. W. BASS, 1941, Chanute Oil Pool, Neosho County, Kansas — a water-flooding operation, *in* A. I. Levorsen, editor, *Stratigraphic type oil fields*: American Association of Petroleum Geologists, p. 57–77.

DONDANVILLE, R. F., 1963, The Fall River Formation, northwestern Black Hills, lithology and geologic history: Wyoming Geological Association - Billings Geological Society Field Conference Guidebook, p. 87–99.

EBANKS, W. J., JR., G. W. JAMES, and N. D. LIVINGSTON, 1977, Evaluation of heavy oil and tar sands in Bourbon, Crawford, and Cherokee counties, Kansas: U.S. Department of Energy, Bartlesville Research Center, BERC/R1–77/20, 110 p.

EISENSTATT, P., 1960, Little Creek Field, Lincoln and Pike counties, Mississippi; Gulf Coast Association of Geological Societies Transactions, v. 10, p. 206–213.

EMERY, L. W., 1962, Results from a multi-well thermal-recovery test in southeastern Kansas: Journal of Petroleum Technology, v. 14, p. 671–678.

ENRIGHT, R. J., 1964, Heavy oil: a billion-barrel prize: Oil and Gas Journal, v. 62, no. 24, p. 41–44.

ETHRIDGE, F. G., 1970, Quantitative petrographic criteria for recognition of environments of deposition: Texas A&M University Ph.D. dissertation, 169 p.

——, and S. A. SCHUMM, 1978, Reconstructing paleochannel morphologic and flow characteristics: methodology, limitations, and assessment, *in* A. D. Miall, editor, *Fluvial sedimentology*: Canadian Society of Petroleum Geologists, Calgary, Alberta, p. 703–721.

HEWITT, C. H., and J. T. MORGAN, 1965, The Fry *in situ* combustion test-reservoir characteristics: Journal of Petroleum Technology (March), p. 337–342.

HAYES, M. O., 1963, Petrology of Krebs Subgroup (Pennsylvanian, Desmoinesian) of western Missouri: American Association of Petroleum Geologists Bulletin, v. 47, no. 8, p. 1537–1551.

HOWE, W. B., 1951, Bluejacket Sandstone of Kansas and Oklahoma: American Association of Petroleum Geologists Bulletin, v. 35, p. 2087–2093.

JACKSON, R. G., 1976, Depositional model of point bars in the lower Wabash River: Journal of Sedimentary Petrology, v. 46, p. 579–595.

JOHNSON, C. T. L., 1973, Environment of deposition of the Pennsylvanian Bartlesville Sandstone, La Bette County, Kansas: Texas A&M University M.S. thesis, 72 p.

KARGES, H. E., Significance of Lower Tuscaloosa sand patterns in southwest Mississippi; Gulf Coast Association of Geological Societies Transactions, v. 12, p. 171–185.

KOLB, C. R., and J. R. VAN LOPIK, 1966, Depositional environments of the Mississippi River deltaic plain — southeastern Louisiana, *in* M. L. Shirley, editor, *Deltas in their geologic framework*: Houston, Texas, Houston Geological Society, p. 17–61.

LEOPOLD, L. B., and M. G. WOLMAN, 1957, River channel patterns: braided, meandering and straight: U.S. Geological Survey Professional Paper 282-B, 85 p.

——, 1960, River meanders: Geological Society of America Bulletin, v. 71, p. 769–794.

——, and J. P., MILLER, 1964, *Fluvial processes in geomorphology*, San Francisco, W. H. Freeman and Company, 622 p.

MARCHER, M. V., and R. G. STEARNS, 1962, Tuscaloosa Formation in Tennessee: Geological Society of America Bulletin, v. 73, p. 1365–1386.

METTLER, D. E., 1966, West Moorcroft Field, Crook County, Wyoming: The Mountain Geologist, v. 3, p. 89–92.

MILES, O. P., JR., 1963, Kummerfeld field: Wyoming Geological Association-Billings Geological Society Joint Field Conference Guidebook, p. 125–128.

MILLER, D. N., JR., 1962, Patterns of barrier bar sedimentation and its similarity to Lower Cretaceous Fall River stratigraphy: Wyoming Geological Association Guidebook, 17th Annual Field Conference Guidebook, p. 232–247.

——, editor, 1968, *Geology and petroleum production of the Illinois basin*: Evansville, Indiana, Illinois and Indiana-Kentucky Geological Societies, 301 p.

MISSISSIPPI GEOLOGIC SOCIETY, 1957, Mesozoic and Paleozoic producing areas of Mississippi and Alabama; v. 1, p. 72.

NUNNALLY, J. D., and H. F. FOWLER, 1954, Lower Cretaceous stratigraphy of Mississippi: Mississippi Geological Survey Bulletin, 79, 45 p.

POTTER, P.E., 1962, Regional distribution patterns of Pennsylvanian sandstones in the Illinois basin: American Association of Petroleum Geologists Bulletin, v. 46, p. 1890–1911.

——, 1963, Late Paleozoic sandstones of the Illinois basin: Illinois State Geological Survey, Report of Investigations 217, 92 p.

PRYOR, W. A., 1973, Permeability-porosity patterns and variations in some Holocene sand bodies: American Association of Petroleum Geologists Bulletin, v. 57, p. 162–189.

RICH, J. L., 1923, Shoestring sands of eastern Kansas: American Association of Petroleum Geologists Bulletin, v. 7, p. 103–113.

RUSSELL, R. D., 1937, Mineral composition of Mississippi River sands: Geological Society of America Bulletin, v. 48, p. 1307–1348.

SCHUMACHER, MADELYN, 1976, Depositional environment of the Bartlesville Sandstone, LaHarpe field, Allen County, Kansas: Texas A&M University M.S. thesis, 82 p.

SCHUMM, S. A., 1968, Speculations concerning paleohydrologic controls of terrestrial sedimentation: Geological Society of America Bulletin, v. 79, p. 1573–1588.

——, 1972, Fluvial paleochannels, *in*, *Recognition of ancient sedimentary environments*: Society of Economic Paleontolgists and Mineralogists, Special Publication 16, p. 98–107.

SCHWALB, H. R., 1968, Subsurface structure map on the base of New Albany black shale or its equivalent, *in Geology and Petroleum Production of the Illinois basin*: Evansville, Indiana, Illinois and Indiana-Kentucky Geological Societies, Pocket.

SCRUTON, P. C., 1950, The petrography and environment of deposition of the Warner, Little Cabin, and Hartshorne Sandstones in northeastern Oklahoma: American Journal of Science, v. 248, p. 408–426.

SCULL, V. J., C. J. FELIX, S. B. MCCALEB, and W. G. SHAW, 1966, The inter-disciplinary approach to paleoenvironmental interpretations: Gulf Coast Association of Geological Society Transactions, v. 16, p. 81–117.

SEARIGHT, W. V., W. B. HOWE, R. C. MOORE, J. M. JEWETT, G. E. CONDRA, M. C. OAKES, and C. C. BRANSON, 1953, Classification of Desmoinesian (Pennsylvanian) of northern Mid-Continent: American Association of Petroleum Geologists Bulletin, v. 37, p. 2747–2749.

SELF, R. P., 1975, Petrologic changes in fluvial sediments in the Rio Nautla drainage basin, Veracruz, Mexico: Journal of Sedimentary Petrology, v. 45, p. 140–149.

SHELTON, J. W., 1973, Models of sand and sandstone deposits: a methodology for determining sand genesis and trend: Oklahoma Geological Survey Bulletin, 118, 122 p.

SMITH, M. W., 1966, Simultaneous underground combustion and water injection in the Carlyle Pool, Iola Field, Kansas: Journal of Petroleum Technology, v. 18, p. 11–18.

SPOONER, H. V., JR., 1964, Basal Tuscaloosa Sediments, East-Central Louisiana: American Association of Petroleum Geologists Bulletin, v. 48, p. 1–21.

STAPP, R. W., 1967, Relationship of Lower Cretaceous depositional environment to oil accumulation, northeastern Powder River Basin, Wyoming: American Association of Petroleum Geologists Bulletin, v. 51, p. 2044–2055.

STEINMETZ, RICHARD, 1972, Sedimentation of an Arkansas River sand in Oklahoma: a cautionary note on dipmeter interpretation: Oklahoma City Geological Society Shale Shaker, v. 23, p. 32–38.

TROTTER, J. F., 1962, Coyote Creek field, Powder River Basin, Wyoming: Wyoming Geological Association, 17th Annual Field Conference Guidebook, p. 297–302.

TRUCHOT, J. F., JR., 1963, The Miller Creek field Crook County, Wyoming: Wyoming Geological Association-Billings Geological Society, Joint Field Conference Guidebook, p. 129–132.

U.S. ARMY CORPS OF ENGINEERS, 1935, Studies of river bed materials and their movement with special reference to the lower Mississippi River: Vicksburg, Mississippi, Waterways Experiment Station paper 17, 161 p.

VISHER, G. S., 1968, Depositional framework of the Bluejacket-Bartlesville sandstone, *in* G.

S. Visher, editor, *Geology of the Bluejacket-Bartlesville sandstone of Oklahoma*: Oklahoma City, Oklahoma, Oklahoma City Geological Society, p. 32–51.

VISHER, G. S., B. S. SAITTA, and R. S. PHARES, 1971, Pennsylvanian delta patterns and petroleum occurrences in eastern Oklahoma: American Association of Petroleum Geologists Bulletin, v. 55, p. 1206–1230.

WANLESS, H. R., 1962, Pennsylvanian rocks of Eastern Interior Basin, *in Pennsylvanian system in the United States*: Tulsa, Oklahoma, American Association of Petroleum Geologists, p. 4–59.

WEIRICH, T. E., 1953, Shelf principle of oil origin, migration, and accumulation: American Association of Petroleum Geologists Bulletin, v. 37, p. 2027–2045.

——, 1968, History of Bartlesville oil sand, *in* G. S. Visher, editor, Geology of Bluejacket-Bartlesville Sandstone Oklahoma: Oklahoma City, Oklahoma, Oklahoma City Geological Society, p. 69–72.

WOMACK, ROBERT, JR., 1950, Brookhaven oil field, Lincoln County, Mississippi: American Association of Petroleum Geologists Bulletin, v. 34, p. 1517–1529.

6

Deltaic Sandstones

Deltas represent the termination of fluvial systems and thus contribute vast amounts of sediment to marine environments. Much of this sediment remains near the site of initial deposition, forming thick sediment bodies. In some cases, marine processes are active during and after the deposition of deltaic sediments, and they modify the form and distribution of the sediments. By these processes, the sediments are moved laterally and downslope to form other coastal or marine sediments. Deltas, then, are only the primary site of sediment deposited in marine basins.

Deltaic sediments have been credited with containing "most of the known accumulations of oil and gas in terrigenous sediments" (Rainwater 1975, p. 3). This claim may be slightly exaggerated, but accounts of petroleum accumulation in deltaic sediments are numerous, particularly in the Tertiary section of the Gulf Coast province.

Despite the undoubted importance of deltaic sandstones as petroleum reservoirs, there are still few descriptions that show details of their cores. Many sandstones credited to deltaic sedimentation may have been the result of modifying processes, and strictly speaking, the reservoir properties of the delta-associated sediments may owe their primary character to processes other than deltaic. Waves, tides, and winds may reshape the deltaic sediments and impose on the sandstones a character not unique to deltas. Therefore, it is the purpose of the following sections to describe selected sequences that owe their character to primary deltaic processes. Not surprisingly, these sandstones bear the imprint of current transport and are, in this sense, the result of fluvial processes.

DELTA PROCESSES

A *delta* is the mass of coarser-grained sediment formed by deposition at the mouth of a river as it enters a body of standing water. This definition suggests the sequence that may be encountered in deltaic deposition. At the base are deposits of the relatively quiet water; these are the *prodelta* sediments and not the deposits derived directly by fluvial currents. Above are coarser sediments of the delta proper; these contain structures that show evidence of current transport. Such structures may be first of the lower flow-regime because the spreading water of the fluvial system dissipates its energy here, causing river-mouth deposition. Higher in the section, there may be structures showing evidence of higher flow-regime transport as the river system builds outward over previously deposited sediments. At the top are the shallow-water to subaerial deposits that represent the final stage of transformation of the preexisting subaqueous environments to the deltaic plain.

The first diagram to suggest this expected sequence (LeBlanc 1975) was presented by Gilbert (1885), and although this diagram represents a special site of deposition, it still presents the essentials of the deltaic sequence (Figure 6–1).

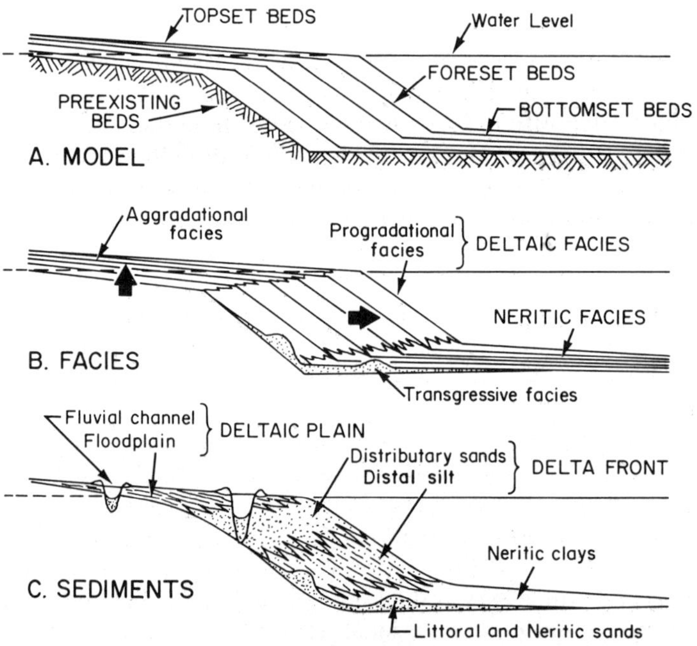

Figure 6–1 Diagrammatic cross sections of depositional units within deltas: (A) the delta concept of Gilbert (1885) showing topset, foreset, and bottomset beds; (B) deltaic and neritic facies from Frazier (1967); (C) sediment types and depositional units of an idealized delta.

The three parts of the delta system are designated *bottomset, foreset,* and *topset* deposits. The bottomset beds are the finest-grained sediment deposited from suspension far from the river mouth, and these form the prodelta sediment. The foreset beds are the subaqueous deposits that are built outward, or *prograde,* by fluvial sedimentation as the river outflow enters the standing body of water, spreads, decreases in velocity, and deposits first its bed load and then, farther from the river mouth, its suspended load. The foreset beds represent rapid deposition of sediment at the submerged delta-front section, and these are therefore the thickest part of the sequence. The topset beds are the shallow subaqueous or subaerial deposits that are built upward, or *aggrade,* to form the delta plain.

The aggrading-topset and prograding-foreset beds represent the delta mass itself, which is dominated by deposition from the decreasing current flow. The bottomset deposits represent sediments of the main water body, not influenced by fluvial processes.

Gilbert's delta was a relatively thin, rapidly prograding mass of coarse-grained sediment that was built outward from the shores of Pleistocene Lake Bonneville, the large body of glacial meltwater that occupied an expanded area around the present Great Salt Lake, Utah. Although this delta was of a special type, the concept of dominant progradation of foreset beds clearly expresses the fundamental processes of delta formation.

Delta processes are, in fact, quite complex from the standpoint of hydrology and associated deposition. Furthermore, these processes are of importance to the commercial use of waterways and have been the subject of extensive study. The fundamental unit of deposition is the river-mouth bar, or *distributary bar;* this is the mound of sediment deposited immediately at the fluvial outlet as the flow spreads laterally. The bar obstructs the outlet and prevents direct access to the river channel, and its location and form are thus of special interest.

The primary river-mouth processes are of four types (Table 6–1) as classified by the nature of flow from the outlet. The first may be called *turbulent jet flow* [Figure 6–2(A)]. High-velocity fully turbulent flow into relatively deep water produces an elongate crescentic or lunate bar with distinct foreset beds of the type described by Gilbert. In jet flow, the inertial force of the turbulent water dominates the form of both flow and bar deposit.

A second flow type is called *turbulent frictional flow* [Figure 6–2(B)]. Moderate- to high-velocity flow into a shallow-water basin results in rapid spreading and deceleration by increased bottom friction, producing a "middle-ground" or shoal bar flanked by diverging channels. These bars result from flow through crevasses, or breached levees, and lateral flow from the main river channel builds *crevasse splays,* or small deltas, into the adjoining swamps or shallow bays.

A third type is called *buoyant flow* [Figure 6–2(C)]. Flow from a deep channel into moderate water depths results in fresh water overriding the denser saline water and in rapid spreading and deceleration of flow. The deposits form a broad crescentic shoaling of coarser-grained sediment succeeded basinward by finer sediment

TABLE 6–1. PRIMARY RIVER-MOUTH PROCESSES

Dominant flow	River-mouth environment	Effluent behavior	Depositional patterns
Turbulent jet flow	High outflow velocities; deep outlet; deep water basinward; small to moderate bed load.	Low spreading angle.	Narrow lunate bar; "Gilbert-type" sequence.
Turbulent frictional flow	Moderate to high velocities; shallow outlet; shallow water basinward; high bed load.	Rapid lateral expansion and deceleration.	Rapid deposition; shoal bar; bifurcating channels; crevasse-splay type.
Buoyant flow	Intermediate outflow velocities; deep outlet; moderate depths basinward; low bed load.	Lateral spreading and thinning; rapid deceleration.	Progressive seaward shoaling; seaward fining; South Pass Mississippi type.
Density flow[a]	High outflow velocities; high suspended load; deep water basinward.	Bottom-flowing density current; rapid spreading.	Little outlet deposition; fan deposits basinward; poorly documented.

SOURCE: Adapted from Wright 1977.

[a]After Bates 1953.

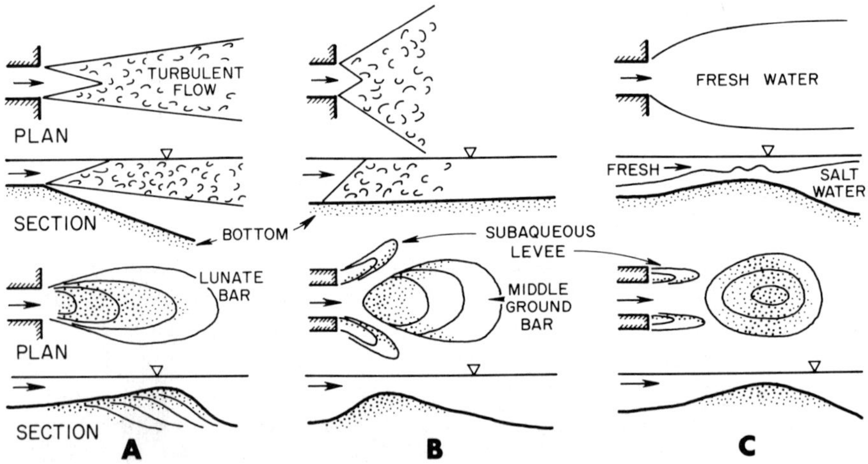

Figure 6–2 Principal types of flow and deposition at river mouths: (A) turbulent jet flow and narrow lunate bar; (B) turbulent frictional flow and broad middle-ground or shoal bar; (C) buoyant flow and broad crescentic bar. [Modified from Wright 1977.]

to produce a proximal river-mouth bar and distal-bar configuration. This type of deposit is that which has formed at the present mouths of the Mississippi River, Southwest Pass, South Pass, and Southeast Pass. In this type of flow, subaqueous levees are well developed, and progradation of bar and levees provides confinement of the deep channel to a straight and relatively narrow channel.

A fourth type of flow is *density flow* or *underflow*. High-velocity discharge with a large suspended load could result in a dense bottom-flowing current, one that would spread rapidly and deposit a thin apron of sediment offshore from the river mouth. The bed load might be left at the outlet, perhaps in a bar similar to the buoyant-flow type, but coarser sediment could also be transported to more distal locations. This type of flow was called *hyperpycnal* by Bates (1953) and was proposed as a common, though poorly documented, process at river mouths. It is still not known whether density flow is a dominant process in any modern delta, but it can be expected wherever high suspended-load streams discharge, probably in combination with the jet- or buoyant-flow types.

The primary river-mouth processes are those that result in the common progradation of coarser current-transported sediment over finer distal sediment to produce an uninterrupted sequence from prodelta below, to distal bar, and then to distributary bar above. This sequence may be modified, however, by marine processes that act on sediment after its deposition from currents. The modifying processes are those of tidal action, wave action, and gravity slumping (Table 6–2). Strong tidal action produces flared sand-filled outlets in which the distributary bar is reworked into linear ridges that are parallel to bidirectional tidal flow. Wave action inhibits outflow, and bar deposition and breaking waves transport sand back to the outlet where it may be distributed laterally by onshore currents, thereby destroying the river-mouth deposits. Gravity acts on deposits of the submarine delta, and aided by pounding of storm waves, the unstable sediment may fail in large slumped masses (Coleman et al. 1974).

TABLE 6–2. MODIFYING RIVER-MOUTH PROCESSES

Process	River-mouth environment	Effluent behavior	Depositional patterns
Tides	High tide range; usually shallow seas.	Intense mixing by tidal currents; bidirectional sediment transport.	Flared outlets sand-filled; tidal ridges parallel to flow.
Waves	Direct exposure to open ocean; steep offshore profile.	Wave action promotes mixing, inhibits outflow.	Crescentic bars; outlet constructed by subaqueous shoals.
Gravity	Moderate to deep basinward; low tide and wave action.	Buoyancy type; lateral spreading and thinning.	Distal deposits slumped; mass gliding.

SOURCE: Modified from Wright 1977.

Examples of delta modification are well known in modern settings (Wright 1977). The modifying process produces a different aspect to the original delta sediment, so much, perhaps, that evidence of the primary mode of deposition may be completely lost. In order to deserve the term *deltaic,* the sediment must preserve, at least in part, the evidence for progradation of fluvial sediment.

MODERN DELTAIC SEDIMENTS

Delta-Front Sequence

The uniform prograding sequence of the delta front is well known in the modern Mississippi. Here the delta front was first described in sufficient detail (Fisk 1961), so that the sequence of sedimentary structures, texture, and composition could be applied readily to other deltas, both modern and ancient.

The modern Mississippi delta is divided into three major units (Figure 6–3). The delta plain ranges from the subaerial to shallow subaqueous area to water depths of 18 ft (5.5 m). Only the narrow levees are emergent, and nearby marsh and bay areas are largely submerged. The delta front is the prograded slope composed of distributary bar, distal bar, and lateral fringing deposits extending to water depths of 240 ft (73 m). The delta front has an average slope of only 0.5° and maximum slope of about 1°, although it is a relatively narrow zone compared to the nearby flat, broad delta plain. The prodelta area is the normal-marine environment of water depths greater than 240 ft (73 m).

A core hole at the Burwood site penetrated nearly 300 ft (91 m) of delta-front section (Figure 6–3). Above the prodelta muds, the section grades upward through a "transition" zone of sandy silt to "clean" sand and then to a thin upper silt of the natural levee. Rapid deposition of this section on delta-fringe and prodelta muds caused buoyant rise of uncompacted clays as "mudlumps," which may penetrate upward to the surface (Morgan, Coleman, and Gagliano 1968).

Texture and composition. Textural gradation in the Burwood test is shown by a plot of median grain size (Figure 6–4). The prodelta clays are succeeded by sandy silt through the section 110–240 ft (34–73 m), which represents the "transition" zone of Fisk (1961) but which is the *distal bar* of modern terminology (Wright 1977). Overlying this unit is very fine to fine-grained sand from 20–110 ft (6–34 m) of the "clean" sand zone in Fisk (1961), which is the distributary-bar sediment that was deposited at the river mouth. The upper levee sediment decreases to coarse silt at the top of the core test. The total delta-front section shows a coarsening-upward sequence from coarse silt to fine sand and a decrease in clay upward to negligible amounts in the distributary bar sands. Composition is represented by modal classes of sand, silt, and clay (Figure 6–4). Based on recent analyses of the sand- and coarse-silt fractions, the quartz content ranges from 75 percent to 93 percent of grain composition with the balance composed of feldspar and rock fragments (Ethridge 1970).

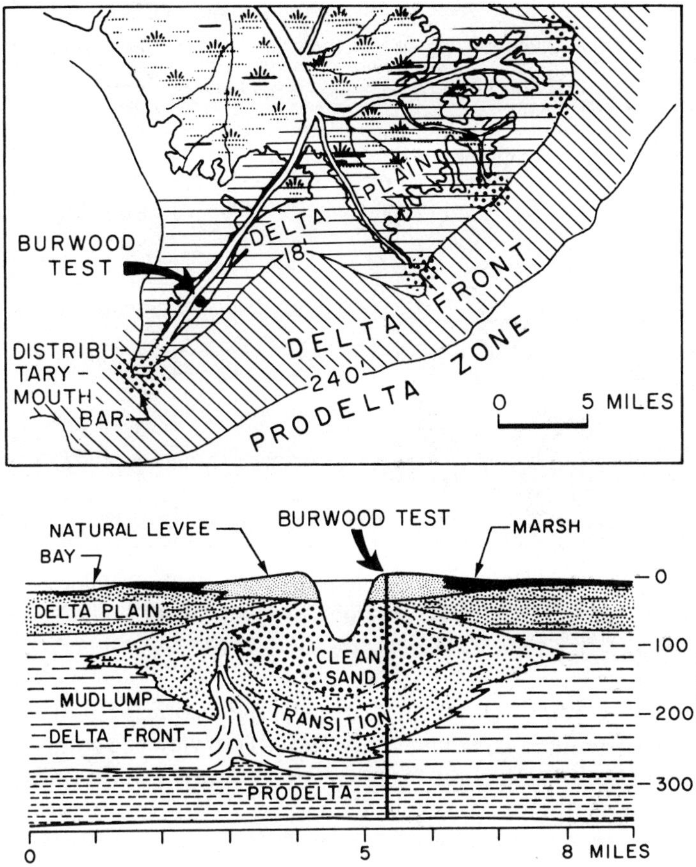

Figure 6–3 Map and cross section of Southwest Pass of the Mississippi River showing the location of the Burwood core test. [From Fisk 1961.]

Increasing upward amounts of finely divided organic material were also noted (Fisk 1961).

Sedimentary structures. Sedimentary structures in the Burwood test also were described and illustrated by Fisk (1961). Bedding is dominated by thin even lamination in the distal bar with increasing numbers of inclined laminae in thin beds upward in the distributary bar. Most of the thin sets represent ripple lamination, and thicker sets with laminae inclined up to 15° are present in the upper distributary bar. Distortion of soft sediment in cores makes the sequence of primary structures somewhat uncertain, but the section appears to be dominated by low flow-regime ripples through much of the sequence and higher flow-regime dune bedding only in the uppermost part (Coleman and Prior 1982).

The Burwood test (Figure 6–4) represents a rather ideal sequence in the center

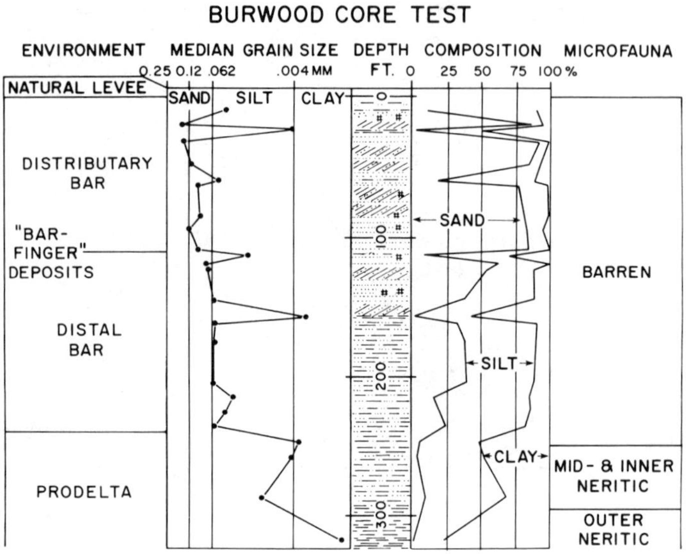

Figure 6–4 Texture and sedimentary structures in the Burwood core test, Southwest Pass, Mississippi River. [Modified from Fisk 1961, Figure 7.] Location of core test in Figure 6–3.

of the prograding delta front. Laterally, the section is characterized by abrupt change to muddy sediments, which led Fisk (1961) to view the composite of successive delta-front deposits as having formed narrow linear bodies of sand, or *bar-finger*, deposits.

Lateral variations. Succeeding studies of the Mississippi have confirmed a complex history of shifting centers of deposition and, consequently, rapid lateral variations in sediment types and vertical succession of deposits (Frazier 1967). Local variations in sequence are the result of progradation by which the delta front is built seaward and thereby provides the base on which the delta plain is formed. The processes of the delta plain may then modify the previous delta-front deposits.

Initial progradation results in the deposition of sand at the river mouth in the form of a distributary bar [Figure 6–5(A)]. Farther from the river mouth, successively finer sediment (silts and clays) is deposited in a broad fan-shaped lens. As the delta builds seaward, the fluvial distributary channels cut deeply into previously deposited delta-front sediment. The interdistributary areas are also filled with fine sediment, especially during periods of flood when the river overflows its banks, constructing natural levees and depositing fine-grained sediment in the adjacent flood plain. The interdistributary areas are commonly well vegetated and are sites of organic accumulation forming thick beds of peat [Figure 6–5(B)]. Thus, the delta-plain sediments build vertically, or *aggrade,* over the previously deposited delta-front section.

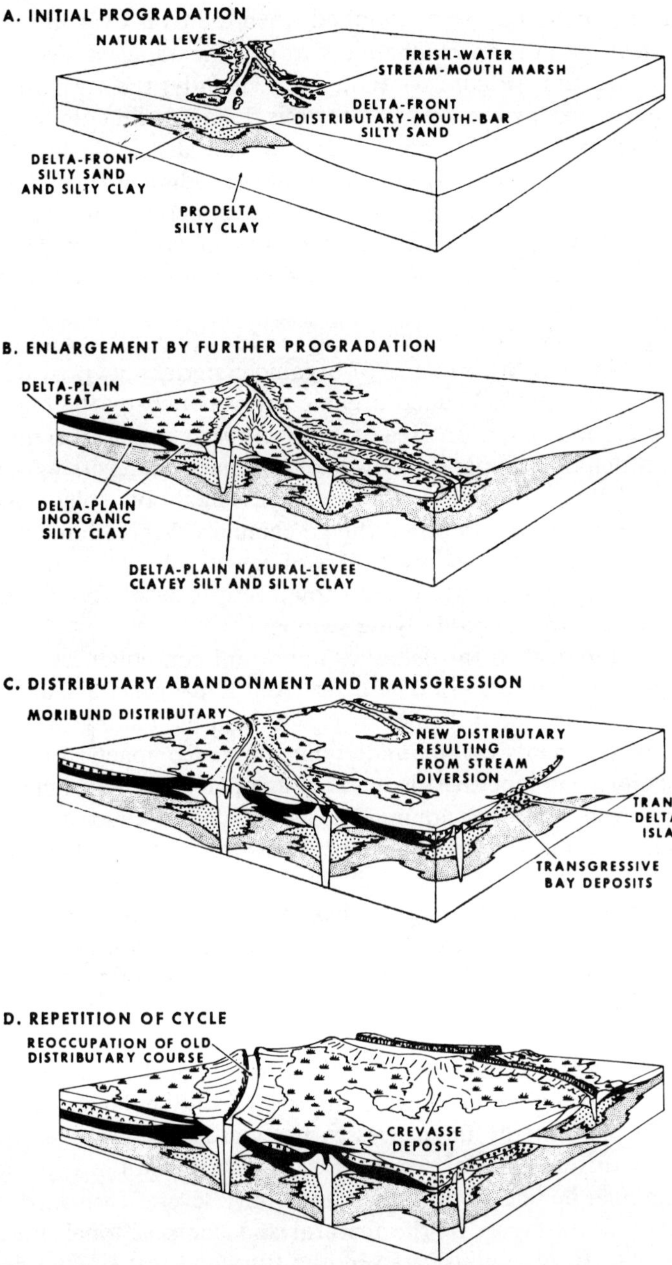

Figure 6–5 Diagrams showing the development of deltaic facies as based on studies of the modern Mississippi delta: (A) initial progradation; (B) enlargement by further progradation; (C) distributary abandonment and transgression; (D) repetition of cycle. [From Frazier 1967.]

Such a delta may be abandoned when the river overflows its banks at an upstream location, cuts a new channel through the adjacent levee, and progrades in a different direction [Figure 6–5(C)]. Subsequently, the river may again change its course and reoccupy the area of previous deltaic sedimentation, cutting new channels and depositing additional amounts of interdistributary sediment with organic accumulations [Figure 6–5(D)]. The resulting sediment then shows a complexly interbedded sequence of delta-front, fluvial channel, and interdistributary marsh deposits. In general, such a sequence grades upward from prodelta marine clays through distal delta sand and silt, and then to distributary bar sand, with flanking levees and marsh deposits.

The delta plain is commonly the site for accumulation of bedded organic materials, and in fact, the presence of organic material may provide the main evidence for deltaic deposition. Bedded organic deposits occupy the highest position in the sequence and may cover an area of relatively wide extent, thereby forming coal beds in ancient deltaic sediments. However, the marsh deposits may not be continuous within the delta-plain section but may be missing in some places along narrow trends that represent the superimposed fluvial channels. Vegetation within the interdistributary marshes accumulates at its place of growth, and therefore the underlying beds of the delta-front sequence often show the effects of roots that extend from the marsh surface into the underlying sediment.

Abandonment of the delta has important consequences for the final sedimentary sequence (Figure 6–5). The delta-front and prodelta sediments are subject to compaction under the influence of gravity and by loading of the rapidly deposited delta-front sediments on the underlying clays. Compaction is followed by marine transgression, and inner neritic muds of an open-bay environment may be deposited directly on the preceding nonmarine deltaic-plain sediments.

During transgression, the uppermost deltaic sediments may be reworked by waves to form clay-free sands. Initially, these sands may be partly emergent and in the form of a narrow "offshore" bar, but further compaction of underlying sediment and resultant transgression may enclose these sands within a blanket of shallow-marine muds. Such transgressive sands, though thin and limited in extent, are not uncommon in the deltaic sequence.

Modification of sequence. Modification of the delta sequence may also take place during active deposition of the delta-front sediment (Coleman and Wright 1975). The main modifying processes are tidal flow and wave action, which impart a distinctly different set of sedimentary structures on the sequence.

The final deltaic section, then, may have several forms. Ideally, these sections all display at least part of the textural and compositional gradation of the prograding delta front as illustrated by the Burwood test (Figure 6–4). But modifications may alter texture and composition as well as structures. If the delta front is subsequently eroded by fluvial flow, the channel-fill sediment may show an abrupt increase in grain size, followed by a change in bed form from dunes to ripples. The composition can be expected to remain essentially uniform because the channel-fill sediment has the same composition as the underlying and previously deposited sediment.

Persistent tidal flow during deposition would impart a dominant rippled bed form on the entire section with little evidence of higher flow-regime. Rippling could be bidirectional, and the sand should be well sorted because the clay and silt would be removed by tidal currents. Some clays remain, however, as thin laminae separating the rippled beds of sand. Persistent wave action on the delta front should impose a horizontally laminated structure on the reworked sediment and should increase the quartz content while completely removing the matrix. The sedimentary structures produced by wave action are most pronounced, whereas changes in texture and composition are less striking, although they do reflect the changes in the processes of sand transport.

A final modification of the delta sequence takes place in the offshore subaqueous environment. The delta-front slope, although only 0.5° to 1°, is unstable, and the sediment is subject to failure and mass movement (Coleman et al. 1974). Instability is caused by high water content aided by methane and carbon dioxide from the decay of organic matter. The basic type of failure is block rotation which involves the movement of sediment masses about 100 ft (30 m) thick along arcuate normal fault planes. Block rotation may be followed by mass gliding of sediment down the slope into deeper water.

Other sequences. Another type of delta has been described in a special setting. In areas of high coastal relief, narrow coastal plains, and deep water basinward, subaerial alluvial fans prograde directly across the shoreline and into the marine environment. The sediment mass has been called a *fan delta* and is characterized by coarse-grained deposits that grade abruptly from fluvial channel to pebbly beach and shallow marine (Wescott and Ethridge 1980). This sequence is limited in areal extent but is common along some coastal areas bordered by highlands.

Delta Morphology

On a regional scale, deltas show a complex of interbedded sediment as a result of shifting sites of delta-front deposition through time. Nowhere is the composite nature of deltas as well known as in the Mississippi (Figure 6–6). Holocene transgression reached a maximum about 5000 yrs BP (before present) in South Louisiana, and remnants of the two oldest deltas are only partly preserved. The next youngest delta is the Teche, which was built outward from the mouth of an ancestral Mississippi that issued near the location of the present Atchafalaya River [Figure 6–6(A)]. Then the Mississippi was diverted southeastward and formed the St. Bernard delta, which extends east from the vicinity of New Orleans [Figure 6–6(B)]. The river was again diverted to near its older site and built the La Fourche delta to the south (Figure 6–6). Finally, the present course of the Mississippi was established and resulted in the southeastward progradation to form the Plaquimines and mod-

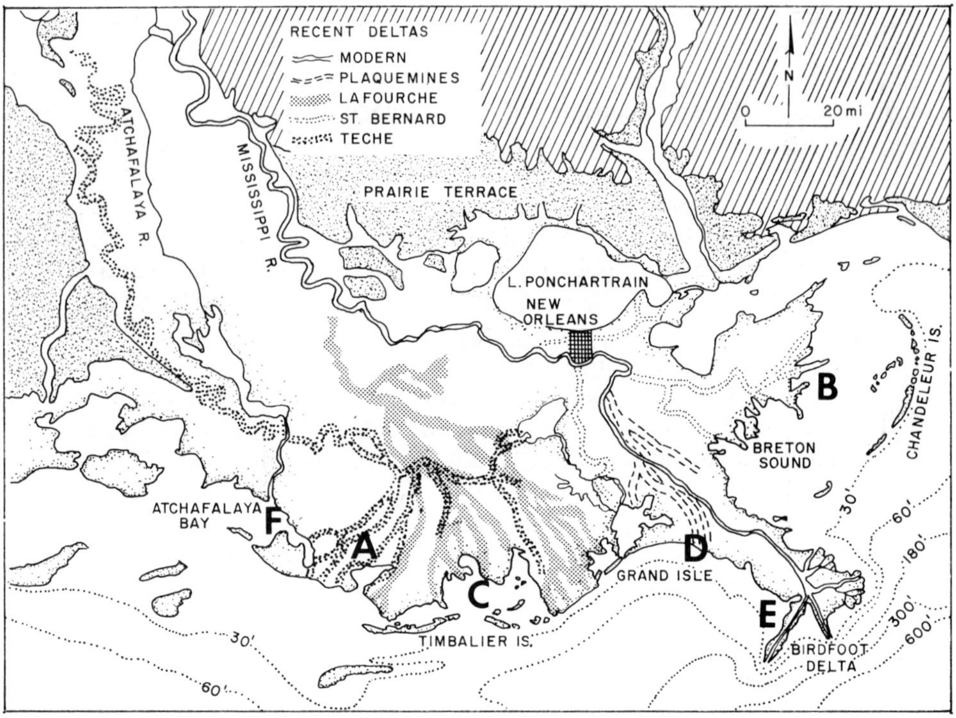

Figure 6–6 Distribution of Holocene deltas of the Mississippi: (A) Teche delta (3800–2700 yrs BP); (B) St. Bernard delta (2800–1700 yrs BP); (C) La Fourche delta (1900–700 yrs BP); (D) Plaquimines delta (1200–500 yrs BP); (E) modern delta (500 to present); (F) Atchafalaya delta (1972 to present). BP = before present. [Modified from Fisk 1955; ages from Kolb and Van Lopik 1966, p. 23.]

ern Balize deltas [Figure 6–6(D) and (E)]. The present passes of the modern or "birdfoot" delta are maintained by dredging, but during annual flooding, about 50 percent of the flow is diverted to the Atchafalaya River at an upstream location. This discharge flows through the Atchafalaya Lake basin and issues at the mouth in Atchafalaya Bay. Since 1972, the Atchafalaya delta has built rapidly Gulfward and now exposes about 5 sq mi (8 km^2) of subaerial deltaic plain (Van Heerden, Wells, and Roberts 1981). This long and complex history of deposition has produced a mass of sediment that has an average thickness of about 200 ft (61 m) and a subaerial extent of 11,000 sq mi (28,600 km^2).

The modern Mississippi delta has prograded rapidly in historic times at the rate of about 3 mi/100 yrs (5 km/100 yrs). Large discharge at times of flood has resulted in diverging flow through four major channels or passes, and the branching distributaries and their deposits have resulted in the splaying or "birdfoot" pattern. Sand deposition in the form of coalescing distributary bars follows the branching trends and coincides with the main channels. Rather than uniform fingerlike exten-

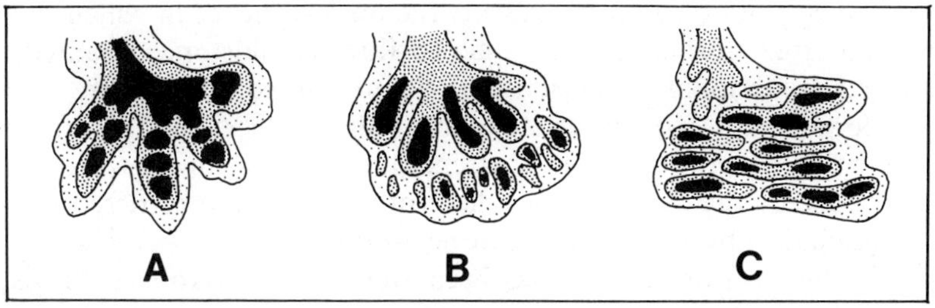

Figure 6–7 Sand distribution in local delta-front sections: (A) distributary-bar pattern of the modern Mississippi; (B) tidal-dominated pattern; (C) wave-dominated pattern. Black areas denote thicker sand. [Adapted from Coleman and Wright 1975.]

sions of sand, the pattern is one of irregular pods that exceed 100 ft (30 m) in thickness and are connected by thinner areas of sand. The intervening sand-poor areas contain large amounts of silt and clay of distal-bar and interdistributary sediments. The resulting distribution of sand has been mapped in detail (Coleman et al. 1974) and is similar to the diagrammatic sand pattern of the distributary-bar type [Figure 6–7(A)].

The distributary-bar pattern of the modern Mississippi delta is characterized by multiple branching trends with a maximum thickness of 80 ft (24 m) [Figure 6–7(A)]. In contrast, the tidal-dominated sand pattern also branches, but this pattern consists of more narrowly linear ridges produced by tidal currents that flowed parallel to the river channels [Figure 6–7(B)]. The wave-dominated pattern is also strongly linear but is oriented parallel to the shore and normal to the direction of river flow [Figure 6–7(C)]. These examples are known by recent core tests in the case of the distributary-bar pattern but are largely inferred in the case of the tidal- and wave-dominated deltas. Nevertheless, they present the variety of sand distributions that can be expected in delta-front environments.

The foregoing description of deltaic deposits is brief and concentrates on the Mississippi because of its well-documented history and sediment sequence (Coleman and Wright 1975). Many other modern deltas have been studied, however, in settings that extend from river-dominated through tidal- and wave-dominated processes (Broussard 1975). These descriptions of sequences are especially important for the interpretation of ancient deltas because they illustrate the variety of processes that work on and modify deltaic sediment.

ANCIENT DELTAS

Deltaic environments are complex, as demonstrated by the variable nature of modern deltaic sequences. Vertical sections of deltaic sedimentary rocks can be expected to display a variety of sandstone types ranging from marine to nonmarine. In fact, vertical heterogeneity can be a key to interpretation.

The deltaic complex presents the only sequence in which there is upward gradation from normal-marine shale to fluvial sandstone. Granted, the section may not be gradational in the sense of uninterrupted deposition. Disconformities may produce abrupt bed contacts and local discordance, but the vertical sequence is the indication of progradation. Current flow is the dominant process in delta deposition. Use of the term *deltaic,* therefore, should be restricted to sections in which progradation by fluvial flow can be established.

The term *delta* often has been applied, or misapplied, to sequences in which the fluvial influence is not observed. For example, deltaic sediments may have been completely reworked by wave action to form a strandplain of beach sandstones, and these deposits have been termed *delta-destructional* in origin, often without direct evidence of deltaic deposition. Such use of the term *delta* is misleading and may suggest a deltaic morphology rather than one produced by the process that last acted on the sediment.

Deltaic sandstones form reservoirs for oil and gas in many parts of the geologic column and in many geologic provinces throughout the world. Indeed, deltaic sandstones have been credited as being the most common and most prolific reservoirs (Rainwater 1975), and undoubtedly sandstones of shoreline origin, including deltas, have been reservoirs of primary importance. Many studies have shown the regional distribution of deltaic deposits, but too few have described the characteristics of local sandstones and their reservoir properties.

The following sections describe deltaic facies in both vertical and lateral sequences of sandstones that are not major oil and gas reservoirs. Nevertheless, these sequences display those characteristics that can be interpreted as of deltaic origin and which may be thought of as representative of deltaic reservoirs.

The first example is that of Lower Cretaceous Fall River Sandstone in the Coyote Creek area, northeast Powder River basin, Wyoming. The section represents a prograding delta front (Figure 6–8) that is a nonreservoir facies in that area but

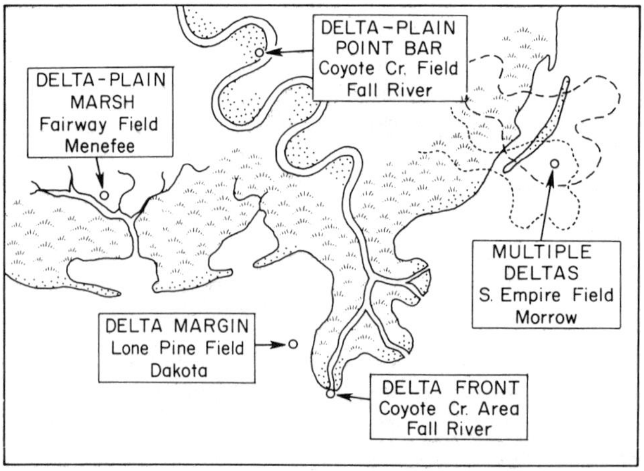

Figure 6–8 Diagram showing relative locations of ancient deltaic sandstones described in the following sections.

is closely associated with fluvial point-bar reservoirs previously described (Chap. 5). The second example is from the Pennsylvanian Morrowan Series in the South Empire field of the deep Delaware basin in southeastern New Mexico. A single cored section displays multiple deltaic facies in vertical sequence. Again, the delta-front sandstones are not the reservoir rocks, but gas production is primarily from the associated fluvial-channel facies. A third example is from the Upper Cretaceous Dakota Sandstone in Lone Pine field of the San Juan basin in northwestern New Mexico. Closely spaced cores provide a picture of rapid lateral changes at the margins of a delta. A fourth example is from the Upper Cretaceous Mesaverde Group in the Fairway field of the San Juan basin, northwestern New Mexico. A core of the Menefee Formation displays a delta-plain section with sandstones of a fluvial system that drained a highly vegetated swamp or marsh environment.

All of these examples show similarities in gross character of the rock sequence, and yet the sections also show significant differences in rock properties, differences that are due to variations that may be anticipated within the deltaic environment. In other words, the examples demonstrate local heterogeneity, and they illustrate that distribution of porosity within the sequence may be difficult to predict.

FALL RIVER SANDSTONE COYOTE CREEK AREA, NORTHEASTERN WYOMING

Sandstones in the Lower Cretaceous Fall River Formation are found in a thin wedge of clastic rocks that extends across the Black Hills uplift from western South Dakota into the Powder River basin of northeastern Wyoming. The sandstones are thin, lenticular, and range from 10 to 50 ft in thickness. Interbedded shales and mudstones grade from dominantly nonmarine on the east to marine in the subsurface on the west. The Fall River sediments were deposited in transitional environments best characterized as deltaic in origin, but the section includes a variety of associated marginal marine and nonmarine environments.

In the subsurface of the northeastern Powder River basin, the Fall River shows two major facies, a more variable *local* facies of sandstones and interbedded shales, and a widespread and more or less consistent *regional* facies. The local facies is that of fluvial point-bar sandstones that form distinctive arcuate reservoir bodies in several small- to moderate-sized oil fields. These reservoirs have been described at Coyote Creek field as sandstones that were formed within large meander loops of a river that cut through previously deposited coastal-plain sediments (Chap. 5).

In the following section, the regional facies is described from a core in a dry hole drilled adjacent to Coyote Creek field. Although this facies does not form petroleum reservoirs in stratigraphic traps, the vertical sequence is typical of delta-front sandstones that may form reservoirs elsewhere on structural closures. In addition, this sequence documents the close association of deltaic sediments with the meander-belt reservoirs of Coyote Creek and similar fields.

Geologic Setting

The Powder River basin of northeastern Wyoming is strongly asymmetric toward the southwest (Figure 5–31). The east flank dips gently southwest at 100 to 200 ft/mi (19 to 38 m/km) across a wide area west of the Black Hills. The basin is bounded on the east by a monocline along which the sedimentary section dips at 10° to 15° and then flattens eastward to form the broad platform of the Black Hills uplift. Across the platform the Lower Cretaceous section is widely exposed in nearly flat outcrops that surround the older exposures of the central Black Hills uplift. It is from this wide exposed area that Fall River sandstones have been described extensively, first because of their stratigraphic importance as part of the Lower Cretaceous section and then because of their economic importance as possible petroleum reservoirs and also as host rocks for uranium (Robinson, Mapel, and Bergendahl 1964; Waage 1959).

Basal Cretaceous formations of the Rocky Mountain region were originally referred to as a basal Lakota conglomerate, an overlying Fuson shale, and an upper Dakota sandstone. At the outcrop in the eastern Black Hills, the Fall River Formation was defined to include the previous Fuson and Dakota units as part of the Dakota Group. Thus, the term *Fall River Formation* is widely used in the Black Hills outcrop, whereas the older name, *Dakota,* is more commonly applied to the upper sandstone by subsurface geologists.

The Fall River Formation is about 150-ft thick in the eastern Black Hills and thins to about 100 ft in the western outcrop and nearby subsurface. Farther westward in the Powder River basin, the Fall River sandstones grade laterally into marine shales, but an equivalent Dakota sandstone is found again on the western margin of the Powder River basin (Figure 6–9). The regional thinning of the Fall River is accompanied by a change from dominantly nonmarine rocks on the east to marginal marine and shallow marine rocks to the west in the subsurface. An eastern source for these sediments is suggested by this lateral change as well as by their content of accessory minerals. The Fall River sandstones are characterized by common staurolite, whereas the Dakota sandstones of central Wyoming are characterized by chlorite (Chisholm 1963; Mackenzie and Ryan 1962).

Fall River sandstones have been described from detailed measured sections along the northwestern perimeter of the Black Hills platform (Dondanville 1963), and the sandstones were characterized as deltaic and marine. The deltaic sections included channel sandstones, "foreset" sandstones, interbedded carbonaceous shales, and blanket sandstones formed by the reworking of abandoned deltas. The marine sections included barrier-island sandstones and lagoonal or marine mudstone and shale. Within this complex of environments, a general progression was discerned in which an earlier delta prograded across the northern part of the outcrop area, and a second delta then was built across the southern part of the area. A third delta lobe extended northwestward, and finally, a fourth delta completely covered the entire outcrop area and reached into the subsurface of the Powder River basin (Dondanville 1963, pp. 96–98). In the subsurface of the Coyote Creek area, the

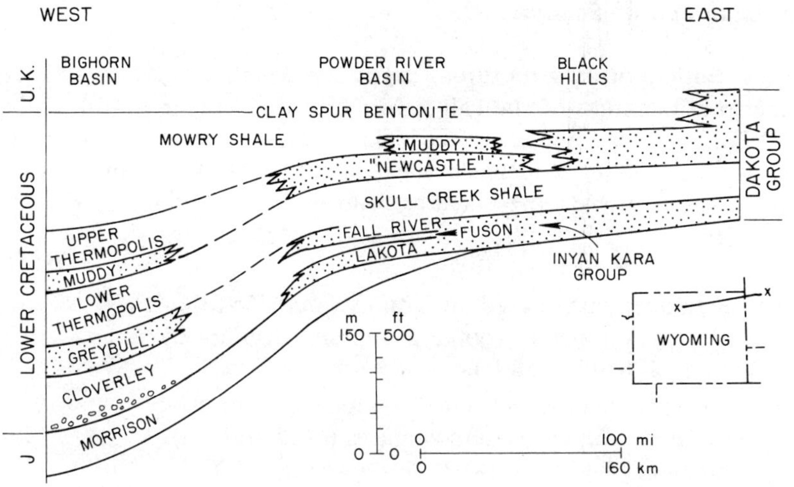

Figure 6–9 Diagrammatic cross section of Dakota Group sandstones from western South Dakota, across the Black Hills and Powder River basin and extending to central Wyoming. [Adapted from Haun and Barlow 1962.]

deltaic facies appears to be dominant and probably represents the fourth and most extensive delta of the outcrop.

Drilling History

Oil was first discovered in the Fall River sandstone at Donkey Creek field in 1956 (Barkley and Gosman 1958). This discovery was the first significant production north of the older Lower Cretaceous Muddy fields, which were located about 50 mi south, and were developed during the period from 1949 to 1951 (Figure 5–31). After Donkey Creek, a dry hole was drilled at the downdip edge of what subsequently became Coyote Creek field. This well, located in Sec. 33–49N–68W, found oil staining in the uppermost Fall River sandstone but recovered only water on drill stem testing (Figure 5–32). Then a second dry hole was drilled, the True Pacific Oil and Gas 1, located in NW NW Sec. 26–49N–68W. This well also found oil staining in the upper Fall River, but again only water was recovered on testing. A third test, located in Sec. 34–49N–68W, was drilled immediately updip from the first dry hole. This well established oil production from the Fall River, and further drilling resulted in Coyote Creek field, a substantial oil accumulation that foretold the possibility of stratigraphic traps over a wide area in the northeast Powder River basin.

The Fall River sandstone was cored in the second dry hole, the True Pacific 1 well, subsequently named the Joseph Watt A–1 location. It is from this core that the character of the regional facies of the Fall River can be designated as deltaic in origin. Its sequence of sedimentary structures is typical of delta-front deposition.

Description of Sandstones

Sedimentary structures. The cored section represents 53 ft (16 m) or nearly the total thickness of the Fall River Sandstone (Figure 6–10). Only the uppermost 10 ft (3 m) were not recovered. This section illustrates the typical character of a prograding deltaic sequence in the regional facies of the Fall River Formation. Three principal units are represented as follows, in descending order:

Unit	*Thickness ft (m)*
1. Sandstone, fine to medium grained, massive to indistinctly laminated, rooted at top and overlain by thin coal (6142–6158 ft).	16 (5)
2. Sandstone, very fine grained, laminated to massive, with numerous truncation surfaces (6158–6177 ft).	19 (5.8)
3. Shale and sandstone, thinly interbedded, sandstones commonly show ripple structure (6177–6195 ft).	18 (5.5)

The lowermost unit (3) is composed dominantly of medium gray shale. Sandstone is found in laminae or in thin beds that range from a few inches to as much as 1 ft (0.3 m) in thickness [Figure 6–11(J), (K), and (L)]. The few thicker sandstone

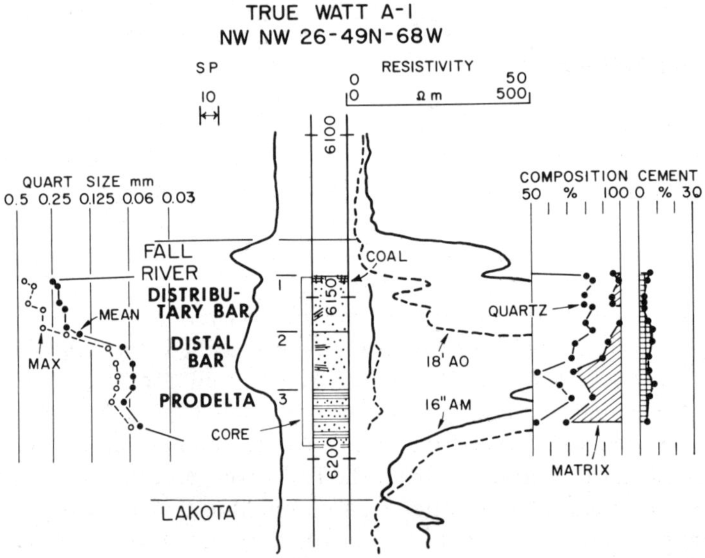

Figure 6–10 Texture, composition, and electric log of Fall River Sandstone in the True Pacific (Watt) A–1 well, Coyote Creek area. Location of well shown in Figure 5-32. Numbers 1, 2, 3 to the left of the depth column refer to units described in text.

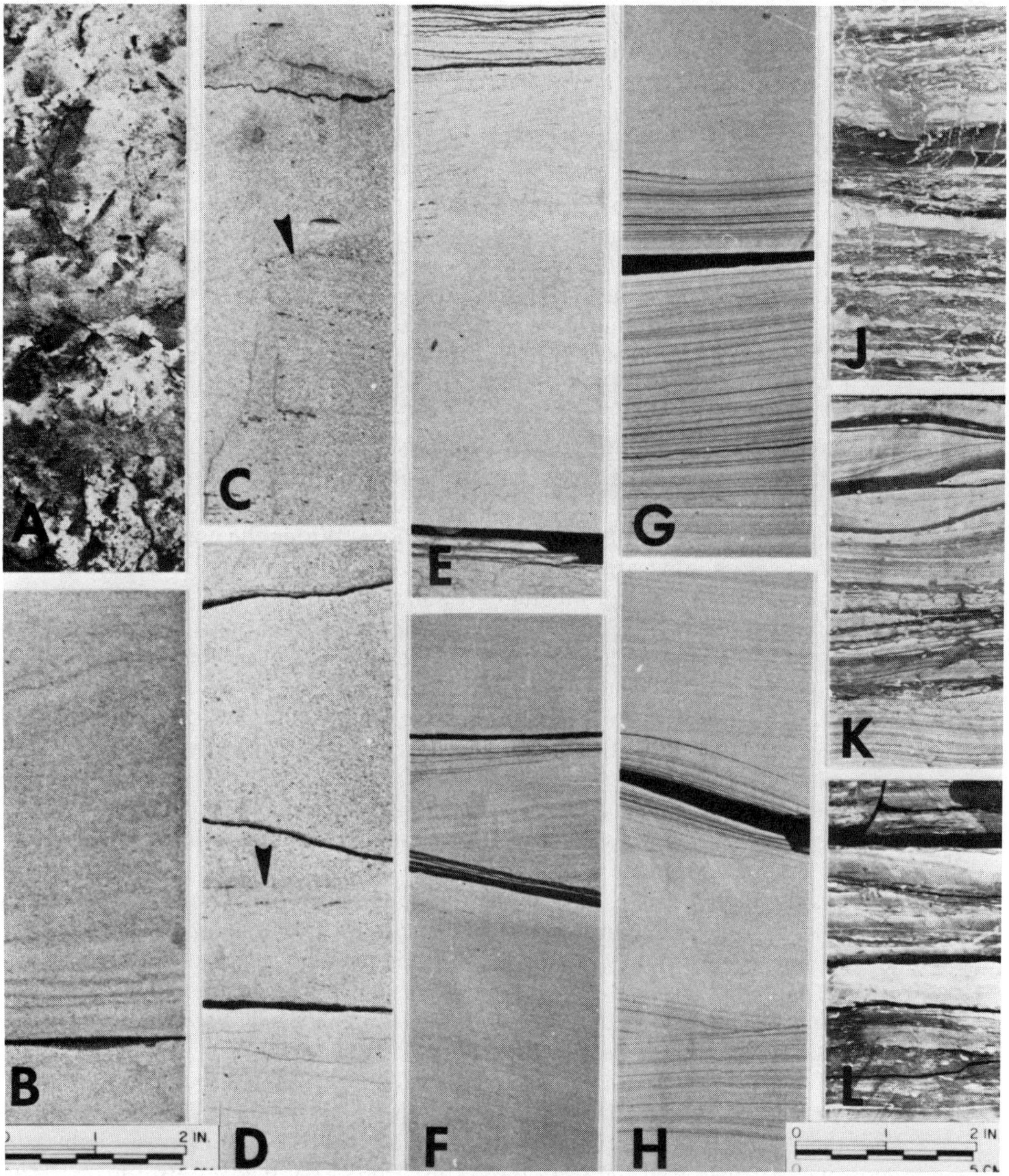

Figure 6–11 Sedimentary structures in Fall River Sandstone, True Pacific (Watt) A–1, Coyote Creek, Wyoming: (A) carbonaceous sandstone with root traces—6143 ft (1872 m); (B) medium-grained sandstone with indistinct laminae—6146 ft (1873 m); (C, D) medium-grained sandstone showing truncation surfaces (arrows)—6152 ft and 6158 ft (1875 m and 1877m); (E) fine-grained sandstone showing thin turbidite bed of the AC type—6161 ft (1878 m); (F)–(H) fine-grained laminated to massive sandstone showing discordant dip surfaces—6169 ft, 6172 ft, and 6174 ft (1880–1882 m); (J) finely bioturbated mudstone—6178 ft (1883 m); (K) interlaminated ripple sandstone and shale—6188 ft (1866 m); (L) silty shale underlying Fall River sandstone—6193 ft (1888 m).

beds are laminated with horizontal, even, parallel laminations that are closely spaced and only a few milllimeters apart. Most of the thinner sandstones are characterized by well-developed ripple structures in lenses that are on the order of 2 to 4 in. (5 to 10 cm) in thickness [Figure 6–11(K)]. The intervening shales show some bioturbation, generally on a fine scale [Figure 6–11(J)]. Thin sand-filled burrow tubes about 1 mm in diameter are oriented more or less vertically and extend through both sandstone and shale laminae. For the most part, bioturbation is not so abundant as to have destroyed the original sedimentary structures, and many thin laminae and lenses of sandstone are preserved.

The next overlying unit (2) is characterized by fine-grained sandstone that is alternately laminated and structureless, forming bed sets that are on the order of ½ to 1 ft (15 to 30 cm) in thickness [Figure 6–11(E)–(H)]. Laminae are even, parallel, and range from closely to widely spaced. Most laminae are horizontal, but many have low dips of 5° to 10°. More rarely, laminae have dips as high as 20°. Despite the average low dip of the laminae, the unit displays many truncation surfaces across which the laminae are slightly divergent or, in some cases, show opposing dips on either side of the truncation surface. Within the unit are a few ripple lenses of sandstone that are as much as 2-in. (5-cm) thick. At several levels there are laminae and thin beds of medium gray shale.

A few thin beds within the unit (2) show a repetition of bedding types from massive or indistinctly laminated sandstone below to an overlying thinly interlaminated sandstone and shale that may show ripple structure of low amplitude [Figure 6–11(E)]. These beds resemble the ordered sequence found in turbidites. Only two such beds were present in the section, and each is approximately ½ ft (15 cm) in thickness.

The uppermost unit (1) has a sharp basal contact on beds below [Figure 6–11(D)], but the surface is essentially horizontal. Other laminae dip at angles up to 20° and commonly truncate underlying laminations [Figure 6–11(C)]. The sandstone is noticeably coarser grained at the base in contrast to underlying sandstones and the unit appears to maintain or slightly increase its grain size upward. The lower half of the unit is distinctly laminated, but laminae dip at low angles of about 5° to 10°. The upper half of the unit is massive to indistinctly laminated, and laminae are commonly widely spaced, on the order of 1 cm apart, and discontinuous [Figure 6–11(B)]. The uppermost part of this unit (2) contains abundant root traces. Irregular thin stringers of carbonaceous material extend vertically through the sandstone [Figure 6–11(A)]. There are also some patches of disseminated carbonaceous material that form distinct dark gray and irregular stains in the otherwise light gray sandstone. At the top of the section is a thin bed of coal that is only 3-in. (7.5-cm) thick. The coal is vitreous on fractured surfaces, but the bed is generally clayey and has a distinct platy fracture.

In summary, the vertical sequence shows increasing scale of sedimentary structures upward from finely rippled sandstone in prodelta mudstone, through gently cross-laminated sandstone, to coarser-grained, cross-laminated sandstone at the top.

Petrography. The sandstones have an average mean grain size of 0.14 mm (fine grained), an average maximum grain size of 0.20 mm, and a standard deviation of only 0.03 mm. Grain size increases upward through the entire cored section from 0.05 mm (coarse silt) in the lowermost unit (3), to 0.25 mm (medium grained) at the top of the section. An abrupt increase in grain size is seen in the uppermost part of the sandstone unit (2) (Figure 6–10). This increase in grain size corresponds to the increasing number of relatively high-angle cross-laminations and truncation surfaces in the lower part of the uppermost unit (1). Within a short interval, grain size increases from 0.07 mm in the upper part of unit 2 to 0.19 mm at the base of unit 3. Mean grain size continues to increase slightly upward within unit 1, and this continued increase indicates further progradation of sediment to the top of the cored section.

The sandstones have an average detrital composition of 74 percent monocrystalline quartz, 7 percent polycrystalline quartz, 1 percent rock fragments, 7 percent other minerals, and 11 percent matrix (Figure 6–10). The content of monocrystalline quartz tends to increase upward in a section until a maximum of 84 percent is reached in the uppermost unit (1). Polycrystalline quartz also increases upward in the section and reaches a maximum of only 6 percent at the top of the unit (2) and thence increases from 10 to 14 percent as a maximum in the uppermost unit (1). The next most abundant component is the clay matrix which also shows a systematic change upward. Thin sandstones in the lowermost unit (3) contain a maximum of 32 percent, and the matrix decreases upward until in the upper part of unit 2, the matrix is less than 10 percent and averages only about 4 percent in the uppermost unit (1). In some samples, the matrix is generally an aggregate of clays, micas, and some organic material. In most cases, however, clay is the dominant component of the matrix. Polycrystalline quartz averages only 7 percent and may be considered rock fragments; other rock fragments are rare. Feldspar content is negligible, and only a few grains of altered plagioclase were noted.

The grain composition is dominated by monocrystalline quartz, especially in the continuous sandstone units 1 and 2. These sandstones can be classified as quartz arenites. Because the rock fragments are dominantly polycrystalline quartz, these sandstones are extreme in their high content of quartz. The dominance of these most stable mineral grains suggests that the original source materials probably were derived from weathering of preexisting sedimentary rocks.

Other detritus includes organic material, zircon, hornblende, and muscovite. The organic material, which is especially abundant in the uppermost sandstone, ranges from opaque to a reddish orange translucent material. Generally, the organic material occurs in groups of bubblelike aggregates between other mineral grains.

Cement is abundant at 20 percent of bulk volume. Nearly all of the cement is silica as grain overgrowths on quartz, and the coarser-grained sandstones, in particular, tend to have a greater abundance of silica overgrowths.

There are only slight differences in composition between the sandstone of the delta-front and equivalent sandstones of the point-bar reservoirs in the Coyote Creek field (Figure 5–36). The delta-front sandstones contain fewer rock fragments and a slightly greater amount of polycrystalline quartz, although these differences are small. The major difference is the overall increase in grain size upward in the delta front with increasing content of quartz and decreasing clay matrix. The texture and composition, therefore, are distinctly different in vertical trends from the Coyote Creek reservoir sandstones.

The textural and compositional changes in the section are not reflected in the well logs (Figure 6–10) because of fresh water in the formation. The dip-trending sandstones of the Fall River are recharged by meteoric water at the outcrop, and fresh water invades the subsurface at least to the Coyote Creek area. The resistivity logs reflect the presence of fresh water, and the spontaneous potential (SP) response is also depressed in the high-resistivity section. Where more saline waters are present, the SP curve shows an increasing upward response that reflects increasing upward grain size (Figure 3–6).

Interpretation of sequence. The total section shows progradation of sandstone with increasing grain size and scale of sedimentary structures upward in the section. The lower unit (3) displays primarily only low flow-regime ripple structure in the interbedded sandstone. The sparseness or fine scale of bioturbation suggests that the section was not normal marine in its environment, but it may have been a restricted bay or lagoonal section. In either case, the lowermost unit (3) represents the prodelta muds.

The overlying unit (2) may represent a distributary-mouth bar composed of sands that were deposited rapidly at the active delta front. The laminated structure and common truncation surfaces suggest a dune phase of sediment transport and deposition. The generally low angle of cross-lamination and the relatively thin bed sets suggest that the bed form was low-amplitude dunes. In other words, the currents that deposited this sand did not reach the uppermost part of lower flow-regime. This interpretation is confirmed by the occasional presence of ripple-bedded sandstone within the generally massive to laminated unit. Deposition was also intermittent as indicated by a few thin laminae of shale.

The uppermost unit (1) represents a continuation of delta-front deposition but of a somewhat higher flow-regime because of increased grain size and a generally higher angle of cross-lamination. Although the unit is coarser grained, the upper sandstones are rather indistinctly laminated. Many laminae dip at relatively low angles, again suggesting that the transporting currents did not attain extremely high velocities.

The distinct differences in texture and bedding upward from unit 2 to unit 1 suggest that the delta-front section may be subdivided into two parts. The lower unit (2) with its finer grain size and generally lower flow-regime structures may represent a distal bar or more distant part of the delta fan. The upper unit (1) with its coarser grain size and somewhat high flow-regime structures may represent the distributary-mouth bar itself.

The shallow subaqueous to subaerial nature of the uppermost section is confirmed by rooting and coal at the top of the cored section. The upper massive sandstones of the underlying unit (2) may owe their homogeneous character to rooting. Although carbonaceous material is lacking in much of the massive sandstone, plant roots of the coal horizon may have extended far below this level and disturbed the sediment of the upper delta front.

The uppermost 10-ft (3-m) sandstone of the Fall River section was not cored, and according to electric logs (Figure 6–10), this bed is a distinct porous unit. The bed appears to be present in other nearby wells, so that it persists for some distance as a distinct unit. Although this sandstone could have a fluvial origin, it may also represent a transgressive-marine deposit produced by reworking of deltaic sands after the local delta system was abandoned. In this case the sandstone would be similar in origin to the barrier-bar sandstones observed at the nearby outcrop (Dondanville 1963).

Of special interest are the thin bed sets found in the distal-bar section; these are composed of massive to indistinctly laminated sandstone overlain by thin rippled sandstone and shale [Figure 6–11(E)]. These beds resemble turbidites of the *AC* type, with a massive *A* division overlain by a rippled *C* division, in the terminology adopted for turbidite sequences (Chap. 9). Similar beds are known from other delta-front sections. The ordered sequence of beds suggests deposition from turbidity currents, the massive division from upper flow-regime currents followed by ripple-form beds from low flow-regime currents. The occurrence of such structures is not unexpected in a distal-bar environment of deposition. Conceivably, turbidity flows could be generated by a sudden influx of sediment-laden water during times of stream flood. A high content of suspended sediment might produce a density contrast of sufficient magnitude to form a bottom-flowing current. Such a flow could carry sediment to the distal parts of the submerged delta fan. Spreading of the unconfined flow would cause rapid deposition of massive sand followed by low flow-regime currents to produce the overlying rippled bed form.

This mode of transport, however, represents a rare event in deposition of distal-bar sediment. Only two ½-ft beds were observed in the total thickness of 19 ft (6 m) for unit 2. Thus, the generation of turbidity flows in the delta-front setting would seem not be a common occurrence.

Porosity and permeability. Although the delta-front section did not produce oil or gas, the upper sandstones have significant porosity and permeability (Figure 6–12). Average porosity for units 1 and 2 is 15.9 percent. Average permeability is 179 md in the distributary-mouth bar sandstones of unit 1, but permeability is reduced to 5.5 md in the distal-bar sandstones of unit 2. The decrease in permeability follows closely the decrease in grain size and the increase in clay matrix from unit 1 to unit 2. No measurements were made for the thin sandstones in unit 3, but their composition and texture suggest values on the order of those in unit 2.

A small amount of residual oil saturation was recorded for the sandstones of unit 1, whereas water saturation was 84 percent. No residual oil saturation was present in the low-permeability sandstones of unit 2.

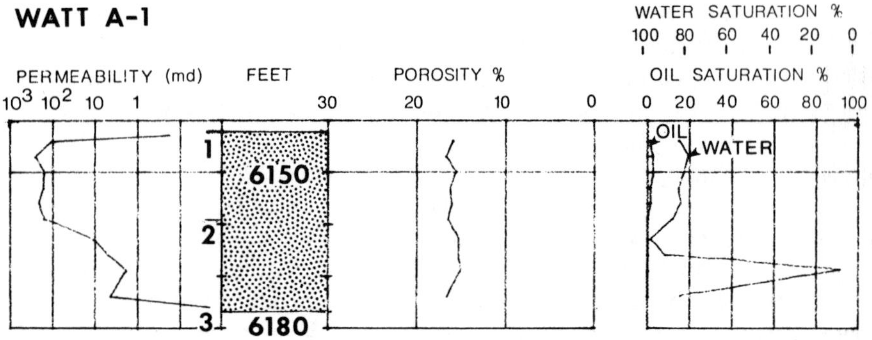

Figure 6–12 Permeabilities, porosities, and fluid saturations in delta-front sandstones of the Fall River Formation, Coyote Creek area. Bed numbers refer to units described in text.

Morphology

There are no published maps that show the local distribution of Fall River sandstones of the deltaic facies. However, correlation of well logs in the area of the cored well suggests that the delta-front sandstones are rather broadly distributed (Figure 6–13).

Correlation of four dry holes extends about 3.3 mi (5.3 km) from northwest to southeast and through the cored section in the True Watt A–1 well. The line of correlation is assumed to be essentially parallel to depositional strike. Maximum thickness of the Fall River is 63 ft (19 m) in the cored well, and the section thins gradually both toward the northwest and the southeast. Thinning appears to take place largely in the delta-front sandstones, which suggests that the original form of the delta-front facies may have been broadly lobate. The uppermost sandstone, of possible transgressive origin, is present in all wells in this strike section, and it shows a slight thickening toward the southeast. However, the transgressive unit is not everywhere present but is missing in wells both updip and downdip from the line of correlation. Therefore, this thin sandstone has an elongate strike trend in the area (Stapp 1967), and its trend and position at the top of the delta-front sequence tend to support its origin by the reworking of the uppermost deltaic section during the succeeding transgression.

Conclusions

The sequence of sedimentary structures in the Fall River sandstone show progradation of distal-bar and distributary-bar sandstones over preceding prodelta muds. The corresponding increase upward in grain size is from coarse silt or finest sand in the distal bar to fine-grained sandstone in the distributary bar. The relatively abrupt change in grain size is similar to that which occurs in the Burwood core test of the modern Mississippi delta (Figure 6–4). The distributary bar is overlain by

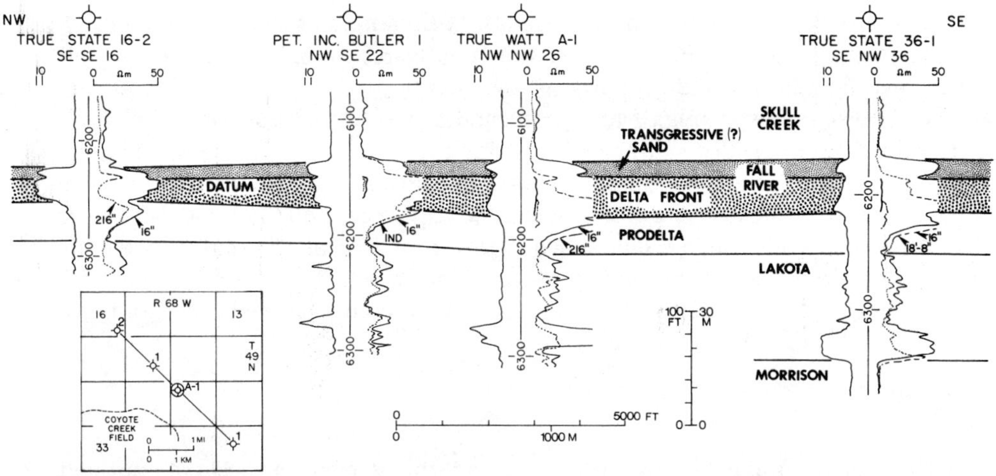

Figure 6–13 Correlation of well logs in the Coyote Creek area that suggests a broad distribution of delta-front sandstones. Cored section was described from the True Pacific (Watt) A-1 well.

clayey coal, which represents the accumulation of organic material in a delta-plain marsh. Therefore, the Fall River sections represent a complete progradation of the delta front from marine below to nonmarine at the top. Abandonment of the delta resulted in marine transgression and reworking of the upper deltaic sediments into strike-trending deposits.

At a late stage in delta progradation, meandering streams of the delta plain eroded deep channels through the delta-front section completely replacing the delta-front sequence in areas such as Coyote Creek and Miller Creek fields. In these producing areas, the reservoir sandstones are point-bar deposits that were deposited within the meandering channels and were later bounded by channel-fill shale. The remaining delta-front section forms a reservoir of limited extent only on low-relief structures.

MORROWAN SANDSTONE, SOUTH EMPIRE FIELD, SOUTHEASTERN NEW MEXICO

The Pennsylvanian section has long been productive of petroleum in the greater Permian basin of western Texas and southeast New Mexico. Pennsylvanian sandstones have produced largely natural gas, whereas carbonate rocks have produced mainly oil. The reservoir sandstones include many different environments of deposition from nonmarine to basinal. Marginal-marine environments are best represented by sandstones of the earlier Pennsylvanian during which there was a gradual marine transgression in a general northward direction across the Permian basin. Regional uplift and erosion followed Pennsylvanian deposition so that the older rocks were truncated beneath the Permian Wolfcampian section. Consequently, the early Pennsylvanian sandstones are not greatly restricted in areal extent and are confined largely to the deep parts of the basins.

Lower Pennsylvanian Morrowan sandstones are natural gas reservoirs in the deep Delaware basin of southeast New Mexico, and many of these marginal-marine sandstones are largely of deltaic origin. An unusual vertical sequence is displayed by a core from the South Empire field. The cored section is 173-ft (53-m) thick and shows parts of at least four deltaic sequences from dominantly shallow-water marine sands, through delta-front sandstones, and finally to fluvial-channel sandstones that were deposited on the delta plain. This section represents successive prograding cycles and lateral facies of the deltaic complex.

Geologic Setting

Early Pennsylvanian Morrowan rocks of southeast New Mexico are confined, for the most part, to the Delaware basin (Figure 6–14), but later Pennsylvanian rocks were deposited over a greater area during continued transgression. The present extent of Pennsylvanian rocks, however, is restricted by pre-Permian uplift and erosion. Total thickness of Pennsylvanian rocks is more than 3000 ft (914 m) in the Delaware basin. The section thins gradually northward across southeast New Mexico, but it is truncated to the west at the Huapache flexure along the eastern margin of the Pedernal uplift. The section also thins abruptly along the faulted west margin of the Central Basin Platform. Early Permian, Wolfcampian sediments are more

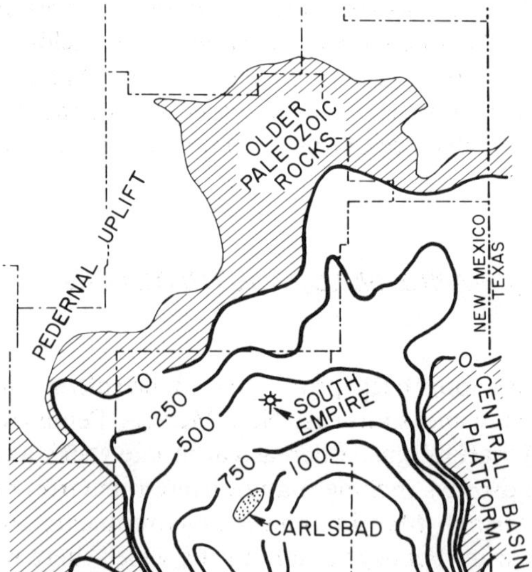

Figure 6–14 Thickness of Morrowan rocks in the Permian basin of southeast New Mexico showing the location of South Empire and Carlsbad fields. Contour interval 250 ft (76 m). [Adapted from Meyer 1966, pp. 32, 35.]

widespread and overlap the truncated margins of the earlier Paleozoic section. Thus, the thickness and depositional patterns in the Pennsylvanian are interrupted by pre-Wolfcampian structures. It is clear that Pennsylvanian sediments were deposited in shoreline and shallow-marine environments. Present limits of earlier Pennsylvanian rocks coincide approximately with depositional limits across a large part of southeast New Mexico.

The Morrowan series has a maximum thickness of about 1300 ft (396 m) in the Delaware basin (Figure 6–14). The Morrowan thins northward at the rate of 25 ft/mi (5 m/km) to a zero edge that trends northeastward across Eddy, southeast Chaves, and Lea counties. According to regional studies (Meyer 1966), Morrowan rocks are dominantly dark gray shale and argillaceous limestone, but significant numbers of fine- to coarse-grained sandstones are present. Sandstone is concentrated within a broad area in the western part of the Delaware basin where net sandstone thickness is greater than 100 ft (30 m). This sandstone area appears to have a southeastward trend, according to a sandstone isolith map (Meyer 1966, p. 35). Near the zero edge of the Morrowan section are some red sandstones that suggest nonmarine conditions at the margins of this depositional system.

During deposition of the Morrowan Series, a major uplift, the Pedernal, was present on the west side of the Delaware basin. Along the central portion of the landmass, Precambrian crystalline rocks were exposed, and extensive areas of Mississippian and older Paleozoic rocks were present on the flanks of the uplift and on its southern extent, called the Otero platform.

Evidence from regional mapping, therefore, suggests that the present edge of the Morrowan Series was near its depositional edge. Weathering and erosion of older rocks in nearby land areas could have supplied an abundance of coarse clastic sediment to the Morrowan shoreline.

Drilling History

The first Pennsylvanian oil in southeast New Mexico was discovered in 1944 on the Central Basin Platform. In succeeding years, exploration resulted in numerous discoveries in Upper Pennsylvanian carbonate rocks, primarily on the northwest shelf. Development of Morrowan gas fields was slow because of the restricted geographic extent of the Morrowan and its location in the deeper parts of the Delaware basin. However, by 1962, Morrowan reservoirs had accounted for cumulative gas production of 48 billion cu ft, mainly in deep stratigraphic traps (Meyer 1966, pp. 84–97). Although the development of early Pennsylvanian gas was delayed, the potential for significant reserves was recognized (Podpochen 1960).

Beginning in 1970 there was renewed interest in exploration and development of deep Morrowan gas. Exploration was stimulated by increases in the price of natural gas in 1970 and 1973 as well as by improved completion techniques. One of the larger field developments of the 1970s was the Carlsbad field (Figure 6–14). Drilling through 1976 (Anderson 1977) resulted in 86 producing wells with a spacing of one well per 320 acres (130 ha). Gas production is largely from Morrowan rocks, but many of the wells have dual completions, mainly in younger zones of the Pennsylvanian.

Development of the South Empire field began in 1970, but the producing area is not extensive. As in other Morrowan fields, the occurrence of porous sandstone is erratic in the section, and reservoirs are unpredictable. The trapping of gas appears to be largely stratigraphic. The interpretation of Morrowan sandstones as deltaic deposits provides a partial explanation for gas accumulation and especially for the nature of the reservoirs.

Description of Sandstones

Sedimentary structures. A cored section in the South Empire field shows interbedded sandstones, siltstones, and shales (Figure 6–15). The sandstones are of two types: (1) fine grained and laminated and (2) coarse grained and steeply cross-laminated. The shales are mostly black and, in part, finely bioturbated. The siltstones or mudstones commonly contain carbonaceous material. In part of the section there are a few thin beds of sandy limestone.

This complexly interbedded section can be subdivided into four sequences. Each begins with black marine shale at the base, grades upward to laminated fine-

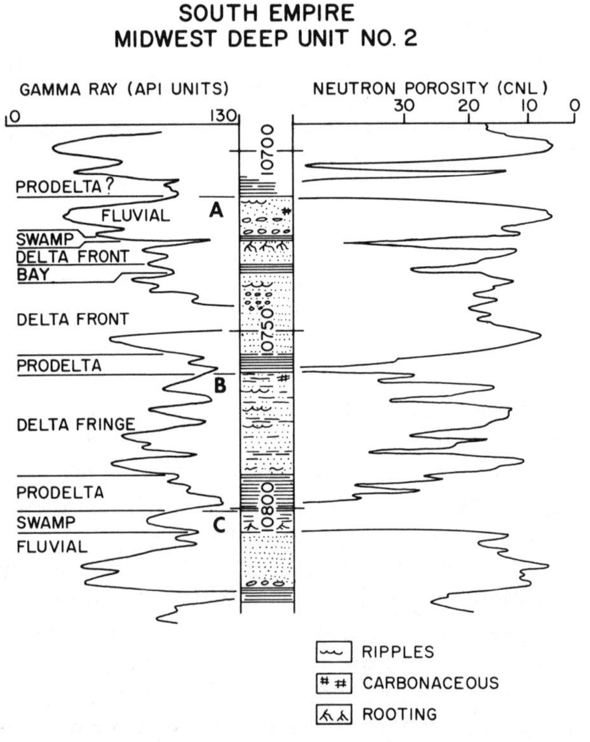

Figure 6–15 Gamma-ray and neutron logs of South Empire Deep Unit 2 well. Letters A, B, C at left of center column refer to bedding divisions described in text.

grained sandstone, then to crossbedded coarse-grained sandstone, and finally to carbonaceous beds at the top. These sequences are designated by letter as follows:

Sequence	*Depth ft (m)*	*Thickness ft (m)*
A	10,715–10,760 (3266–3280)	45 (13.7)
B	10,760–10,800 (3280–3292)	40 (12.2)
C	10,800–10,870 (3292–3313)	70 (21.3)
D	10,870–10,910 (3313–3325)	40 (12.2)

These sequences correspond reasonably well to subdivisions displayed by the gamma-neutron logs (Figure 6–15). Only the top part of sequence (D) was cored, but the upward gradation from dominantly shale at the base to sandstone at the top can be inferred from the gamma-ray log. The overlying sequences C, B, and A show a similar increase in sandstone upward according to log response. However, the gamma log shows unusually high readings within the lower (A) sequence from 10,743–10,750 ft (3274–3276 m), similar to that of black shale at greater depths, but the reason for the anomalous readings is not known. Nevertheless, the major sedimentary sequences based on the succession of rock types observed in the core agree well with the log.

The lowermost (D) sequence contains coarse-grained fluvial sandstone at the top. This sandstone is 10-ft thick, and grain size fines upward. In a hand specimen, conspicuous crystal faces are seen, indicating abundant silica overgrowths on quartz grains. The rock has micro-vuggy porosity that is visible between other areas that are tightly cemented. An underlying thin shale is black and carbonaceous.

Sequence C is composed largely of black shale. In the middle of the unit is a laminated sandstone in beds about 4 to 6-ft (1.2 to 1.8-m) thick. The laminae are even, nearly horizontal, but some dip up to 12°. A few thin beds show ripple structure. Also interbedded with the shale and sandstone are thin beds of sandy limestone about 2-ft (0.6-m) thick. These beds contain fossil fragments, many of which are crinoids and brachipods, and also some oolites. The limestones are largely micritic, however, and are classified as wackestones. An upper coarse-grained sandstone is 12-ft (3.6-m) thick, decreases in grain size upward, and contains some pebbles and coaly fragments. The uppermost section of the sequence is siltstone and shale, which is 11-ft (3.4-m) thick. The rock is dark gray to greenish black, and contains plant fragments, a few coaly laminae, and stringers of carbonaceous material that represent bioturbation by rooting.

The overall aspect of sequence C, therefore, is that of an upward gradation from marine shale, through delta-front laminated sandstones, and then to fluvial-channel sandstones that are overlain by nonmarine delta-plain sediments. The sequence represents a continuous progradational cycle of deposition. The dominance of marine shale with some limestone through the middle of sequence C suggests that this section was probably at the distal margin of the delta lobe. The abrupt appearance of fluvial sandstone at the top indicates that streams of the delta plain cut deeply into the delta-front section at a late stage in the depositional cycle.

The overlying sequences B and A show similar progradational cycles, but the sections contain more sandstone. Details of sedimentary structures are also well illustrated in the upper two units. Unit B begins with a finely bioturbated black shale [Figure 6–16(Q)], which is 9-ft (2.7-m) thick. Overlying the black shale is sandstone 4-ft (1.2-m) thick, in which bedding is highly contorted [Figure 6–16(P)], but the next overlying bed is horizontally bedded, bioturbated black shale [Figure 6–16(O)]. Because adjacent units are not folded, the thin contorted sandstones may represent a block of sediments that slumped from the delta front into a deeper marine environment.

The succeeding sandstone is 15-ft (4.6-m) thick, shaly at the base, and sparsely bioturbated. The major part of this sandstone is characterized by small-scale ripple bedding in which the thickness of ripple lenses is on the order of 2 cm and may reach a maximum of about 4 cm [Figure 6–16(M) and (N)]. The rippled sandstone is succeeded by black shale that is 7-ft (2-m) thick and contains lenses of sand that are highly disturbed by distinct bioturbation [Figure 6–16(L)]. The shallow-marine character of this shale probably indicates an open bay because it is immediately succeeded by nonmarine mudstone. This carbonaceous unit is 3-ft (1-m) thick and contains coal fragments, root traces, and thin ironstone lenses [Figure 6–16(K)]. The next overlying section is normal-marine shale. Therefore, the bedding types of sequence B suggest that sand prograded into a normal-marine environment and was carried by currents of low velocity, probably to the distal margins of a delta fan. The subaqueous delta fringe was succeeded by bay and swamp deposits when the delta lobe was abandoned. Sequence B has considerably more sandstone than the underlying sequence C, but it lacks the fluvial channel sandstone at the top.

Sequence A begins with black shale that is 3-ft (1-m) thick, finely bioturbated, and normal marine in aspect [Figure 6–16(J)]. The shale is succeeded by a major unit of sandstone that is 25-ft (8-m) thick and finely laminated with parallel, even, and continuous laminae that dip at 5° to 10° [Figure 6–16(D)–(H)]. Several truncation surfaces are seen [Figure 6–16(E) and (H)]. Only a minor amount of ripple

---

Figure 6–16 Sedimentary structures in Morrow sandstones, Midwest Deep Unit 2, South Empire field: (A) bioturbated sandstones churned by burrowing—10,716 ft (3266 m); (B) coarse-grained fluvial sandstone; note thin lens of coal in (c)—10,726 ft (3269 m); (C) bioturbated sandstone; churned appearance and carbonaceous stringers represent rooting; note large root trace (arrow)—10,728 ft (3270 m); (D)–(H) delta-front sandstone, mostly well laminated; note truncation surface in (E) and (H) and thin pebbly sandstone in (G)— 10,732 ft, 10,739 ft, 10,740 ft, 10,741 ft, and 10,743 ft (3271–3274 m); (J) black, silty, finely bioturbated marine shale—10,761 ft (3280 m); (K) sandy siltstone with coal fragments and root traces—10,765 ft (3281 m); (L) bioturbated siltstone and shale, probably representing open bay sediments—10,770 ft (3284–3285 m); (M) and (N) ripple-laminated sandstone of the delta-fringe facies—10,775 ft and 10,779 ft (3284–3285 m); (O) bioturbated marine sandstone and shale—10,787 ft (3288 m); (P) contorted sandstone, probably representing slumped sediments—10,789 ft (3288 m); (Q) finely bioturbated marine shale—10,797 ft (3291 m).

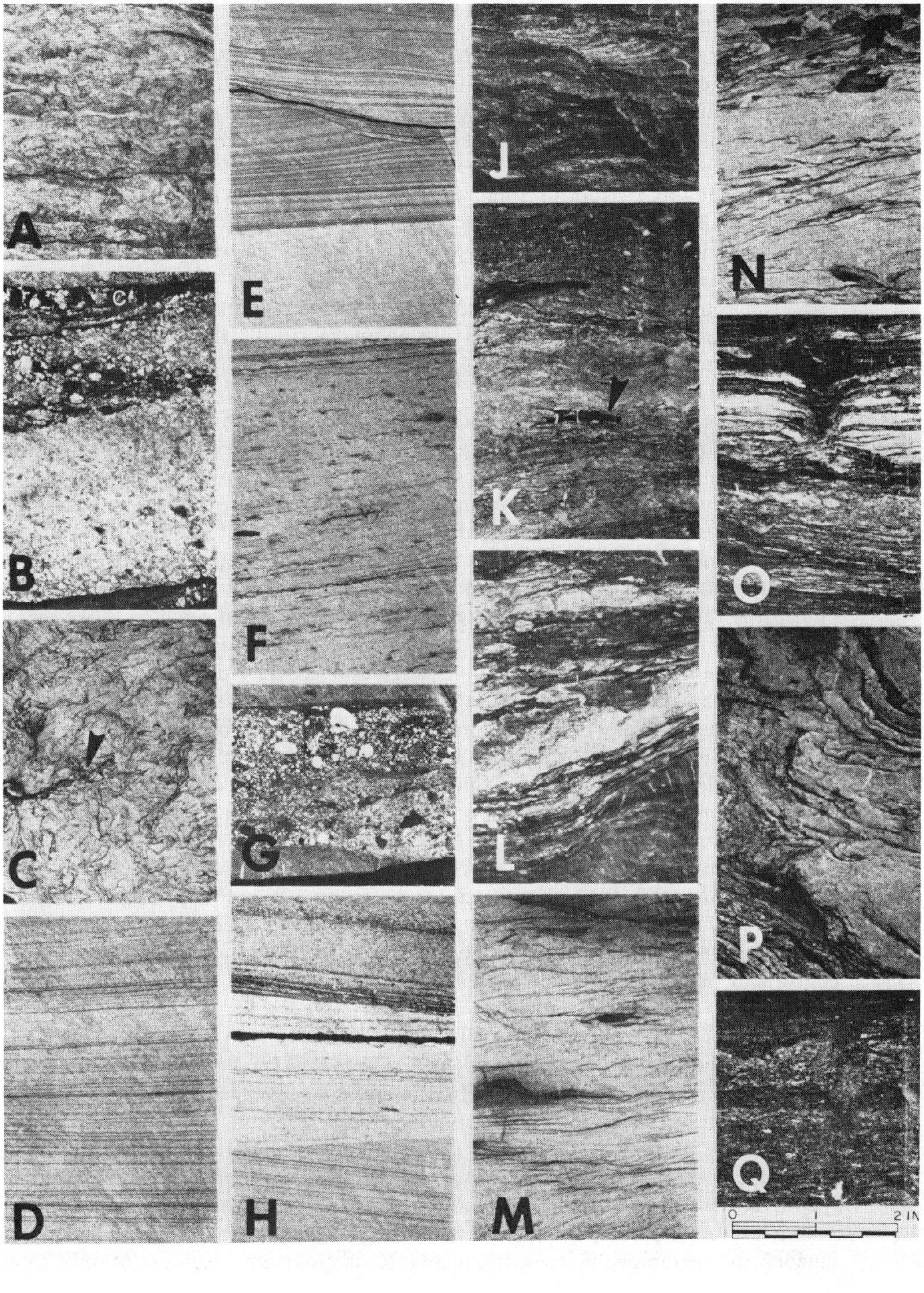
A
B
C
C
D
E
F
G
H
J
K
L
M
N
O
P
Q
0
1
2 IN

bedding is present, and near the base and middle of the unit are a few laminae and thin beds of coarse-grained to pebbly sand [Figure 6–16(G)]. Within the dominant sandstone unit is a black shale 2-ft (0.6-m) thick, which is immediately succeeded by a thinly laminated sandstone 7-ft (2-m) thick [Figure 6–16(D)]. The top of this sandstone contains stringers of carbonaceous material and appears to be highly churned, probably by rooting [Figure 6–16(C)]. Next above is a black shale that is 1-ft (0.3-m) thick and highly carbonaceous. This shale probably represents sediment of a poorly drained swamp that supported the vegetation responsible for the root traces in the underlying sand.

The uppermost sandstone of sequence A is coarse grained and crudely bedded with even and parallel laminae that dip at angles of 10° to 20°. The sandstone contains some small pebbles with up to a 4-mm diameter and thin beds and fragments of black coaly laminated shale [Figure 6–16(B)]. The upper part of this sandstone is irregularly laminated and has a churned appearance that denotes bioturbation [Figure 6–16(A)]. No carbonaceous material is present, and therefore the churned appearance probably was due to burrowing by marine organisms. This bed represents the top of the core, and borehole logs (Figure 6–15) suggest that the next overlying bed is black shale, probably indicating a return to normal-marine conditions and the beginning of yet another cycle of deposition.

The bedding units of sequence A, therefore, appear to represent delta-front sandstones that were subject to currents of low velocity that resulted in well-laminated sand of relatively low dip. In fact, the bedding is most similar to the distal delta-front sandstones of the Fall River Formation [Figure 6–11(F)–(H)]. Laminae of coarse grains and pebbles indicate proximity to an active fluvial source. The delta-front sandstones are overlain immediately by nonmarine shale and by fluvial sandstones that represent channeling into the delta-front sequence by streams of the delta plain. The delta was finally abandoned and subject to compaction, which resulted in transgression and deposition of overlying normal-marine shales.

Petrography. The fluvial sandstones are coarse grained, and grain size tends to decrease upward. Mean grain size ranges from 1.26 mm (very coarse grained) near the base of the fluvial sandstone to 0.16 mm (fine grained) in rippled sandstone at the top of the fluvial section (Figure 6–17). Maximum grain size is generally very coarse and ranges from about 1.6 to over 3 mm.

In contrast, delta-front sandstones are very fine grained, and the average grain size is in the range of only 0.07 to 0.10 mm (Table 6–3). Maximum grain size is fine grained on the average, except for rare pebble zones. The delta-front sandstones do not show a systematic textural variation in vertical sequence.

The sandstones have a relatively high quartz content that averages about 86 percent in the delta-front sandstones (Table 6–3). The quartz is composed of monocrystalline grains only. Rock fragments are the next most abundant components of the fluvial sandstones and average about 11 percent. These rock fragments are dominantly polycrystalline quartz grains with only a minor number of finely crystalline igneous and metamorphic rock fragments. Rock fragments decrease to only 2 per-

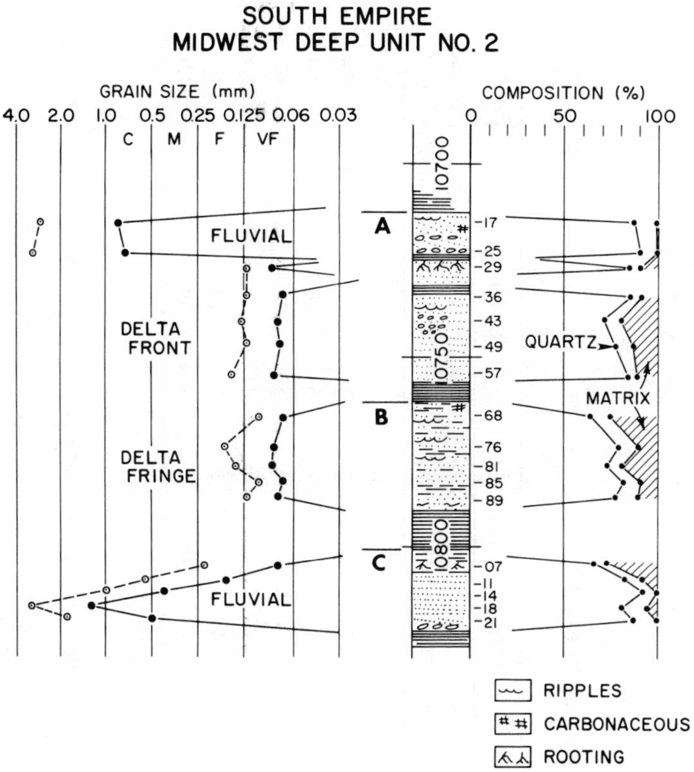

Figure 6–17 Composition, texture, and sedimentary structures of Morrow sandstone section, Midwest Deep Unit 2, South Empire field. Letters to the left of the center column refer to bedding sequences described in text.

cent in the delta-front sandstones. Feldspar is present in only very small amounts. Matrix is most abundant in the delta-front sandstones and is composed largely of silt-size particles and some clay minerals. Other minerals are present in small amounts and include a trace of mica, some glauconite grains, and more rarely, small particles that appear to be oolites composed of siderite. Shell fragments are also found in trace amounts, particularly in the lower half of the section.

Cement minerals form an important part of the bulk volume and average about 20 percent. Silica cement is dominant as overgrowths on quartz grains, and carbonate cement is less common. The lowermost delta fringe section contains the most carbonate. (Table 6–3).

Texture and composition help to confirm the depositional sequence inferred from sedimentary structure. The very fine size of delta-front sandstones seems to imply that sand was transported by currents of low velocity, probably for a considerable distance from the fluvial source. Coarser grains and pebbles are found only in the laminated sandstones of unit A, suggesting that these delta-front sand-

TABLE 6–3. AVERAGE PROPERTIES OF MORROW SANDSTONES, SOUTH EMPIRE FIELD, NEW MEXICO

			Quartz size[a]			Detrital composition[b]					Cement[c]		
Depth (ft)	Environ-ment	Number of samples	Mean (mm)	Max (mm)	σ (mm)	Qz (%)	F (%)	Rx (%)	Oth (%)	Mx (%)	(% of total)	Permea-bility (md)	Porosity (%)
10,717–10,725	Fluvial	2	0.78	2.92	0.17	88	tr	11	tr	1	22	<0.1	3.9
10,729–10,757	Delta front	5	0.07	0.13	0.10	80	2	2	3	13	20	<0.1	5.8
10,768–10,789	Delta fringe	5	0.08	0.13	0.08	74	1	2	7	16	18	<0.1	4.8
10,807	Swamp	1	0.08	0.23	0.06	66	-	tr	6	28	13	<0.1	4.2
10,811–10,821	Fluvial	4	0.58	1.64	0.16	86	-	10	tr	4	18	<0.1	5.4
10,847	Delta fringe	1	0.10	0.14	0.08	88	4	2	5	1	52	-	-
10,870–10,882	Fluvial	3	0.68	2.15	0.14	85	1	13	tr	1	19	308.0	11.9

SOURCE: Petrography by T. L. Burnett 1970.

[a]Long-axis measurements; σ = standard deviation.

[b]Qz = monocrystalline quartz, F = feldspars, Rx = rock fragments including polycrystalline quartz, Mx = matrix, Oth = other minerals, tr = trace (less than 1 percent).

[c]Cement = silica overgrowths and calcite.

stones were closer to the river mouth and were subject to influx of coarser detritus at times of flood.

The composition of both fluvial-channel and delta-front sandstones is similar, and this similarity suggests a common source. Quartz content averages 74 percent or more, whereas feldspar is small in amount, and polycrystalline quartz is the dominant type of rock fragment. This composition seems to indicate that the source terrain was composed largely of weathered sedimentary rocks.

Interpretation of sequence. The Morrowan section displays an apparently mixed sequence of marine to nonmarine rock types. The entire section represents different views of the transitional environments in the deltaic complex. There is an ordering in this sequence, however, that permits reconstruction of the several deltas through time (Figure 6–18).

The sandstones of the delta front may be visualized as comprising individual lobate bodies composed of distal-bar sandstones succeeded by distributary-mouth sandstones. Building of a single lobe was confined to a local area until the distributary was abandoned and a new delta lobe was constructed at an adjacent site. Any vertical sequence, then, may penetrate several lobes at different positions within or lateral to the axial parts of the lobes. Such a composite vertical sequence has been constructed for the Mississippi delta (Coleman and Gagliano 1964), and a similar picture may be established for the Morrowan section.

The lowest (D) sequence was not completely cored. Only the uppermost fluvial

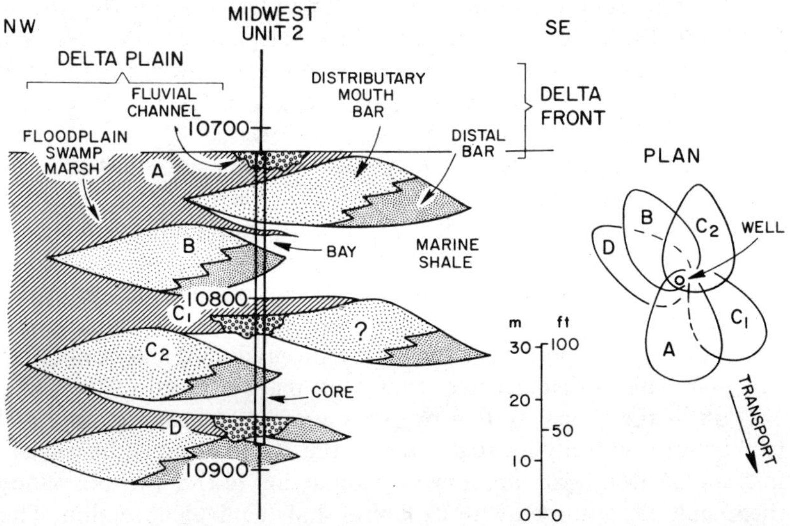

Figure 6–18 Interpretation of lateral relationships in deltaic facies of the Morrowan sandstone, Midwest Deep unit 2, South Empire field, showing hypothetical cross section at left and diagrammatic plan view of delta lobes at right.

channel was recovered. Based on log interpretation, the channel is underlain by a thin sandstone, presumably some part of a delta-front section (Figure 6–18).

The next-higher C sequence is largely marine shale with a thin section of rippled sandstone and some clastic limestone. A fluvial channel rests abruptly on the marine shale and is overlain by swamp deposits. This succession may be explained as parts of two delta-front lobes, the lower one designated C1, a more distal section, and an upper one designated C2, a delta-plain section that followed a delta-front unit.

The overlying B sequence shows a more sandy distal section of the delta front. Marine shale grades upward to sandy mudstone and then to ripple-bedded sandstone. The unit has considerably more sand than the underlying C1, but the dominant ripple structure indicates low flow-regime of bottom currents in a location that was still significantly removed from the active distributary mouth. When this distributary was abandoned, the delta-front sequence was compacted and transgressed by the marine environment. First formed was bioturbated mudstone, probably of an open-bay environment. Then followed a marsh or swamp of the delta plain in which was formed the highly vegetated, carbonaceous, and rooted claystones. Finally, the entire section was again transgressed by marine shale.

The uppermost A unit shows the distributary-mouth bar itself. Numerous low-angle cross-laminae and truncation surfaces indicate the active fluvial system. A few thin zones of coarse-grained to pebbly sand testify to the proximity of the higher-energy fluvial source. The bar was succeeded by a thin bioturbated mudstone that probably represents an interdistributary bay, then by another thin laminated sandstone, which is highly rooted near the top, and again by carbonaceous clay. These deposits represent a transition to the delta plain that overlay the abandoned delta front. Finally, a fluvial channel of the delta plain was cut deeply into the swamp or marsh deposits of the flood plain. Above the cored section, there was again a return to the marine environment, as judged from the response of the borehole logs (Figure 6–15).

This complex system of delta deposits was not stacked in exact vertical sequence, as shown in the hypothetical cross section. Most likely, there were also lateral shifts of the delta-front lobes upward through the sequence, as suggested by the plan view (Figure 6–18). The C units, especially, probably represent lateral sections because of the lack of a regular succession and the abruptness of the fluvial channel at the top of the unit. The overlying B and A units were more directly in line with the transport direction of the prograding delta-front environments.

The lateral displacement that took place upward in the section suggests deflection of the fluvial system from the previous transport route of the C sequence to the more centrally located route of the B and A units. The shift may have been in response to a lesser amount of compaction in the previous sandy delta front of the C unit as compared with its lateral shaly equivalent section. Thus, the B and A units were offset from previous transport routes.

Similarly, this complex sequence might not have been deposited continuously. Abandonment of earlier delta-front systems may have been followed by long periods of delta construction elsewhere before the location was occupied later by distributary systems. Lateral displacement in the despositional systems were also responsible for a large amount of variability in laterally equivalent sections, as illustrated in logs from adjacent wells.

Morphology

The validity of the reconstructed environments (Figure 6–18) cannot be demonstrated at the South Empire field. Well control is not sufficient to map the details of facies distribution because field wells are widely spaced and not numerous. Therefore, the depositional patterns are better observed in an area where subsurface control is more abundant.

The distribution of Morrowan sandstones is well illustrated in the Carlsbad field (Figure 6–14), located about 20 mi (32 km) south of the South Empire field. Subsurface control is relatively abundant and is provided by 86 producing wells and 16 dry holes over an area of more than 50 sq mi (130 km^2). Although no cores were examined from the Carlsbad area, it may be assumed that the Morrowan section had an origin similar to the deltaic sequence at South Empire.

The morphology of Morrowan sandstones is shown by net sandstone thickness within two principal divisions of the section (Andersen 1977, pp. 23, 26). The maps (Figure 6–19) were constructed by measuring thickness of sandstone that showed readings of 50 API units or less on gamma-ray logs. Net thickness of sandstone was mapped for two divisions, a lowermost unit about 250 ft (76 m) in thickness and an overlying unit about 200 ft (61 m) in thickness. According to well logs, these Morrowan subdivisions consist of cyclic units similar to those at the South Empire Deep Unit 2 (Figure 6–18). In each of the Morrowan zones there appear to be several distinct sequences that begin with prominent shaly units at the base, grade upward into interbedded sandstones and shales, and have increasing numbers of sandstones upward. Each of these sequences is 100 ft (30 m) or more in thickness and equivalent to at least two of the sequences designated in the South Empire core.

The resulting maps, therefore, represent sandstone distributions within more than one progradational unit. Nevertheless, the sandstone thicknesses display a preferred orientation similar to that which might be expected for sandstones within a deltaic complex.

In the lower Morrowan unit, net sandstone has an average thickness of about 20 ft (6 m) and reaches a maximum of nearly 60 ft (18 m) (Figure 6–19). The trends of thicker sandstone are in a southeastward direction, essentially parallel to structual dip. These trends may represent concentrations of fluvial-channel sandstones that were eroded into the previously deposited delta-front sandstones of earlier stages of seaward progradation.

Greater thickness of sandstone is shown in the upper Morrowan (Figure 6–19). Sandstones are generally more than 40 ft (12 m) in thickness and reach a maximum of 140 ft (43 m). The southeastward trends are still clearly seen, but concentrations of sandstone have a more lobate pattern. This configuration may reflect

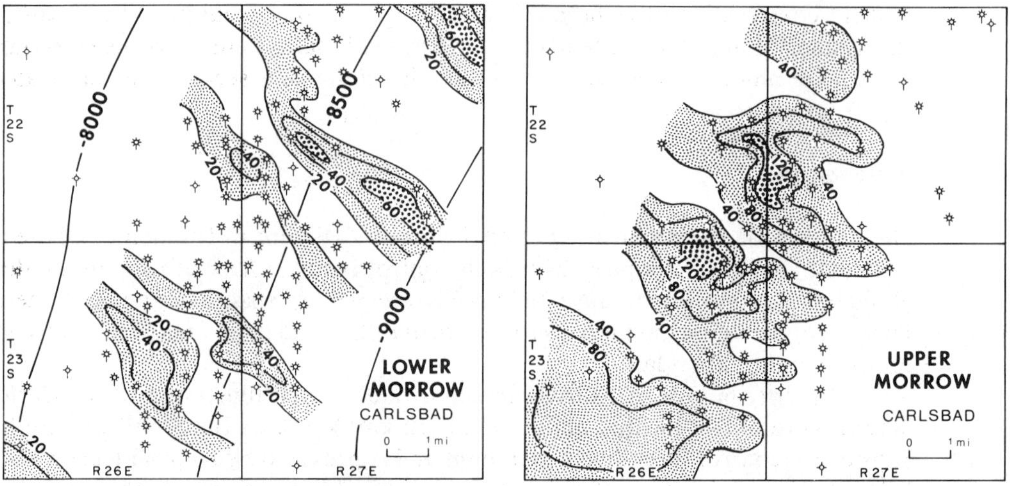

Figure 6–19 Structure and distribution of sandstone in the Morrowan at the Carlsbad field. Lower Morrow map (left) shows generalized structure contours at 500-ft (152-m) intervals and net sandstone thickness at 20-ft (6-m) intervals. Upper Morrow map (right) shows net sandstone at 40-ft (12-m) intervals. [Adapted from Andersen 1977, pp. 22, 23.]

accumulations of delta-front sandstone, but because of the composite nature of the map, trends of later channel sandstones are superimposed on the earlier-deposited delta lobes.

Trends within the lower and upper zones are not coincident. Concentration of channellike bodies in the lower unit appear to be offset from the larger concentrations of sandstone in the upper unit. The reason for this may be differential compaction over fluvial sandstones of the lower unit, which caused later deposition to avoid the areas of previous channels.

Control on gas accumulation is not clearly shown by the distribution of sandstones. Wells are concentrated along strike, presumably controlled in part by the presence of weak structural closure in the area of the discovery well (Andersen 1977, p. 22). Further drilling to date has continued largely along the strike and clearly shows that gas accumulation is independent of structure. Consequently, there is little subsurface control in the direction of structural dip to establish whether sandstones persist in either an updip or downdip direction.

In general, it can be expected that sandstone will be lost in a downdip direction, which is also the direction of depositional dip, because the seaward margins of the ancient deltas may pass rapidly to shale. Similarly, sandstone may be lost in an updip direction, or at least may become thinner and more erratically distributed, because landward from the delta front, sandstone concentrations may be found only as narrow, sinuous fluvial channels.

Despite the compound nature of the sandstone maps, the deltaic patterns of sandstone distribution are well illustrated. It is not known whether gas accumulations at the Carlsbad field are concentrated within the more porous fluvial sandstones or whether delta-front sandstones also contribute substantially to gas production. Judging from the development of porous sandstone within the South Empire core, the fluvial channels probably provide the best reservoirs.

Porosity and Permeability

Stratigraphic variation in the Morrowan sequence determines the variations in reservoir properties. In general, only fluvial channels have sufficient porosity and permeability to form adequate reservoir sandstones. Even these channels, however, are not always highly permeable.

In the South Empire Unit 2 well, the fluvial channel in the D sequence was perforated and flowed gas at the rate of 15.3 million cu ft/day (428,000 m^3/day). This section has an average permeability of 342 md and an average porosity of 12.8 percent (Figure 6–20). Permeabilities range from 25 and to 715 md, and the relatively high permeability of this section appears to be largely the result of diagenesis. Quartz grains have well-developed overgrowths of silica, but silicification of the sandstone is irregular. Areas of more complete cementation are found in patches, between which are smaller regions of little cementation. The irregular distribution of cement gives the sandstone a small-scale vuggy appearance as if cementation was followed by leaching and removal of some cement.

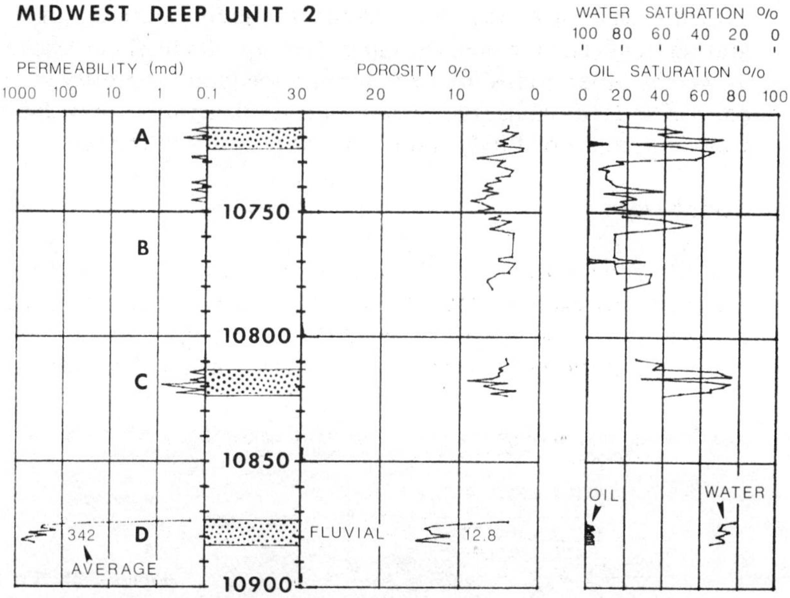

Figure 6–20 Porosity and permeability of Morrowan sandstones, Midwest Deep Unit 2, South Empire field. Letters at left side of depth column refer to bedding sequences described in text.

The upper fluvial sandstones of the A and C sequences have low permeabilities in the range of 0.1 and to 0.8 md and porosities of from 3 percent to 9 percent (Figure 6–20). The quartz grains also have silica overgrowths, but the sandstones are uniformly well cemented and lack the highly vuggy aspect of the lowest fluvial sandstone. Apparently diagenesis of these sandstones did not proceed to the ultimate stage reached by the gas-productive sandstone. The delta-front sandstones have the lowest permeabilities of less than 0.1 md.

Diagenesis in the delta-front sandstones can be explained as a result of flow of connate waters (Fisher et al. 1967). During deposition, the greater head of fresh water in the distributary channels caused displacement of salt water from the delta-front sands, but after burial the flow was reversed (Figure 6–21). Water from mud compaction was forced undip as burial proceeded, and salt water displaced the fresh water in the delta front as well as in the fluvial sands immediately updip. The waters of compaction carried small amounts of dissolved silica, perhaps only 40 ppm, but continued flow through a long period of burial transported enough silica to produce a high degree of cementation. Finally, the delta-front and fluvial systems were highly cemented by silica overgrowths, and permeability was reduced to low values.

At a later stage, the organic material in the marine shales began to convert to natural gas under the influence of higher temperatures at depth. The conversion also produced carbon dioxide, carbonic acid, and organic acids. This transformation probably was accompanied by the development of high fluid pressure as the result of converting solid organic matter to gas, and the high pressures, in turn, caused fracturing of the rock and further updip migration of fluids. When the migrating fluids encountered the sandstones, the acidic fluids leached the silica cement, as well as other grains, and produced a secondary porosity in which the migrating gas was lodged. Thus, the pressure of gas generation provided the mechanism for flow, migration of fluid, and enhanced porosity in the reservoirs.

Conclusions

The Morrowan section at South Empire shows, in vertical section, at least four different prograding deltaic sequences. Although the total aspect of the section is complex in its variety of interbedded rock types, the sequences are consistent with the interpretation of shifting deltaic environments. The morphologies of the sand-

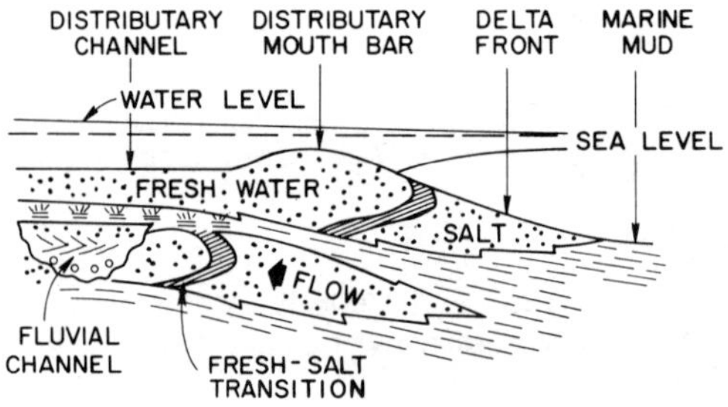

Figure 6–21 Freshwater invasion of distributary-mouth bar and delta-front sands during deposition (above), and replacement by updip flow of salt water during burial and compaction (below).

stones reflect narrow sinuous trends of fluvial channels as well as more broadly lobate patterns of the delta-front section. Natural gas is trapped stratigraphically, but it is not clear whether the morphology of the sandstones provides the primary control on gas accumulation or whether diagenetic processes are the primary factors in the erratic development of permeability. The evidence from the South Empire core seems to favor a complex cementation history as the primary control on gas accumulation. Other littoral deposits, such as barrier-island sandstones, might be present along strike of the Morrowan shorelines.

DAKOTA SANDSTONE, LONE PINE FIELD, NORTHWESTERN NEW MEXICO

The Upper Cretaceous Dakota sandstone at Lone Pine field shows a typical deltaic sequence from basal bioturbated marine mudstones to overlying fluvial current-transported sandstones. However, the delta-front sediments were highly bioturbated and probably were deposited more slowly than the normal delta-front sequence. Therefore, the section has been interpreted as a delta-margin sequence that was deposited at a location lateral to the major influx of fluvial sediment (Berg 1979). The Dakota reservoir is a heterogeneous sandstone that posed many problems during development because thin permeable zones are erratic in occurrence, and distribution of oil and gas does not conform to structural closure. Interpretation of cores helped to explain the nature of the reservoir and the local accumulation of oil and gas.

Geologic Setting

Lone Pine field is located on the south flank of the San Juan basin in northwestern New Mexico (Figure 6–22), a typical intermontane basin of the Rocky Mountain foreland province. The basin is strongly asymmetric toward the northeast, and the gently dipping south flank is uniform and without significant structural deformation. The northern boundaries of the basin, however, are marked by monoclines along which the Upper Cretaceous sedimentary rocks dip at angles of 20° to 30°.

The total sedimentary section of the basin is 15,000 ft (4575 m) in thickness (Peterson et al. 1965). A significant part is composed of Upper Cretaceous shales and sandstones, which comprise a maximum thickness of 6500 ft (1982 m). This section is dominated by deltaic and shoreline deposits that prograded northeastward into a shallow marine shale environment. The sandstones and shales of the nonmarine and transitional environments form the thick wedges of the Gallup and Mesaverde sequences. These coarser clastic wedges thin toward the northeast and interfinger with thick marine shale of the Mancos and Lewis formations.

The central and deeper part of the San Juan Basin has been an area of major gas production from the Mesaverde Group in the greater Blanco field. In the same area, significant reserves of natural gas are also found in sandstones of the Dakota

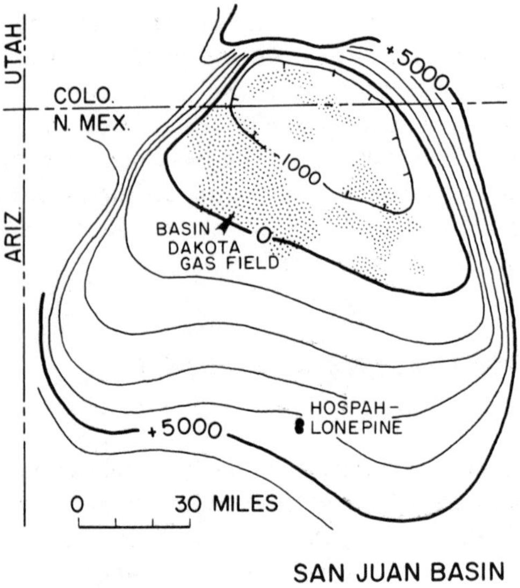

Figure 6–22 Structure map of the San Juan basin, northwestern New Mexico, showing location of Basin Dakota gas field (stippled) and Hospah-Lone Pine area. Structure on base of Greenhorn limestone marker; contour interval 1000 ft (305 m). [Adapted from Peterson et al. 1965, Figure 4, p. 2084. Reprinted from Berg 1979 and published with the permission of the American Association of Petroleum Geologists.]

Group at the base of the Upper Cretaceous section (Deischl 1973). In contrast to the deeper basin, the southern part lacks major accumulations of petroleum. In this area, most of the Upper Cretaceous section is eroded, and only a thin section of lower Mancos Shale, Gallup Sandstone, and underlying Dakota Sandstone remains in the subsurface. A few structural closures of small areal extent provide structural traps for oil and gas. The Hospah-Lone Pine anticline is one of these.

The thickness of the Dakota Formation ranges from 100 to 300 ft (30 to 92 m), generally thickening from west to east. In the outcrop, the vertical sequence of rock types shows an overall transgression from nonmarine below to dominantly marine above (Owen 1973, p. 39). Coarse-grained channel sandstones and carbonaceous beds at the southern margin of the basin (Figure 6–23) form a thinner but continuous basal unit. Overlying interbedded sandstones and marine shales form a transition to the marine shales of the Mancos.

Immediately above the Dakota at the outcrop is a fossiliferous bed of regional extent called the *Gryphaea* subzone (Young 1960), and in the subsurface, calcareous shales of the lowermost Mancos at the same level are called the *Greenhorn Limestone*. The Greenhorn is an excellent marker for structural mapping (Figure 6–22) and serves as a time horizon to which the Dakota sandstones can be referred. As the Dakota thins westward, the lower carbonaceous beds converge on the Greenhorn or *Gryphaea* subzone (Figure 6–23). This relation suggests that the Dakota is younger toward the west and that the upper sandstones represent short-term regressions within the overall transgressive sequence. These sandstones have been termed *littoral deposits* (Young 1960, p. 176) or *coastal sandstones* (Owen 1973).

In the subsurface of the Hospah-Lone Pine area, the interbedded sandstones and shales of the Dakota are found in a section that is about 280 ft (85 m) thick (Figure 6–24). The lower part of this section probably is equivalent to the channel

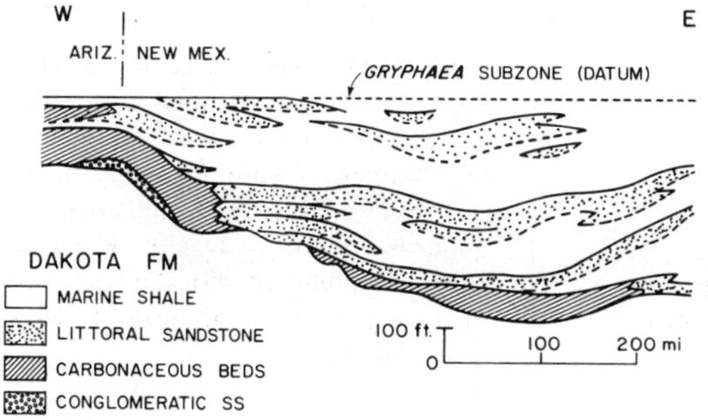

Figure 6–23 Diagrammatic cross section of Dakota Group at the outcrop of the south margin of the San Juan basin, New Mexico, showing interfingering of littoral-margin sandstone with marine shale. [Adapted from Young 1960, p. 174. Reprinted from Berg 1979 and published with the permission of the American Association of Petroleum Geologists.]

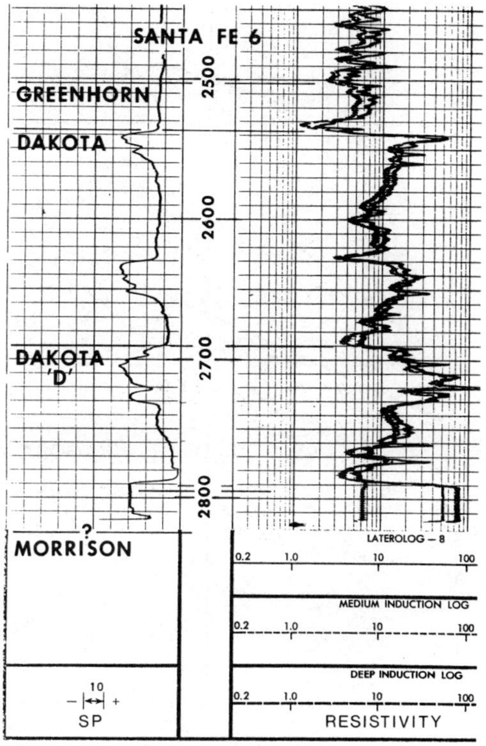

Figure 6–24 Electric log of Dakota Group, Santa Fe 6 well, Lone Pine field, showing stratigraphic position of D sandstone reservoir. Location of well (SF6) shown in Figure 6–25. [From Berg 1979 and published with the permission of the American Association of Petroleum Geologists.]

sandstones of the outcrop, and the upper part is equivalent to the interbedded sandstones and shales of the upper Dakota. The subsurface Dakota is thinner to the west and thickens eastward, generally conforming to the lateral stratigraphic change illustrated at the outcrop (Figure 6–23). The sandstones are variable, however, and net sandstone thickness may range from only a few feet to more than 200 ft (61 m). At the Lone Pine field, the net sandstone thickness is about 70 ft (21 m) in the major part of the Dakota section, and sandstones comprise only about 25 percent of the section.

Drilling History

The Hospah-Lone Pine area (Figure 6–25) has produced oil for many years. Hospah field was discovered in 1924 by a well drilled on the crest of the anticline mapped

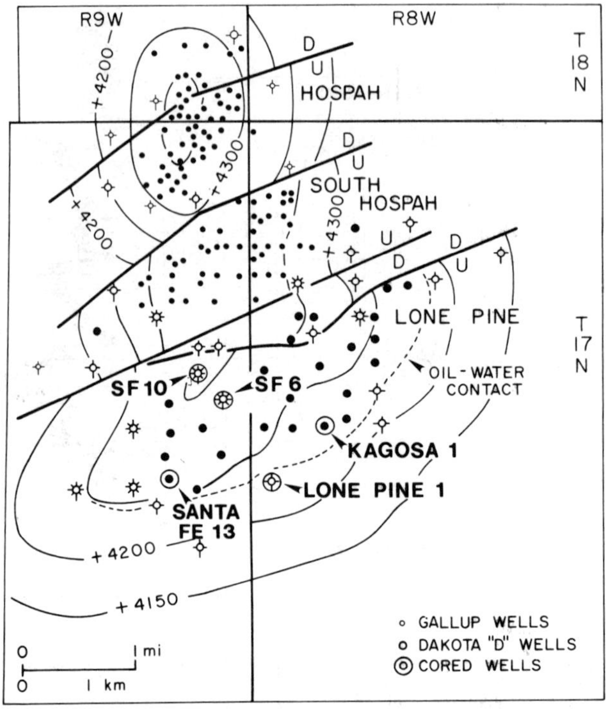

Figure 6–25 Structure on top of Dakota D sandstone, Hospah-Lone Pine area, adapted from a map by Tenneco Oil Company. Contour interval 50 ft (15 m). Cores of Dakota D sandstone were obtained from circled wells. [From Berg 1979 and published with the permission of the American Association of Petroleum Geologists.]

at the surface (King and Wengerd 1957). Oil was produced from the uppermost Gallup sandstones at a depth of 1550 ft (473 m). Although production rates were low, Hospah field had produced a total of 5 million bbls (800,000 m^3) of oil by 1974. The Hospah structure was apparently limited by a transverse fault to the southeast that was believed to have about 200 ft (61 m) of throw. Further drilling in the area perhaps was delayed by the presence of this fault as well as by the loss of porous sandstone toward the southeast and by a few early dry holes south of the field. Not until 1968 was South Hospah field developed in a lower member of the Gallup sandstone. Finally, the Lone Pine Dakota field was discovered in 1970 by further drilling to the south across another fault zone (Figure 6–25).

Oil at Lone Pine was found in the middle of the Dakota Formation about 1000 ft (305 m) below the Gallup reservoirs. The producing horizon is referred to as the Dakota D sandstone (Figure 6–24), and the gross thickness is 40 ft (12 m). Porous sandstones are more numerous and thicker on the east part of the Lone Pine structure, and porosity decreases toward the west and northwest. Therefore, oil production on the Hospah-Lone Pine structure is limited largely by stratigraphic change within the reservoir sandstones from uppermost Gallup sandstone to the north, to lower Gallup sandstone in the middle section, to Dakota D sandstone on the south.

The Lone Pine field has produced only about 1 million bbls (159,000 m^3) of oil. Early development was complicated by the erratic occurrence of porous zones in the relatively thin reservoir. Water was produced with oil in many wells, and it soon became apparent that the oil-water contact did not conform to the structural configuration. As an aid in the evaluation of the Dakota D reservoir, a study of cores was undertaken during the period of development drilling. The major part of the Dakota D section was present in each of the cores from five wells (Figure 6–25), and the cores were examined in sequence from east to west in the wells Kagosa 1, Lone Pine 1, Santa Fe 6, Santa Fe 13, and Santa Fe 10.

Description of Sandstones

Sedimentry structures. The first core from the Kagosa 1 well consists of sandstone in a gradational sequence, but the section can be divided into three units characterized by different sedimentary structures:

Unit	*Thickness ft (m)*
1. Sandstone, fine to medium grained, cross-laminated in part with rare burrows.	8 (2.4)
2. Sandstone, fine grained, moderately bioturbated.	14 (4.3)
3. Sandstone, fine grained, clayey, highly bioturbated.	18 (5.5)
Total	40 (12.2)

The lowermost unit (3) is itself gradational upward from a highly bioturbed mudstone that consists of sand-filled burrow tubes in a matrix of sandy and silty clay. The mudstone has a highly churned appearance but retains a generally horizontal aspect to the bedding. The overlying sandstone of unit 3 retains the highly churned appearance of horizontal bioturbation but contains less clay and is a dominantly sandy section [Figure 6–26(F)]. The next overlying unit (2) shows a decrease in clay content upward [Figure 6–26(C),(D), and (E)]. The section still contains significant bioturbation with a generally horizontal aspect in the lower part of the section [Figure 6–26 (E)] but has an increasing number of vertical burrows in the upper part [Figure 6–26(C)]. Within unit 2, thin zones about ½ ft or less in thickness have less clay, few burrows, and show low-angle cross-laminae that dip at about 10° [Figure 6–26(D)]. These thin zones have sharp basal contacts and contain a few small shale clasts aligned parallel to lamination [Figure 6–26(D) (arrow)].

The uppermost unit (1) is massive to crossbedded, contains very little clay, and only a few vertical burrows [Figure 6–26(A) and (B)]. Crossbedding is not distinct because of the coarser grain size and clay-free nature of the sandstone, but where observed, the laminae dip at angles of 10° to 20° [Figure 6–26 (B)]. Bioturbation is common only in the uppermost 1 ft (0.3 m) of the unit.

The sequence observed in the Kagosa 1 core is gradational upward from bioturbated inner-neritic mudstone, through increasing amounts of sand, less clay, and less bioturbation, to a cleaner crossbedded sandstone at the top. The transition up-

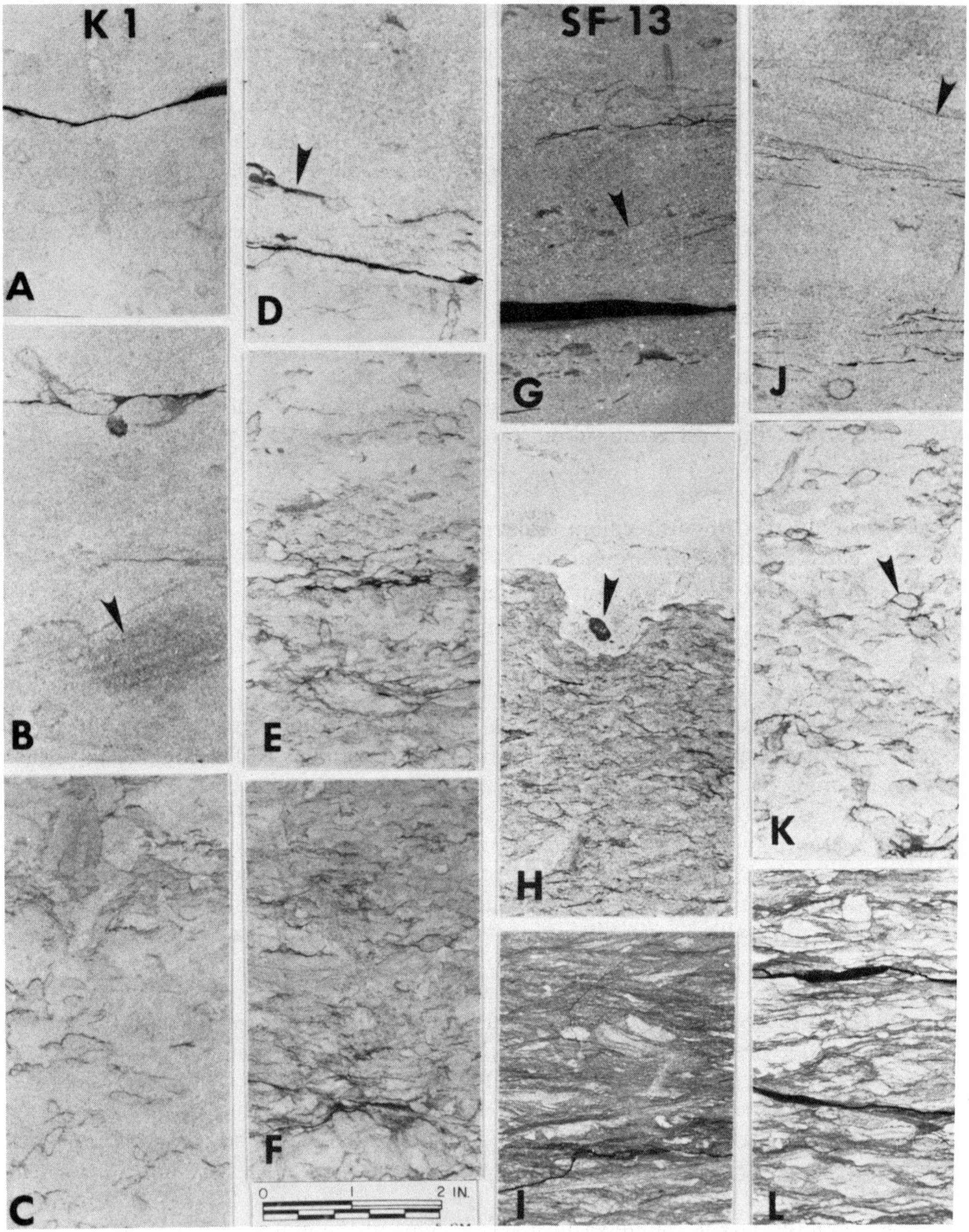

Figure 6–26 Sedimentary structures in Dakota D sandstones, Kagosa 1 (A–F) and Santa Fe 13 (G–L) cores, Lone Pine field: (A) medium-grained sandstone with vertical burrow—2715 ft (828 m); (B) crossbedded sandstone (arrow), partly burrowed—2718 ft (829m); (C) biotur-

ward from marine to current-transported sand can only be interpreted as a prograding deltaic section.

The most unusual aspect of this sequence is the large amount of bioturbation that persists upward through the lower part of the section, and although less abundant in the upper part, bioturbation is a persistent feature. The abundance of bioturbation, as compared with other deltaic sequences, must indicate that deposition was not rapid. Sand supply was intermittent, and sufficient time elapsed between periods of sand influx so as to enable benthic organisms to reoccupy and thoroughly probe the bottom sediment. Increasingly rapid deposition of sand upward, however, probably accounts for the rare bioturbation in the uppermost cross-laminated sandstones.

The second core was from the Lone Pine 1 test, located about one-half mile southwest of the Kagosa 1 well (Figure 6–25). The section was similar in most respects to that in the Kagosa 1 but with a significant difference. The uppermost crossbedded sandstone showed a distinct fining upward in grain size, and steeply dipping laminae of 15° to 20° were more abundant. These characteristics suggest that the unit represents a stream channel eroded into the upper part of the delta-front section.

The next core examined was from the Santa Fe 13 well, and the section contained many of the same rock types but in a mixed sequence, as in the following summary:

Unit	*Thickness ft (m)*
1. Shale, dark gray, silty, massive.	6 (1.8)
2. Sandstone, very fine grained, ripple-bedded.	3 (0.9)
3. Mudstone, sandy, bioturbated to shale, black, carbonaceous, laminated.	3 (0.9)
4. Sandstone, fine to medium grained with pebble zone, crossbedded.	13 (4.0)
5. Mudstone, highly bioturbated.	7 (2.1)
6. Sandstone, fine grained, crossbedded.	6 (1.8)
7. Sandstone, fine grained, bioturbated.	4 (1.2)
8. Mudstone, highly bioturbated.	4 (1.2)
Total	46 (14.0)

bated sandstone—2722 ft (830 m); (D) crossbedded sandstone with small shale clasts (arrow)—2723 ft (830.5 m); (E) bioturbated sandstone—2728 ft (832 m); (F) bioturbated clayey sandstone—2737 ft (835 m); (G) coarse-grained sandstone with shale clasts (arrow)—2791 ft (851 m); (H) coarse-grained sandstone with pebble (arrow) loaded on bioturbated sandstone—2799 ft (854 m); (I) bioturbated mudstone—2805 ft (856 m); (K) bioturbated sandstone with small horizontal burrow tubes (arrow)—2814 ft (858 m); (L) bioturbated mudstone—2817 ft (859 m).

The lower units 6, 7, and 8 form a gradational sequence that becomes increasingly sandy and less bioturbated upward [Figure 6–26 (J), (K), and (L)]. This sequence is similar to that observed in the entire Kagosa 1 core except that the lower sandstone section is only 10-ft (3-m) thick. This lower section represents a small prograding delta, probably deposited as a crevasse splay that originated by levee breakthrough from an adjacent fluvial channel.

The succeeding unit 5 is a bioturbated mudstone, more shaly below and increasingly sandy above [Figure 6–26 (H) and (I)]. The base of the mudstone is in sharp contact with the underlying crossbedded sandstone, and although the section becomes more sandy upward, the unit remains clayey and thoroughly bioturbated.

The contact with the overlying unit (4) is abrupt [Figure 6–26 (H)]. The base of the coarser-grained sandstone contains small quartzose pebbles, and the sand intrudes into the underlying bioturbated sandstone by loading, especially adjacent to the pebbles [Figure 6–26 (H), (arrow)]. The coarser sandstones of unit 4 also contain distinct laminae that dip from 10° to 20°, but these crossbedded units are in sets that rarely exceed 1 ft (0.3 m) in thickness. The sandstones also contain scattered granules and small shale clasts [Figure 6–26 (G) (arrow)]. These sandstones appear to represent deposits of fluvial channels that were cut deeply into the previous delta-front secton. Deposition was rapid and took place before compaction of the underlying section, as demonstrated by loading of the sandstone into the highly bioturbated mudstone below [Figure 6–26 (H)].

Fine-grained deposits succeed the fluvial sandstones. The next overlying unit (3) is bioturbated mudstone that grades upward into black carbonaceous shale. Apparently this section was deposited as a result of compaction and submergence of the underlying deltaic sequence. Transgression first formed a shallow neritic environment that was soon succeeded by a more stagnant reducing environment of black mud. The overlying unit (2) is well-sorted coarse siltstone that shows well-developed ripple structure. This thin siltstone is partly folded and contains thin dikes of intruded shale as if the silt had been deposited rapidly on underlying muds and deformed by sinking into the soft sediment. The presence of ripple bedding indicates low flow-regime and suggests that the siltstone was probably an overbank deposit, or levee, from an adjacent stream.

The uppermost unit (1) is a dark gray silty shale that is massive and lacks bioturbation. This section marks the top of the Dakota D section and is of unknown origin, but it may also represent an overbank deposit, probably a shallow-water area of the flood plain.

The next core was obtained from the Santa Fe 6 well near the top of the structure (Figure 6–25, SF 6). The cored section is similar to that of the Santa Fe 13 and includes a lower thin prograding delta sequence followed by a second upward gradation from bioturbated mudstone to sandstone. However, the overlying crossbedded sandstone of channel-fill origin is only 5-ft (1.5-m) thick and is succeeded by bioturbated sandstone that becomes increasingly clayey upward. This upper section, when compared with equivalent units in the Santa Fe 13 (units 4 and 5), contains a lesser thickness of current-transported sands. The loss of sand suggests that this section was located laterally to the main route for sediment transport across the area.

The final core was obtained from the Santa Fe 10, the highest well on the structure (Figure 6–25, SF 10). The cored section is generally similar to those of the Santa Fe 13 and 6 wells, but the upper fluvial-channel deposits are missing.

The lower section becomes increasingly sandy and less bioturbated upward and is identical to the lower part of the Santa Fe 13 core and, again, represents a thin crevasse splay. The succeeding units represent another gradational sequence that lacks the evidence for current transport. Bioturbated mudstone rests in sharp contact with the underlying crossbedded sandstone, and the amount of sand increases upward in the section, but the unit is intensely bioturbated on a very fine scale.

A succeeding sandy section is also bioturbated, but the scale of burrowing increases. The sandstone contains long vertical sand-filled burrows that are 2 cm or more in diameter, and one large burrow was traced through a thickness of 1.5 ft (0.5 m) through the core. These large burrows were probably made by burrowing crustacea, perhaps crabs, that penetrated deeply into the soft sediment. Most of the horizontal burrows also are of large diameter. This facies completely replaces the equivalent crossbedded and more finely burrowed sandstones in the nearby Santa Fe 6.

The uppermost units in Santa Fe 10 are similar to the equivalent units (2 and 3) in Santa Fe 13. Bioturbated mudstone is succeeded by massive black shale and overlain by contorted and rippled siltstone. These upper units again indicate a succession from open bay, to restricted bay, and then to levee.

In summary, the vertical sequence shown in the Kagosa 1 core can only be interpreted as a prograding delta-front section. The lowermost bioturbated mudstones represent inner-neritic deposits, and these are gradational upward through bioturbated sandstones to the uppermost crossbedded sandstones that indicate sediment transport in fluvial channels. The additional four cores show rapid lateral variation within the deltaic complex. Thinner prograding delta-front sections partly incised by fluvial channels are replaced by distal delta-margin sediments of bioturbated sandstones that retain no evidence of the nearby fluvial system.

Petrography. The sandstones have an overall mean grain size of 0.18 mm (fine grained), and grain size varies with bedding (Table 6–4). Crossbedded sandstones have the largest mean grain size of 0.27 mm (medium grained). Mean grain size decreases to 0.17 mm (fine grained) in burrowed sandstones and to 0.12 mm (very fine grained) in more highly burrowed sandstones. The rippled sandstones have the finest grain size of 0.08 mm (very fine grained).

Textural variation with bedding is clearly shown in vertical sequence (Figures 6–27 and 6–28). Mean size increases upward through the succession from highly burrowed to less burrowed and to crossbedded sandstones as, for example, in the Kagosa 1 core (Figure 6–27). However, within the thicker crossbedded sandstones, there is a decrease upward in mean size as, for example, in the upper crossbedded

TABLE 6–4. PETROGRAPHIC SUMMARY FOR DAKOTA D SANDSTONES, LONE PINE FIELD

			Quartz size[a]		Detrital composition[b]				
Facies	Bedding type	Number of samples	Mean (mm)	Max (mm)	Qz (%)	F (%)	Rx (%)	Mx (%)	Cement[c] (% of total)
Levee	Rippled	2	0.08	0.12	70	10	7	13	16
Fluvial	Crossbedded	10	0.27	0.54	69	11	15	5	11
Delta front	Bioturbated	14	0.17	0.34	71	10	13	6	8
Outer delta front	Highly bioturbated	7	0.12	0.26	65	8	13	14	10
Means		33	0.18	0.37	69	10	13	8	10

[a]Long-axis measurements of monocrystalline quartz.

[b]Qz = monocrystalline quartz, F = feldspars, Rx = rock fragments, Mx = matrix.

[c]Cement is calcite.

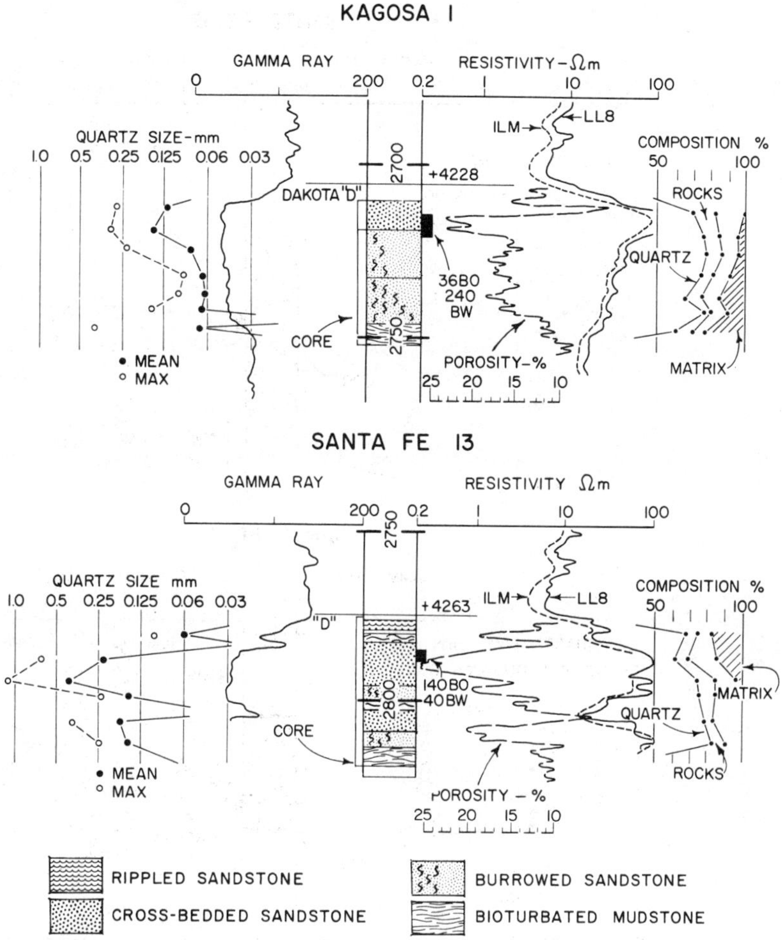

Figure 6–27 Texture, composition, and borehole logs of Dakota D sandstone in Kagosa 1 (upper) and Santa Fe 13 (lower) wells, Lone Pine field. Location of wells is shown in Figure 6–25. [From Berg 1979 and published with the permission of the American Association of Petroleum Geologists.]

sandstones in Kagosa 1 and Santa Fe 13 (Figure 6–27). This change supports the conclusion that the thicker and coarser-grained sandstones are fluvial-channel deposits. Thinner crossbedded sandstones, as in the lower part of Santa Fe 13 (Figure 6–27) and Santa Fe 6 and 10 (Figure 6–28), are gradational upward from burrowed sandstones, and mean grain size shows no reversal in trend.

Average detrital composition is 69 percent monocrystalline quartz, 10 percent feldspar, 13 percent rock fragments, and 8 percent matrix (Table 6–4). Crossbedded sandstones and less burrowed sandstones tend to have higher quartz and lower matrix content.

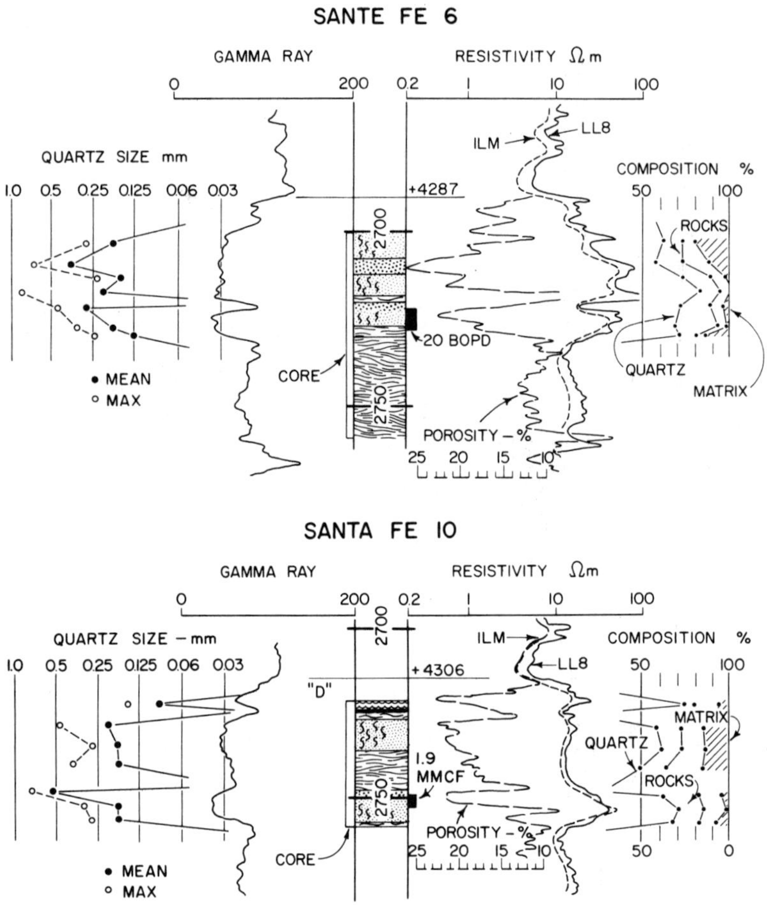

Figure 6–28 Texture, composition, and borehole logs of Dakota D sandstone in Santa Fe 6 (upper) and Santa Fe 10 (lower) wells, Lone Pine field. Symbols indicating bedding types are explained in Figure 6–27. Location of wells is shown in Figure 6–25. [From Berg 1979 and published with the permission of the American Association of Petroleum Geologists.]

The sandstones have a relatively high content of stable grains, which indicates that the source sediment was probably derived from the weathering of older sedimentary rocks. In addition to monocrystalline quartz, the rock fragments are composed largely of polycrystalline quartz and chert. In a few samples of crossbedded sandstone, rock fragments include weathered ooids, indicating a sedimentary source. Feldspars are mainly orthoclase and microcline.

Calcite cement is present in the average amount of 10 percent of bulk volume (Table 6–4). The occurrence of calcite is patchy, but in a few samples it forms a pervasive groundmass. For example, the uppermost crossbedded sandstone in Kagosa 1 has 35 percent calcite cement, which accounts for a thin zone of reduced porosity near the top of the section (Figure 6–27). Only a trace of silica cement is present as overgrowths on quartz grains.

The petrography is consistent with a deltaic origin for the sandstones. Grain size increases upward with change in bedding type, indicating the prograding nature of the delta-front sequence. The content of quartz, feldspar, and rock fragments is similar to that expected for the primary fluvial source of sediment.

Interpretation of sequence. The variable successions observed in the five cores may be viewed as lateral equivalents to give an integrated view of facies change within the deltaic sequence (Figure 6–29). The major rock types may be interpreted as follows. Bioturbated mudstones are the normal shallow neritic deposits of the prodelta facies at the base of the section, and the same rock type also represents open-bay deposits that are found on top of the deltaic sequence. Burrowed sandstones represent a greater influx of sand along the delta front. Crossbedded sandstones are of two types. The first is gradational upward from delta-front sandstone and probably represents deposits of the distributary-mouth bar. The second type, more common in the upper part of the section, is in sharp contact with the delta-

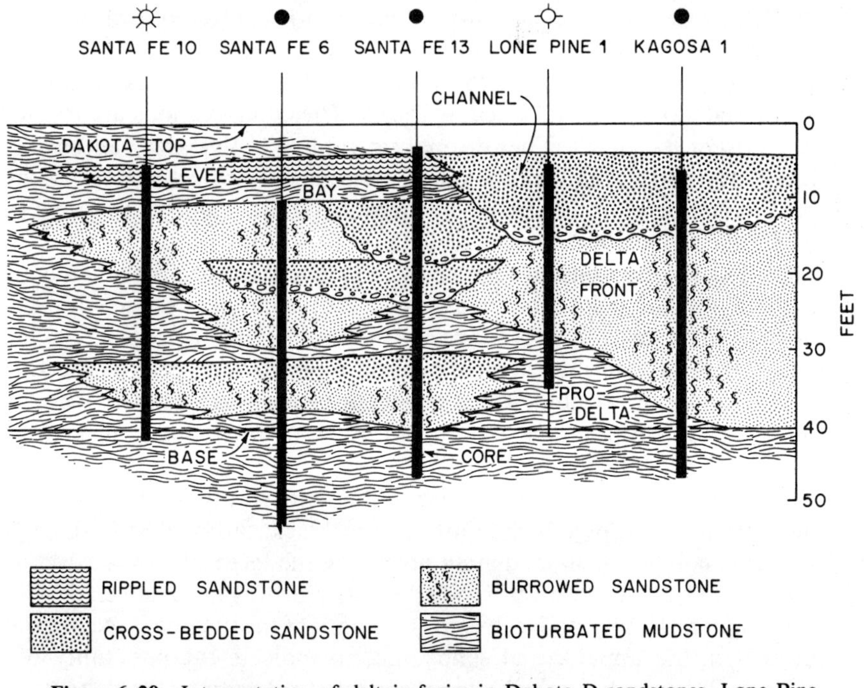

Figure 6–29 Interpretation of deltaic facies in Dakota D sandstones, Lone Pine field. Wells projected into northwest-southwest section with no horizontal scale. Locations shown in Figure 6–25. [From Berg 1979 and published with the permission of the American Association of Petroleum Geologists.]

front sandstone and represents fluvial-channel fills. Finally, the rippled siltstones at the top of the sequence represent overbank levee deposits lateral to the uppermost fluvial channels.

The Kagosa 1 section is the thicker and more uniform sequence and is dominated by delta-front sandstone, but coarser crossbedded sandstones at the top may represent fluvial-channel fill. The sandstones thin toward the Lone Pine section and part of the lower delta-front sandstone is replaced by prodelta mudstone. The Santa Fe 13 core shows that the single deltaic sequence of the Kagosa location was replaced by two cycles of deltaic deposition. The lower, thinner sequence represents a small delta or perhaps crevasse splay, and the upper, thicker section is composed entirely of fluvial channels, which are cut deeply into prodelta mudstones. The lower delta persists through the Santa Fe 6 to the Santa Fe 10.

In the Santa Fe 6 section, the upper delta-front sandstones again are dominant and enclose a single fluvial channel. In the Santa Fe 13, there are two fluvial channels, but only the lower channel reaches the Santa Fe 6. Above this channel is an increasing amount of bioturbated sandstone in Santa Fe 6, which anticipates the change to the more distal location of the upper section in Santa Fe 10.

The coarse bioturbation in the upper delta in the Santa Fe 10 section is significant to the environmental interpretation. Large vertical burrows of crabs suggest that water depths were shallow and confirm the inferred location of the section as lateral to the main delta system.

The most remarkable characteristic of the total Dakota D sandstone is the abrupt lateral variation in deltaic facies. The cored sections are located only 0.5 mi (0.8 km) or less apart. The regional strand line in this area can be assumed to have had a northwest trend. If the cored sections are projected to a northwest-trending profile, as shown in Figure 6–29, the rapid lateral changes can then be expected to have taken place in distances of 0.25 mi (0.4 km) or less.

The second unusual feature of this delta sequence is the large amount of bioturbation in the delta-front sandstone. Bioturbation is expected in prodelta neritic muds, but in delta-front sandstones the rapid influx of coarser sediment normally prevents the establishment of a benthic fauna. In the case of the Dakota D sandstone, it may be concluded that sediment influx was relatively slow. Between times of flood, a benthic fauna reoccupied much of the delta-front area and deeply reworked the sediment.

A third distinguishing feature of this delta section is the lack of bedded organic deposits in the topset beds. Only a thin black carbonaceous shale is evidence of reducing conditions immediately preceding the levee siltstone. Although the uppermost 4-ft (1.2-m) section of the Dakota D unit was not recovered in any core, there is no high resistivity on electric logs that might indicate coal. Neither are root traces present in the upper cored sandstones to indicate the possibility of an overlying swamp. Therefore, it may be concluded that this part of the delta was continuously submerged, probably dominantly marine, and far removed from the site of subaerial, delta-plain deposition.

An additional explanation is possible to account for the high degree of bioturbation in delta-front sandstones and the general lack of organic deposits. That is, the entire Dakota D section at Lone Pine may represent only the distal margins of a larger delta complex that is present to the southeast of the field. This alternative is supported by a thickening of the delta-front sequence toward the Kagosa 1 location where the section is dominantly sandstone through nearly the entire Dakota thickness. Although the main complex has not been drilled, it can be expected to have been the site of more rapid deposition under direct influence of the distributary channels. Furthermore, it may contain more crossbedded sandstones that represent the active channel-mouth bars of the delta. Therefore, the Lone Pine section may be only a lateral feather edge of the delta system in an area that received less sediment and was largely submerged in shallow-marine waters. This explanation also accounts for the thin topset beds and the lack of organic deposits in the uppermost Dakota.

Porosity and permeability. Rapid vertical and lateral changes in the deltaic section explain the erratic distribution of more permeable sandstones. As might be expected, the highest permeabilities are associated with current-transported fluvial sandstones, which are coarser grained and contain small amounts of clay and bioturbation.

The delta-front sequence of Kagosa 1 shows an upward increase in permeability and porosity (Figure 6–30) as clay content decreases upward and grain size increases. The best revervoir section is the coarser fluvial sandstone at the top of the core where average values reach a maximum of 613 md and 22.8 percent. Calcite cement in the uppermost fluvial section reduces permeability and porosity to low values.

Permeability and porosity in the Santa Fe 13 core are variable but correspond closely to bedding types (Figure 6–30). The lower delta-front and upper fluvial sections have maximum values of 146 md and 20.8 percent.

The Santa Fe 10 core shows the lowest average values of permeability and porosity (Figure 6–30). The lower delta-front sandstones have a maximum permeability of 67 md and a porosity of 18.5 percent. The upper delta-front sandstones have a maximum average permeability of only 9 md and an average porosity of 22.3 percent. These low values reflect the large amount of bioturbation and higher clay content of this distal delta section. Furthermore, this sandstone had no trace of oil saturation and a high value of water saturation of 74 percent. Therefore, this section would probably be productive only of water, although its location is at the highest position on the structure.

Morphology

Distribution of reservoir facies can be estimated from borehole logs because of the close correspondence of log response with rock properties. Particularly useful are the gamma-ray and density logs.

Characteristic log response is clearly shown in the Kagosa 1 section in which

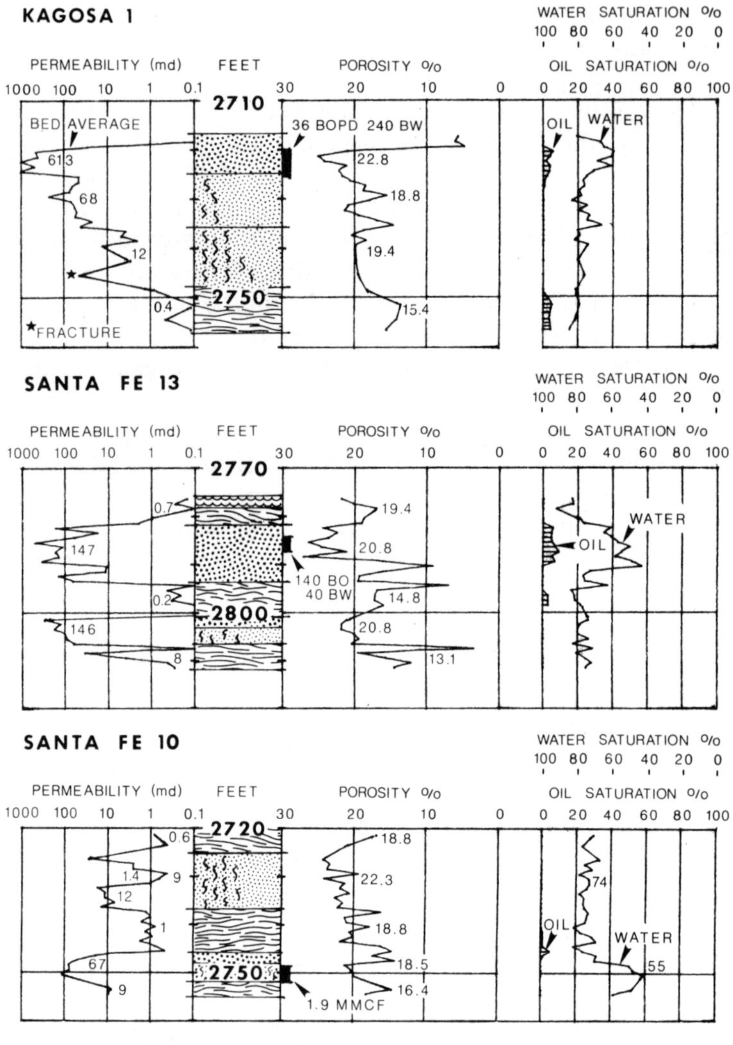

Figure 6-30 Permeabilities, porosities, and fluid saturations in selected cores, Lone Pine field. Bedding types designated by symbols are explained in Figure 6-27. Core depths are not corrected to borehole logs. [From Berg 1979 and published with the permission of the American Association of Petroleum Geologists.]

the principal facies are found in a single continuous vertical sequence (Figure 6-27). The bioturbated mudstone, a nonreservoir facies at the base of the section, shows the highest gamma readings and lowest porosity because of the high clay content. The delta-front sandstones, highly burrowed below and less burrowed above, show decreasing gamma readings upward as clay content decreases. Through the same section, porosity shows a marked increase over the bioturbated mudstone and is somewhat variable. The fluvial crossbedded sandstone at the top of the section shows the lowest gamma readings and the highest porosities.

Similar response of gamma and density logs is shown for the other cored sections (Figures 6–27 and 6–28), but some seemingly anomalous readings are apparent. For example, in Santa Fe 10, the highly burrowed nonreservoir sandstones of the upper delta front show relatively high porosities of 20 percent to 22 percent, similar to the range of porosities in the fluvial crossbedded sandstones. However, the section still can be distinguished on the basis of gamma readings that suggest a higher shale content in the delta-front sandstone and lower shale content in the fluvial sandstones.

The most porous sandstones are those of fluvial channels, and these can be mapped across the field area (Figure 6–31). Fluvial sandstones that exceed 15 ft (4.6 m) of net thickness form a northeast-trending body along the flank of the structure. This body probably is composed of more than one channel unit as indicated in the section of Santa Fe 13 (Figure 6–29). Unfortunately, none of the cored wells has a maximum section of net pay, and the interpretation cannot be confirmed in detail, but the trend of thicker sandstone suggests that channels may be concentrated in the direction of stratigraphic dip that was generally northeast.

Oil and Gas Column

The five cores of the Dakota D sandstone not only reveal the local distribution of porous sandstones, but they also explain the uneven distribution of oil and gas across the Lone Pine structure. It can be assumed that the cored sections are representative of facies change within the field as well as of fluid production related to these facies.

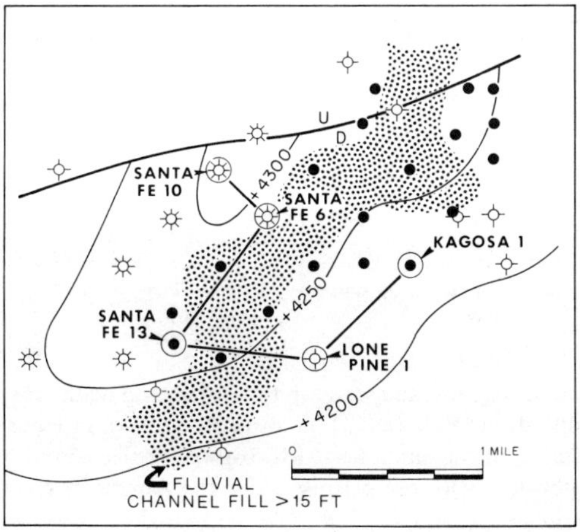

Figure 6–31 Distribution of thicker fluvial-channel sandstones across Lone Pine field. Coarse-dot pattern shows trend of net porous sandstone greater than 15 ft (4.6 m). [Based on well log interpretation by David E. Pollard and presented with permission of INTERCOMP, Houston, Texas, from an unpublished report prepared for Southern Union Company. [From Berg 1979 and published with the permission of the American Association of Petroleum Geologists.]

Then the relative structural positions of the cores permit an estimate of the vertical heights of the oil and gas columns (Figure 6–32).

The lower delta-front section produced only gas at the highest location, Santa Fe 10, and oil with much dissolved gas at the next highest location, Santa Fe 6. The upper delta-front section had high water saturation in the Santa Fe 10 (Figure 6–30) but lower water saturation and a trace of oil saturation at the Santa Fe 6. These relationships suggest the presence of a separate gas column that may have a vertical height of about 20 ft (6 m) (Figure 6–32).

The upper fluvial-channel section in Santa Fe 13 produced oil with some water, but a nearby dry hole recovered only water. This well, the Santa Fe 16, was located about one-quarter mile south of Santa Fe 13, and the highest porosity in the Dakota D was at an elevation of +4240 ft (+1293 m). These relationships suggest that the oil column is about 20 ft (6 m) in the southwest part of the field.

However, oil is found at lower levels to the northeast. For example, oil was produced at Kagosa 1 from perforations as low as +4215 ft (+1286 m), whereas only water was produced in Lone Pine 1 at about +4220 ft (+1287 m). These relationships suggest that an additional oil column of about 25 ft (7.6 m) is present because of the tilt toward the northeast.

The total vertical column of oil and gas, therefore, is on the order of 65 ft (20 m). There is, however, a seemingly anomalous occurrence of water in the upper

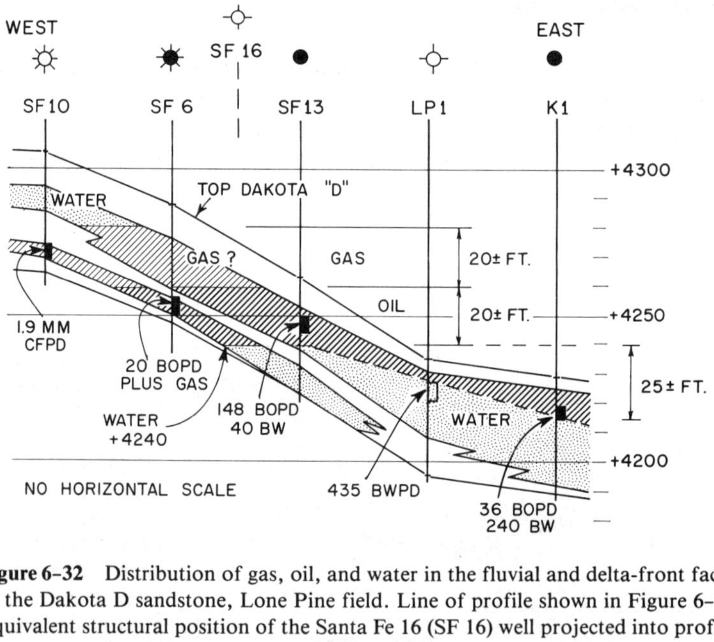

Figure 6–32 Distribution of gas, oil, and water in the fluvial and delta-front facies of the Dakota D sandstone, Lone Pine field. Line of profile shown in Figure 6–31. Equivalent structural position of the Santa Fe 16 (SF 16) well projected into profile. [From Berg 1979 and published with the permission of the American Association of Petroleum Geologists.]

delta-front section at the highest structural position in Santa Fe 10. An explanation for this water occurrence lies in the reduced permeability of this facies and its increased capacity for trapping oil and gas because of greater capillary pressure. This possibility may be tested by calculating the hydrostatic oil column that could be held by this facies, according to methods previously described (Berg 1975).

By selecting appropriate values for rock and fluid properties, the oil and gas column may be calculated. A minimum value of 9 md (Figure 6–30, Santa Fe 10) can be used for the barrier facies. If it is assumed that the total hydrostatic column is half gas and half oil, then the average density of the combined column is 0.45 g/cm^3. Based on these values, the total oil and gas column that can be held by the water-wet facies is 35 ft (10.7 m). This value agrees well with the estimated hydrostatic column of about 40 ft (12 m). Thus, the calculation confirms the trapping capacity of the water-wet facies.

The additional 25 ft (7.6 m) of oil column may be accounted for by hydrodynamic flow. The regional hydrodynamic gradient in the Dakota is extremely low and its direction is generally eastward, as estimated from an unpublished map (Berry 1959). Control for this map was sparse in the Hospah area, and the scale of the map was small. Nevertheless, the potentiometric gradient can be estimated to be on the order of only 5 to 10 ft/mi (1 to 2 m/km). Based on the lower of these values, the tilt of the oil-water contact should be about 12.5 ft/mi (2.4 m/km) or about 20 ft (6 m) between the Santa Fe 13 and Kagosa 1 wells. The observed tilt is about 25 ft (8 m), which agrees with the additional oil column that might be expected by the potentiometric gradient. Consequently, it may be concluded that hydrodynamic flow accounts for the observed tilt and added oil column on the east part of the Lone Pine structure.

The total oil and gas column is the sum of the hydrostatic and hydrodynamic columns, or 55 ft (17 m), as compared with the estimated maximum column of 65 ft (20 m). Therefore, the calculations (Berg 1979) tend to confirm the nature of the trap and stratigraphic control on distribution of oil and gas.

Conclusions

Complex lateral variations in the delta-margin sequence result in porosity and permeability changes that control the distribution of oil and gas. These variations are typical of deltaic sections and were, in fact, predictable from the initial interpretation of depositional conditions in the first core. Optimum recovery of oil and gas at Lone Pine field will depend on detailed interpretation of the reservoir section and mapping of thin porous units within the productive area. Subsequent reservoir interpretations must rely on borehole logs, but these interpretations will be supported by the facts of sandstone properties observed in the cored sections. Other reservoirs of deltaic origin can be expected to show similar complexities. Successful recovery of oil and gas in such reservoirs can be achieved only by a detailed knowledge of the rock sequence.

MENEFEE SANDSTONE, FAIRWAY FIELD, NORTHWESTERN NEW MEXICO

Interbedded sandstones and shales from the Menefee Formation represent a deltaic-plain environment, probably similar to that of the freshwater Atchafalaya basin on the deltaic plain of the Mississippi in south Louisiana. Two Menefee oil wells were completed in the Fairway field, but reservoir sandstones are thin, laterally restricted, and probably of fluvial origin. Although initial production from the reservoir was significant, production rapidly declined, and the wells were not successfully offset. Thus, the sandstone represents an insignificant reservoir that yielded a low return of producible oil. However, the cored section is of geologic interest because it illustrates not only the delta-plain sandstones but also the occurrence of coal in two modes, one formed in place and the other of detrital origin. In fact, the nature of the total section made possible the prediction that the reservoir sandstones would be greatly restricted in areal extent.

Geologic Setting

The San Juan basin of northwestern New Mexico contains large gas reserves in Upper Cretaceous rocks of the Mesaverde Group. The structure of the basin is distinctive (Figure 6–33). The south flank of the basin dips gently northeastward at

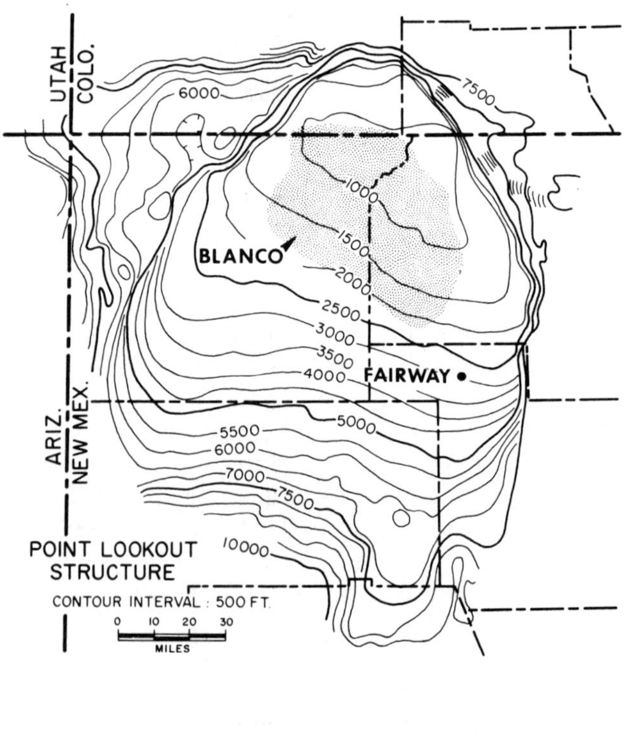

Figure 6–33 Structure map of the San Juan basin, northwestern New Mexico, contoured on top of the Point Lookout Formation, Mesaverde Group. Contour interval 1000 ft (305 m). Blanco gas field (stippled) produces from the Mesaverde reservoir. Cored well is indicated as Fairway field.

rates that average only 100 ft/mi (19 m/km) to a structural low on the Colorado-New Mexico boundary. The other three sides of the basin consist of steep monoclines across which the dip of the beds ranges from 10° to as much as 30°. The deep central part of the basin is occupied by the Blanco gas field. This area is well known as a closed potentiometric low, which has been referred to as a *hydrodynamic trap* (Berry 1959). The Fairway field is located just updip from the Blanco gas field, and, unfortunately, it was not within the gas-producing trend.

The Upper Cretaceous Mesaverde Group is composed of three formations, which are, in ascending order, (1) the Point Lookout Sandstone, (2) the Menefee Formation, and (3) the Cliff House Sandstone (Figure 6–34). The Mesaverde Group exceeds 3000 ft (915 m) in thickness on the outcrop across the southwest flank of the basin, but it thins to about 1500 ft (457 m) downdip in the area of the Blanco gas field. The Mesaverde wedge is surrounded by marine shales, the thick Mancos Shale below, and the Lewis Shale above. The Mesaverde wedge represents a regressive and transgressive sequence of deltaic sediments. The bounding sandstones of the Point Lookout and Cliff House have thicknesses that range from 50 ft to 100 ft (15 m to 30 m) and are modified delta-front sandstones. These sandstones are the major gas-producing horizons in the Blanco field (Allen 1955), and trends within the field area are relatively narrow and strike in a northwest direction (Hollenshead and Pritchard 1962). These trends are parallel to the regional depositional strike and are at right angles to the northeastward thinning of the Mesaverde wedge. Therefore, the sandstones have been interpreted as deltaic sands that were reworked

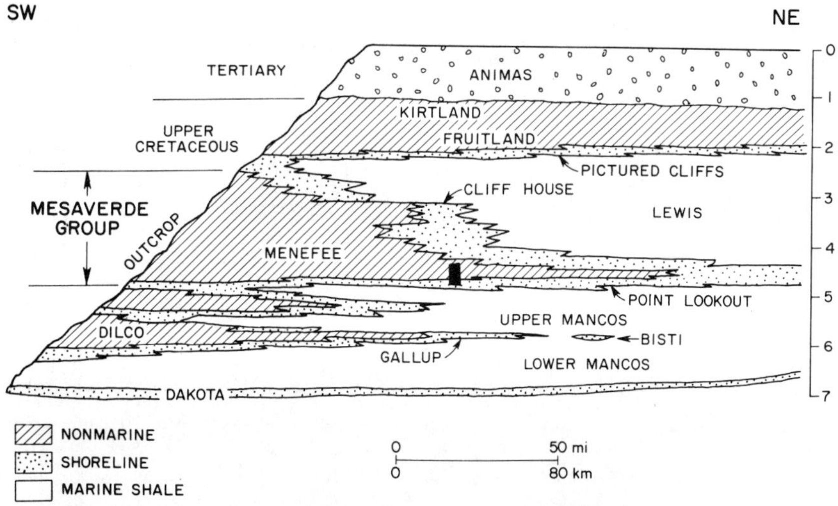

Figure 6–34 Stratigraphic diagram of the Upper Cretaceous section, San Juan basin, northwestern New Mexico. Approximate location of the cored section in the Menefee formation is indicated by the short vertical bar. [Diagram adapted from Silver 1950.]

by waves into strike-trending shoreline deposits. Individual sandstones are lenticular in cross section and form partly overlapping shingled reservoir bodies in both the transgressive and regressive sections.

The Menefee Formation forms the greater part of the Mesaverde wedge, thickening southwestward toward the basin flank. The Menefee is composed dominantly of shale with interbedded thin sandstones. At the outcrop, the section also contains coal beds that have been mined by stripping, but in the downdip subsurface, the coal beds are thin and limited in areal extent.

The depositional aspect of the Mesaverde Group, therefore, is that of a broad deltaic plain bordered by relatively narrow shoreline deposits. The nature of the delta-plain sequence is well illustrated by a core from the Menefee section.

Drilling History

The discovery well for the Fairway field was the Tesoro-North American Parlay 1, Sec. 29–22N–3W, Sandoval County (Figure 6–33). A second well, the Parlay 2, was completed in January 1972, for 300 bbls/day (48 m^3/day) of oil from a thin sandstone in the lower Menefee Formation (Figure 6–35). Unfortunately, subsequent offset wells were unsuccessful. A core of lower Menefee was obtained in the Parlay 2, and the base of the core was only 35 ft (11 m) above the top of the Point Lookout

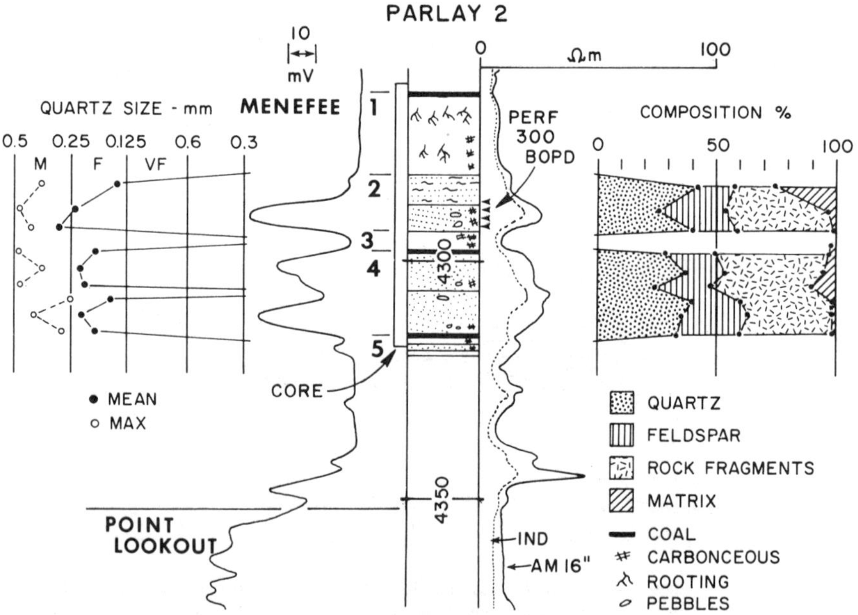

Figure 6–35 Electric logs, sedimentary structures, texture, and composition of the lower Menefee section, Fairway field, Sandoval County, New Mexico.

Sandstone. Because of its stratigraphic position, the cored section probably represents a depositional location that was only a few miles landward from the shoreline.

Description of Sandstones

Sedimentary structures. The cored section contains interbedded sandstones and shales that can be divided into the following units:

Unit	*Thickness ft (m)*
1. Claystone, greenish gray, rooted, overlain by thin coal.	17 (5.2)
2. Sandstone, with inclined laminae, overlain by interlaminated sandstone and claystone.	12 (3.6)
3. Claystone, dark gray, carbonaceous.	4 (1.2)
4. Sandstone, in two beds, with thin coals at base and top.	17 (5.2)
5. Claystone and sandstone interbedded.	5 (1.5)
Total	55 (16.8)

The sandstones are characterized by even, parallel, continuous laminae that have dips of about 10° [Figure 6–36 (G) and (H)]. Lamination is distinct because of concentrations of finely divided carbonaceous material and conspicuous dark rock fragments. The lower sandstone unit (4) is divided into two beds separated by thin horizontal laminae with no development of ripples. Some rounded shale pebbles occur in all sandstones as well as thin angular shale clasts [Figure 6–36 (F)]. The uppermost sandstone unit (2) grades upward into interlaminated shale with thin ripple lenses of sandstone [Figure 6–36 (E)]. The upper part of this interbedded unit shows contortion by soft-sediment deformation [Figure 6–36 (D)].

Two thin beds of coal are adjacent to the lower sandstone unit (4). The coals contain small blebs of fine silt and clay [Figure 6–36 (J)], and both coals probably represent accumulations of detrital organic material. The basal coal is underlain by laminated claystone that shows no evidence of rooting; the uppermost coal is in sharp contact with the sandstone below, which also shows no evidence of rooting. Apparently, these coals were formed from organic material that accumulated at the bottom of a shallow drainage channel, which subsequently was filled by sand, and this was followed by an accumulation of organic material in the abandoned channel.

The uppermost unit (1) has a highly churned appearance characteristic of bioturbation by rooting [Figure 6–36 (A) and (B)]. In the lower part of the unit are thin irregular streaks of carbonaceous material that probably represent root traces [Figure 6–36 (B)]. At the base of the unit is a large mass of sideritic claystone that contains white calcite-filled cracks that appear to radiate outward from the center

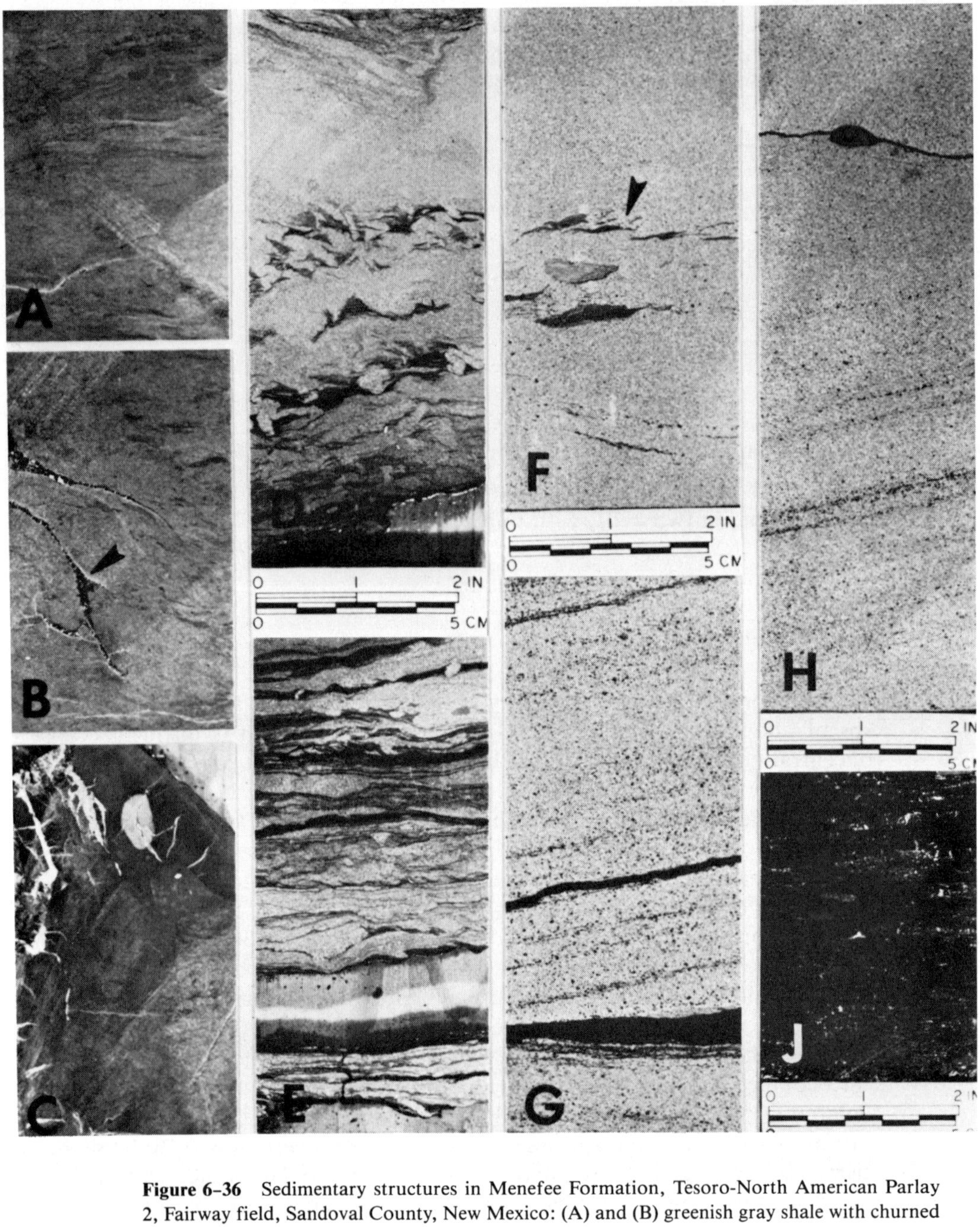

Figure 6–36 Sedimentary structures in Menefee Formation, Tesoro-North American Parlay 2, Fairway field, Sandoval County, New Mexico: (A) and (B) greenish gray shale with churned appearance due to rooting; note large root traces in (B) (arrow) filled with carbonaceous material—4267 ft and 4271 ft (1301 m and 1302 m); (C) part of large siderite concretion with septarian cracks filled with white calcite—4276 ft (1303 m); (D) sandstone with interlaminated clay, slumped and contorted—4281 ft (1305 m); (E) thin ripple lenses of sandstone and interlaminated greenish gray clay—4285 ft (1306 m); (F) indistinctly laminated sandstone with squeezed shale clasts (arrow)—4287 ft (1307 m); (G) and (H) laminated sandstone with abundant dark grains and carbonaceous laminae, dip 12°—4302 ft and 4307 ft (1311 m and 1313 m); (J) thin coal at base of sandstone; small white blebs are fine sand and silt—4312 ft (1314 m).

of the mass [Figure 6–36 (C)]. The mass probably represents a large concretionary body, probably a "septarian" concretion.

At the top of the section is a thin coal, which evidently formed in place. The organic material accumulated from plants that extended their roots deeply into the underlying claystone, and rooting resulted in the highly churned appearance of the "underclay."

Other than rooting in the uppermost unit (1), bioturbation is absent in the section. A few small tubelike structures of sand are present in the interbedded claystone and sandstone units (2 and 3). Although some of these structures may be burrows, the majority appear to be the result of soft-sediment deformation by loading sand into soft claystone, or by subsequent deformation of the sediment.

Petrography. The sandstones are uniformly fine grained with a mean size of 0.20 mm (fine grained) and an average maximum size of 0.38 mm (medium grained). Both sandstone units (2 and 4) show a decrease in mean grain size upward (Figure 6–35). Maximum size ranges up to 0.47 mm, a significantly large value, and feldspar grains and rock fragments appear to be generally larger than quartz. In composition, the sandstones have an unusually low quartz content of 34 percent and large amounts of feldspar and rock fragments (Table 6–5). Rock fragments include igneous rocks, limestone fragments, polycrystalline quartz grains, and chert. Chert grains are the most abundant at an average 15 percent of the total detrital composition. The sandstones can be characterized as feldspathic litharenites.

Porosity and permeability. The uppermost sandstone unit (2) has the best porosity and permeability (Figure 6–37), but the two lower superposed sandstones (unit 4) have greatly reduced permeability and porosity. The reason for this difference appears to be in the amount of calcite cement, which has a low value of 5 percent of the upper sandstone and an average amount of 12 percent to 27 percent in the lower sandstones. Calcite cement may have been derived largely from solution and reprecipitation of limestone rock fragments. Chert grains also have affected the porosity and appear to have been distorted during compaction and squeezed into adjacent interstitial voids. Partial solution of chert grains may also have provided a small amount of silica, which forms thin overgrowths on quartz grains in some samples.

Interpretation

The sandstones are all of fluvial origin, and this interpretation is based largely on the low quartz content, the fining upward grain size within beds, and the laminated character of the sandstones. Dip of the laminae, however, is relatively low and suggests a dune phase of sediment transport in which the bed form was of low amplitude. It is concluded, therefore, that transport was by relatively low-velocity currents.

TABLE 6–5. AVERAGE PROPERTIES OF MENEFEE SANDSTONES, TESORO-NORTH AMERICAN PARLAY 2, SANDOVAL COUNTY, NEW MEXICO

	Quartz size[a]			Detrital composition[b]					Cement[c]		
Depth (ft)	Mean (mm)	Max (mm)	σ (mm)	Qz (%)	F (%)	Rx (%)	Cht (%)	Mx (%)	(% of total)	Permeability (md)	Porosity (md)[d]
4282-91	0.22	0.40	0.07	36	21	15	18	10	5	14.	21.8
4296-03	0.20	0.43	0.06	31	20	30	13	6	12	0.42	15.4
4306-13	0.18	0.30	0.05	35	25	23	16	1	27	0.34	12.6

SOURCE: Petrography by Ahr 1972.

[a]Long-axis measurements; σ = standard deviation.

[b]Qz = monocrystalline quartz, F = feldspars, Rx = rock fragments, Cht = chert and polycrystalline quartz, Mx = matrix.

[c]Cement includes silica as grain overgrowths and calcite.

[d]Average permeability and porosity from total core.

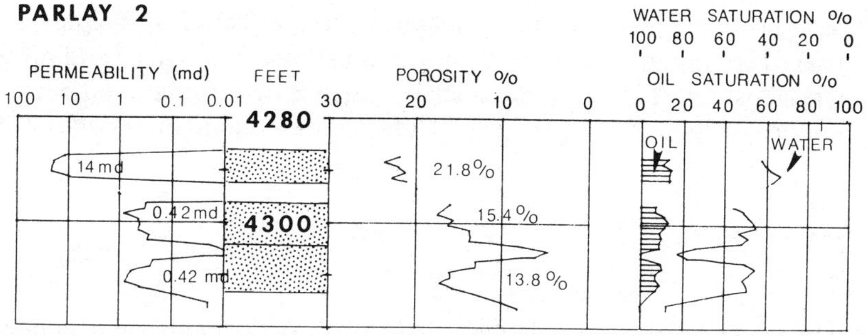

Figure 6–37 Permeability, porosity, and fluid saturations in Menefee sandstones, Tesoro-North American Parlay 2.

The total cored section displays most features of the delta-plain environment. Delta-plain sediments represent the subaerial and shallow subaqueous deposits that are formed as topset beds on the main mass of delta sediment. Sandstones are mainly fluvial in origin, and fine-grained siltstones and shales represent overbank deposits lateral to stream channels. The overbank areas are commonly poorly drained and highly vegetated. Thus, the delta-plain environment is one of the principal sites for coal deposits.

Characteristics of the modern deltaic plain of the Mississippi have been described in the Atchafalaya Basin (Coleman 1966). This freshwater lake and adjoining swamp are part of the drainage system of the Mississippi River, and large areas are flooded each spring. The alternate flooding and draining produce distinctive features of the sediment sequence.

The uppermost coal of the Menefee represents a poorly drained swamp. The modern swamp environment has a stable depth of about 3 ft (1 m) and contains dense woody vegetation and aquatic plants (Coleman 1966). Bottom conditions are anaerobic with hydrogen sulfide production. Sediment is black highly organic clay that contains large plant fragments and thin woody peat beds. Bioturbation by rooting is abundant and results in destruction of lamination. Accessory minerals are small pyrite globules, vivianite, a hydrous ferrous phosphate, and small calcium carbonate and iron carbonate modules.

The massive claystone unit (1) represents a well-drained swamp. This highly vegetated area is inundated only during floods, and consequently the clays lack a high organic content and are gray to light gray in color. Bioturbation by rootlets is common but not intense, so lamination is preserved. Accessory minerals include abundant calcium carbonate in large and small nodules with some small iron oxide nodules.

The thin sandstones of the Menefee represent the fill of sluggish gathering channels that drained the flood basin. The lowermost sandstone unit (4) with adjacent detrital coals indicates that the channels were draining a highly vegetated swamp. These coals cannot be expected to have wide lateral extent because they were probably restricted to the meandering channelways. In contrast, the uppermost coal overlying the rooted claystone unit (1) formed in place and probably has a wide lateral extent. Thus, the presence or absence of rooting in adjacent sediments is an important indicator of the extent of bedded organic materials.

Conclusions

The Menefee section was deposited in a deltaic-plain environment. Fluvial sandstones represent shallow channels that drained a highly vegetated swamp, probably similar to the modern Atchafalaya Basin of the Mississippi deltaic plain. The channel-fill sandstones are thin and probably have a narrow, highly sinuous, and dip-trending distribution. Although larger and higher-energy channel deposits may occur locally, the composition of the sandstone is such that reservoir characteristics cannot be expected to change greatly in the immediate area.

Better reservoir sandstones probably occur seaward in a downdip direction. In fact, the vertical sequence suggests the lateral relationships of sandstones (Figure 6–35). The underlying Point Lookout Sandstone is a shoreline deposit formed by reworking of sediments supplied by the deltaic-plain streams. The Menefee section was deposited near sea level, but it is difficult to predict the proximity of the marine environment; it may be several miles or several tens of miles downdip to the ancient delta-front environment of more porous strike-trending sandstones. Although the fluvial sandstones of the Menefee are poor reservoirs, the sandstones themselves suggest the presence of better reservoirs in a seaward direction.

SUMMARY

Deltas are complex bodies of sediments deposited at the mouths of rivers. The uniform prograding delta-front section is recognized by the vertical sequence that is gradational upward from shallow marine at the base to fluvial-transported sands and thence to nonmarine beds above. Grain size increases upward, and composition is typically that of the immediate fluvial source sediment. The sequence may be modified because of the complex shifting depositional conditions and progradation of deltaic-plain sediments over the previous delta front. A common result of progradation is that fluvial channels may have cut deeply into the previous sediment to produce an inverted sequence of sedimentary structures and texture characteristic of deposition in fluvial channels. The delta-front section may be modified by wave action or tidal currents during deposition, and finally, marine transgression over the delta sequence may have reformed the upper beds into thin marine deposits. The resulting delta complex, therefore, may show rapid vertical and lateral change in sediment type.

Higher porosities and permeabilities are commonly associated with the current-transported sediments, either those of the distributary bar or of the fluvial channels. The overlying transgressive sandstones may also have excellent reservoir properties but in relatively thin units. Rapid lateral change within the deltaic sequence results in low predictability for the occurrence of reservoir sandstones. Deltaic reservoirs commonly occur in structural traps, as in the Gulf Coast province, but stratigraphic traps also may have developed, especially within the meandering channel deposits of the deltaic plain.

REFERENCES CITED

ALLEN, R. W., JR., 1955, Stratigraphic gas development in the Blanco Mesaverde pool of the San Juan basin: Four Corners Geological Society Guidebook, p. 144–149.

ANDERSEN, ROD, 1977, Carlsbad field, *in The oil and gas fields of southeastern New Mexico:* Roswell, New Mexico, Roswell Geological Society Symposium (Supplement), p. 21–28.

BARKLEY, C. J., and R. F. GOSMAN, 1958, Donkey Creek area, Crook County, Wyoming: Wyoming Geological Association Guidebook, p. 174–179.

BATES, C. C., 1953, Rational theory of delta formation: American Association of Petroleum Geologists Bulletin, v. 37, p. 2119–2162.

BERG, R. R., 1968, Point-bar origin of Lower Cretaceous Fall River Sandstone, Powder River basin, Wyoming: American Association of Petroelum Geologists Bulletin, v. 52, p. 2116–2122.

———, 1975, Capillary pressures in stratigraphic traps: American Association of Petroleum Geologists Bulletin, v. 59, p. 939–956.

———, 1979, Oil and gas in delta-margin facies of Dakota sandstone, Lone Pine field, New Mexico: American Association of Petroleum Geologists Bulletin, v. 63, p. 886–904.

BERRY F. A. F., 1959, Hydrodynamics and geochemistry of the Jurassic and Cretaceous systems of the San Juan Basin, northwestern New Mexico and southwestern Colorado: Stanford University Ph.D dissertation, 192 p.

BROUSSARD, M. L., editor, 1975, *Deltas, models for exploration:* Houston, Texas, Houston Geological Society, 555 p.

CHISHOLM, W. A., 1963, The petrology of Upper Jurassic and Lower Cretaceous strata of the Western Interior: Wyoming Geological Association-Billings Geological Society Guidebook, p. 71–86.

COLEMAN, J. M., 1966, Ecological changes in a massive freshwater clay sequence: Gulf Coast Association of Geological Societies Transactions, v. 16, p. 159–174.

———, and S. M. GAGLIANO, 1964, Sedimentation in the Mississippi River deltaic plain: Gulf Coast Association of Geological Societies Transactions, v. 14, p. 67–80.

———, J. M. SUHAYDA, T. WHELAN, and L. D. WRIGHT, 1974, Mass movement of Mississippi River delta sediments: Gulf Coast Association of Geological Societies Transactions, v. 24, p. 49–68.

———, and L. D. WRIGHT, 1975, Modern river deltas: variability of processes and sand bodies, *in* M. L. Broussard, editor, *Deltas, models for exploration:* Houston, Texas, Houston Geological Society, p. 99–149.

———, and D. B. PRIOR, 1982, Deltaic environments of deposition: American Association of Petroleum Geologists Memoir 31, p. 139–178.

DAVIS, R. E., and G. A. IZETT, 1958, Keyhole Sandstone Member of the Fall River Formation, northern Black Hills, Wyoming and South Dakota: American Association of Petroleum Geologists Bulletin, v. 42, p. 2745–2750.

DEISCHL, D. G., 1973, The characteristics, history, and development of the basin Dakota gas field, San Juan basin, New Mexico, *in Cretaceous and Tertiary rocks of the southern Colorado Plateau:* Four Corners Geological Society Memoir, p. 168–173.

DONDANVILLE, R. F., 1963, The Fall River Formation, northwestern Black Hills, lithology and geologic history: Wyoming Geological Association-Billings Geological Society Guidebook, p. 87–99.

ETHRIDGE, F. G., 1970, Quantitative petrographic criteria for recognition of environments of deposition: Texas A&M University Ph.D. dissertation, 169 p.

FISHER, W. L., L. F. BROWN, A. J. SCOTT, and J. H. MCGOWEN, 1967, *Delta systems in the exploration for oil and gas, a research colloqium:* Austin, Texas, University of Texas Bureau of Economic Geology, p. 58–59.

FISK, H. N., 1955, Sand facies of Recent Mississippi delta deposits: 4th World Petroleum Congress Proceedings Section I/C, p. 1–21.

———, 1961, Bar-finger sands of Mississippi Delta, *in* J. S. Peterson, and J. C. Osmond, editors, *Geometry of sandstone bodies:* Tulsa, Oklahoma, American Association of Petroleum Geologists, p. 29–52.

FRAZIER, D. E., 1967 Recent deltaic deposits of the Mississippi River—their development and chronology: Gulf Coast Association of Geological Societies Transactions, v. 17, p. 287–311.

GILBERT, G. K., 1885, The topographic features of Lake shores: U.S. Geological Survey Annual Report No. 5 (1883–84), p. 69–123.

HAUN, J. D., and J. A. BARLOW, JR., 1962, Lower Cretaceous stratigraphy of Wyoming: Wyoming Geological Association Guidebook, p. 15–22.

HOLLENSHEAD, C. T., and R. L. PRITCHARD, 1961, Geometry of producing Mesaverde sandstones, San Juan basin *in* J. A. Peterson and J. C. Osmond, editors, *Geometry of sandstone bodies:* Tulsa, Oklahoma, American Association of Petroleum Geologists, p. 98–118.

KING, V. L., and S. A. WENGERD, 1957, The Hospah oil field, McKinley County, New Mexico: Four Corners Geological Society Guidebook, p. 155–168.

KOLB, C. R. and J. R. VAN LOPIK, 1966, Depositional environments of the Mississippi River deltaic plain—southeastern Louisiana, *in* M. L. Shirley, editor, *Deltas in their geologic framework:* Houston, Texas, Houston Geological Society, p. 17–61.

LEBLANC, R. J., 1975, Significant studies of modern and ancient deltaic sediments, *in* M. L. Broussard, editor, *Deltas, models for exploration:* Houston, Texas, Houston Geological Society, p. 13–85.

MACKENZIE, F. T., and D. J. RYAN, 1962, Cloverly-Lakota and Fall River paleocurrents in the Wyoming Rockies: Wyoming Geological Association Guidebook, p. 44–61.

MEYER, R. F., 1966, Geology of Pennsylvanian and Wolfcampian rocks in southeast New Mexico: New Mexico Bureau of Mines Memoir 17, 123 p.

MORGAN, J. P., J. M. COLEMAN, and, S. M. GAGLIANO, 1968, Mudlumps: diapiric structures in Mississippi delta sediments, *in Diapirism and diapirs:* American Association of Petroleum Geologists Memoir 8, p. 145-161.

OWEN, D. E., 1973, Depositional history of the Dakota sandstone, San Juan basin area, New Mexico, *in Cretaceous and Tertiary rocks of the southern Colorado Plateau:* Four Corners Geological Society Memoir, p. 37-51.

PETERSON, J. A., A. J. LOLEIT, C. W. SPENCER, and R. A. ULLRICH, 1965, Sedimentary history and economic geology of San Juan basin: American Association of Petroleum Geologists Bulletin, v. 49, p. 2076-2119.

PODPOCHEN, F. W., 1960, Lower Pennsylvanian gas exploration, Eddy County, southeastern New Mexico, *in Oil and gas fields of southeastern New Mexico:* Roswell Geological Society Symposium, p. xxii-xxiii.

RAINWATER, E. H., 1975, Petroleum in deltaic sediments, *in* M. L. Broussard, editor, *Deltas, models for exploration:* Houston, Texas, Houston Geological Society, p. 3-11.

ROBINSON, C. S., W. J. MAPEL, and M. H. BERGENDAHL, 1964, Stratigraphy and structure of the northern and western flanks of the Black Hills uplift, Wyoming, Montana, and South Dakota: U.S. Geological Survey Professional Paper 404, p. 23-40.

SILVER, CASWELL, 1950, The occurrence of gas in the Cretaceous rocks of the San Juan basin, New Mexico and Colorado: New Mexico Geological Society Guidebook, p. 109-123.

STAPP, R. W., 1967, Relationship of Lower Cretaceous depositional environment to oil accumulation, northeastern Powder River basin, Wyoming: American Association of Petroleum Geologists Bulletin, v. 51, p. 2044-2055.

VAN HEERDEN, I. L., J. T. WELLS, and H. H. ROBERTS, 1981, Evolution and morphology of sedimentary environments, Atchafalaya delta, Louisiana: Gulf Coast Association of Geological Societies Transactions, v. 31, p. 399-408.

WAAGE, K. M., 1959, Stratigraphy of the Inyan Kara group in the Black Hills: U.S. Geological Survey Bulletin, 1981-B, p. 11-90.

WESCOTT, W. A., and F. G. ETHRIDGE, 1980, Fan-delta sedimentology and tectonic setting, Yallahs Fan Delta, southwest Jamaica: American Association of Petroleum Geologists Bulletin, v. 64, p. 374-399.

WRIGHT, L. D., 1977, Sediment transport and deposition at river mouths: Geological Society of America Bulletin, v. 88, p. 857-868.

YOUNG, R. G., 1960, Dakota group of the Colorado Plateau: American Association of Petroleum Geologists Bulletin, v. 44, no. 2, p. 156-194.

7

Coastal Sandstones

The coastal environment of deposition is that of the interdeltaic shoreline. It includes not only the shoreline itself but also the associated brackish to freshwater environments immediately landward and extends to the shallowest marine at the shoreface or littoral zone. Thus, the coastal environment represents the transition from nonmarine to marine. In terms of sands and sandstones, the beach and shoreface are the most important deposits, and the dominant process is wave action. Yet other sands in bays, lagoons, and estuaries form deposits related to the shoreline environment. Therefore, the coastal zone is used here rather broadly to include all sands that may have accumulated adjacent to, if not directly at, the shoreline.

SHORELINE PROCESSES

Waves approaching the shoreline from the open sea build up and then spill or break toward the shoreline, continually transporting great amounts of water onto the beach. The energy of breaking waves is also responsible for carrying large amounts of sand to the shoreface. In addition, the wave action sets up a littoral current by which sands are carried long distances parallel to the shoreline.

The large waves, or swell, of the open ocean begin to build up into solitary waves as the shallowing bottom retards the progress of bottom water (Figure 7–1). Further shallowing increases bottom friction, and as the uppermost water breaks toward the shore, water is carried up the beach face in the swash zone. Returning water flows beneath the swash at the bottom and is again caught up in the next line of breaking waves. On a steeply sloping beach face, the line of breaking waves is

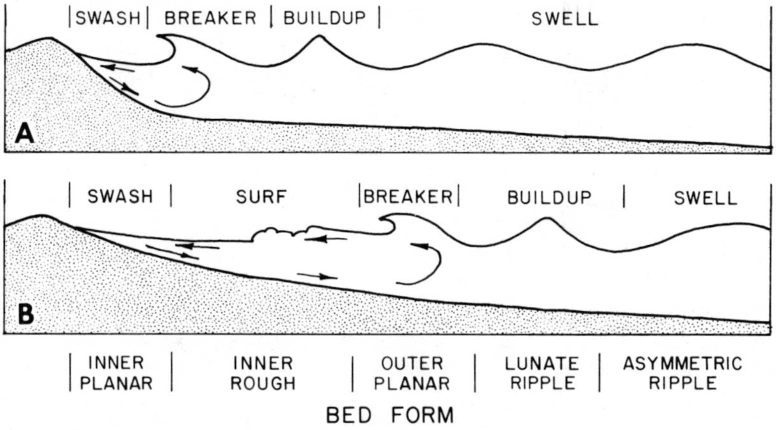

Figure 7–1 Characteristics of waves and bed forms in the nearshore area: (A) steeply sloping beach face; (B) gently sloping beach face. [Wave diagram from Ingle 1966; bed forms from Clifton, Hunter, and Phillips 1971.]

close to shore. On a gently sloping beach face, the line of breaking waves is farther from the shoreline, and the water spills into a rolling surf zone.

On a sandy coastline, characteristic bed forms are developed in the nearshore area. The orbital motion of water particles in the waves at the surface is modified to a tangential motion at the bottom. This velocity of water movement sets the sand grains in motion, first to form ripples at lower velocities and then to form a plane-bed phase at higher velocities (Figure 2–14). The velocities of bottom water increase shoreward, and a regular sequence of bed forms is developed in this direction [Figure 7–1(B)]. In somewhat deeper water under the ocean swell, asymmetric ripples are developed, and under the zone of wave buildup, the ripple crests are discontinuous and form lunate patterns. Below the breaker and swash zones, the higher water velocities develop planar bedding, while beneath the surf zone, the bed form is rough and irregular with small ridges and troughs.

Most nearshore waters are characterized by lines of breaking waves that are interrupted by the rip currents of return flow (Figure 7–2). The breakers commonly advance at some angle to the shoreline, either because of the direction of the prevailing winds or because the wave advance is refracted by bottom irregularities. In this case, water is translated laterally beneath the surf and swash zones, and a littoral current is generated by the alternate advance and return of water. The alternate inshore and offshore transport results in a sinusoidal or meandering form for the littoral current, and the finer sand is moved parallel to the shoreline, whereas the coarser sand is left behind. Thus, longshore transport in the surf and swash zones may produce a textural gradation, fining in the direction of the littoral current. If the supply of sand is fine enough in relation to wave energy, the sand may be transported completely away from the immediate source, such as at the mouth of a river.

Farther offshore, the water is also moving parallel to the shoreline due to the

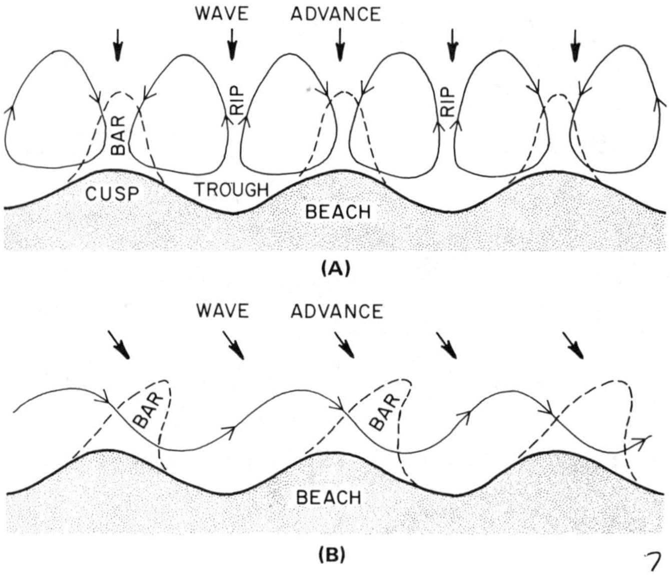

Figure 7–2 Diagrams of currents in the nearshore area: (A) breakers move water into the swash zone, and return flow is concentrated as rip currents; (B) longshore currents generated by alternating inshore and offshore transport of water. [(a) from Ingle 1966; (b) from Sonu 1972.]

action of prevailing winds and aided, perhaps, by tidal flow. This longshore current is called the *coastal current* (Ingle 1966), and on many coasts it is confined largely to the upper layers of the water. When this current is amplified by tidal action, it is also capable of transporting large amounts of sand.

The interaction of waves and sand on the beach produces a well-sorted sediment. Grain size depends on the immediate source of sand, the energy of the approaching waves, and the slope of the shoreface (Figure 7–3). On the exposed shore of the western United States, the average waves have high energy, largely because of the long fetch of swell across the open Pacific Ocean, whereas on the shore of the eastern United States, the average waves have lower energy. The differences in grain size between the shorelines is largely a function of the sand supplied to the beaches, but the median size concentrated in the swash zone is related to the slope of the shoreface. When the slope is high, maximum grain size is much larger on the high-energy coast than on the low-energy coast. For the same medium grain size, the slope is much lower on the high-energy coast. This relationship is explained by the coarser-grained sediment having a greater permeability, which allows more of the return flow to be carried within the bottom sediment. Infiltration of water on the coarser-grained beach results in a reduced capacity of the return flow to transport larger grain sizes. Thus, the slope of the beach face is actually a function of the median grain size supplied to the beach (Komar 1976, p. 303).

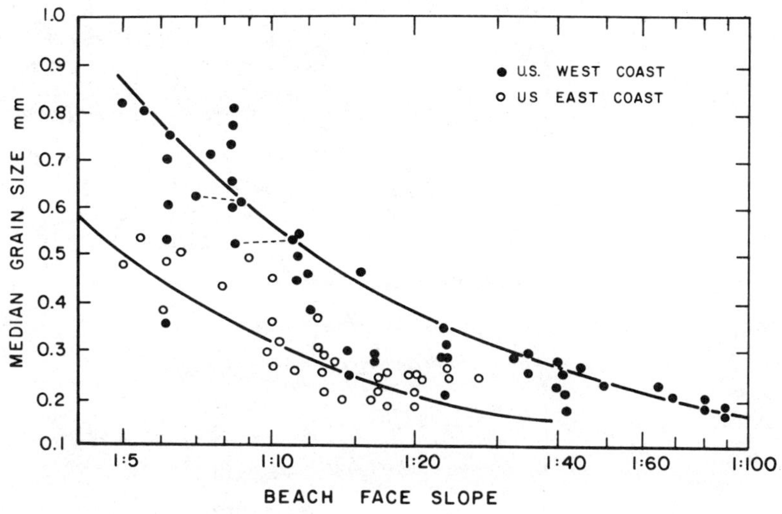

Figure 7–3 Relation of median grain size of sand to slope of the beach face. [Data for the West Coast of the United States from Bascom 1951 and for the East Coast from Wiegel 1964. Modified from Komar 1976.]

The importance of the grain size-slope function (Figure 7–3) is that high-energy beaches can be relatively coarse grained. Furthermore, the seaward termination for a coarser-grained beach could be more abrupt than for a finer-grained beach. This information might be important, in a qualitative way, in explaining the morphology of some beach sands, say in ancient shoreline sections. It would not be possible, however, to predict the beach-face slope from grain size alone, for a large range in medium size is present on both high- and low-energy beaches.

MODERN COASTAL SANDS

Sandy coastlines may be characterized as of two types: (1) those with beach ridges attached to the shoreline and (2) those with barrier islands separated from the shore by a lagoon or bay. In the beach-ridge type, a succession of beach sands may be amalgamated to form a broad, sandy flat or strand plain. In the barrier-island coast, these sands are concentrated in narrow trends that are often parallel to the coast for many miles. Both types of deposits were formed by the same processes of wave action so that their internal textures and structures are the same.

Barrier-Island Sequence

The modern barrier islands of the Texas Gulf Coast are well known from the study of closely spaced core holes along transverse profiles (Figure 7–4). The internal se-

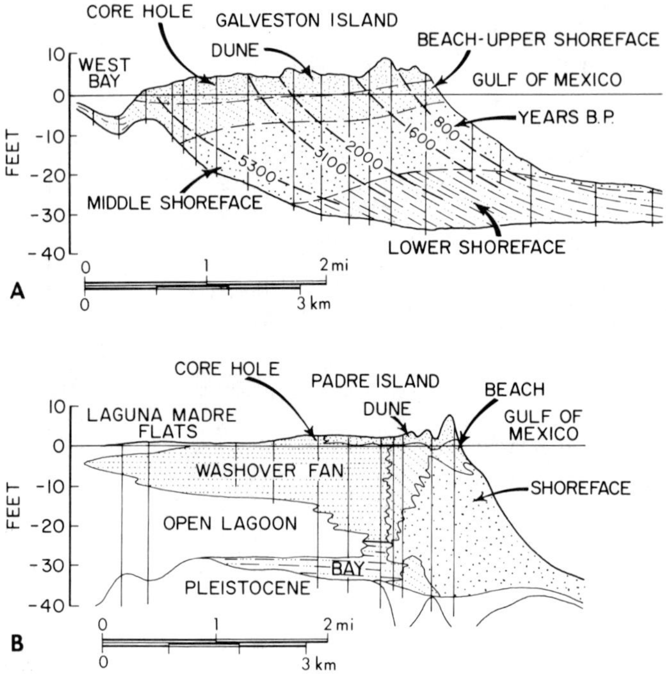

Figure 7–4 Cross sections through barrier islands of the Texas Gulf Coast: (A) Galveston Island—time lines in years before present are based on radiocarbon dating; (B) Padre Island on the south Texas coast. Galveston Island has a shallow lagoon called West Bay, whereas Padre Island is bounded by the broad Laguna Madre. [(A) after Bernard et al., (B) after Fisk 1959.]

quence is illustrated by the barrier island at Galveston in southeast Texas (Bernard et al. 1970, p. 202):

Unit	*Thickness ft (m)*
1. Eolian and beach sand, fine to very fine grained, weathered, massive, with dark pseudolaminae near the base.	6 (1.8)
2. Beach and upper-shoreface sand, fine to very fine grained, well laminated, with shells.	6 (1.8)
3. Middle-shoreface sand, very fine grained, burrowed, laminated, and thin to thick bedded with shells.	18 (5.5)
4. Lower-shoreface sand, shelly, burrowed, laminated, and interbedded with silt-clay.	6 (1.8)
Total	36 (10.9)

The lower shoreface represents the inner-neritic environment of highly bioturbated sand, silt, and clay. The middle shoreface is clean sand that is worked by wave action only during periods of storms. In the period of normal wave action between storms, the bottom is populated with benthic organisms that burrow into the sediment and tend to disrupt the sedimentary structures. Therefore, the middle-shoreface section is one of generally massive appearance, although relict laminae may be preserved. The beach and upper shoreface are subject to continual wave action, and the sand is finely bedded with thin, nearly horizontal laminae. The eolian dune sand is carried from the beach face across the barrier island by winds and by surge during major storms. The eolian sands are vegetated by grass, vines, and small shrubs whose roots extend downward to the shallow water table. Bioturbation by rooting tends to destroy the sedimentary structures, and the sand has a massive appearance.

Padre Island on the south Texas coast has a different distribution of facies (Figure 7–4). Middle-shoreface and beach sands are restricted in lateral extent, and the section is dominated by a thick washover fan composed of fine-grained sands that were deposited behind the barrier island in the adjacent lagoon. The washover deposits are partly eolian in origin and partly the result of sand carried over the barrier island and into the lagoon by the storm surge of hurricanes. The washover sands are finely laminated to rippled and may be partly bioturbated. The washover fans extend as large arcuate bodies into the lagoonal area. Fan sediments may contain thin algal mats. Because of the low tidal range, the washover fans are commonly exposed and may contain gypsum crystals because of the arid climate. The eolian dune sands of the barrier island also prograde across the exposed fans. Deposits directly related to the storm surge of hurricanes are only about 1-ft (0.3-m) thick and are cross-laminated because of the current transport across the fan (Hayes 1967). The fans prograded shoreward over muds of the open lagoon that commonly contain small banks of oyster shells.

Texture and composition. Grain size and composition from Galveston Island cores show an increase in mean size upward and an increase in quartz content (Figure 7–5). Mean grain size of quartz increases from very fine to fine grained at the top of the section. The content of the monocrystalline quartz increases abruptly to 60 percent or more above the highly bioturbated and clayey lower shoreface. Quartz continues to increase in amount upward, whereas feldspars and rock fragments decrease. Clay matrix is common only in the lower part of the section. Shell fragments are relatively abundant throughout the section but are not included in the detrital composition.

The subenvironments of the barrier island tend to segregate according to composition and mean grain size (Figure 7–6). The eolian-dune, beach, and upper-shoreface sands have the highest content of quartz, which reaches a maximum of 87 percent. The middle-shoreface sands have a range of 68 percent to 76 percent with a corresponding increase in rock fragments and feldspars. The lower-shoreface

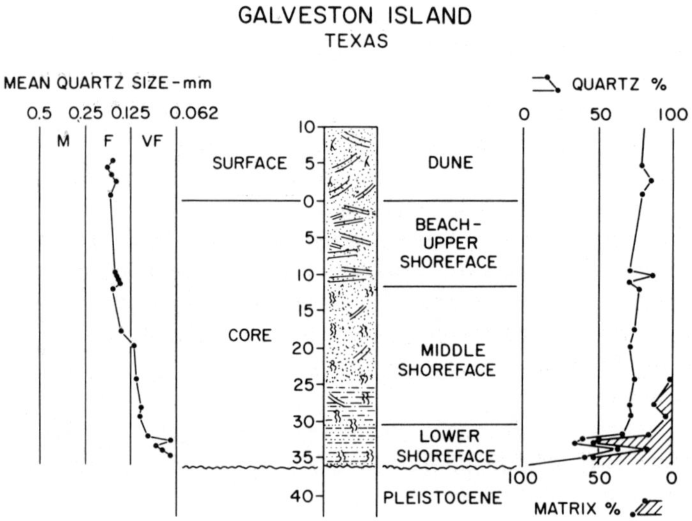

Figure 7–5 Grain size and composition of Galveston Island sand from a core hole along the Eight Mile Road section (Figure 7–7). Mean size of quartz increases upward, and quartz is dominant above the lower shoreface. [Data from Ethridge 1970.]

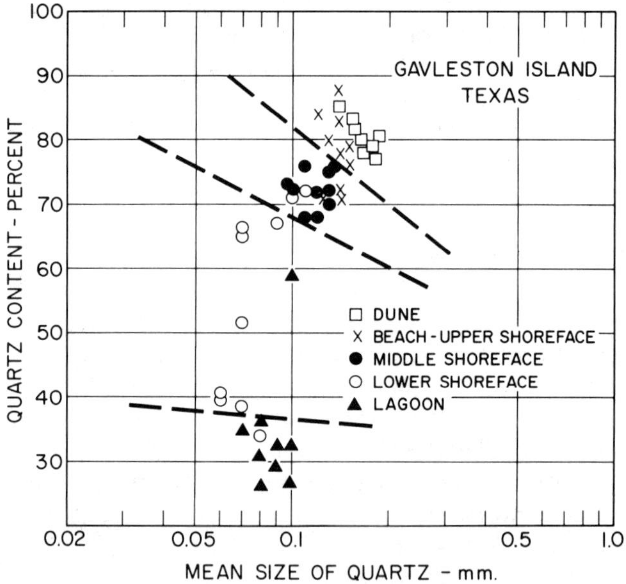

Figure 7–6 Content of monocrystalline quartz as a function of mean grain size for Galveston Island, southeast Texas. Subenvironments of the barrier island tend to segregate according to composition and mean size.

sands show a decreased quartz content and an increase in clay matrix. The lagoonal deposits are largely muds.

The high quartz content is the result of wave action on the shoreface, which removes clay matrix, reduces the amount of feldspar and rock fragments, and enriches the sands in monocrystalline quartz. Source sediment is of fluvial origin and has an average quartz content in the range of 40 percent to 60 percent with abundant feldspars and rock fragments. The fluvial sands also tend to have a large mean grain size. These feldspathic and lithic sands are greatly modified along the beach, and enrichment in monocrystalline quartz is characteristic of beach sediments.

Morphology. The morphology of the barrier islands is typically narrow and linear, parallel to the shoreline. Galveston Island, for example, is only 2 mi (3.2 km) wide, but it is nearly 30 mi (48 km) in length (Figure 7–7). At its northeast end, Galveston Island is separated from Point Bolivar by a major channel that is currently maintained by dredging. Maximum sand thickness is about 50 ft (15 m), and thickness gradually diminishes toward the southwest along the coast. The island is bounded on the southwest by another major channel, San Luis Pass.

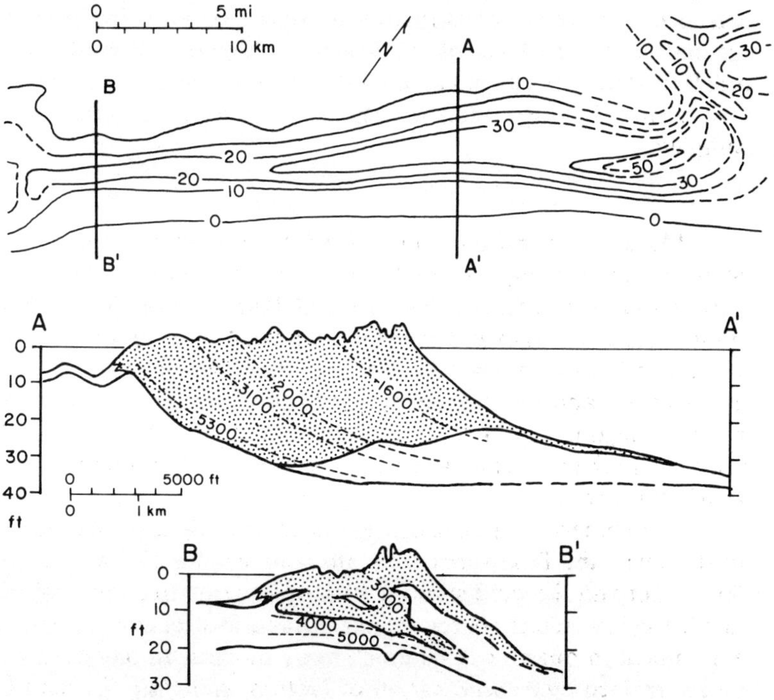

Figure 7–7 Thickness of sand in Galveston Island, southeast Texas, and generalized cross sections at Eight Mile Road (AA′) and West Line (BB′). Time lines in years before present (BP) based on radiocarbon dating. [From Bernard et al. 1970.]

Longshore drift is toward the southwest, and sand has been transported from the northeast. In fact, the High Island area, located northeastward from Point Bolivar, is an area of active erosion and shoreline retreat, which exposes the underlying Pleistocene Beaumont Clay. Sand accumulation began at the northeast end of Galveston Island about 5500 years before present (BP), and the island has prograded seaward by the accretion of sand at the shoreface. Southwestward extension of Galveston Island was accomplished largely by spit accretion, and the accumulation of sand at the southwestern end of the island began about 4000 years before present (Figure 7–7). Time lines shown in the cross sections are based on radiocarbon dating (Bernard et al. 1970). Longshore currents continue to transport sand in a southwesterly direction, constructing similar barrier islands along the middle Texas coast. Toward the south, however, longshore drift is toward the north from Mexico, and the longshore currents converge in the vicinity of Padre Island. A large amount of sand accumulates here, much of which is carried inland across Laguna Madre to form the dune fields of the south Texas coastal plain.

The process of erosion and transport of sand along the Texas coast emphasizes an important aspect of the barrier islands. Beaches of sand everywhere are ephemeral features. Their formation depends on a balance between sand supply, wave energy, and longshore drift. It is well known that if a beach system becomes unbalanced, say by the construction of groins or piers, an excess of sand begins to accumulate at one location while erosion begins at another. In the long term, all beach sands are in transit and are subject to erosion. Thus, if sand supply is reduced, for example, by the diminished capacity of rivers, the barrier islands may disappear completely.

Modification by erosion. The barrier islands of the East Coast of the United States are undergoing more active erosion than those on the Texas coast. The typical barrier-island sequence is that of the Delaware coast (Kraft 1971). These sands are transgressive in origin and were built landward over previous marsh deposits (Figure 7–8). The Delaware coastal sands are somewhat coarser grained than those of the Texas coast, and mean grain size is in the range of 0.25 to nearly 0.5 mm. Composition, again, is greatly enriched in quartz at the beach face where the content of feldspar, rock fragments, and clay is reduced, and monocrystalline quartz may reach nearly 90 percent.

Enrichment in quartz content is characteristic of shoreline sands. In the case of the Texas and Delaware coasts, the source sediment has a quartz content of about 50 percent and the sand shows a striking increase in quartz content. If, on the other hand, the immediate source sediment has a low quartz content, say 25 percent, the enrichment in quartz will be much more modest. In any case, the composition will be determined by a combination of factors, including the nature of the source sediment, the wave energy of the nearshore area, and the length of time the sands are in motion under wave action or in transport by longshore currents.

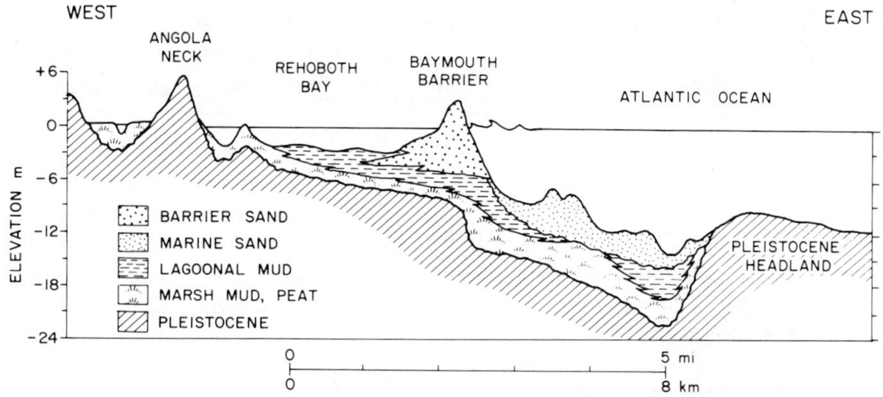

Figure 7–8 Barrier-island sands of the Delaware coast were deposited by transgression over previous marsh deposits and are now being actively eroded. [After Kraft and John 1979.]

Strand-Plain Sequence

The strand-plain coastline shows a broad distribution of shoreline sands. Such a section has been described in the coastal zone of Nayarit, northwestern Mexico (Curray, Emmel, and Crampton 1969). Here the beach ridges have coalesced to form a sand sheet that ranges from 5–15 km in width and is bordered on the landward side by extensive mangrove marshes and more limited lagoons (Figure 7–9). The sheet sands extend for at least 100 km along the coast and are, therefore, a significant body of sand. The marsh lagoon sequence contains washover sands as well as alluvial deposits. The strand-plain section is a composite transgressive-regressive sequence. Deposition began 7000 years before present with a rise in sea level that produced a rapid shoreline advance over lagoon-marsh sediments, fol-

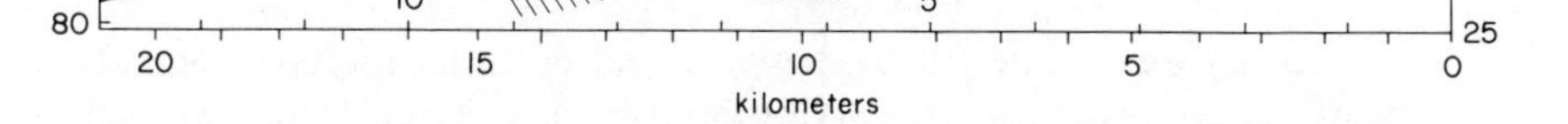

Figure 7–9 Cross section of modern coastal deposits of the Nayarit area, northwest Mexico, showing a transgressive sequence from marsh-lagoon to beach sands. [After Curray, Emmel, and Cramption 1969.]

lowed by regression or progradation beginning about 4500 years before present, during which time longshore drift brought larger amounts of sand to the area. These sands accreted at the shoreface to form narrow beach ridges with intervening marshy swales.

The shoreface sands are fine grained, and the beach-ridge and dune sands are medium grained. Stratification was described as fair to poor with only crude lamination sometimes visible (Curray, Emmel, and Crampton 1969, pp. 76–78). The sands are composed of 40 percent quartz, 40 percent volcanic rock fragments, 10 percent potassium feldspar, and 10 percent plagioclase. The composition of the fluvial source sediments is unknown, and the degree of quartz enrichment cannot be determined.

Other Coastal Environments

The interdeltaic coastal zone contains other depositional environments in which sand is less common than mud. These environments include the chenier plain, estuaries, and mud flats. Some coastlines receive little sand to be concentrated in the form of beaches. Sands may be lacking either because of sand-poor source sediment or because of low energy for sediment transport. Estuaries, for example, are located landward from the shoreline and are largely protected from wave action. Only where tidal currents are strong is sand transported into the mouth of estuaries.

Chenier plain. Isolated sand ridges are found in the marshy coastal plain of southwestern Louisiana. Here the sand ridges rise nearly 10 ft (3 m) above the surrounding marsh, and their well-drained character allows the growth of oak trees, hence the term *chenier ridges* from the French word for "oak" (*chêne*). The supply of sand to the area is small because most of the bed load of the Mississippi River is deposited at the delta mouth and not transported westward by coastal currents.

The chenier plain is a prograding sequence that is built by a succession of beach ridges and their adjoining coastal marshes (Figure 7–10). The sand ridges are narrow and only about 2000 ft (610 m) in width, but they extend parallel to the shoreline for distances of 5 mi (8 km) or more. Presumably, these were originally thin beach deposits that extended across the mouths of bays that were subsequently filled by muddy sediment. The sand ridges have a thickness of only 10 ft (3 m) and are generally highly quartzose like other Gulf Coast beaches, and grain size also increases upward. However, these sands also contain abundant shell fragments, which greatly reduce the grain content.

Estuaries. The fill of drowned valleys may include deposits of sand, but in general, estuaries form a relatively sand-poor section. However, estuarine sediments may underlie barrier-island and strand-plain deposits. The estuarine sequence is one of transgression, so the section may begin with fluvial deposits of the preceding alluvial valley. Succeeding deposits form a transition of tidal flats, to bay muds, and finally to the shallow-marine mudstones. Within this sequence, sands may be

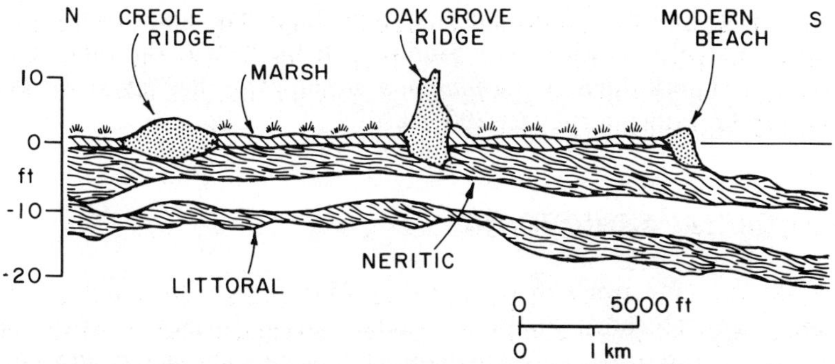

Figure 7–10 Diagrammatic section of the chenier plain of southwestern Louisiana showing narrow beach ridges isolated within marsh deposits. [Adapted from Byrne, Leroy, and Riley 1959 and Gould and McFarlan 1959.]

deposited in the form of small deltas, tidal-flat channels, or linear tidal bars. In general, sands are not extensive in the estuary sequence, and when present, they may form narrow linear bodies that are oriented parallel with the axis of the drowned valley.

Few modern estuarine sections have been documented, and these commonly show only fluvial sands concentrated at the base. Estuarine fill of the Chesapeake Bay (Hack 1957) and the Texas Gulf Coast (Wilkinson and Byrne 1977) are sequences of this type. Thin sands reworked by tidal currents into elongate bars are known in other modern estuaries (Allen 1971). Still, the volume of sand is relatively low in the total sequence unless succeeded by beach deposits.

Tidal flats. Extensive areas of mud and sand may be present on open coastlines as well as within estuaries. The tidal-flat sequence is well known from studies of coastlines bordering the North Sea. Along the coast of the Netherlands, broad tidal flats up to 7 km in width border the shoreline (Reinick 1972). The total thickness of deposits is equal to the tidal range, which is 2 to 4 m. Below the supratidal salt marsh are mud flats with sand concentrations only in ebb-flow channels. Increasing amounts of sand are found seaward to the level of low tide. The mud flats are characterized by interbedded muds and ripple lenses of sand that form a distinctive sequence characterized as *flaser bedding*. Sand flats are also dominated by low flow-regime structures.

Thicker tidal-flat deposits are present in the estuary known as the Wash on the northeast coast of England (Evans 1965). The maximum tidal range in the Wash is 22 ft (6.7 m), and tidal sediment has a thickness of the same range. This thicker section is dominated again by muds or mixtures of sand and mud. As in the Netherlands coast, sand is concentrated only in narrow ebb channels. In both areas, the mud-flat deposits may be highly bioturbated because they support an abundant fauna, largely of mollusks that burrow within the shallow sediment.

Despite their sand-poor character, tidal flats may form thin but extensive deposits by progradation. For example, in the Wash, the tidal-flat deposits form a coastal plain known as the *fenlands,* a plain that has advanced seaward at least 15 mi (25 km) during the last 400 years.

ANCIENT COASTAL SANDSTONES

Reservoir sandstones of coastal origin are present in many ancient sections of diverse ages. The most commonly recognized sandstones are of shoreline origin, either barrier-island or strand-plain deposits. They are characterized by high grain content, often enriched in quartz, and relatively free of matrix. When cores are available, the shoreline sandstones show a characteristic sequence of sedimentary structures similar to those of the modern Texas coast (Davies, Ethridge, and Berg 1971 and Dickinson, Berryhill, and Holmes 1972). When cores are lacking, shoreline sandstones can often be identified by their narrow linear trends parallel to the ancient coastline.

Sandstones described in the following sections are from the Lower Cretaceous Muddy Formation in the Powder River basin in Wyoming and Montana. The Muddy consists of interbedded nonmarine to marine sandstones and shales in an overall transgressive sequence. The Recluse field shows an estuarine sequence from a basal fluvial-channel fill to overlying transgressive marine sandstones (Figure 7–11). The Bell Creek field displays a barrier-island sandstone and adjacent lagoonal deposits. The Hilight field has a coastal-marsh section succeeded by shoreline sandstones that were partly eroded or reworked during transgressions. These examples illustrate the variety of sandstones and morphologies that can be encountered in the coastal zone.

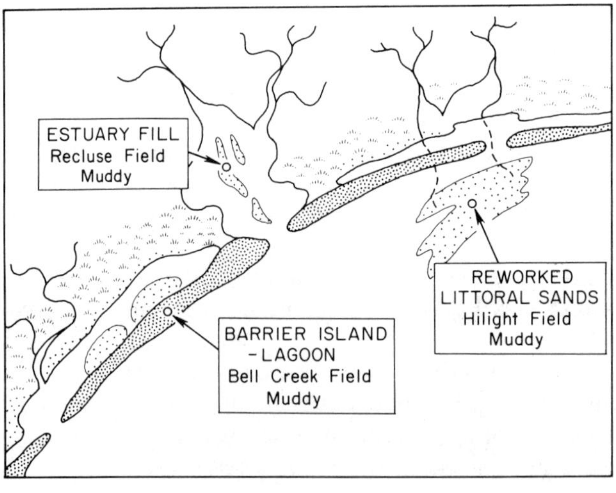

Figure 7–11 Diagrammatic map showing relative locations of ancient coastal interdeltaic sandstones described in the following sections.

MUDDY SANDSTONE, RECLUSE FIELD, NORTHEASTERN WYOMING

Lower Cretaceous reservoir sandstones of the Muddy Formation are found at four levels in the Recluse field, and the sandstones are of two types: (1) narrow sinuous bodies represent the fill of fluvial channels, and (2) small ovate bodies represent shallow-marine deposits. These fluvial and marine sandstones are the initial deposits of the Muddy transgression. They are closely associated in vertical sequence and demonstrate the proximity of both depositional environments to the shoreline. The thinly interbedded reservoir sandstones and shales form a complexly layered reservoir system within the field area.

Geologic Setting

The Powder River basin (Figure 7–12) is an asymmetric syncline that is bounded on the east by the Black Hills platform. West of a narrow monocline, the basin dips southwestward at rates of only 150 ft to 200 ft/mi (28 to 38 m/km) toward the deep basin axis that is bordered on the west by the Bighorn Mountains and Casper uplift, and on the south by the Laramie Range and Hartville uplift. Oil production from the Muddy Formation was first established in the Newcastle area during the 1920s and then in the Fiddler Creek and Clareton fields during the late 1940s. Major field development did not come until 1967 when the Bell Creek field was discovered in southeastern Montana (McGregor and Biggs 1968) and was soon followed by the

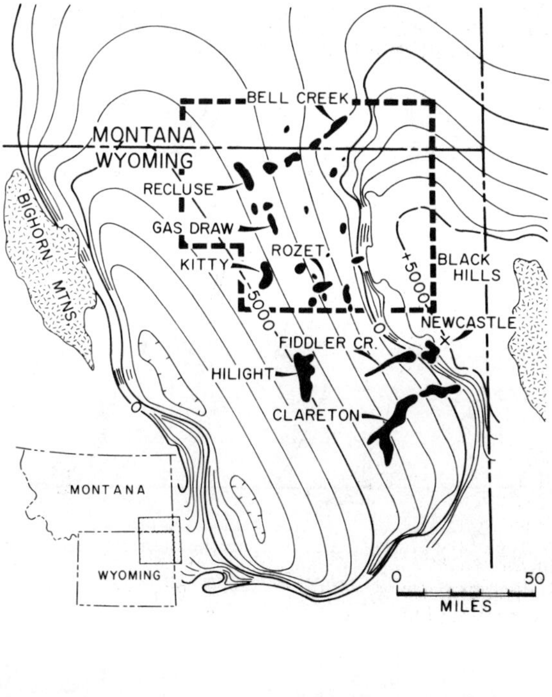

Figure 7–12 Structure map of the Powder River basin, northeastern Wyoming and southeastern Montana, showing location of oil fields in the Lower Cretaceous Muddy Formation. Contours on top of Lower Cretaceous Fall River (Dakota Formation); contour interval 1000 ft (300 m). Heavy dashed line shows area of Figures 7–13 and 7–15.

Recluse field discovery in Wyoming. Subsequent drilling established production in many small- to medium-sized fields through a broad trend that extended southward along the basin flank. A final major development in the trend was the discovery of Hilight field in 1971.

The lenticular coastal sandstones of the Muddy form stratigraphic traps for oil in all of the fields (Figure 7–13). The fields are located in a uniform regional dip that trends generally to the southwest but locally to the northwest as at Bell Creek field. Trapping is by capillary-pressure differences as a result of lateral changes in pore size from reservoir sandstones to finer-grained nonreservoir deposits (Berg 1975). An important factor in the trapping of oil is downdip hydrodynamic flow within the Muddy aquifer system, and a strong hydrodynamic effect has been demonstrated at several fields (Berg 1976*a*; Larberg 1980; Lin 1981).

It can be estimated that recoverable reserves of oil may approach 400 million barrels in all fields on the east flank of the Powder River basin. Cumulative production from the larger fields has now exceeded 200 million barrels (32 million m^3) (Table 7–1), and some reservoirs will be amenable to both secondary and tertiary

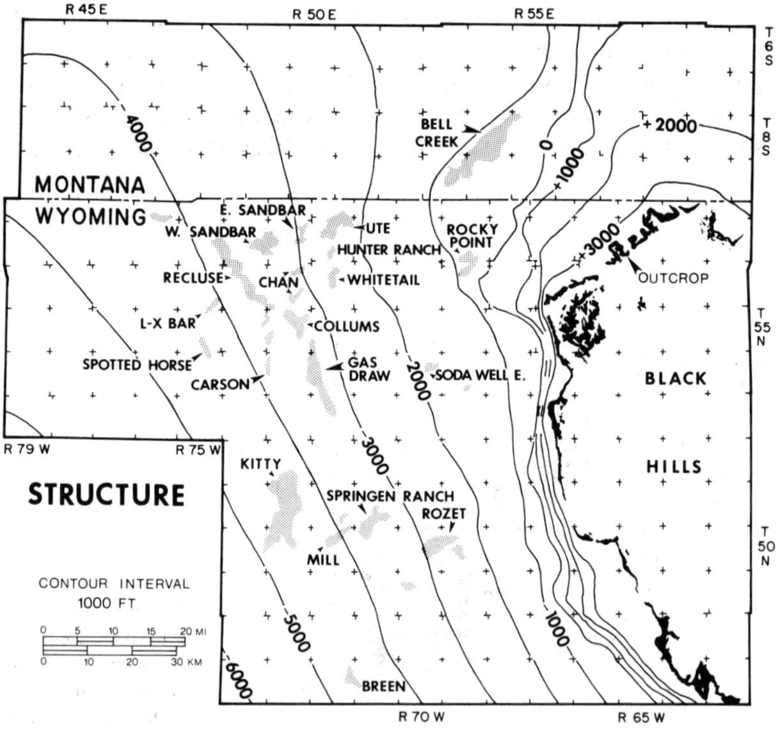

Figure 7–13 Structure on top of the Muddy Formation, northeastern Powder River basin, and location of Muddy stratigraphic oil fields. Contour interval 1000 ft (300 m). [Map drawn by L. D. Recker.] Map area shown in Figure 7–12.

TABLE 7-1. OIL PRODUCTION FROM SELECTED MUDDY FIELDS, NORTHEASTERN POWDER RIVER BASIN, MONTANA AND WYOMING

Name	Discovery date (mo/yr)	Producing wells (Total Jan. 1, 1977)		Production[a] (Annual 1976)	Production[a] (Cumulative Jan. 1, 1977)
Bell Creek	6/67	196	180	8.75	86.3
Hilight	2/69	195	176	3.82	57.6
Recluse	8/67	58	46	8.38	20.8
Gas Draw	8/68	83	47	1.8	19.7
Rozet	4/59	28	28	0.70	15.7
Kitty	8/65	175	161	0.50	15.1
Springen Ranch	11/68	43	28	0.70	8.9
Collums	2/69	42	23	0.20	5.4
Sandbar, East	1/68	24	12	0.10	4.2
Sandbar, West	2/68	16	11	0.06	3.3
Rozet East	6/61	8	8	0.16	1.7
Mill	3/69	22	22	0.01	1.5
Gillette	10/62	11	0	0	1.3
Recluse SE	12/68	5	4	0.02	0.4
L-X Bar	2/73	7	7	0.10	0.6

SOURCE: Data from International Oil Scouts Association, *Yearbook 1977.*

[a] Production in million barrels of crude oil.

recovery methods. Given the relative complexity of deposition and the thinness of individual sandstones, these coastal deposits form a remarkably prolific reservoir system.

The Muddy Formation has a variable thickness that averages only about 60 ft (18 m), and it consists of interbedded sandstones, siltstones, and shales. The formation can be divided into six zones (Figure 7–14) as based on marker beds in the lower part and thin bentonites in the upper part. These horizons can be correlated widely with considerable confidence so that the six zones represent time-rock units within which the distribution of depositional environments can be determined rather accurately. Reservoir sandstones are present at all levels, primarily in the lower zones to the west, and in the upper zones across most of the area.

The underlying Skull Creek Shale is a platy black marine claystone that was widely deposited throughout the Powder River basin. After Skull Creek deposition, the seas withdrew northwestward and exposed the Skull Creek Shale to subaerial erosion. The erosional surface is documented by the regional discordance of its top with the overlying Muddy, and locally by the loss of marker beds beneath the Muddy.

Regional thickness. Thickness of the Muddy Formation ranges from less than 20 ft (6 m) on the east to a maximum of 140 ft (43 m) on the west across the main producing area (Figure 7–15). The general westward thickening is not uniform, however, because of the unconformity at its base. The Muddy thickness represents successive filling of irregular topographic relief that was developed on the top of

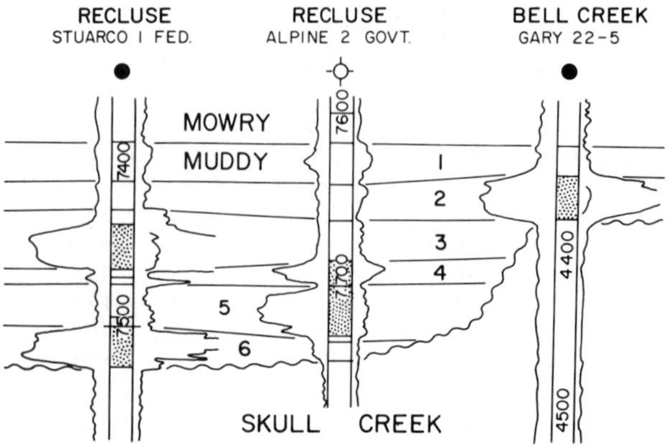

Figure 7–14 Correlation of electric logs showing six zones within the Muddy Formation from Bell Creek field on the northeast to the vicinity of Recluse field on the southwest. Location of Bell Creek and Recluse fields given in Figure 7–12. No horizontal scale.

the exposed Skull Creek shale (Stone 1972). In other words, the Muddy thickness mirrors underlying topography. Prominent trends of thicker Muddy reflect a dendritic drainage pattern on this surface. Tributary valleys from both the northeast and southeast converge near the central part of the area, turn abruptly north, and then continue westward. Another prominent drainage system is present in the south-central part of the map area.

Relief on the Skull Creek surface is gentle, and slopes seldom exceed 20 ft/mi (4 m/km). The Muddy interval thins toward the southeast and is presumed to be absent across a broad area that trends southwestward into the basin. The Muddy section reappears farther south in the Hilight field (Figure 7–12), and thus the area of missing section represents a local high that was not submerged by the Muddy sea. It has been suggested that the trends of Muddy valleys, as well as this arch, were controlled by northeast-trending structural lineaments that reflect basement structures along which there was recurrent movement to influence depositional patterns in both older and younger sections (Slack 1981).

Muddy sequence. The initial deposits of the Muddy were fluvial sandstones within the deeper drainage channels. Transgression of the Muddy shoreline then proceeded eastward across the unconformity surface, probably inundating the irregular topography to form an indented coastline of broad shallow estuaries and bays. Within the distal parts of the estuaries, older sediments were reworked by waves to produce transgressive marine sands that trend parallel to the drowned valleys. Near the shoreline, fluvial channels supplied sediment from the east to feed small deltas. The interdeltaic areas also consisted of swamps and marshes, and locally, wave action produced shoreline sands of limited extent. As the Muddy shoreline transgressed farther eastward, the previous coastal deposits were submerged, and a series of beach sands and barrier islands were deposited along a southwestward-trending shoreline. During this time, sands were supplied to the coastal area by a major delta system that persisted to the northeast (Waring 1976). Finally,

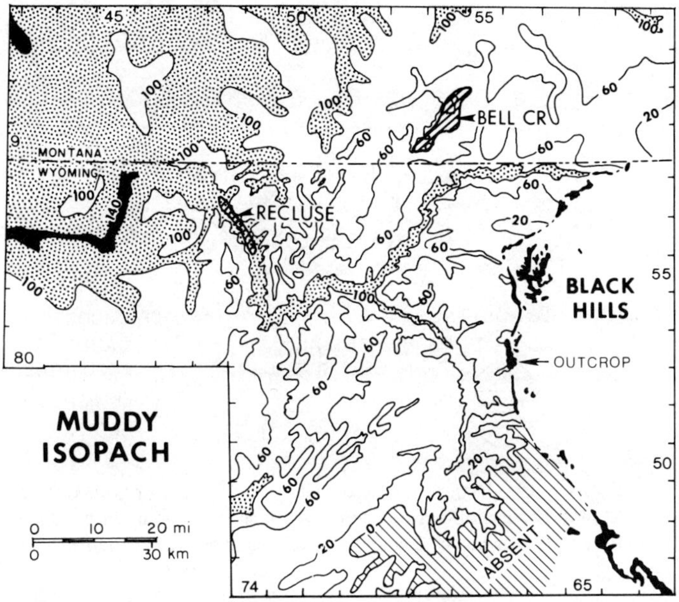

Figure 7–15 Thickness of the Muddy Formation in the northeastern Powder River basin. Contour interval variable in feet. [Map adapted from Stone 1972.] Location of map area shown in Figure 7–12.

transgression extended far eastward across the entire area and buried the Muddy coastal deposits beneath a blanket of marine mud. Marine conditions persisted to form the silicious shales and interbedded bentonites of the overlying Mowry Formation (Figure 7–14).

Within the overall transgressive sequence, sands of different origin were deposited during successive time intervals. The distribution of environments within the individual zones can be shown diagrammatically (Figure 7–16). The initial fluvial-channel deposits are present in the sixth zone at Recluse field (Berg 1976*a*). Succeeding transgressive estuarine sands are present in the fourth and fifth zones at Recluse, and deltas and small fluvial systems are present nearby at Gas Draw field (Lin 1981) and at Kitty field (Larberg 1980). The third zone deposits represent a minor regression wherein small fluvial channels and marsh deposits were widespread with limited areas of beach sand at the shoreline. Renewed transgression eastward, followed by a short stillstand, resulted in a series of barrier islands in the second zone at Bell Creek field (Berg and Davies 1968), as well as other shoreline sands to the southwest. Further transgression nearly covered the area with marine deposits, but a broad tidal flat persisted at Bell Creek in the first zone, and shoreline sands were deposited to the southeast at Rozet and Hilight fields (Figure 7–12).

The distribution of contemporaneous environments (Figure 7–16) is highly generalized and applies chiefly to the fields named. Other field areas may show

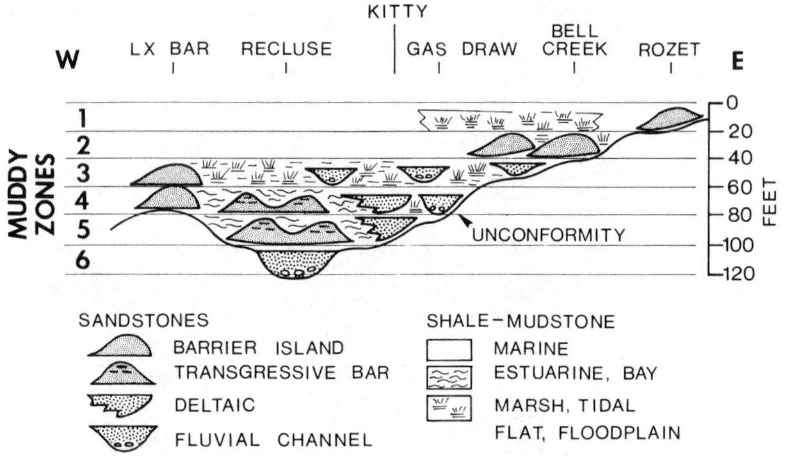

Figure 7–16 Diagrammatic distribution of contemporaneous environments in the Muddy Formation across the northern part of the main producing area, northeast Powder River basin. Location of oil fields shown in Figure 7–12.

somewhat different, but equally complex, interbedding of coastal deposits, both marine and nonmarine.

The Muddy sandstones at Recluse field were deposited during the slow and fluctuating transgression of the Muddy seas across the unconformity surface on top of the underlying Skull Creek shale. The local depositional setting is illustrated by the total Muddy thickness in the area (Figure 7–15). A prominent drainage system on top of the Skull Creek shale is indicated by thicknesses greater than 100 ft (30 m) of the Muddy Formation. Maximum thickness of the Muddy in the Recluse field area is about 120 ft (36 m), which represents nearly a maximum thickness for the Muddy in this area. Reservoir sandstones lie within this thick area and represent the initial deposits of the transgressing Muddy sea. Sandstones representing the sixth, fifth, and fourth Muddy zones are distributed along the trend of maximum Muddy thickness. During the initial transgression, the Recluse area must have been a broad shallow estuary that received sediment transported along the main drainage routes to the marginal marine environment.

Drilling History

The discovery well at Recluse field was completed in July 1967, only one month after the discovery of oil at Bell Creek in southeastern Montana. Rapid development of both fields helped to accelerate the search for Muddy oil production in the northeastern Powder River basin. Production at Recluse extends for about 13 mi parallel to the present regional strike (Woncik 1972). The field is narrow and only 1 to 1.5 mi (0.6 to 2.4 km) in width, but the producing area covers more than 8000 acres (3240 ha). The average net pay thickness is about 20 ft (6 m), and cumulative production to January 1, 1976, was approximately 20 million barrels (3.2 million m^3) of oil from 58 wells (Table 7–1).

Because of their lenticular nature, the Muddy reservoir sandstones are highly variable in occurrence within the producing area. In the central and southern parts of the field, a well-defined oil-water contact is generally present, and the trap is formed by loss of sandstone in an updip direction. In the north part of the field, however, there is no apparent oil-water contact, and the sandstones abruptly disappear in both updip and downdip directions. Furthermore, permeable but water-bearing sandstones occur laterally and even updip from the producing area.

The nature of the reservoir sandstone is known from cores of producing wells and dry holes. Full-diameter cores were obtained from four wells in the north part of the field; representative samples and thin sections were available from three additional wells.

Description of Sandstones

Sedimentary structures. A composite section of the Muddy Formation shows the following units:

Unit	*Thickness ft (m)*	
Third Muddy		
1. Shale, dark greenish gray, carbonaceous root traces, sandstone locally present, fine grained, inclined laminae.	20	(6)
Fourth Muddy		
2. Mudstone, medium gray, sandy, highly bioturbated.	2	(0.6)
3. Sandstone, fine grained, fines upward, laminated, horizontal to 10° dip.	10	(3.0)
Fifth Muddy		
4. Mudstone, medium gray, sandy, highly bioturbated.	8	(2.4)
5. Sandstone, fine grained, fines upward, laminated, horizontal to 10° dip.	25	(7.6)
Sixth Muddy		
6. Shale, brown to greenish gray, sandy.	1	(0.3)
7. Sandstone, medium grained, fines upward, carbonaceous fragments and thin beds, inclined laminae dip 20° to 30°.	28	(8.5)
Total	94 ft	(28.6)

The section is underlain by black platy shale of the Skull Creek Formation.

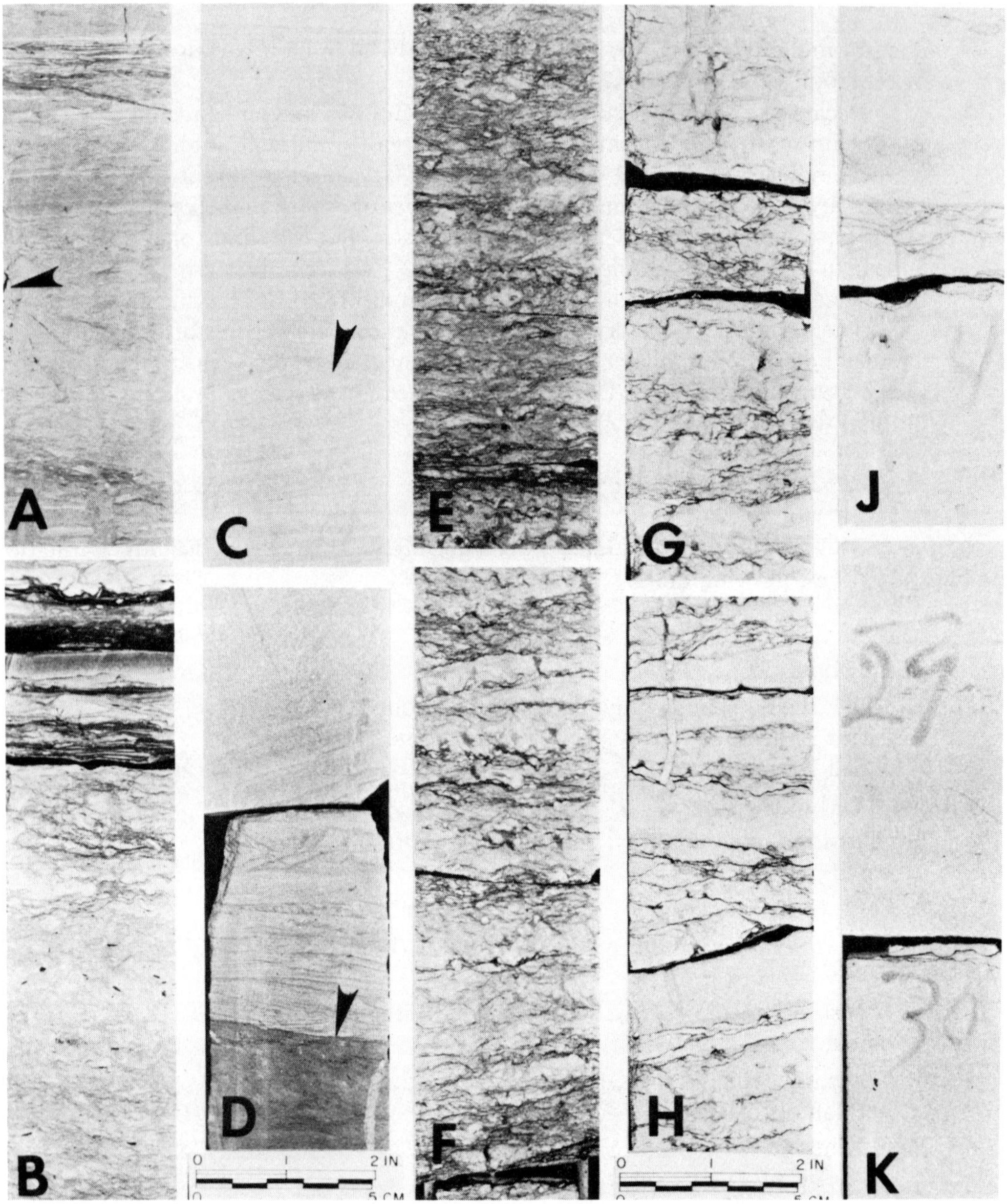

Figure 7–17 Sedimentary structures in third, fourth, and fifth Muddy sandstones in Alpine Parker 2 well, Recluse field: (A) Third Muddy, massive claystone partly bioturbated by rooting; note carbonaceous root trace (arrow); 7688 ft (2343 m); (B)–(D) Fourth Muddy sandstone; (B) Mudstone bioturbated by marine organisms and overlain by black shale and silt; 7690 ft (2344 m); (C) Fine-grained, laminated sandstone; note low angle of lamination and truncation surface (arrow); 7700 ft (2347 m); (D) Fine-grained, laminated sandstone showing sharp basal contact on underlying mudstone (arrow); 7703 ft (2348 m); (E)–(K) Fifth Muddy sandstone; (E) and (F) Highly bioturbated sandstone and shale which becomes increasingly shaly upward; 7705 and 7708 ft (2348 and 2349 m); (G) and (H) Bioturbated, fine-grained sandstone with thin shale laminae; 7711 and 7715 ft (2350 and 2351 m); (J) and (K) Fine-grained sandstone, massive to indistinctly laminated, with few shale laminae and minor bioturbation; 7718 and 7723 ft (2352 and 2354 m).

The lowermost sixth Muddy unit (7) is medium grained and conspicuously coarser than the succeeding units. Inclined laminae are common throughout the section, and laminations commonly dip 20°, and more rarely up to 30°, indicating a dune phase of sediment transport. Only in the uppermost 2.0 ft (0.6 m) of the unit (7) are there wavy nonparallel laminae that indicate a ripple phase of transport. Carbonaceous material is abundant throughout the section, both as fragments and in thin beds. Near the base, scattered carbonaceous fragments up to 1 cm in diameter are present with a few clasts of dark gray shale. Thin beds of coaly black shale up to 1 cm thick are also found at several levels. A thin shale unit separates the Sixth Muddy sandstone from the overlying unit. The decreasing scale of sedimentary structures upward denotes decreasing flow regime and deposition as a fluvial channel-fill deposit.

The overlying fourth and fifth Muddy sandstones are similar in character, and both are strikingly different from the underlying sixth Muddy. The sandstone units (3 and 5) are fine grained, laminated in the lower part, and generally massive above. The sandstones have sharp basal contacts on underlying shales [Figure 7–17 (D)]. Laminations near the base are plane, even, parallel, and either horizontal or inclined at 10° [Figure 7–17 (C) and (D)]. The overlying massive sandstone [Figure 7–17 (J) and (K)] contains a few vertical sand-filled burrows, and thin shale laminae increase in abundance upward [Figure 7–17 (G) and (H)]. The sandstones grade into overlying sandy mudstone, which becomes increasingly shaly and bioturbated upward [Figure 7–17 (B), (E), and (F)]. The bioturbation is an intensely churned mixture of sand and clay with a few larger rounded sand-filled burrows. This bioturbation is typical of normal-marine environments.

The total aspect of the fourth and fifth Muddy sandstones is that of deposition in shallow-marine waters of increasing depth. Laminations at the base of the section denote wave action, and the overlying massive character probably indicates homogenization by bioturbation. Increasing clay and bioturbation upward indicate submergence in shallow-marine waters. The two sections represent, therefore, transgressive marine sandstones.

The upper third Muddy unit (1) shows a return to partly emergent nonmarine conditions. The section is dominantly shale, massive, dark greenish gray, and only slightly sandy to silty in thin discontinuous laminae. Vertical carbonaceous stringers are common throughout the section and represent root traces [Figure 7–17 (A)]. Much of the shale has a highly churned aspect, which denotes intense bioturbation by rooting extending through at least 10 ft (3 m) of the section. However, no coaly horizon is present to mark the level of plant growth. Laterally, thick sandstones appear abruptly in the third Muddy unit (1). Where present, the sandstones are fine grained and contain inclined laminae in the lower parts and ripple laminae above. These sandstones, which locally attain a thickness of 23 ft (7 m), indicate the fill of fluvial channels. The third Muddy, therefore, represents a marsh environment that was locally drained by fluvial channels.

The overlying first and second zones of the Muddy were not observed in cores.

However, the character of these zones on electric logs suggests a dominant shale section and a probable return to marine conditions through the remainder of Muddy deposition.

Petrography. The sixth Muddy sandstone is characterized by a coarser grain size than the overlying zones (Figure 7–18). In the ARCO Bow and Arrow 3 core, the mean quartz size is 0.25 mm (medium grained), and grain size fines upward. Maximum grain size exceeds 0.50 mm (coarse grained). The overlying fifth Muddy sandstone is fine grained and has an average mean size of 0.15 mm and a maximum size of 0.26 mm.

Both the sixth and fifth Muddy zones have a relatively high quartz content of 72 percent and 80 percent, respectively (Table 7–2). In both sandstones, feldspar is found in amounts of 1 percent or less. Rock fragments are composed of polycrystalline quartz, metaquartzite, and chert. Matrix consists of clay minerals, finely divided organic material, and other fine-grained minerals. Muscovite and heavy

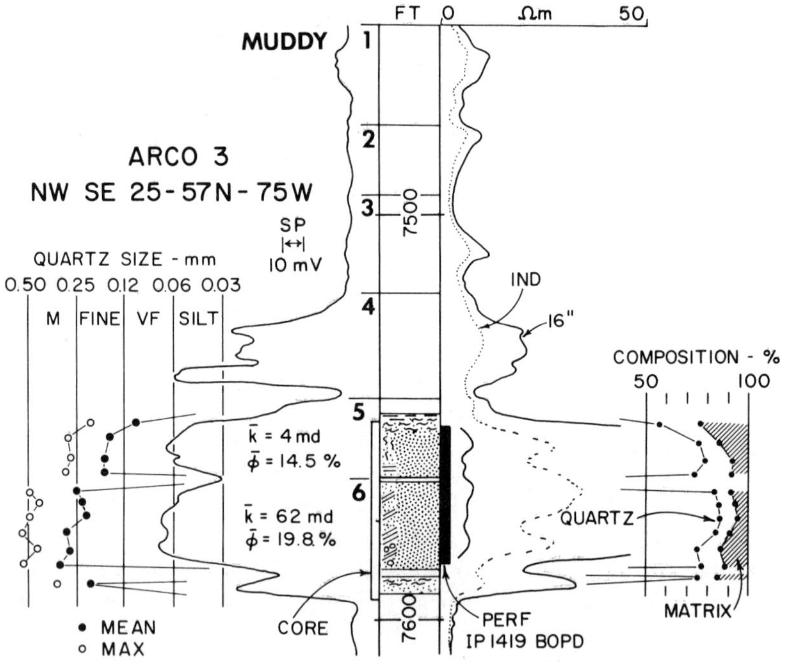

Figure 7–18 Electric logs, texture, and composition of the fifth and sixth Muddy sandstones in the ARCO Bow and Arrow 3 core. Location of well shown in Figure 7–25.

TABLE 7-2. PROPERTIES OF MUDDY SANDSTONES, RECLUSE FIELD, WYOMING

Well	Zone	Number of samples	Quartz size[a]			Detrital composition[b]					Cement[c]	Permeability (md)	Porosity (%)
			Mean (mm)	Max (mm)	σ (mm)	Qz (%)	F (%)	Rx (%)	Oth (%)	Mx (%)	Silica (% of total)		
Stuarco Federal 1	Third	8	0.17	0.31	0.05	85	—	7	—	8	3	90	16.4
Phillips Greenough 1	Third	4	0.16	0.28	0.04	86	—	6	1	7	1	—	—
Alpine Parker 2	Fourth	3	0.11	0.21	0.04	74	1	9	2	14	3	2.0	15.1
Apache State 1	Fourth	7	0.16	0.28	0.04	78	1	13	2	6	3	1.3	12.5
Alpine Parker 2	Fifth	7	0.18	0.30	0.04	72	—	14	2	12	4	17	16.0
Apache State 1	Fifth	5	0.20	0.35	0.05	87	1	8	—	4	9	403	19.5
ARCO Bow and Arrow 3	Fifth	4	0.15	0.26	0.04	72	1	11	2	14	3	4	14.5
ARCO Bow and Arrow 3	Sixth	7	0.25	0.48	0.07	80	—	9	1	10	3	62	19.8

SOURCE: Petrography from D. L. Stoudt 1974.

[a] Long-axis measurements; σ = standard deviation.

[b] Qz = monocrystalline quartz, F = feldspars, Rx = rock fragments including polycrystalline quartz, Mx = matrix, Oth = other minerals.

[c] Silica as grain overgrowths.

minerals are also found in trace amounts. Cement is largely silica as grain overgrowths, but some samples contain a significant amount of calcite as interstitial cement.

The main compositional difference between the fifth and sixth Muddy zones is in the relatively large amount of carbonaceous material present in the lower sandstone. Fragments and thin beds of coal and coaly shale are present in the sixth sandstone but are only rarely present at the base of the fifth sandstone. In addition, the fifth sandstone contains more abundant clay matrix, which increases upward with increasing bioturbation.

The fifth Muddy sandstone shows a great variation in composition laterally. For example, a high content of monocrystalline quartz of 87 percent is present in the Apache State 1 core (Figure 7–19). High quartz content is accompanied by a low matrix content, which results in the highest permeability and porosity of any Muddy zone. Slabbed cores were not available for description of sedimentary structures in the fifth Muddy sandstone. However, the average grain size is 0.20 mm (fine grained), and grain size decreases upward, all within the fine-grained class. Therefore, this sandstone is assumed to be similar to the fifth Muddy sandstone in the ARCO Bow and Arrow 3 core and is significantly finer grained than the sixth Muddy sandstone (Figure 7–18).

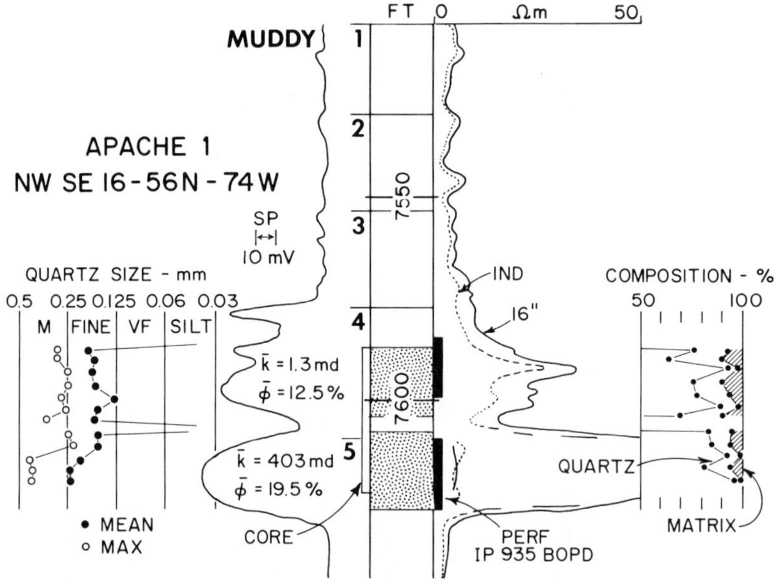

Figure 7–19 Electric logs, texture, and composition of the fourth and fifth Muddy sandstones, Apache State 1 well, Recluse field. Location of well shown in Figure 7–25.

The fifth Muddy sandstone in the Alpine Parker 2 core is similar to that in the Bow and Arrow 3 (Figure 7–20). The sandstone has a mean size of 0.18 mm (fine grained), and grain size decreases upward. Monocrystalline quartz averages 72 percent, and quartz content is highest at the base of the fifth Muddy sandstone and decreases upward as clay matrix increases. This change corresponds to the increase upward in the amount of shale and the degree of bioturbation. The overlying fourth Muddy sandstone shows no textural gradation but has an increasing quartz content upward. Bioturbation is confined to the uppermost part of the unit, and the transition to the succeeding mudstone is more abrupt.

Sandstone in the third zone is present in only a few wells, but a thick section was cored in the Stuarco Federal 1 well (Figure 7–21). Mean grain size is 0.17 mm (fine grained), and size decreases upward. Quartz content averages 85 percent with a matrix of only 87 percent. This composition closely resembles the fourth and fifth Muddy sandstones, but a core slab was not available for observation of sedimentary structures. A core from a nearby well, however, showed a coarse-grained basal unit with shale clasts and coal fragments; the succeeding section was composed largely of thin rippled sandstones with scattered shale clasts. The nature of the sequence suggests channel fill, probably of fluvial origin.

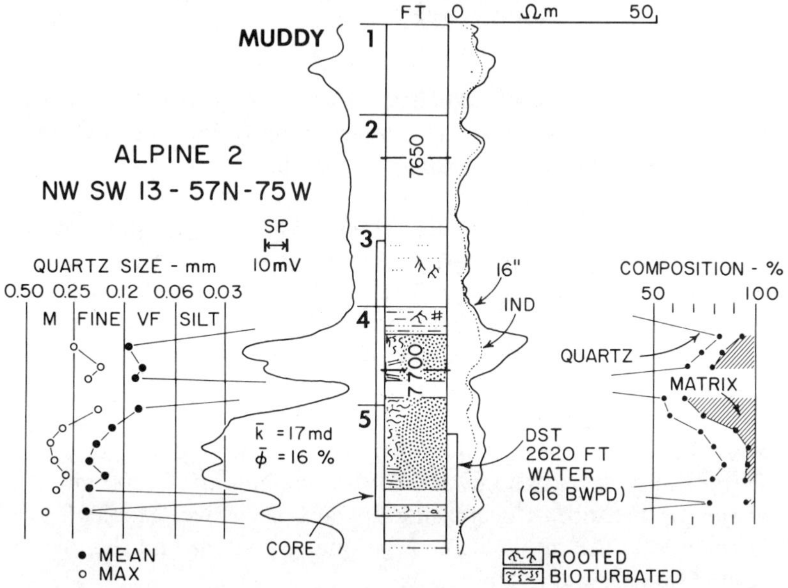

Figure 7–20 Electric logs, texture, and composition of the fourth and fifth Muddy sandstones, Alpine Parker 2 well. Location of well shown in Figure 7–25.

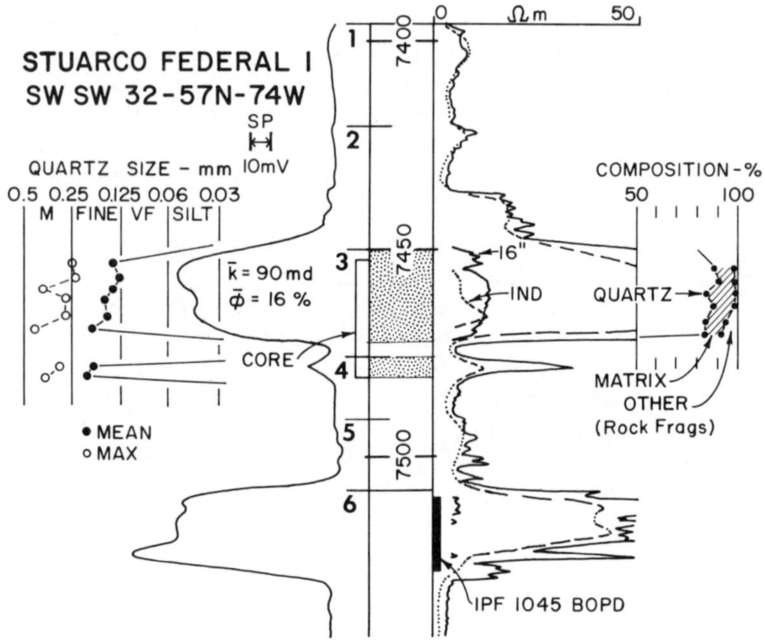

Figure 7–21 Electric logs, texture, and composition of the third Muddy sandstone, Stuarco Federal 1 well, Recluse field. Location of well shown in Figure 7–25.

Porosity and permeability. Reservoir properties of the Muddy sandstones are highly variable in the field area. Composition and texture play an important role in controlling porosity and permeability.

The sixth Muddy sandstone has an average permeability of 62 md and an average porosity of 19.8 percent (Figure 7–22). Permeability decreases from a maximum of 368 md at the base to 15 md at the top, a change that is controlled by decreasing upward grain size. The overlying fifth Muddy sandstone in the Bow and Arrow 3 core has an average permeability of 1.3 md and an average porosity of 12.5 percent, a reduction caused in large part by its finer grain size as compared with the sixth Muddy sandstone.

The fifth Muddy sandstone has the highest average permeability of 403 md and an average porosity of 19.5 percent in the Apache State 1 core (Figure 7–22). Permeability decreases upward from a maximum of 1200 md at the base to 80 md at the top because of decreasing upward grain size. The high average permeability is apparently the result of the high quartz content and the low matrix of the sandstone (Table 7–2).

The fifth Muddy sandstone of the Alpine Parker 2 core has an average permeability of 17 md and an average porosity of 16 percent (Figure 7–23). A drill-stem test of this sandstone recovered water at the rate of 600 bbls/day (95 m^3/day), although the well was located only about 2000 ft (610 m) from oil production at the

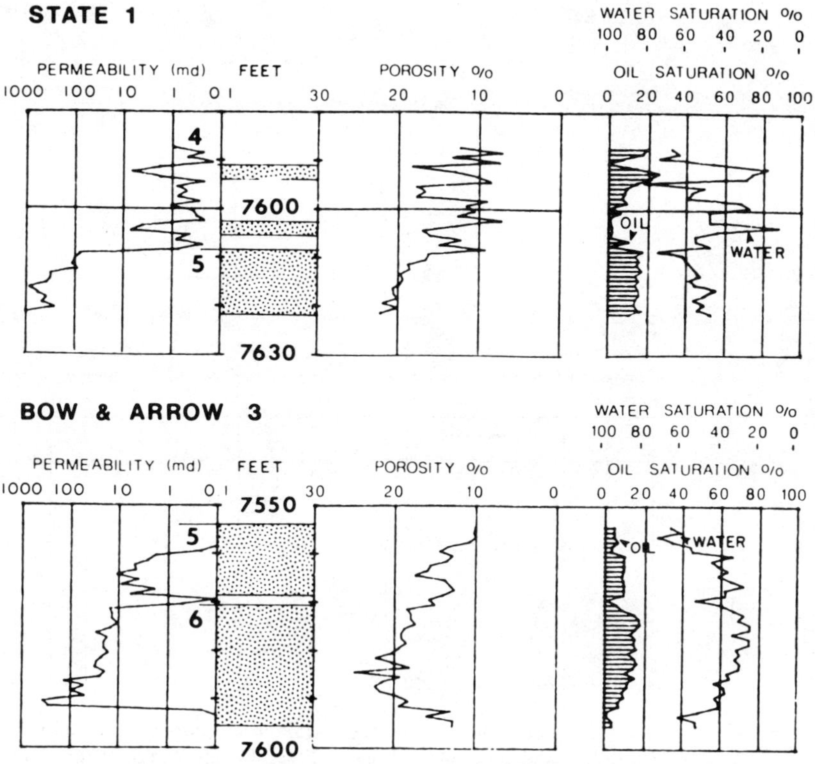

Figure 7–22 Permeability, porosity, and fluid saturations of the fourth, fifth, and sixth Muddy sandstones in the Apache State 1 and ARCO Bow and Arrow 3 wells.

same level to the south. Apparently, the fifth Muddy sandstone has greatly reduced permeabilities and no communication between these locations.

The fourth Muddy sandstone is also highly variable in its reservoir properties, but only low average permeabilities are recorded in the available cores (Figures 7–22 and 7–23). This sandstone, however, is a principal reservoir in the south part of the field and presumably has better reservoir properties.

The third Muddy sandstone has an average permeability of 90 md and an average porosity of 16.4 percent in the Stuarco Federal 1 core (Figure 7–23). Unfortunately, the sandstone has a restricted areal extent and is not a major reservoir unit in the field.

Morphology

The interpretations of the Muddy sandstones are confirmed by their distributions in the field area. The sixth sandstone forms a narrow sinuous body, which is present in only part of the field area [Figure 7–24 (A)]. This pattern is typical of fluvial

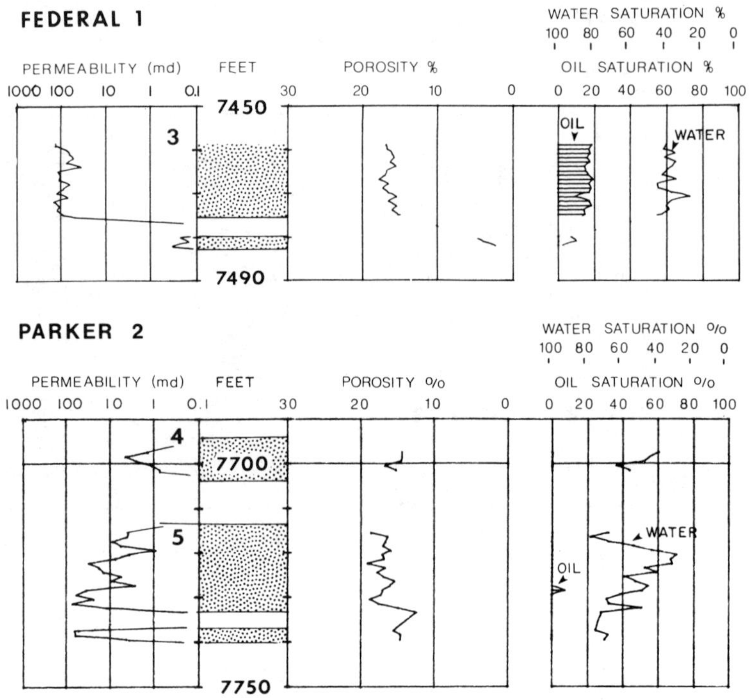

Figure 7–23 Permeability, porosity, and fluid saturations of the third, fourth, and fifth Muddy sandstones in the Stuarco Federal 1 and Alpine Parker 2 wells.

sandstones. Maximum thickness is 20 ft (6 m), and average width is about 2000 ft (610 m). Where it is present, the sixth sandstone is a major reservoir unit as in the ARCO Bow and Arrow 3 and Stuarco Federal 1 wells (Figures 7–18 and 7–21).

The overlying fifth sandstone is more widely distributed and has a distinctive pattern. The sandstone is present along the strike for the entire length of the field [Figure 7–24 (B)]. It is not uniform but appears to be a composite sandstone body composed of coalescing lenses with widths of 0.5 mi (0.8 km) and lengths up to 1 mi (1.6 km). The general elongation of the composite body is the result of deposition within the shallow estuary eroded into the underlying Skull Creek shale (Figure 7-15). The fifth sandstone shows the effects of wave action. Low-angle laminations are prominent in the lower part of the fifth sandstone and denote washing by waves. Bioturbation in the upper part of the fifth sandstone indicates that shallow-marine conditions were present in the estuary.

The fifth sandstone shows considerable variation on a local scale. The section in the Apache State 1 well (Figure 7–19) is highly quartzose and free of matrix. On the other hand, some sections are highly bioturbated and clayey as in the Alpine Parker 1 well (Figure 7–24, SE SW 13–57N–75W) where the basal 2-ft (0.6-m) section of the fifth sandstone is clean and quartzose and is overlain by high bioturbated mudstone of low porosity and permeability.

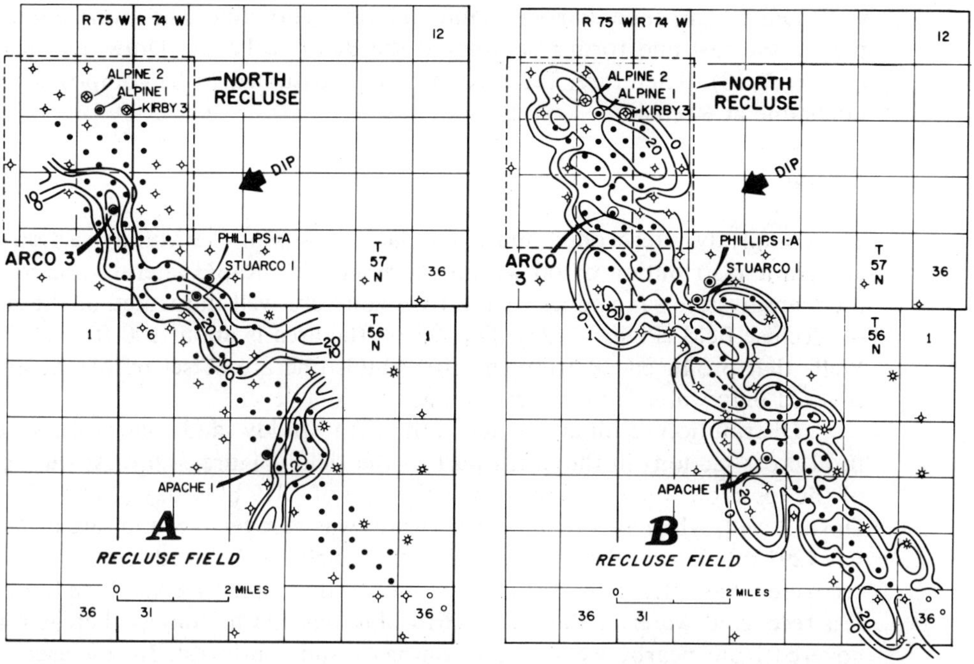

Figure 7–24 Gross thickness of sixth Muddy sandstones (A) and Fifth Muddy sandstones (B) at Recluse field. Contour interval 10 ft (3 m). Cores or samples were examined from circled and named wells. The north Recluse area is shown in Figure 7–25. [Maps from Berg 1976*a* and published with the permission of the Wyoming Geological Association.]

The fourth sandstone has a distribution that is similar to the underlying fifth sandstone. Average thickness is about 20 ft (6 m), and the unit is formed of coalescing lenses throughout the extent of the producing area. Low-angle lamination shows the effect of wave action, but some higher-angle laminations and truncation surfaces suggest that part of the section was deposited under conditions of low flow-regime. The fourth sandstone is more thinly bedded than the fifth in most wells and has a larger amount of carbonaceous material, which suggests, perhaps, the contribution of fluvial sands to the estuarine system.

The third Muddy zone shows a return to nonmarine conditions from the previous shallow-marine estuary. Common rooted mudstones suggest deposition in tidal flats and marshes. Sandstones are rarely present, and thicker sandstones on the order of 20-ft (6-m) thick are present only in the area near the Stuarco Federal 1 and Phillips 1–A wells (Figure 7–24). The sandstones are fine grained and contain common ripple structure, which suggests that these channel-fill deposits may have been washed by tidal currents. During deposition of the third zone, the shoreline was not far distant because barrier-island sands were deposited at the same time at the L-X Bar field located about 4 mi (6 km) to the southwest.

Succeeding deposits of the second Muddy zone represent renewed transgression. During this time, barrier-island and other shoreline sands were deposited farther to the east and form reservoirs at the Bell Creek, Gas Draw, and Kitty fields (Figure 7-13). Therefore, the second zone is assumed to be composed of marine mudstone at Recluse.

Oil Column

The complexly interbedded and lenticular deposits of the Muddy section result in an erratic distribution of oil within the reservoirs. Rapid lateral variations within the transgressive sandstones of the fifth zone are well illustrated at the north end of Recluse field (Figure 7-25). Dip across the area is about 100 ft/mi (19 m/km). Wells that produced oil from the fifth sandstone are offset by dry holes that recovered only water from sandstones at the same level.

Distribution of oil and water in the fifth Muddy sandstone is shown by structural cross sections in the north part of the field (Figure 7-26). At the updip part of the reservoir, oil was produced at the rate of 1819 bbls/day (289 m^3/day) in the Davis 2 well, while at the same level, water was recovered at the rate of 1843 bbls/day (293 m^3/day) in the Davis 3 well. Other wells at the north end of the field also recovered water from the fifth sandstone. In the Alpine Parker 2 well, a drill-stem test recovered water at the rate of 616 bbls/day (98 m^3/day), whereas oil was recovered from nearby wells to the southeast and southwest. In this area, the total oil column is 130 ft (40 m) as shown by the levels of highest and lowest oil production along the two cross sections (Figure 7-26).

The distribution of oil and water within the fifth Muddy reservoir is the result

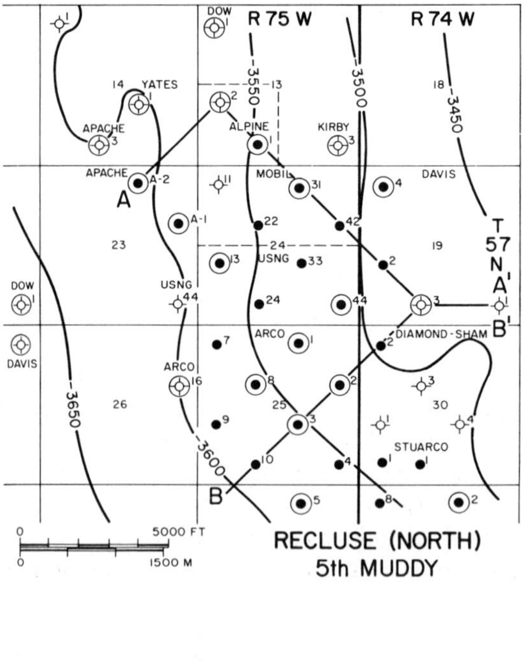

Figure 7-25 Structure on top of the fifth Muddy sandstone in the north part of Recluse field. Contour interval 50 ft (15 m). Location of map area shown in Figure 7-24; lines of cross sections AA′ and BB′ shown in Figure 7-26. Drill-stem tests were run in the circled wells. [Map from Berg 1976*a* and published with the permission of the Wyoming Geological Association.]

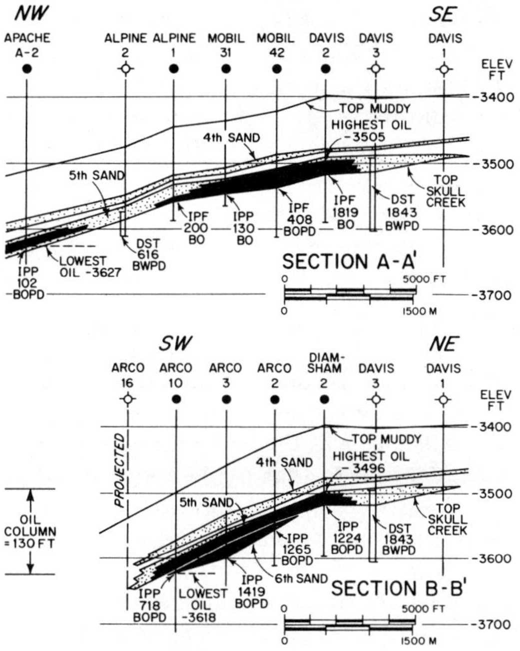

Figure 7–26 Cross sections AA′ and BB′ in the north part of Recluse field showing water recovery adjacent to oil production and an oil column greater than 100 ft (30 m). Location of cross sections shown in Figure 7-25. BOPD = bbls oil/day; BWPD = bbls water/day. [From Berg 1976*a* and published with the permission of the Wyoming Geological Association.]

of rapid lateral changes. A core from the Alpine Parker 1 well (Figure 7–25, SE SW 13) showed a highly bioturbated section of reduced permeability, and oil was produced only from the lowermost 2 ft (6.6 m) of the zone. Similar abrupt changes in the reservoir were detected near other wells by the interpretation of pressure buildup in drill-stem tests (Berg 1976). For example, pressure buildup in the well, USNG 44 (Figure 7–25, SW SW 24), indicates an abrupt reduction in permeability only a few hundred feet from the borehole and, presumably, a change to highly bioturbated sandstone. Apparently, the reservoir is heterogeneous on a local scale as suggested by the composite morphology of the sandstone [Figure 7–24 (B)].

Conclusions

Muddy sandstones represent interbedded nonmarine and marine sediments that were formed adjacent to the shoreline. The local depositional setting was characterized by a shallow estuary in which the initial deposits were fluvial channel-fill sands. Transgression caused submergence of the area by shallow-marine waters in which sands were transported by low flow-regime currents and washed by wave action. These processes produced a distinctive vertical sequence that represents a transgressive marine sandstone. These sandstones have abrupt basal contacts on underlying shale and are massive to laminated in lower parts with a high content of monocrystalline quartz. Clay matrix and bioturbation increase upward as water depth increased during transgression. Two periods of deposition of transgressive sandstones were followed by nonmarine deposits of a widespread tidal flat or marsh that contained narrow, sinuous, fluvial channels of limited areal extent. The area was finally transgressed by normal-marine deposits.

The transgressive marine sandstones are highly variable locally in their reservoir properties, probably due to the irregular distribution of bioturbation and clay matrix. The variable nature of the sandstones results in a complex reservoir unit in which the distribution of oil and water is erratic.

MUDDY SANDSTONE, BELL CREEK FIELD, SOUTHEASTERN MONTANA

An excellent reservoir of shoreline origin is found in the Lower Cretaceous Muddy Formation at Bell Creek field, northeastern Powder River basin (Figure 7–12). This sandstone, the Bell Creek barrier island, was the first reservoir to be recognized as displaying a complete sequence identical to that of the modern barrier islands of the Texas Gulf Coast (Berg and Davies 1968). Although the section is thin, the succession of shoreface environments can be recognized, including an uppermost eolian facies that attests to the emergent nature of the sand body during deposition. High porosity and permeability result from the high quartz content of the reservoir and its relatively shallow depth of burial. These characteristics have resulted in optimum oil recovery by primary and secondary methods and suggest that a large additional amount of oil may be recovered by tertiary methods.

Geologic Setting

The Bell Creek field is located off the northwest flank of the Black Hills uplift in southeastern Montana (Figure 7–13). Regional dip across the area is northwest at 100 ft/mi (19 m/km). The total Muddy section is thin, and only the first and second zones are present (Figure 7–14). Reservoir sandstone is found only in the second zone of the Muddy, which rests unconformably on the underlying Skull Creek shale. The thin Muddy section of only 40 ft (12 m) denotes that the area remained high, exposing the Skull Creek shale, while sands were being deposited farther to the west at Recluse field (Figure 7–17). After deposition of the third zone at Recluse, there was a general eastward transgression of the Muddy sea across the areas of previous deposition. When the shoreline reached the Bell Creek area, there was a pause or stillstand in the overall transgression. Sand continued to be supplied to the area by fluvial deltaic systems located just northeast of Bell Creek, and these sands were carried by longshore drift toward the southwest to accumulate as shoreline sands at Bell Creek. Some sand bypassed the Bell Creek area and was transported as far as the Gas Draw and Kitty fields (Figure 7–13), where thin sandstones are also present in the second Muddy zone. More sand of high quartz content, however, accumulated in the Bell Creek area.

Thickness of sandstone in the second Muddy shows two elongated bodies that partly overlap in the center of the field (Figure 7–27). The sandstones reach a maximum thickness of about 30 ft (9 m) and have an average thickness of about 20 ft (6 m). The sandstone shows a barrier-island sequence in cored sections (Berg and Davies 1968).

At least two distinct periods of barrier-island deposition can be inferred from the map. The first is a large, somewhat ovate body that forms the main part of the field in T8S-R54E. This sandstone has broad arcuate extensions of sand toward the northeast and southwest in which the sandstone thins to about 10 ft (3 m). The second body is located just northwest and appears to be somewhat thinner and more narrow and linear in its trend. This sandstone represents a second period of barrier-island deposition by progradation of the shoreline toward the northwest. This later sandstone body extends farther toward the southwest and forms an extension of the main Bell Creek field.

Drilling History

Bell Creek field was discovered in 1967 as the result of an extensive exploratory drilling program for Lower Cretaceous stratigraphic traps (McGregor and Biggs 1968). At the time of drilling, there was only a small area of oil production from the Muddy sandstone at the Ranch Creek field located about 1 mi (1.6 km) west of

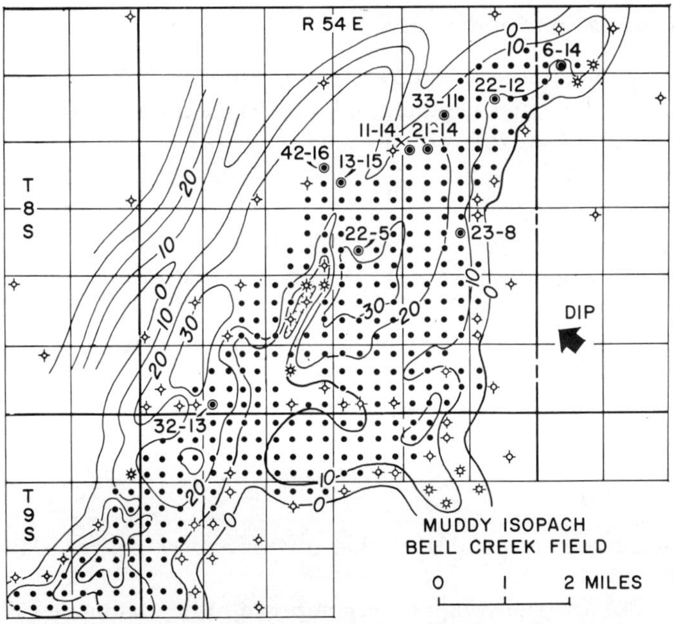

Figure 7–27 Thickness of Muddy sandstones at Bell Creek field, Montana, showing linear barrier-island sandstones partly overlapped in the center of the field. Regional dip is toward the northwest at 100 ft/mi (19 m/km). [Modified from Berg and Davies 1968.]

the south end of what was to become the Bell Creek field (Figure 7–27). Only four wells were present at Ranch Creek, and total production was about 80 bbls/day (13 m^3/day). No other oil production had been found in the entire southeastern part of Montana.

The Bell Creek discovery well was located in the south part of the field area (Figure 7–27, NE NE 33–8S–54E). Rapid development resulted in completion of over 200 producing wells, and oil production reached 170,000 bbls/month (27,000 m^3/month) by the middle of 1968. During field development, two distinct oil-water contacts were present in the north part of the field, representing separate accumulations within the two linear sand bodies (Figure 7–27). In the area of overlap, wells were completed in extensions of either the updip or downdip sandstones as in 15–8S–54E, whereas farther to the southwest a separation of the two sandstone bodies resulted in several dry holes in the middle of the field area as in 28–8S–54E (Figure 7–27).

At an early stage of field development, cores were obtained from ten wells throughout the field area. These wells are circled and numbered on the field map (Figure 7–27). The cores were representative of thick sections within the two sandstone bodies as well as of the area of overlap and thin extensions of the reservoir to the east and south. Interpretations of these cores resulted in the first description of the barrier-island sequence at Bell Creek field (Berg and Davies 1968).

Description of Sandstones

Sedimentary structures. The Muddy sandstone shows a sequence similar to that of the barrier island, and a typical section is shown in a core from the well Boekel 21–14, as follows:

Unit	*Thickness ft (m)*
1. Eolian sandstone, very fine grained, massive.	5 (1.5)
2. Beach and upper-shoreface sandstone, fine grained, laminated.	5 (1.5)
3. Middle-shoreface sandstone, fine grained, massive or with discontinuous laminae.	10 (3)
4. Lower-shoreface mudstone, shaly, highly bioturbated.	3 (0.9)
Total	23 (7)

This sequence is identical to that of the modern barrier island at Galveston Island (Figure 7–4).

The lowermost unit (4) is composed of interlaminated siltstone and shale [Figure 7–28 (D)]. Laminations are wavy and regular to irregular. The section is highly bioturbated but contains some distinct silt-filled burrow tubes up to 5 mm in diameter. Although bioturbation is intense, the section has a general horizontal aspect to the bedding.

The overlying middle-shoreface unit (3) is fine grained and mostly structureless or massive [Figure 7–28 (C)]. The section contains a few continuous or discontinuous laminae. There is also some indistinct mottling, which suggests that the original bedding may have been destroyed by burrowing organisms, and thus the general massive appearance was probably the result of homogenization by burrowing. Distinct burrow tubes are absent because the sandstone is well sorted and generally lacks a fine-grained matrix.

The upper-shoreface unit (2) is fine grained and mostly well laminated [Figure 7–28 (B)]. Laminae are even, parallel, and horizontal or gently inclined. Some laminae are gently curved and represent ripple structure.

The uppermost eolian unit (1) is generally massive but may contain a few discontinuous laminae [Figure 7–28 (A)]. The section also contains some indistinct mottling and rare, thin, burrowlike traces that contain a slight concentration of matrix. These traces may represent bioturbation by rooting, but no organic material is preserved.

Other cores in the field area show the same sequence of sedimentary structures through a section that is, on the average, about 25-ft (7.6-m) thick. The individual units may be variable in thickness and are everywhere gradational. The upper eolian unit (1) and the lower-shoreface unit (4) maintain rather uniform thicknesses, but the upper-shoreface unit (2) appears to thicken in some wells at the expense of the underlying massive middle-shoreface unit (3). The lowermost part of the middle-shoreface unit (3) contains some distinct bioturbation and clay matrix, and the gradation upward from the lower shoreface is generally abrupt.

The barrier-island sequence is found in cores within the main body of sandstone (Figure 7–27), in which the total thickness is greater than 15 ft (4.6 m). In all cores, the section is dominated by the massive middle-shoreface unit (3), but the thinner upper-shoreface and eolian units (1 and 2) are commonly present. The main body of sandstone, therefore, represents a sequence deposited at the shoreline that had a widespread subaerial dune facies signifying its true emergent character.

Updip from the main body of sandstone is the lagoonal facies. In this area, the section is less than 15 ft (4.6 m) in thickness, and much of the sandstone is distributed in a broadly arcuate pattern (Figure 7–27). The complete lagoonal facies was recovered in one core at the northeast end of the field, the Gary 6–14 well, and showed the following units:

Unit	*Thickness ft (m)*
1. Sandstone, very fine grained, white clayey matrix.	2 (0.6)
2. Sandstone, fine grained, quartzose.	2 (0.6)
3. Sandstone, very fine grained, ripple to horizontally laminated, interlaminated shale, bioturbated.	4 (1.2)
4. Mudstone, highly bioturbated.	2 (0.6)
Total	10 (3)

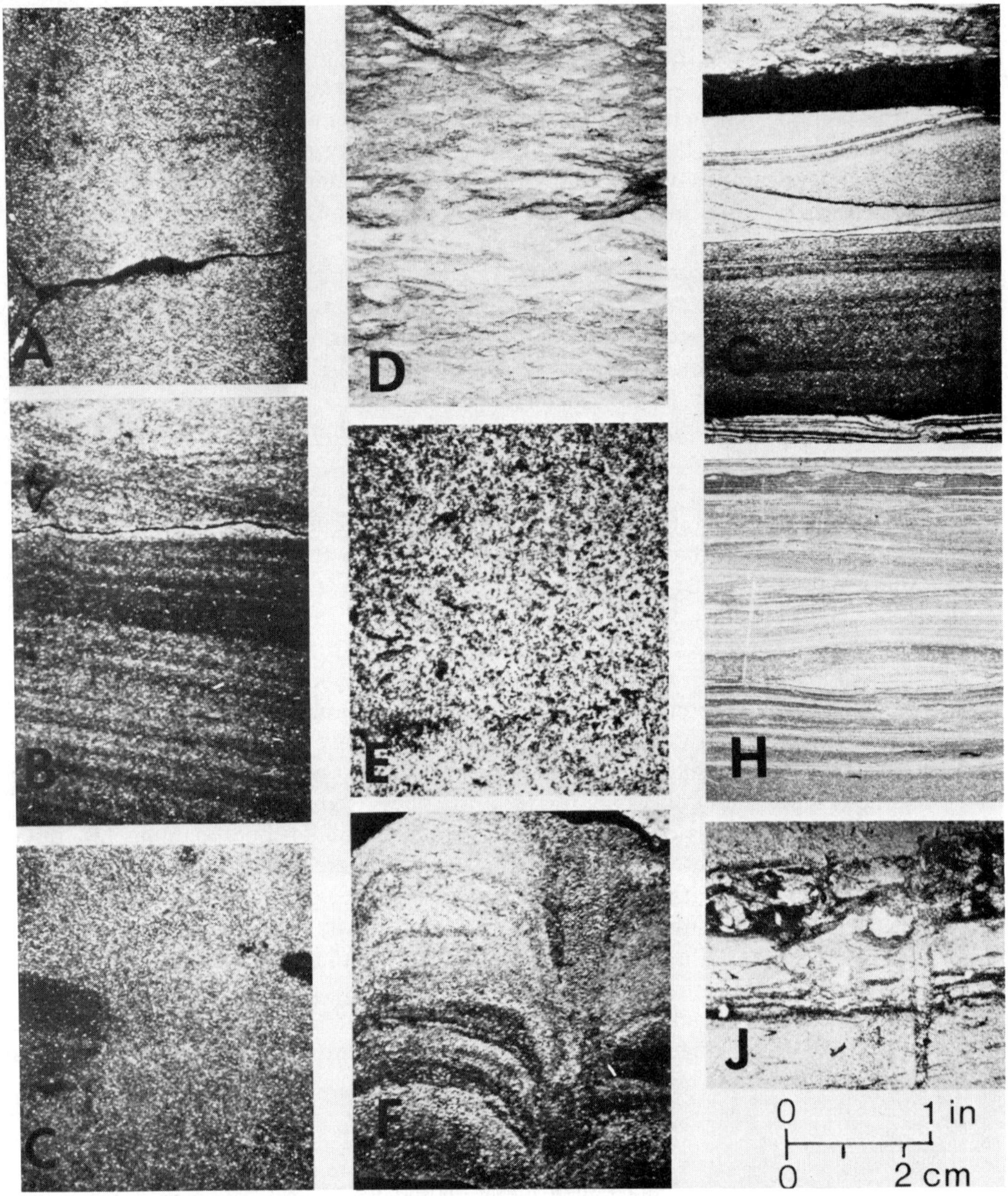

Figure 7–28 Sedimentary structures in Muddy sandstone, Bell Creek field; (A)–(D) barrier-island sequence—Boekel 21–14 well; (A) massive eolian sandstone—4675 ft (1425 m); (B) laminated upper-shoreface sandstone—4678 ft (1426 m); (C) massive middle-shoreface sandstone; the darker round mottles probably represent burrows—4690 ft (1430 m); (D) bioturbated lower-shoreface mudstone—4695 ft (1431 m); (E)–(J) lagoonal sequence—Gary 6–14 well; (E) bentonitic sandstone with small blebs of siderite—4410 ft (1344 m); (F) laminated quartzose sandstone with large mollusk (?) burrow—4412 ft (1345 m); (G) laminated and rippled sandstone—4414 ft (1345 m); (H) laminated silty claystone—4416 ft (1346 m); (J) bioturbated clayey sandstone—4418 ft (1347 m).

The lowermost mudstone unit (4) represents a normal-marine, lower-shoreface or open-bay environment that preceded deposition of the fine-grained lagoonal sandstones above [Figure 7–28 (E–J)].

The overlying unit (3) is thinly bedded and heterogeneous in character [Figure 7–28 (G) and (H)]. Beds up to about 10 cm in thickness are alternately rippled, horizontally laminated, or highly bioturbated. Throughout the section are thin laminae of dark shale. These beds denote deposition in shallow lagoonal waters, largely of suspension deposits that were alternately agitated by wave or current action and, at times, disturbed by burrowing organisms.

A thin bed of clean sandstone forms the next overlying unit (2), and this friable sandstone is not always recovered in cores of the lagoonal facies. Where it is seen in the Gary 6–14 core [Figure 7–28 (F)], the sandstone contains no clay, and bedding is disrupted by large vertical burrows, perhaps formed by large mollusks that rested within the shallow bottom sediment. The clean quartzose nature of this sandstone suggests that it is a washover facies composed of sands from the emergent barrier island to the west, sands that were transported by winds or storm waves into the protected lagoonal area. This thin unit is the principal reservoir sandstone within the lagoonal section on the updip side of the field.

The uppermost sandstone unit (1) has a clayey ash-white matrix and grades upward to an overlying bentonite bed that is about 6 ft (1.8 m) in thickness [Figure 7–28 (E)]. The bentoite bed contains random root traces of carbonaceous material. The final episode, then, of barrier-island deposition was the fall of volcanic ash that covered the lagoonal area as well as the barrier-island sediments at the shoreline. It was, perhaps, this volcanic ash that prevented erosion of the uppermost barrier-island sediments during the succeeding marine transgression.

Typical sections dominated by either barrier-island or lagoonal sandstones are illustrated by the bedding sequence and log response in the two contrasting facies (Figure 7–29). The barrier-island section is a continuous rather uniform sandstone as shown by the Boekel 21–14 well. The section is a single unit about 25-ft (8-m) thick and is characteristic of many wells in the thicker sandstone area of the field (Figure 7–27). The equivalent lagoonal sandstones are thin-bedded and shaly as shown in the Gary 6–14 well. This section is only 10-ft (3-m) thick and is characteristic of wells in the thinner sandstone area on the updip side of the field. Wells in the transition area, that is, from 10–20-ft (3–6-m) thick, generally show a combination of thin-bedded shaly sandstones overlain by a more continuous uniform sandstone.

Variations in the barrier-island sequence are found in the area of partial overlap between the two major sandstone bodies. Sections in this area commonly have porous sandstones at the top and the base of the Muddy section separated by a middle unit of fine-grained clayey sandstone or mudstone.

A core from the area of overlap comes from the Boekel 13–15 well (Figure 7–30). This section shows the following units:

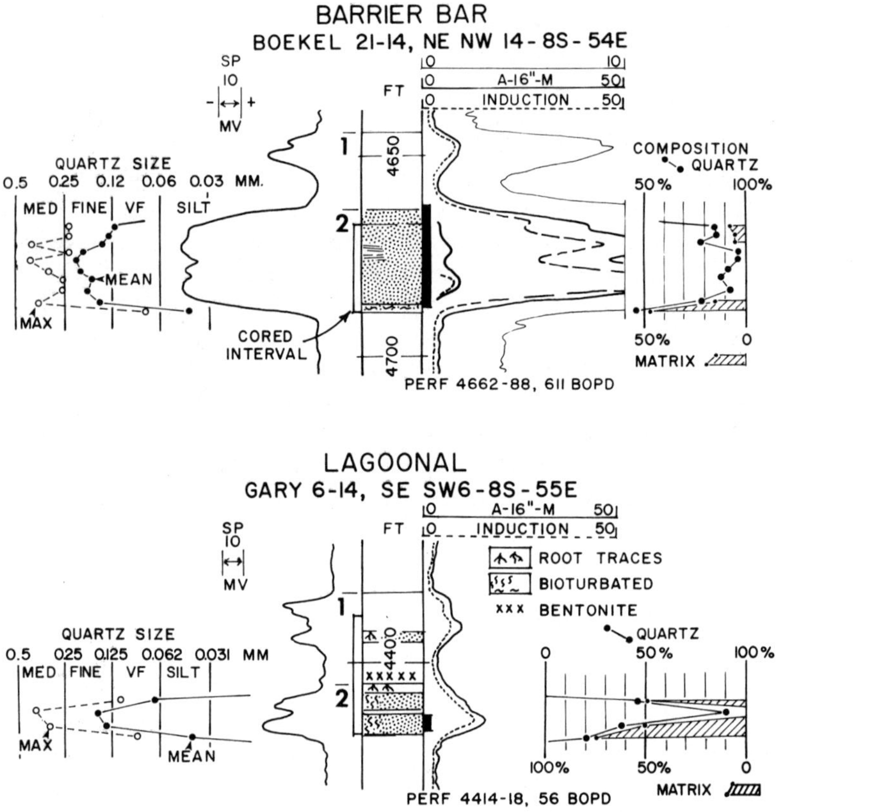

Figure 7–29 Electric logs, texture, and composition of typical barrier-island and lagoonal sections, Bell Creek field. Location of wells shown in Figure 7–27. Sedimentary structures for these wells are shown in Figure 7–28.

Unit	*Thickness ft (m)*
1. Washover sandstone, massive to rippled.	8 (2.4)
2. Lagoonal siltstone, clayey, rippled	3 (1.0)
3. Middle-shoreface sandstone, mostly massive, some ripple lamination.	14 (4.3)
4. Lower-shoreface mudstone, bioturbated.	2 (0.6)
Total	27 (8.1)

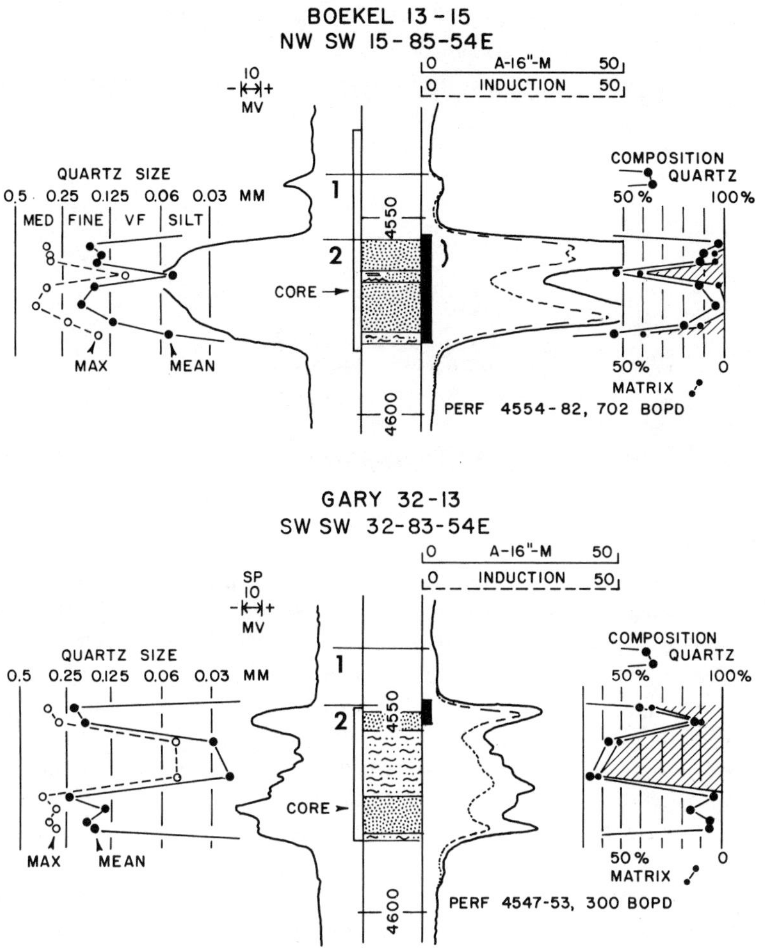

Figure 7–30 Electric log, texture, and composition of sections in the area of overlap between the two barrier-island sandstone bodies. Location of wells shown in Figure 7–27.

The lower sandstone unit (3) represents the seaward extension of the main barrier-island sandstone farther updip. The sandstone is mostly massive but contains some ripple lamination. The sandstone is highly quartzose, but it contains some shale laminae and bioturbation at the base where it grades into the underlying lower-shoreface unit (4).

The upper sandstone unit (1) represents washover sandstone above lagoonal siltstones (2). This section is the landward equivalent of the barrier-island sandstone to the west. The upper sandstone unit (1) is massive or ripple-laminated and highly quartzose, and the underlying siltstone contains some ripple lamination as well as a large amount of clay matrix. In this well, both the upper and lower sandstone units (1 and 3) were oil-productive.

Another section in the area of overlap is from the Gary 32–13 well located in the southwest part of the main Bell Creek field (Figure 7–30). In this case, the full core was not available, and the section was described from the small core plugs, as noted below:

Unit	*Thickness ft (m)*
1. Washover sandstone, massive to laminated.	6 (2)
2. Lower-shoreface mudstone, highly bioturbated.	16 (4.8)
3. Middle-shoreface sandstone, massive.	9 (2.7)
4. Lower-shoreface mudstone, bioturbated.	2 (0.6)
Total	33 (10.0)

The lower sandstone unit (3) is massive, highly quartzose, and represents the toe of the barrier island to the east. The succeeding section is dominated by lower-shoreface mudstone deposited in the open-marine environment seaward from the barrier-island section. The upper sandstone unit (1) represents the landward extension of the barrier-island sandstone to the west. The thick underlying lower-shoreface unit (2) suggests that there was a considerable period of open-marine deposition in this area before the second barrier-island sandstone prograded toward the southwest. In this well, only the upper sandstone unit (1) produced oil, whereas the lower sandstone unit (3) was below the oil-water contact of the barrier-island section to the east.

Lateral variations in the barrier-island sequence are shown by correlation of logs across the field (Figure 7–31). The logs selected are from cored wells, and they are shown in their positions relative to changes within the barrier-island section. Uniform continuous sections of barrier-island sandstones in wells 22–5 and 42–16 represent the updip and downdip sandstone bodies, respectively. The area of overlap between the two barrier-island sections is shown by well 13–15 (Figure 7–30), in

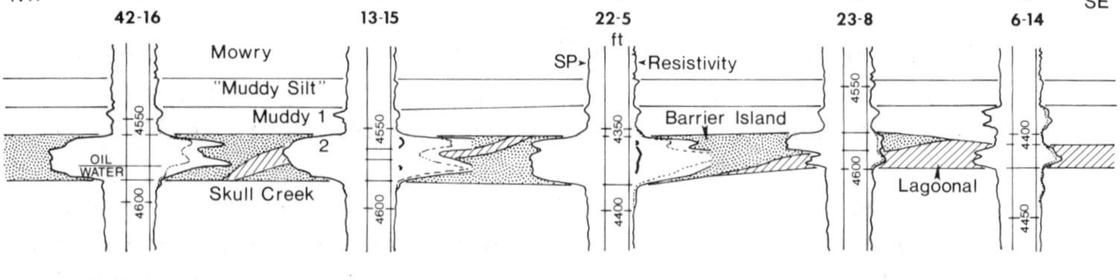

Figure 7–31 Correlation of electric logs from cored wells across Bell Creek field showing overlap of two barrier-island sandstones and adjoining lagoonal facies. No horizontal scale.

which the upper part of the sandstone represents washover from the downdip barrier-island section, whereas the lower sandstone unit represents middle-shoreface sandstone near the toe of the updip barrier-island section. The two sandstones are separated by a thin lagoonal siltstone. Eastward from the barrier island, the transition is shown in well 23–8 in which washover sandstones overlie fine-grained lagoonal sandstones. The abrupt change to a dominant lagoonal section is illustrated in well 6–14.

Petrography. The continuous barrier-island sequence shows an increase in grain size upward through the middle- and upper-shoreface units, commonly with a slight decrease in grain size in the uppermost eolian unit, as shown in the Boekel 21–14 well (Figure 7–29). Mean grain size through the barrier-island sandstone is 0.16 mm (fine grained). In contrast, grain size within the lagoonal section is variable as shown in the Gary 6–14 well (Figure 7–29). Mean grain size is 0.06 mm (very fine grained) but ranges from coarse silt to fine sand. In the more variable sections, the barrier-island sandstones are generally fine grained including their extensions of washover sandstones, whereas lagoonal and lower-shoreface sections are significantly finer grained (Figure 7–30).

The barrier-island section also has the highest content of monocrystalline quartz, which averages about 90 percent in the middle- and upper-shoreface sections (Figure 7–29). In these sections, matrix may be present in only trace amounts and is generally 10 percent or less in the eolian of other middle-shoreface sandstones. Quartz content decreases as matrix increases in the lagoonal section (Figure 7–29). In the Gary 6–14 well, the highest quartz content and grain size occur in the thin bed of clean sandstone that represents washover from the barrier-island section into the lagoonal section.

Cement consists of trace amounts of silica overgrowths on quartz grains in the barrier-island sandstones. Calcite cement is found in significant amounts only in the lagoonal siltstones.

An unusual cement is opaline silica, which occurs erratically in the upper part of the barrier-island sandstone. This cement has apparently been derived by solution of silica from the bentonite that lies immediately above the second Muddy sandstone. The hydrous nature of this cement causes a decrease in resistivity at the top of the section, which is noted in many wells as, for example, in the Boekel 21–14 induction log (Figure 7–29). The amount of cement is small, and despite the decrease in resistivity, oil saturation remains high.

An interesting compositional variation is found in a barrier-island section located in a well about 5 mi downdip from the field. A core of the second and third Muddy zones showed sandstones that display middle- and upper-shoreface structures (Waring 1976). These sandstones contain a white clay matrix in the amount of about 25 percent, although the grain content is dominated by monocrystalline quartz. The clay is authigenic kaolinite, which is formed by alteration of feldspars (Webb 1974). Apparently, the clay occurs only in water-saturated sandstones and is believed to be a late alteration product whose formation is promoted by hydrodynamic flow of connate water. Similar authigenic clays in quartz-rich sandstones are known from cores of other Muddy sandstones in the northeast Powder River basin.

Porosity and permeability. The barrier-island sandstones have the highest values of permeability and porosity because of the clean quartzose nature of the rock. Average permeability is 5300 md, and average porosity is 30 percent in the Boekel 21–14 well (Figure 7–32). Maximum permeabilities in this core ranged up to 13 darcys in the middle- and upper-shoreface units. The overlying eolian unit shows a slight reduction to about 500 md because of finer grain size. In the zone of overlap

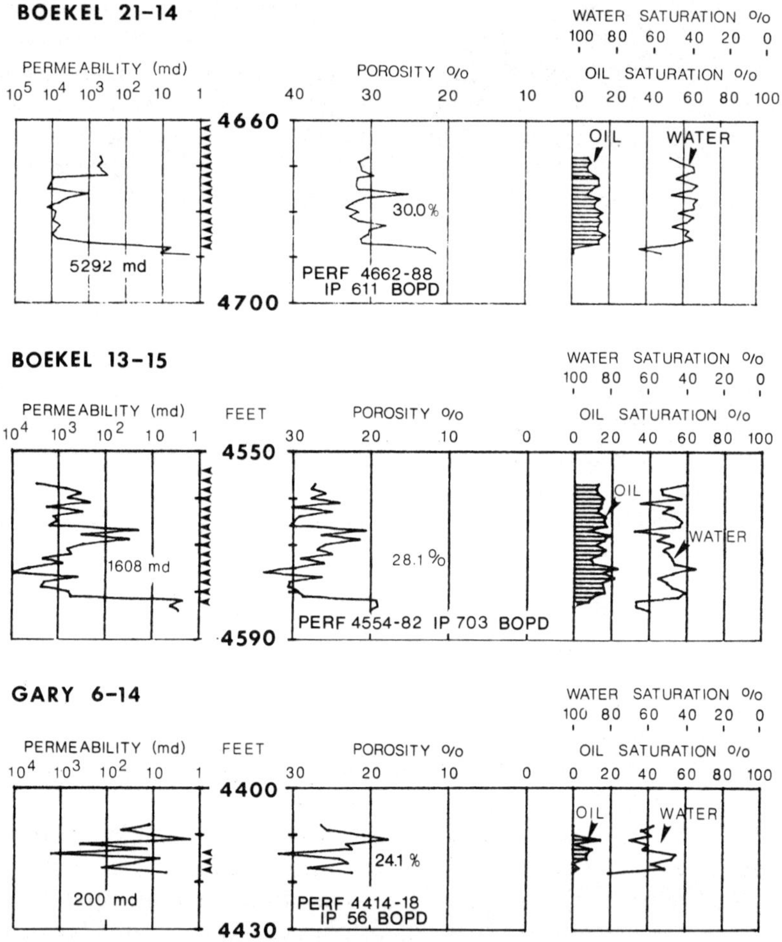

Figure 7–32 Permeabilities, porosities, and fluid saturations in barrier-island and lagoonal sandstones, Bell Creek field. Texture and composition for the sections are illustrated in Figures 7–29 and 7–30. Note change in scales.

between the two barrier islands, permeabilities are also high and in the range of 1000 to 10,000 md in the middle-shoreface section at the base of the Muddy in the Boekel 13–15 well. Permeabilities are somewhat reduced in the washover sandstones of the upper Muddy and are greatly reduced in the intervening lagoonal siltstones. In the lagoonal section of the Gary 6–24 well, permeability decreases to an average of only 200 md. The producing interval in this well includes a 2-ft (0.6-m) section of washover sandstone that has a high quartz content and a maximum permeability of 1500 md.

Production from the Bell Creek barrier-island sandstone has been outstanding because of the excellent continuity of high porosity and permeability in the sandstone. Through 1979, cumulative production was 86.3 million bbls (13.7 million m^3) or about 440,000 bbls per well (70,000 m^3/well). Unit operations began in 1970, and water was injected on the downdip side of the reservoir (Burt, Haddenhorst, and Hartford 1975), thereby halting the decline in primary oil production and increasing recovery significantly during succeeding years. The performance of individual wells was also outstanding during the water flood. For example, the thin washover sandstone in the Gary 6–14 well has produced over 90,000 bbls (14,300 m^3) from only 4 ft (1.2 m) of section. Other thinner sections on the updip side of the field have produced equally impressive amounts of oil.

The excellent nature of the reservoir promises a third period of production by chemical flooding. A micellar-polymer recovery project was begun in 1975, and the initial performance indicates that an additional 54 million bbls (8.6 million m^3) might be recovered by this method (Goldberg and Morganthaler 1976).

Conclusions

The barrier-island sands accumulated at Bell Creek field during a pause in the overall transgression of Muddy seas. The sands accumulated on an outlier of Skull Creek shale, and sand may have been derived in part by reworking of previously deposited deltaic sediments that were present in adjacent low-lying areas. Another possible source was thin deltaic sediments in an active fluvial transport system that existed about 10 mi to the northeast (Waring 1976). Sands were carried southwestward by longshore currents, and continual reworking by wave action produced a sediment high in quartz content and nearly lacking in matrix. Progradation of the shoreline produced two partly overlapping barrier islands, the earliest of which was bounded by a broad lagoon that received washover sands. The barrier island sands grade southwestward into more clayey sands that were submerged and only occasionally washed by storm waves.

The resulting barrier-island reservoir has been highly productive because of its high grain content, lateral continuity, and relatively homogeneous internal structures. Shallow depth of burial and lack of significant cement also contribute to the storage capacity of the reservoir. Other shoreline sandstones, even those more deeply buried, should form the best reservoirs of the coastal environment.

MUDDY SANDSTONE, HILIGHT FIELD, NORTHEASTERN WYOMING

Discovery of the Hilight oil field in 1969 was spectacular for its high flow rates from thin sandstones of relatively low permeability in the Lower Cretaceous Muddy Formation. The field subsequently developed into the largest Muddy producing area of recent years as the original Hilight field coalesced with the Grady and Jayson fields to the north. The discovery also extended production far to the south of the original success of Muddy exploration in 1967 at Bell Creek and Recluse fields (Figure 7–12).

At Hilight it was soon apparent that the Muddy's thin sandstones and associated coaly shales represented a depositional sequence that was distinctly different from the thicker sandstones of fluvial and marine origin that were found in the earlier fields to the north. Consequently, this study was completed (Berg 1976*b*) to explain the distribution of sandstones in the field area and to aid in understanding Muddy depositional patterns in adjacent areas.

Local Stratigraphy

The Muddy Formation consists of interbedded sandstones, siltstones, and shales and has a maximum thickness of 70 ft (21 m) in the north half of the producing area, as shown by the well Inexco Evans 3 (Figure 7–33). The Muddy thins abruptly within short distances to the nearby wells—Kirby Gov't 1, Inexco Mills 6, and Mills 11. Also, there is a rapid change in facies within the section from 25 ft (8 m) of porous sandstone in the Evans 3 to thinner sandstones to the south and to siltstone and shale to the north at Kirby Gov't 1.

Four zones in the field area were designated in descending order as the A, B,

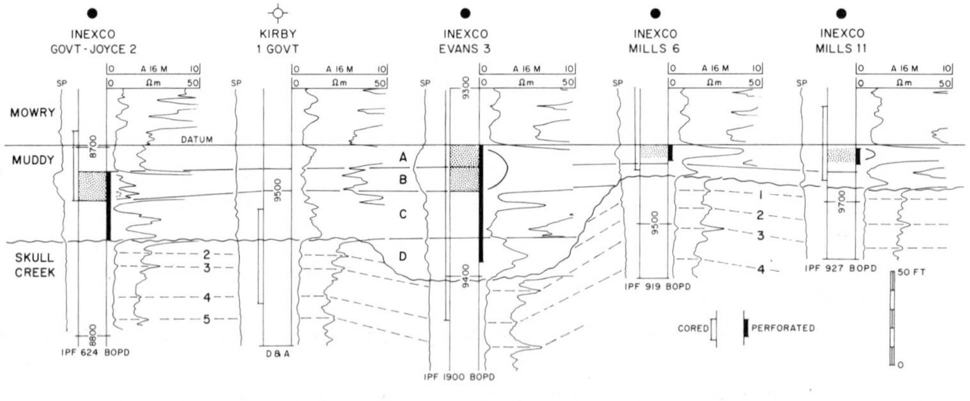

Figure 7–33 Zones of the Muddy sandstone in wells of the Hilight area. Correlation of resistivity curves suggests erosion at the top of Skull Creek Shale. Location of wells shown in Figure 7–34. No horizontal scale. [From Berg 1976*b* and published with the permission of the Rocky Mountain Association of Geologists.]

C, and D zones (Prescott 1970). These zones are distinctive horizons on the electric logs, and they can be correlated with some confidence within the area by means of gamma-ray logs. However, the distributions of the zones are quite different. The lowest zone (D) is confined to a narrow band across the north half of the field, whereas higher zones are more extensive, and the A zone occurs widely across all of the area. These zones correspond, respectively, to the first through fourth zones of the Muddy that have been defined earlier for the northeastern Powder River basin (Figure 7–14), but for the Hilight area the original letter designations are retained.

The Muddy is underlain by black platy marine shales of the Skull Creek Formation and is overlain by siliceous marine shales of the Mowry Formation. Key beds can be correlated within the Skull Creek by means of the amplified short-normal resistivity curves (Figure 7–33). These beds, numbered one through five, are present where the Muddy is thin, but the upper beds are missing where the Muddy is thick. These correlations confirm that an erosional surface of low relief was developed on the Skull Creek shale.

Drilling History

The field discovery in February 1969 was the Inexco Oil Company Isenburger 1, SE SW Sec. 26, T45N, R71W (Figure 7–34), which when completed was producing 1008 bbls/day (160 m^3/day) from a 3-ft (1-m) perforated zone. Spacing was soon established at one well for each 160 acres, and development was spurred by many completions that ranged to as high as 7960 bbls/day (1266 m^3/day). Oil gravity was 30° API in the north part of the field to 45° API in the south. The separate fields at Jayson, Grady, central Hilight, and South Hilight were soon joined to form the greater Hilight area (Figure 7–34), and by January 1972, there were 305 wells that produced 23,500 bbls/day (3736 m^3/day). Primary recovery was estimated to be 43 million barrels of oil (Prescott 1970, p. 105), and secondary recovery was expected to add another 40 million barrels.

Slabs of full-diameter cores were examined from six wells in the Hilight area (Figure 7–34). Cores were selected to represent contrasts in the reservoir section.

Description of Sandstones

Sedimentary structures. Muddy sandstones represent three depositional environments: beach, lagoonal, and fluvial. Beach sandstones are thicker, highly quartzose, and laminated sandstones that were deposited in marine shoreline environments. Lagoonal sandstones are thinner, moderately quartzose, and borrowed sandstones that were deposited shoreward from the beaches. Fluvial sandstones are thin, poorly quartzose, and contain large amounts of clay and carbonaceous fragments. Fluvial sediments were deposited largely in sluggish streams that drained highly vegetated marshy areas. All of these sandstones are associated, both laterally and vertically, with highly carbonaceous shales and siltstones.

Beach sandstones are present in both the A and B zones of the upper Muddy

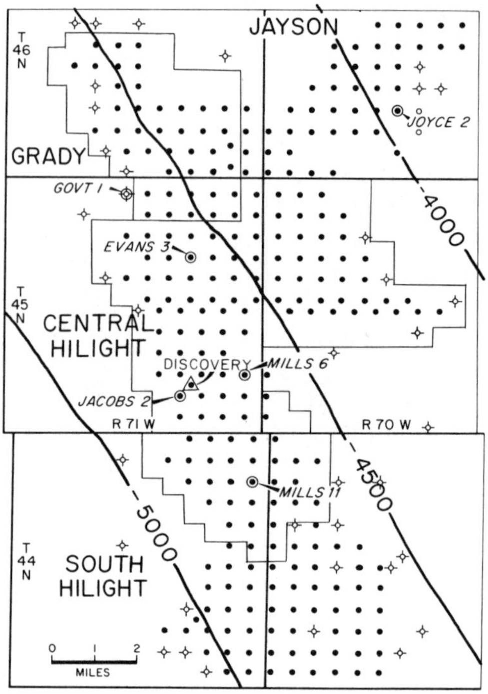

Figure 7–34 Hilight area structure on top of Muddy Sandstone. Contour interval 500 ft (152 m). Cores studied were from circled and named wells. Producing area extends 2 mi south of map. [From Berg 1976*b* and published with the permission of the Rocky Mountain Association of Geologists.]

in the Evans 3 core (Figure 7–35). Bedding is characterized by laminations that are inclined mostly at low angles, but a few laminae dip as much as 20°. Mean size of quartz is 0.21 mm (fine grained), and texture is uniform with no significant vertical change. Quartz content averages 91 percent, and clay matrix is 3 percent. Secondary overgrowths are abundant on quartz grains. The high quartz content, good sorting, and dominantly low dip of laminae suggest that these sandstones were deposited in a wave-washed environment at the shoreface.

The A and B sands in the Evans 3 core are separated only by a thin zone of wavy carbonaceous laminae. The top of the underlying B sandstone has a few vertical carbonaceous stringers that represent root traces. Hence, the B sand probably was exposed and vegetated before deposition of the A sand.

Lagoonal sandstone is present in the A zone in the Jacobs 2 core (Figure 7–36). Laminations generally dip at low angles of 5° to 10°, but a few laminae have dips as high as 20°. Bedding is similar to that in the beach sandstones, but bed sets appear to be relatively thin. Furthermore, lagoonal sandstones are quite different in composition. Mean size of quartz is 0.12 mm (very fine grained), quartz content is only 55 percent, and matrix is 25 percent. The bedding of these sandstones appears to represent a high-energy wave environment, but the finer grain size and high content of matrix suggest that these sands may have been washed from the beach into a shoreward environment of quiet water in which clays were deposited from suspension.

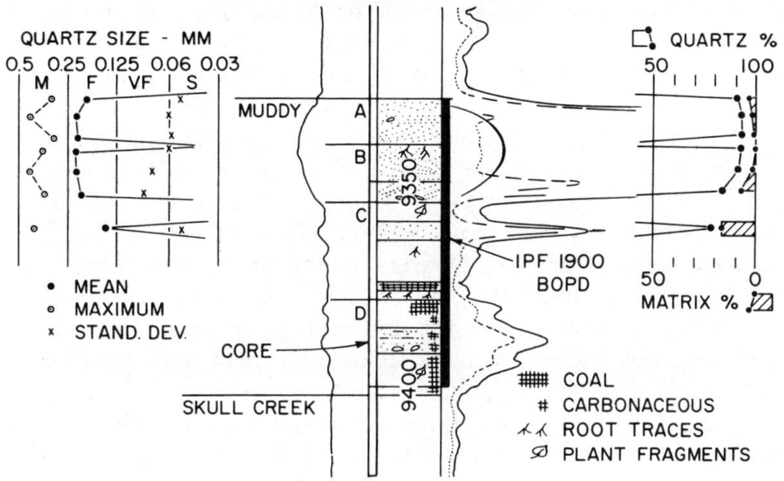

Figure 7-35 Petrography of Muddy sandstones in Inexco Evans 3 core. Location of well shown in Figure 7-34. [From Berg 1976*b* and published with the permission of the Rocky Mountain Association of Geologists.]

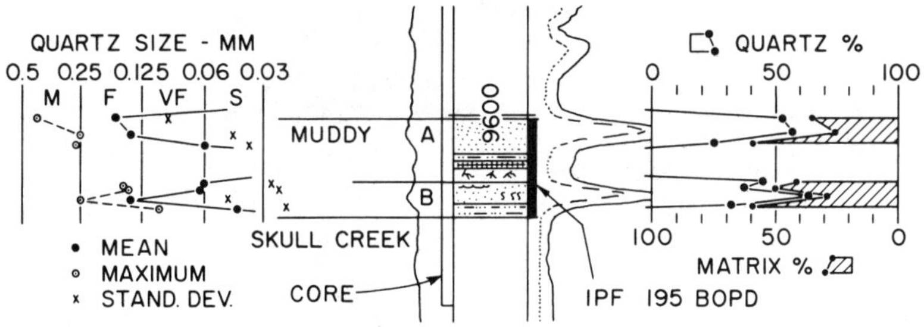

Figure 7-36 Petrography of Muddy sandstones in Inexco-King Jacobs 2 core. Location of well shown in Figure 7-34. [From Berg 1976*b* and published with the permission of the Rocky Mountain Association of Geologists.].

Lagoonal sandstone is also present in the B zone in the Jacobs 2 core (Figure 7-36). The unit is laminated, partly ripple-bedded, and characterized by large vertical burrows that occur as tubes and irregular masses of bioturbated sand. Mean size of quartz is 0.08 mm (very fine grained), quartz content is 44 percent, and clay matrix is 49 percent. The dominance of ripple structure and high clay matrix suggest an environment of quiet water behind the beach zone. Although bioturbation is present, the unit is dominantly well laminated and not highly burrowed and clayey as might be expected in a neritic environment.

The lagoonal sandstones of the A and B zones are separated by dark gray rooted claystone that contains a thin bed of black coaly shale. These deposits indicate that the underlying lagoonal sandstones were succeeded by vegetated marshes before being covered by succeeding washover sands.

The distinction between beach and lagoonal sandstones is less clear to the northeast at Jayson. In the B zone of the Joyce 2 core (Figure 7–37), a 17-ft (5.2-m) sandstone is generally massive, but a few laminae dip at low angles to as much as 15°. Mean size of quartz is 0.17 mm (fine grained), quartz content is 74 percent, and matrix is 21 percent. Quartz content ranges from a high of 87 percent at the base to 60 percent at the top, and clay matrix increases upward. Carbonaceous material is conspicuous in the unit with small chunks of coal near the base and carbonaceous fragments, plant stems, and small coaly pebbles in the top 4 ft (1.2 m). In addition, the upper part of the sandstone has a disturbed appearance, probably due to rooting. The entire unit may have originated as a series of washover sands that were intermittently vegetated, and primary structures that might have indicated the mode of sand transport were largely destroyed by rooting. The mean grain size is somewhat coarser than other lagoonal or littoral sandstones, however, and may indicate the fluvial influence. Nevertheless, the moderate amount of quartz and relatively high matrix suggest that the lower part of the sandstone is most likely of lagoonal origin rather than of the high-energy zone of the beach, whereas the uppermost part may be of fluvial origin.

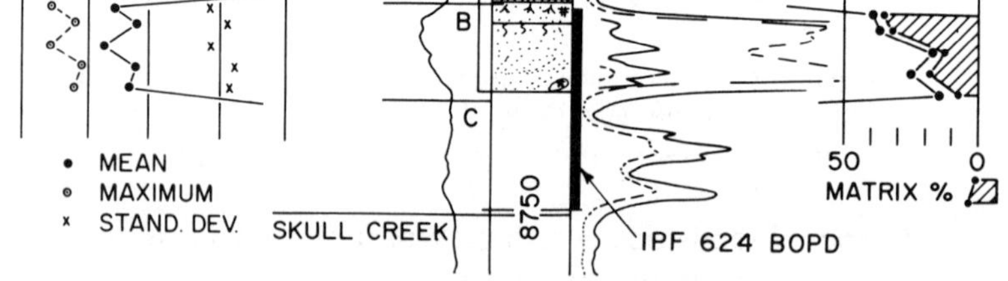

Figure 7–37 Petrography of Muddy sandstones in Inexco Joyce 2 core. Location of well shown in Figure 7–34. [From Berg 1976*b* and published with the permission of the Rocky Mountain Association of Geologists.]

Fluvial sandstone and mudstone are present in the C zone in the Kirby Gov't 1 core (Figure 7–38). Mean size of quartz is 0.13 mm (fine grained), maximum grain size is 0.36 mm (medium grained), and grain size decreases upward. Quartz content is 32 percent, clay matrix is 61 percent, and rock fragments are 7 percent. The sandstones range in composition from highly clayey sandstone to very sandy mudstone, and clay content increases upwards as grain size decreases.

The lower part of the fluvial sandstone is gray, indistinctly cross-laminated, and contains abundant angular fragments of carbonaceous material and coaly shale. These fragments average 2 mm in size but range up to about 1 cm. The upper part of the fluvial sandstone grades to greenish gray sandy mudstone, which is massive and contains irregular, vertical, carbonaceous root traces. These channel-fill deposits are succeeded by green rooted claystone and by highly carbonaceous black shales or clayey coals that contain abundant impressions of plant stems.

The upward succession of thin fluvial sandstone to sandy mudstone, rooted claystone, and coal suggests that the sequence was deposited in low-lying coastal marshes or swamps that were not far removed from the shoreline. The fluvial sandstones probably were deposited in low-energy streams that formed sluggish drainage channels through the marshes.

Fluvial and marsh sediments also occur in the other zones of the Muddy. In the Evans 3 core (Figure 7–35), the D zone consists of three distinct sequences of fluvial sandstone to mudstone, rooted claystone, and coal. Identical sequences occur

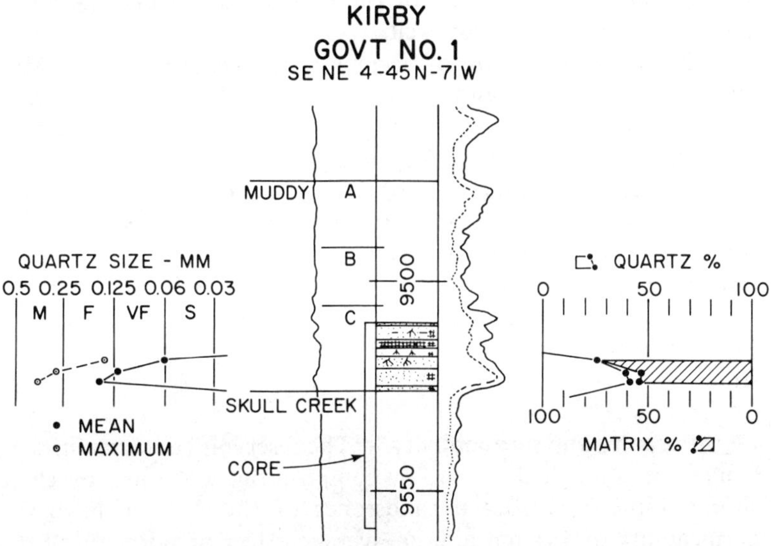

Figure 7–38 Petrography of Muddy sandstones in Kirby Gov't 1 core. Location of well shown in Figure 7–34. [From Berg 1976*b* and published with the permission of the Rocky Mountain Association of Geologists.]

in the A and B zones in the Mills 6 and Mills 11 cores (Figure 7–34). The similarity of these sections is striking and suggests that the same environment persisted at various levels throughout deposition of the entire Muddy section.

A less carbonaceous facies is present in the C zone in the Evans 3 core (Figure 7–35). Here a lower section of green sandy claystone is overlain by reddish brown massive sandstone. Although the underlying claystones are abundantly rooted, there are no highly carbonaceous beds or coals in the sequence. Thus, the C zone in this core represents a better-drained facies of the marsh environment. The sandstone of this section also differs from other fluvial sandstones. Mean size of quartz is 0.14 mm (fine grained), quartz content is 79 percent, and matrix is 17 percent. The quartzose composition probably indicates deposition in a more active fluvial channel.

Petrography. Muddy sandstones at Hilight also show compositional differences that are generally consistent with depositional environment (Table 7–3). Beach sandstones are highest in quartz content for their grain size, and lagoonal sandstones are less quartzose, finer grained, and contain increased amounts of clay matrix. Fluvial sandstones are relatively low in quartz content and have moderate to high matrix and significant amounts of rock fragments. Fluvial mudstones are characterized by high clay matrix as well as by abundant carbonaceous fragments.

One fluvial sandstone of the C zone (Table 7–3, Evans 3) has a high quartz content of 79 percent and low rock fragments of 3 percent. This sandstone showed massive bedding with only a few clay laminae and was called "fluvial" because of its location within a section of rooted clays and lying laterally to poorly drained marsh deposits in a nearby well.

Rock fragments are of various types but are characteristically composed of polycrystalline quartz and chert. Other rock fragments are relatively abundant in fluvial sandstones, and these may consist of fine-grained igneous and metamorphic rocks and minor limestone fragments.

Cements are commonly of two types. In the more quartzose beach and lagoonal sandstones, silica overgrowths on quartz grains are common. Fluvial sandstones have some grain overgrowths but generally have a more abundant mixed cement of calcite mottles, intergranular chert, and iron oxides.

In general, the correlation of environment, texture, and composition for Hilight sandstones agrees well with the relationships established for other Muddy sandstones.

Porosity and permeability. The reservoir characteristics of the Muddy sandstones are controlled in part by composition, and thus, by environment of deposition (Table 7–3). Beach sandstones of the A zone have the highest average permeability of 141 md and an average effect of porosity of 16.4 percent (Figure 7–39). The B zone has the next highest average permeability of 99 md and a porosity of 17.1 percent. Other quartzose sandstones have similar high permeabilities and porosities, although silica overgrowths on quartz grains result in a compact inter-

TABLE 7-3. SUMMARY OF MUDDY SANDSTONE PROPERTIES IN THE HILIGHT FIELD

			Average	Quartz size[a]			Detrital composition[b]					Cement[c]			Effective
Well	Environment	Zone	depth (ft)	Mean (mm)	Max (mm)	σ (mm)	Qz (%)	F (%)	Rx (%)	Oth (%)	Mx (%)	SiO_2 (% of total)	Oth	Permeability (md)	porosity (%)
Evans 3	Beach	A	9336	0.21	0.42	0.06	93	tr	4	1	2	11	—	141.	16.4
Evans 3	Beach	B	9350	0.21	0.42	0.07	89	1	6	1	3	9	—	99.	17.1
Jacobs 2	Lagoonal	A	9604	0.12	0.44	0.06	55	6	9	4	26	26	—	1.7	7.8
Jacobs 2	Lagoonal	B	9618	0.08	0.25	0.03	44	2	4	3	46	12	—	0.13	10.0
Joyce 2	Lagoonal	B (lower)	8723	0.17	0.38	0.06	82	tr	5	tr	12	16	—	72.	14.3
Mills 6	Fluvial	B	9455	0.18	0.45	0.06	46	5	16	0	33	—	1	(not available)	
Mills 11	Fluvial	B	9690	0.17	0.45	0.08	61	4	23	1	11	—	18	0.14	6.3
Joyce 2	Fluvial	B (upper)	8713	0.17	0.38	0.06	61	tr	4	1	33	—	8	0.22	11.5
Gov't 1	Fluvial	C	9523	0.13	0.36	0.07	32	tr	7	tr	61	—	—	<0.1	8.0
Evans 3	Fluvial	C	9358	0.14	0.39	0.05	79	tr	3	tr	17	—	10	0.13	7.4

[a] Long-axis measurements; σ = standard deviation.

[b] Qz = monocrystalline quartz, F = feldspars, Rx = rock fragments including polycrystalline quartz, Mx = matrix, Oth = other minerals, tr = trace (less than 1%)

[c] SiO_2 = silica as grain overgrowths, Oth = calcite mottles, intergranular chert, and minor iron oxide.

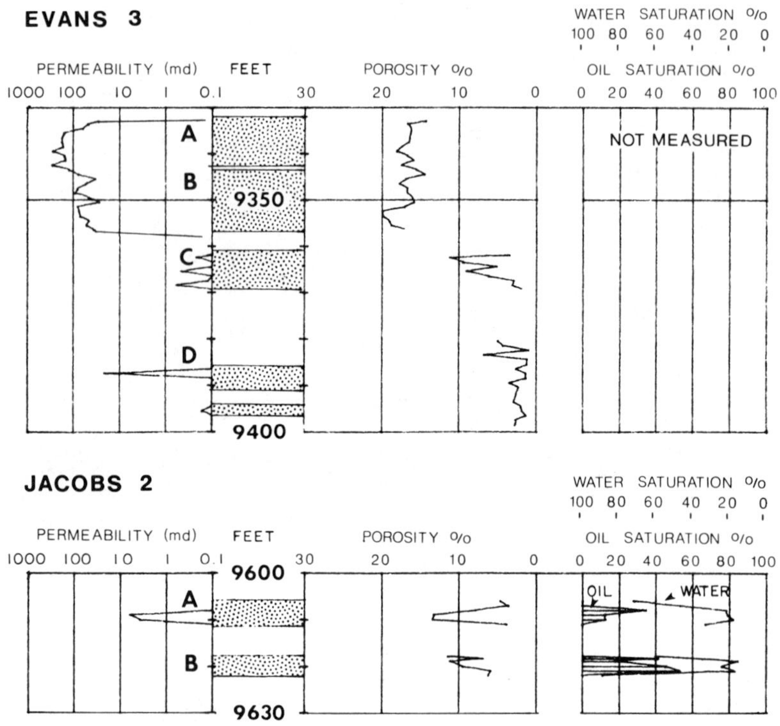

Figure 7–39 Permeability and porosity in the Muddy sandstone in the Inexco Evans 3 and Inexco-King Jacobs 2 cores, Hilight field.

locking grain fabric. Lagoonal and fluvial sandstones have lower permeabilities and porosities because of greater amounts of clay matrix as in the A and B zones in the Jacobs 2 core (Fig. 7–36). Calcite cement is abundant in some samples, thereby reducing permeabilities and porosities to lower values.

The A and B zones have the widest distribution, and the productivity of these horizons appears to be controlled in part by the depositional environment. The highest rates of initial production, commonly greater than 2000 bbl/day (318 m^3/day), were found in the central Hilight area where beach sandstones are present, but significant flow rates were also found in areas of lagoonal and fluvial sandstones, although many wells in these areas have flows of 500 bbl/day (80 m^3/day) or less. In most wells, perforated intervals include the total Muddy section (coaly marsh deposits as well as sandstones), and the productivity of individual zones cannot be ascertained. Because sandstones are thin and irregular in distribution, high flow rates in early wells were attributed to fracturing. Indeed, a number of high-angle natural fractures were found in several cores, but these were mostly filled with a finely crystalline white calcite. The contribution of fractures to flow may not be significant.

Interpretation

Distribution of environments. The depositional environments of Muddy sandstones have consistent distributions within the several zones (Figure 7–40). The C and D zones are largely marsh deposits, and their clayey sandstones are of fluvial origin. The A and B zones show greater lateral variation from a beach origin in the northern part of the central Hilight area (Evans 3) to lagoonal areas farther south where the sandstones are separated by marsh deposits (Jacobs 2). Still farther to the south and east, the A and B sandstones are of fluvial origin in a dominantly marsh section (Mills 6 and 11). To the north in the Grady Unit, the nature of the A and B zones is not known from cores, but they are presumed to be largely of shallow-marine origin. The overall sequence, then, appears to be transgressive from nonmarine below to littoral marine in the upper Muddy.

Depositional history. The environmental interpretations from cores can be integrated with subsurface mapping to confirm the depositional history previously inferred for the Muddy in the Hilight area (Prescott 1970). The total Muddy section is relatively thin, and its thickness is inversely reflected by the underlying Skull Creek shale. An isopach map of the Skull Creek (Figure 7–41) shows thinning toward the west from 140 ft to nearly 100 ft (43 m to 30 m), and across the same area, the Muddy thickens from a minimum of 10 ft to more than 70 ft (3 m to 21 m). The top of the Skull Creek is interpreted to be an erosional surface upon which the Muddy was deposited, and the Skull Creek isopach thus reflects pre-Muddy topography. The branching thin areas in the north and south parts of the area represent drainage routes toward the west. The relief on this surface is exceedingly gentle.

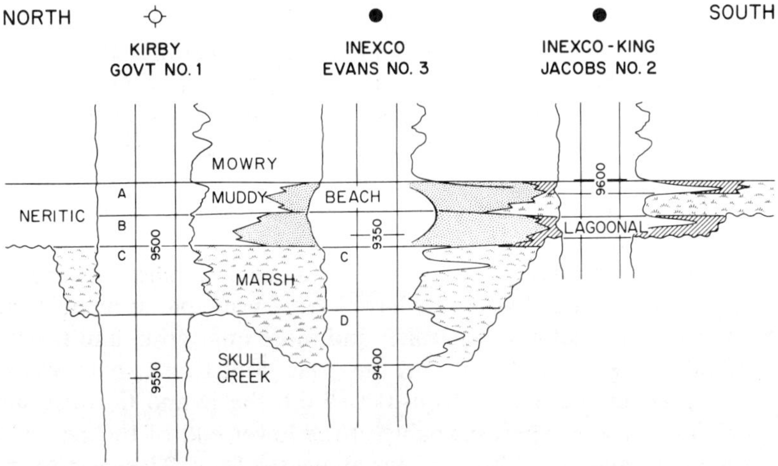

Figure 7–40 Distribution of environments in the Muddy Formation, central Hilight area. Location of wells shown in Figure 7–34.

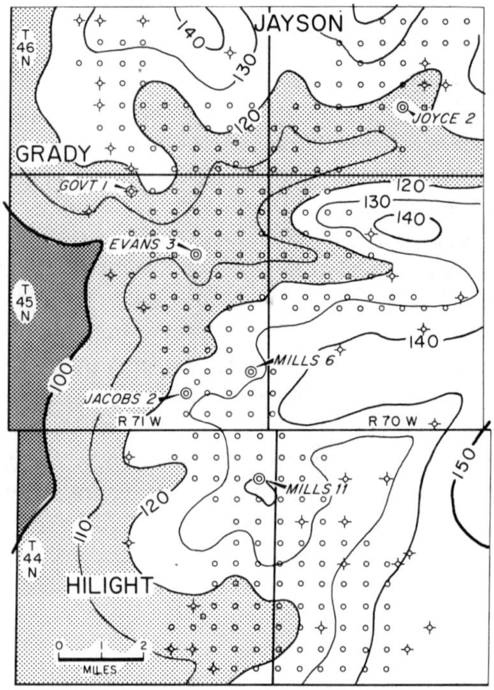

Figure 7–41 Isopach map of the Skull Creek Shale, which may be interpreted as representing erosional topography on which the Muddy was deposited. Contour interval 10 ft. [Adapted from Prescott 1970, Figure 4, p. 95; from Berg 1976*b*; and published with the permission of the Rocky Mountain Association of Geologists.]

Slopes of about 20 ft/mi (4 m/km) can be inferred locally for the surface in the northwest part of T45N, R70W, but for most of the area, slopes are no greater than 10 ft/mi (2 m/km) and are generally less.

Muddy deposition was initiated on this erosional surface in a narrow band that coincides with the principal drainage area across the north part of the Hilight unit (Figure 7–42, D zone). Thin fluvial sandstones that average less than 10 ft (3 m) in thickness occupy the valley, and their carbonaceous nature and associated rooted and coaly claystones suggest that streams were bordered by well-vegetated marshes or swamps. At this time the Skull Creek shale was still exposed over a wide area.

Succeeding deposits of the same type filled this drainage area more completely during deposition of the C zone (Figure 7–42, C zone). Fluvial sandstones were still thin, indicating poor drainage of the adjacent marshes, but the main sandstone area around the Evans 3 well may have been a principal drainage channel because the C sandstone in that core is cleaner and more quartzose than fluvial mudstones of the adjacent marshes. During this time the Skull Creek shale was still exposed over a broad area. The marsh deposits of both the D and C zones suggest that the area was one of low relief, probably at the lower end of the general transport route of sediment, and very likely not far above sea level. The sandstone areas do not show definite meander patterns, probably because of wide spacing of control points, but the sandstone has a generally sinuous form.

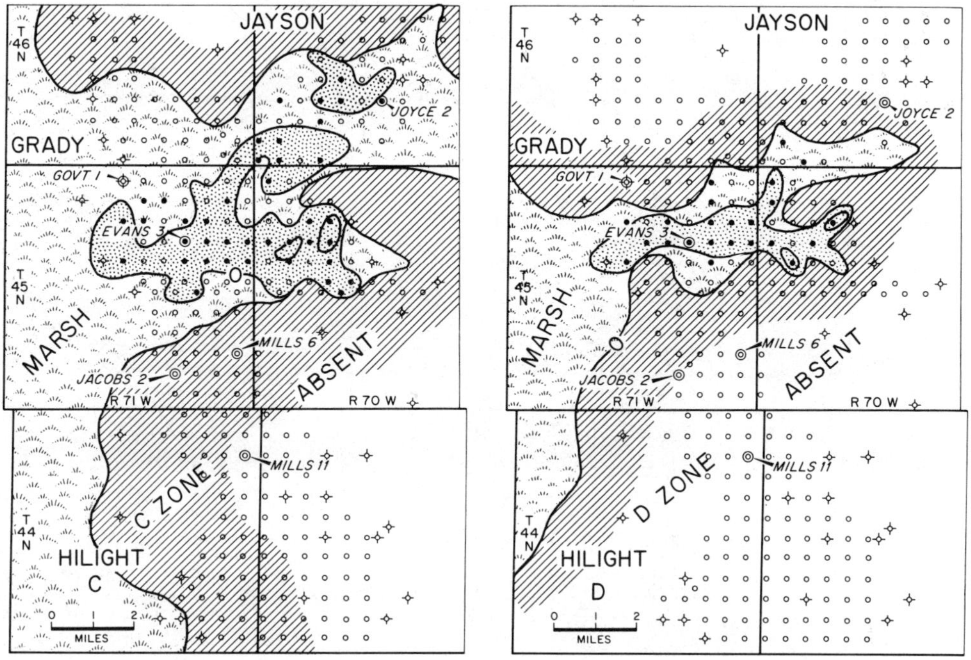

Figure 7-42 Paleogeography of the Hilight field during deposition of the C and D zones of the Muddy Formation. Sandstone areas shown by dotted pattern, and contour interval of 10 ft (3 m). Oil well symbols indicate production for each zone. [Maps adapted from Prescott 1970, Figures 5-8, pp. 97-103; from Berg 1976*b;* and published with the permission of the Rocky Mountain Association of Geologists.]

Marine conditions are indicated for the B zone [Figure 7-43, (B)] because of the beachlike character of sandstone in the Evans 3 core (Figure 7-35). The local trend of the shoreline at this time would appear to have been northeastward because washover and other lagoonal sandstones are present in the Jacobs 2 core to the south, but the sections in the Mills 6 and Mills 11 cores of the same area (but slightly eastward) are still dominated by fluvial sandstones and interbedded marsh deposits. In the northeast part of the area at Jayson, the section in the Joyce 2 core (Figure 7-37) is less definitive, but more highly quartzose sandstone at the base of the B zone gives way to clayey sandstones and then to rooted and highly carbonaceous sandstones above. This section still represents a dominantly nonmarine environment, but there is no evidence from subsurface control that a body of shoreline sands existed in the vicinity. Neither are shoreline deposits indicated by the area just downdip from the Evans 3 well, largely because subsurface control is lacking.

The area to the north in the Grady unit is one of very thin but rather widely distributed sandstone. No cores were examined in this area to indicate the mode of sediment transport, but the section here may be presumed to be marine also. The

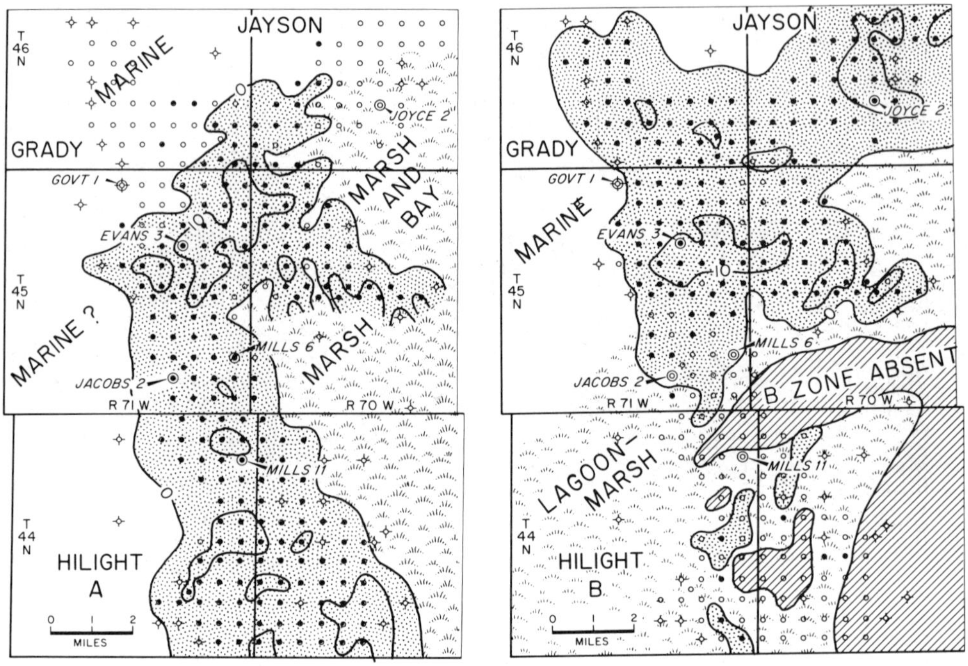

Figure 7–43 Paleogeography of the Hilight field during deposition of the A and B zones of the Muddy formation. Sandstone areas shown by dotted pattern, and contour interval of 10 ft (3 m). Oil well symbols indicate production for each zone. [Maps adapted from Prescott 1970, Figures 5–8, pp. 97–103; from Berg 1976*b;* and published with the permission of the Rocky Mountain Association of Geologists.]

absence of a well-defined shoreline trend may be the effect of continued transgression of the seas, which may have resulted in reworking of previous deposits and spreading of sands under the influence of prevailing longshore currents. Consequently, the present distribution of sands does not clearly indicate the former shoreline. During deposition of B zone sediments, the area of exposed Skull Creek shale was greatly restricted, and Muddy sediment was spread over a wide area.

During the final phase of Muddy deposition (Figure 7–43), beach sands of the A zone were spread in a generally northeastward trend across the central Hilight unit, but no sand was deposited immediately to the northwest in the Grady area. A lagoonal environment persisted, however, at the Jacobs 2 location, whereas marsh environments were still present immediately east at Mills 6 and Mills 11, and also to the northeast at Jayson in the Joyce 2. In the latter area, more normal marine sediments soon covered the previous marsh.

The area was then entirely submerged by marine waters in which the siliceous shales of the Mowry were deposited. The abrupt change to Mowry shales with little or no gradation suggests that the final transgression was rapid.

Morphology. The thin sandstones lack strong orientations. Trends of the reservoir sandstones are broadly linear, dip-trending for the fluvial sandstones of zones C and D but strike-trending for the beach and lagoonal sandstones of zones A and B. Within the upper zones, however, the widely distributed sandstones have more subtle internal trends. In the central Hilight Unit, B sandstones greater than 10 ft (3 m) in thickness have an eastward trend, and A sandstones have a northeast trend. The direction of Muddy transgression must have been generally toward the east, but nowhere is a consistent pattern developed between sandstone distribution and facies, as might be expected for shoreline deposits.

Part of the mapping problem may be in the wide spacing of wells, which makes the definition of individual sandstones a most difficult task. However, another interpretation of the Hilight area (Curry 1972) based on resistivity mapping displays a strong northeast lineation of reservoir sandstones in T44N, R70, 71W. In the central Hilight area (T45N, R70, 71W), there is a strong east-west lineation of sandstones because of the presence of fluvial deposits in the C and D zones, but superimposed is a prominent northeast lineation produced by sandstones of the A and B zones. These relations are suggestive of stronger trends and greater lateral separation of reservoir sandstones.

The Hilight section is similar to the modern Delaware coast (Figure 7–9) where Holocene transgression has produced a sequence that extends upward from marsh through lagoon to beach sands. This sequence is particularly well displayed through the bay-mouth barriers (Kraft 1971, Figure 15, p. 2146). The Delaware coast, however, is being actively eroded, and the sedimentary section is incomplete. It appears that further transgression will leave only a partial record, if any, of the previous coastal environments, which will be covered, finally, by a thin transgressive sand (Kraft 1971, pp. 2152–53). In order to preserve such a sequence, rapid transgression would be necessary, possibly aided by thicker and more extensive littoral sand that might not be completely eroded.

The Muddy environment may have closely resembled the strand-plain and marsh-lagoon coastal zone of Nayarit, northwest Mexico (Figure 7–9). Here beach ridges have coalesced to form a sand sheet that ranges from 5 to 15 km wide and is bordered landward by extensive mangrove marshes and more limited lagoons. The marsh-lagoon sequence also contains washover sands as well as alluvial deposits.

A major difference, however, is the orientation of upper Muddy sandstones, which have a northeasterly trend rather than an expected north-south trend, normal to the generally eastward advance of the Muddy sea. The northeast trend may have been produced by reworking of littoral sands during transgression, partly eroding the strand plain but preserving and modifying much of the previous sediment. The upper Muddy sandstones have retained contrasts in composition and bedding that permit interpretation of different environments of deposition. Therefore, the sandstones appear to owe their configurations to erosive forces that shaped previously existing sands during transgression, rather than to constructional forces that produced new sand bodies in a shallow neritic environment.

Conclusions

The Muddy section shows areally restricted marsh deposits that are succeeded by thin but widespread littoral-marine sandstones of beach, lagoon, and marsh origin. The overall sequence represents a transgression to the east across an unconformable surface on the underlying Skull Creek shale. Paleogeography during early Muddy deposition probably was a low-relief coastal environment that was bordered by marshes with sluggish streams. Later, short-term progradation of the shoreline produced highly quartzose beach sands that became the principal oil reservoirs. Subsequent rapid transgression over the strand-plain sequence reshaped the littoral sands and produced northeast-trending bodies that now lie at an angle to the assumed north-south trend of the advancing shoreline. Similar sandstone distributions can be expected in other transgressive sections.

SUMMARY

Barrier islands are the major concentrations of sandstone in the coastal interdeltaic zone. The vertical sequence is (1) lower-shoreface, inner-neritic mudstone, (2) middle-shoreface, massive sandstone, (3) beach and upper-shoreface, laminated sandstone, and (4) eolian-dune sandstone at the top. Grain size increases upward, and the sandstones have a high content of quartz, decreased amounts of feldspar and rock fragments, and little matrix. Barrier-island and other shoreline sandstones are inserted in marsh, estuarine bay, and normal-marine coastal deposits. Other sandstones in the sequence are fluvial-channel and transgressive-marine deposits. The transgressive sandstones are laminated at the base and have increasing bioturbation upward. Grain size decreases upward, and quartz content also decreases upward as matrix increases.

Barrier-island sandstones are thin, narrow, linear bodies that are oriented parallel to the ancient coast. Porosities and permeabilities are high because of dominant grain content. Transgressive sandstones are smaller and linear-to-ovate bodies that are internally variable because of bioturbation and matrix, which also reduce porosities and permeabilities.

REFERENCES CITED

Allen, G. P., 1971, Relationship between grain size parameters distribution and current patterns in the Gironde Estuary (France): Journal of Sedimentary Petrology, v. 41, no. 1, p. 74–88.

Bascom, W. N., 1951, The relationship between sand size and beach-face slope: American Geophysical Union Transactions, v. 32, p. 866–874.

Bernard, H. A., C. F. Major, Jr., B. S. Parrott, and R. J. LeBlanc, Sr., 1970, Recent sediments of southeast Texas: University of Texas Bureau of Economic Geology Guidebook 11, p. 16.

BERG, R. R., 1975, Capillary pressures in stratigraphic traps: American Association of Petroleum Geologists Bulletin, v. 59, p. 939–956.

———, 1976a, Trapping mechanisms for oil in Lower Cretaceous Muddy sandstone at Recluse field, Wyoming: Wyoming Geological Association Guidebook, p. 261–272.

———, 1976b, Hilight Muddy field, Lower Cretaceous transgressive deposits in the Powder River basin, Wyoming: The Mountain Geologist, v. 13, no. 2, p. 33–45.

———, and D. K. DAVIES, 1968, Origin of Lower Cretaceous Muddy Sandstone at Bell Creek field, Montana: American Association of Petroleum Geologists Bulletin, v. 52, p. 1888–1898.

BURT, R. A., F. A. HADDENHORST, and J. C. HARTFORD, 1975, Review of Bell Creek waterflood performance, Powder River, Montana: Journal of Petroleum Technology (December), p. 1443–1449.

BYRNE, J. V., D. O. LEROY, and C. M. RILEY, 1959, The Chenier Plain and its stratigraphy, southwestern Louisiana: Gulf Coast Association of Geological Societies Transactions, v. 9, p. 237–259.

CLIFTON, H. E., R. E. HUNTER, and R. L. PHILLIPS, 1971, Depositional structures and processes in the nonbarred high-energy nearshore: Journal of Sedimentary Petrology, v. 41, p. 651–670.

CURRAY, J. R., F. J. EMMEL, and P. J. CRAMPTON, 1969, Holocene history of a strand plain, lagoonal coast, Nayarit, Mexico, *in* A. A. Castanares and F. B. Phleger, eds., *Coastal lagoons, a symposium:* Mexico, D. F., Universidad Nacional Autonoma de Mexico, p. 63–100.

CURRY, W. H., III, 1972, Resistivity mapping of sandstone stratigraphic traps: Wyoming Geological Association Earth Science Bulletin, v. 5, no. 4, p. 3–11.

DAVIES, D. K., F. G. ETHRIDGE, and R. R. BERG, 1971, Recognition of barrier environments: American Association of Petroleum Geologists Bulletin, v. 55, p. 550–565.

DICKINSON, D. A., H. L. BERRYHILL, JR., and C. W. HOLMES, 1972, Criteria for recognizing ancient barrier coastlines: Society of Economic Paleontologists and Mineralogists Special Publication 16, 192–214.

ETHRIDGE, F. G., 1970, Quantitative petrographic criteria for recognition of environments of deposition: Texas A&M University Ph.D. dissertation, 169 p.

EVANS, GRAHAM, 1965, Intertidal flat sediments and their environments of deposition in the Wash: Geological Society of London Quarterly Journal, v. 121, p. 109–245.

FISK, H. M., 1959, *Padre Island and the Laguna Madre Flats, coastal South Texas:* 2nd Coastal Geography Conference, April 6–9, Baton Rouge, Louisiana, Louisiana State University, p. 103–152.

GOLDBERG, ARNOLD, and C. E. MORGENTHALER, 1976, Micellar-polymer oil recovery demonstration in the Bell Creek field, Montana: Energy Research and Development Administration Symposium on Enhanced Oil and Gas Recovery Proceedings, v. 1, p. A5/1-A5/26.

GOULD, H. R., and E. MCFARLAN, JR., 1959, Geologic history of the Chenier Plain, southwestern Louisiana: Gulf Coast Association of Geological Societies Transactions, v. 9, p. 261–270.

HACK, J. T., 1957, Submerged river system of Chesapeake Bay: Geological Society of America Bulletin, v. 68, p. 817–830.

HAYES, M. O. 1967, Hurricanes as geological agents; case studies of hurricane *Carla,* 1961, and *Cindy,* 1963: University of Texas Bureau of Economic Geology Report of Investigations 61, 56 p.

HORNER, D. R., 1951, Pressure buildup in wells: 3rd World Petroleum Congress Proceedings Section II, E. J. Brill, Leiden, p. 503–521.

INGLE, J. C., JR., 1966, The movement of beach sand: *Developments in Sedimentology,* v. 5, Amsterdam, Elsevier, 221 p.

INTERNATIONAL OIL SCOUTS ASSOCIATION, 1977, International oil and gas development yearbook: v. 47, part 2.

KRAFT, J. C., 1971, Sedimentary facies patterns and geologic history of a Holocene marine transgression: Geological Society of America Bulletin, v. 82, p. 2131–2158.

———, and C. J. JOHN, 1979, Lateral and vertical facies relations of transgressive barrier: American Association of Petroleum Geologists Bulletin, v. 63, p. 2145–2163.

KOMAR, P. D., 1976, *Beach processes in sedimentation:* Englewood Cliffs, N. J., Prentice-Hall, Inc., 429 p.

LARBERG, G. M., 1980, Depositional environments and sand morphologies of the Muddy sandstones at Kitty field, Powder River basin, Wyoming: Wyoming Geological Association Guidebook, p. 117–135.

LIN, J. T. C., 1981, Hydrodynamic flow in Lower Cretaceous Muddy sandstone, Gas Draw field, Powder River basin, Wyoming: The Mountain Geologist, v. 18, p. 78–87.

MCGREGOR, A. A., and C. A. BIGGS, 1968, Bell Creek field, Montana: rich stratigraphic trap: American Association of Petroleum Geologists Bulletin, v. 52, p. 1869–1887.

PRESCOTT, M. W., 1970, Hilight field, Campbell County, Wyoming: Wyoming Geological Association Guidebook, p. 89–103.

REINICK, H. E., 1972, Tidal flats, *in* J. K. Rigby, and W. K. Hamblin, editors, *Recognition of ancient sedimentary environments:* Society of Economic Paleontologists and Mineralogists Special Publication 16, p. 146–159.

SLACK, P. B., 1981, Paleotectonics and hydrocarbon accumulation, Powder River basin, Wyoming: American Association of Petroleum Geologists Bulletin, v. 65, p. 730–743.

SONU, C. J., 1972, Field observation of nearshore circulation and meandering currents: Journal of Geophysical Research, v. 77, p. 3232–3247.

STONE, W. D., 1972, Stratigraphy and exploration of the Lower Cretaceous Muddy Formation, Northern Powder River basin, Wyoming and Montana: The Mountain Geologist, v. 9, no. 4, p. 355–378.

STOUDT, D. L., 1974, Depositional environments of Lower Cretaceous Muddy sandstone, Recluse area, Campbell County, Wyoming: Texas A&M University M.S. thesis, 118 p.

WARING, JULIANA, 1976, Regional distribution of environments of the Muddy sandstone, southeastern Montana: Wyoming Geological Society Guidebook, p. 83–96.

WEBB, J. E., 1974, Relation of oil migration to secondary clay cementation, Cretaceous sandstones, Wyoming: American Association of Petroleum Geologists Bulletin, v. 58, no. 11, p. 2245–2249.

WIEGEL, R. L., 1964, *Oceanographical engineering:* Englewood Cliffs, N. J., Prentice-Hall, 532 p.

WILKINSON, B. H., and J. R. BYRNE, 1977, Lavaca Bay-transgressive deltaic sedimentation in central Texas estuary: American Association of Petroleum Geologists Bulletin, v. 61, no. 4, p. 527–545.

WONCIK, JOHN, 1972, Recluse field, Campbell County, Wyoming, *in* Stratigraphic oil and gas fields: American Association of Petroleum Geologists Memoir 16, p. 376–386.

8

Shelf Sandstones

The modern marine-shelf environment ranges in water depth from the lower shoreface at about 20 ft (6 m) to the outer margin of the continental shelf at about 600 ft (200 m). It is a marine environment, perhaps not continually stirred by breaking waves, but one, nevertheless, in which the bottom sediment is affected periodically by waves and by intermittent current flow. The primary physical processes that agitate the bottom sediment are: (1) wave motion, (2) tidal currents, (3) storm currents, and (4) turbidity currents. As a result of these processes, sands have accumulated on the modern shelves in a variety of forms, mostly as sand waves or ridges that rise above the surrounding sea bottom. These deposits are not known in great detail because of the difficulties of access to the marine environment. Nevertheless, such features do provide some knowledge whereby the nature of ancient shelf sands may be predicted.

Sandstones of marine-shelf origin are important reservoirs for oil and gas. Examples of these sandstones demonstrate that some may have formed as sand ridges, but the deposition of most ancient shelf sands seems to have been as more subtle moundlike features or was controlled by bottom topography. The resulting reservoirs are widely variable in their shape, thickness, and internal properties.

SHELF PROCESSES

The oceans of the world are in continuous motion. Not only surface waves but internal waves and currents agitate the seas and transport sea water. Water movements, in turn, cause sediment to be entrained and transported, sometimes for great distances. These fluid movements have produced distinctive distributions of sand across the shelves and in the deeper basins of the modern ocean.

The origins of oceanic fluid motions are ultimately astronomical forces applied to the atmosphere and to the hydrosphere. These forces are solar radiation and gravitational attraction, and the motions that they cause can be classified according to their origins (Weggel 1972). The direct application of *astronomical forces* causes tides, tidal currents, and solar heating of the water, thereby producing density differences and currents (Table 8–1). The direct application of *meteorological forces* causes wind waves and wind-induced currents such as longshore and rip currents. Also pressure differences caused by storms generate storm surge along the shorelines and thereby generate currents, largely by the return flow of the storm surge.

There is also indirect application of astronomical forces, which includes the *impulsive forces* generated within the earth by the release of stress. Earthquakes and submarine landslides can cause tsunami waves and density currents by suspending fine sediment to produce turbidity flows.

The physical processes that cause mass transport of fluid are reasonably well known, both theoretically and by observation. The forces that act in the water column, however, are greatly modified as the sea bottom, or boundary layer, is approached. Therefore, the forces that act directly on bottom sediment are less well understood, but it is clear that there are several major processes that are responsible for sand accumulation. These include wave motion, tidal currents, storm currents, turbidity currents, and other currents.

Wave Motion

The action of waves on bottom sediment can be significant, even in relatively deep water. Progressive waves at the surface set water particles in orbital motion, and the height of the orbit is equal to the wave height (see Figure 2–12 in Chapter 2). The orbital paths decrease below the surface until at a depth of one-half the wave length, the motion is negligible. When waves move into shallow water, the particle orbits become tangent to the bottom, and the horizontal velocity may be great

TABLE 8-1. CLASSIFICATION OF OCEANIC FLOWS ACCORDING TO DRIVING FORCES

Astronomical forces	Meteorological forces	Impulsive forces
Solar and lunar attraction:	Wind stresses:	Stress release:
Tides (longwaves)	Wind waves	Tsunamis
Tidal currents (nearshore)	Currents (longshore, rip)	Turbidity currents
	Pressure systems:	
Solar radiation:	Storm surge	
Density currents	Currents	

SOURCE: Weggel 1972.

enough to erode and suspend sand grains. The effect on bottom sediment is determined by maximum bottom velocities, which can be estimated as a function of water depth and wave height (Figure 2–13). Even modest wave heights can move sand grains at depths of 100 ft (30 m), and storm waves with heights of 50 ft (15 m) can erode sand at a depth of 300 ft (91 m).

In shallow water near the shoreline, the effect of waves is to set up longshore and coastal currents; but in deeper water of the shelf, there is little transport of sand by wave action.

The major effect of storm waves is to put sediments into suspension. When the wave heights decline, sand may be deposited while clays remain in suspension and subject to transport by currents. The net effect of wave motion, then, is a winnowing process that cleans and sorts the bottom sediment.

The importance of storm waves as a sedimentary process raises the question of their frequency of occurrence. Waves generated by hurricanes are of particular interest to petroleum engineers for the design of offshore drilling platforms, and statistical studies have been made to predict wave heights (Bea 1975). Major storms in the Gulf of Mexico have occurred at the rate of 1.06 per year since 1871, and reliable information can be obtained on 43 storms since 1900. Most of these produced wave heights greater than 25 ft (8 m), and heights on the order of 40 ft (12 m) have been common. Based on these data, hindcasts were made for maximum wave heights (Figure 8–1). The results show that wave heights of 70 to 80 ft (21 to 24 m) may be expected every 100 to 200 years (Bea 1975), and other studies, based

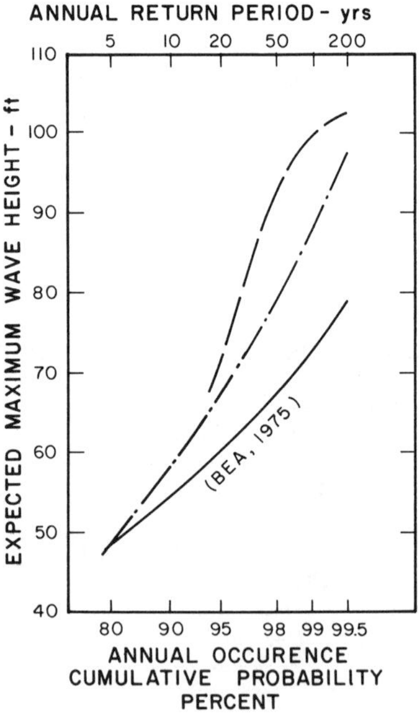

Figure 8–1 Expected maximum wave heights generated by hurricanes in the Gulf of Mexico. Solid curve is based on hindcast studies of storms during the period 1900 to 1969; other curves represent estimates of earlier studies. [From Bea 1975, Figure 14, p. 1170.]

on somewhat fewer data, have estimated maximum heights in the range of 60 to 100 ft (18 to 30 m). All studies indicate that wave heights of 50 ft (15 m) are rather common and can be expected to occur about every 5 years.

Of equal importance is the area affected by the storms. The hindcast studies (Figure 8–1) have referred to the Continental Shelf from Alabama to South Texas, almost the entire area of the Gulf Coast province of Tertiary clastic deposition. Therefore, these results may be applied with some confidence to depositional conditions during the Tertiary and perhaps earlier periods.

From a geological standpoint, the effects of storm waves on a shallow shelf are significant. In a geographic location similar to the shelf in the Gulf of Mexico, sedimentation rates are rather low. As an approximation, the Holocene sediment thickness averages about 75 ft (23 m) and was deposited during about 10,000 years. The sedimentation rate is 0.038 ft/5 yrs (0.012 m/5 yrs), which means that a major storm could have affected about every $\frac{1}{2}$ in. or 1 cm of sediment.

Tidal Currents

Tides are large waves caused by the gravitational attraction of the moon and sun on the earth's surface. The tidal bulge of surface water is symmetrical; that is, there is an attraction of water mass on the side toward the moon and an opposite bulge on the reverse side of the earth. Thus, as the moon rotates about the earth, sea level at a given point rises and falls twice a day. The wave length of the tidal bulge is one-half the circumference of the earth, which is many times greater than the average ocean depth, so that the depth to length ratio is small. Therefore, the tidal bulge can be considered a "shallow-water" wave (Chap. 2).

This simple view of the tidal bulge, however, is complicated by other factors. In the open ocean, the Coriolis force deflects the tidal flow from its expected path. On the continental shelf, the shoaling water also impedes and deflects the flow. On approach to the shoreline, the tidal flow is again deflected, and at the shoreline, the effect may be either dissipated or magnified depending on the coastal configuration. Thus, the tidal force has a nonuniform effect on the sea bottom or shoreline sediment.

In relatively small isolated ocean basins, such as the Gulf of Mexico, the tidal range is about 1 ft (0.3 m), and tidal effects are negligible. In other basins, such as the North Sea, the tidal range is much greater because the tidal bulge from the open North Atlantic is confined to a smaller area. When the tides are concentrated in narrow estuaries, the tidal range can be 22 ft (7 m) as in the Wash on the northeast coast of England or a maximum of 40 ft (12 m) as in the famous Bay of Fundy in Canada.

The tidal wave generates significant currents when the wave is confined in some smaller oceans. For example, in the North Sea the tidal flow is exceedingly strong, and surface currents of 1 m/sec or more flow across the shelf parallel to shore, reversing direction according to the flood and ebb flow. Again, the tide can be considered a "shallow-water" wave, and the maximum bottom velocity produced by such a wave is given by Eq. (2–61) in Chapter 2. For the North Sea, the tidal height may be about 1 m, and it can be calculated that at a depth of 10 m, the bottom velocity is 50 cm/sec, and at a depth of 30 m, the bottom velocity is 28 cm/sec. These velocities are sufficient to erode and transport a large range of sand sizes (Figure 2–1) and to produce significant bed forms on the shelf.

Storm Currents

Another important mechanism for the erosion and transport of sand is that of wind-induced currents, especially during storms. A related and more pronounced effect is that of major storms that produce a piling of water in coastal areas called a *storm surge.* In both cases, the return flow can result in bottom currents of significant velocities.

Onshore winds of modest force produce bottom currents on the shallow shelf. For example, currents were measured (Murray 1972) in 15 m of water northeast of the Mississippi delta during passage inland of a storm that had maximum wind speeds of 25 knots (30 mi/hr, 48 km/hr) to the northwest. Velocity profiles were recorded and resolved into north-south and east-west components (Figure 8–2).

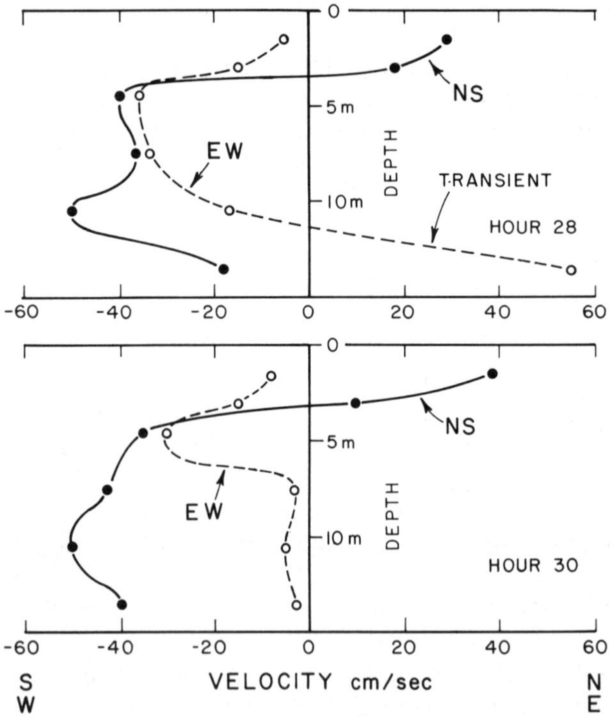

Figure 8–2 Velocity profiles at a location 25 km northeast of the Mississippi delta in a water depth of 15 m during passage of a storm with onshore winds of 25 knots (30 mi/hr, 48 km/hr). East and north components are taken as positive; west and south are negative. [From Murray 1972, p. 133.]

Typical observations near the time of maximum winds showed that surface water had a strong northerly flow of 30 to 40 cm/sec. Most of this current was deflected parallel to shore as it reached shallow water. At the same time, deeper waters had a strong southerly flow of 40 to 50 cm/sec, decreasing to 20 to 40 cm/sec near the bottom. These velocities contain a significant tidal component that contributes as much as 15 cm/sec to the observed velocities. Thus, the wind-driven surface flow is accompanied by bottom flow in an offshore direction, which is added to the tidal flow to produce an erosive current.

A greater effect is produced by the approach of major storms. The surge generated by hurricanes may result in a mound of water as high as 20 ft (6 m) in coastal areas, and the return flow near the bottom can result in velocities estimated at 5 percent of wind speed (Murray 1970). This means that minimum hurricane-force winds of 75 mph (120 km/hr) may produce return flows of 168 cm/sec, and winds of 150 mph (240 km/hr) may produce flows of 335 cm/sec. These powerful currents may be responsible for the transport of large amounts of sand into deeper water.

Turbidity Currents

Return flow of storm surge has another major effect, that of producing turbidity currents, or density currents, in which greater density is due to suspended sediment. The exact mechanism whereby the turbidity currents are generated is not known, but the results of turbidity flow have been documented on the shelf.

The Texas shelf off Padre Island south of Corpus Christi was sampled after the passage of hurricane *Carla* in 1961 (Hayes 1967). Cores of bottom sediment contained a thin graded bed that ranged from 1 to 10 cm (0.4 to 4 in.) in thickness (Figure 8–3). The beds were composed of very fine-grained sand and coarse silt that graded upward to fine silt and clay, and they rested with a sharp basal contact on underlying pre-*Carla* muds. Their character suggested rapid deposition from turbidity currents. These turbidities were recorded in water depths of between 60 and 120 ft (18 to 36 m), and greater thicknesses were observed offshore from the larger channels that were cut through the barrier island by the hurricane surge. This relationship suggested that the return flow of water piled along and behind the barrier island was responsible for the deposits.

Carla passed inland about 100 mi (160 km) north of the Padre Island area with wind speeds estimated to be 175 mph (280 km/hr), and in the area of Padre Island, maximum winds were on the order of 90 mph (144 km/hr). The accompanying storm surge was a maximum of 22 ft (6.7 m), and many coastal areas experienced a surge of 10 ft (3 m).

Wind speeds of 90 mph (144 km/hr) could generate return flow of 2 m/sec according to the studies of Murray (1970). Therefore, these currents were sufficient to erode and entrain all sands and muds present in the coastal zone. The sediments thus suspended probably formed turbidity currents that moved across the shelf for distances greater than 17 mi (27 km). Deposition took place due to a decrease in

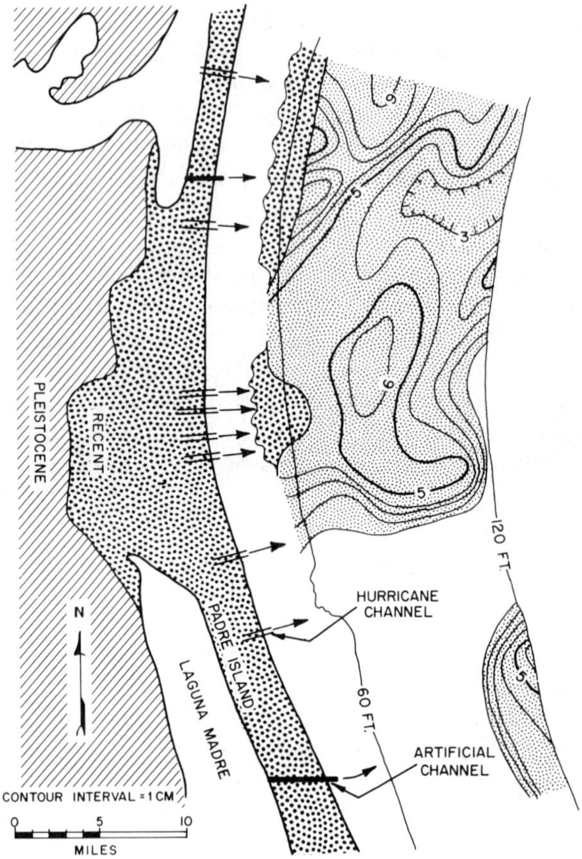

Figure 8–3 Thickness of turbidites deposited in water depths of 60 to 120 ft (18 to 36 m) off Padre Island, lower Texas coast, by the effects of hurricane *Carla* in 1961. Contour interval 1 cm. [Adapted from Hayes 1967.]

velocity as the currents spread across the shelf. Deposits were thicker where the slope of the sea floor was least, whereas submerged ridges had no deposits, suggesting a topographic control on deposition (Hayes 1967, p. 41).

The shelf turbidites generated by hurricane *Carla* are unique in the oceanographic literature, and one may well ask whether the process was not also unique. It has been said that later attempts to sample these turbidites recovered only bioturbated silt and mud rather than graded beds, and so it appears that benthic organisms quickly reoccupied the areas of deposition. This conclusion may explain why shelf turbidites have not been more widely reported.

Turbidity currents generated by major storms may be an important mechanism for the transport of sand to the shelf. The frequency of these modern storms (Figure 8–1) also suggests that this process could have been important in the geologic past. In fact, storm-generated density currents have been called upon to explain thin sandstones in some ancient shelf deposits (Berg and Powell 1976; Snedden and Kersey 1982).

Other Currents

The open oceans beyond the shelf have large-scale permanent circulation systems that may at times affect the outermost shelf. These currents are confined to the upper or surface layers and commonly have an effect no deeper than 100 to 200 m. The ocean circulation closely resembles that of atmospheric circulation, and the basic patterns are those of large closed systems called *gyres.* For example, in the Northern Hemisphere, there are subpolar and subtropical gyres bounded on the south by equatorial currents. The gyre currents have velocities of only 3 to 6 km/day (3.5 to 7 cm/sec). The major patterns are modified locally by large eddies, called *Ekman spirals,* and by movement of surface waters normal to the dominant wind direction and toward the centers of the basin where circulation again proceeds outward toward the margins of the gyre by so-called geostrophic flow.

At some places the major circulation systems develop jet currents of high velocity. In the Northern Hemisphere, these currents form on the eastern continental margins and are called, from the ocean viewpoint, *western boundary currents.* Such currents are the Kuroshiro current of the western Pacific and the Gulf Stream of the western Atlantic Ocean. Velocities range from 40 to 120 km/day (46 to 140 cm/sec), and yet their principal effect is confined to the surface layer and extends to no more than a depth of 200 m. It is possible that jet currents such as the Gulf Stream can affect the outer continental shelf periodically because the currents are known to shift laterally. Still, velocities decrease rapidly with depth, and so it is concluded that the jet currents are not a major factor in shelf transport and deposition.

In summary, the major processes of wave action and current transport make the shelf environment one of continually agitated waters, sometimes violently by storms and at other times more gently by persistent currents. The marine shelf is an environment of both transport and deposition of sand, and shelf processes cause sand accumulation in distinct bodies and, thus, in the form of potential reservoir rocks.

MODERN SHELF SANDS

Sands are distributed in various patterns on the modern continental shelves. In many areas the shelf is a mud environment, but sands may be concentrated in large waves or ridges that have linear trends. These sands are conspicuous because of their topographic relief, and they have been studied in many areas of the world. Various interpretations have been offered for their origins, but there are obstacles to interpretation. Texture and composition may be known from bottom samples or short cores, but the sequence of sedimentary structures commonly is not described, probably because the unconsolidated nature of the sediment either obscures or has not allowed preservation of structure. Nevertheless, the morphology and general character of these features indicate the types of sand bodies that can be expected in the shelf environment. The following examples show only three types, and these illustrate some of the processes and resulting morphologies that may be expected in ancient sections.

Sabine-Heald Banks

Concentrations of sand in the modern Gulf of Mexico form broad linear banks off the coast of southeast Texas. Two of these features, known as the *Sabine* and *Heald banks,* are located from 20 to 30 mi (32 to 48 km) from the shoreline and west of the mouth of the Sabine River, which forms the boundary between Texas and Louisiana (Figure 8–4). The area is in water depths of 50 to 60 ft (15 to 18 m), and the banks rise only about 20 ft (6 m) above the sea floor. Thus, they are subtle features of low relief.

The general sedimentary sequence has been determined by shallow seismic pro-

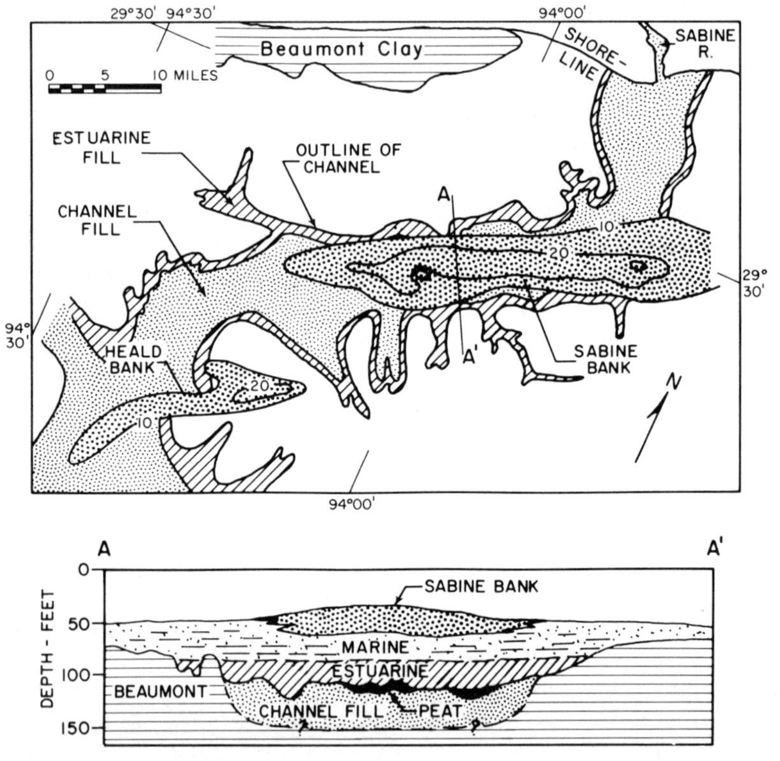

Figure 8–4 Holocene sediments in the Sabine-High Island area, Gulf of Mexico. Heald bank and Sabine bank are sand shoals overlying the ancient Sabine River channel cut in Pleistocene Beaumont clay. Map shows thickness of bank sands greater than 10 ft (3 m) and consisting of more than 50 percent sand or shelly sand. [Adapted from Nelson and Bray 1970, Plate V.]

filing and has been supported by bottom samples and short cores (Nelson and Bray 1970). The ancient Sabine River channel is incised in Pleistocene Beaumont Clay. The channel extended southward from the modern shoreline, then turned abruptly westward for about 40 mi (64 km), and again turned southward. The deepest portion of the channel is about 75 ft (23 m) below the general level of the Beaumont surface. One long core showed the channel to be filled with coarse fluvial sands overlain by a discontinuous layer of peat and succeeded by estuarine deposits. The succeeding Holocene sediments are largely muds that were deposited during the transgression that followed Pleistocene glaciation.

Distribution of Holocene sediments was the main object of the study (Nelson and Bray 1970). These sediments could be divided into an older and younger section on the basis of seismic reflections. Bank areas are composed of sand and shelly sand. The sands are fine grained and show, on the average, 83 percent of quartz and feldspar with minor amount of shell fragments, silt, and clay. Shelly sands are still relatively quartzose with 71 percent of quartz and feldspar, but shell fragments are 18 percent of bulk composition. The offbank sediments are muddy sands that are 37 percent silt and clay, and locally there are concentrations of sandy shell that are composed of 50 percent of whole shells and fragments.

The thickness and composition of upper Holocene sediments were mapped in significant detail, and the morphology of the sand deposits can be reconstructed from these maps (Nelson and Bray 1970, Plate V). Areas of more than 50 percent sand or shelly sand, and greater than 10 ft (3 m) in thickness, show broad linear deposits that coincide with the banks. The low ridges are parallel to the shoreline, and the largest feature, Sabine bank, has an average width of about 6 mi (10 km) and a length that exceeds 35 mi (56 km). Average thickness is 20 ft (6 m) and maximum thickness is 30 ft (10 m).

The method used to construct the map (Figure 8–4) raises some uncertainties with regard to the true amount of sand in the banks. Undoubtedly, sand distribution is probably less uniform than shown in the reconstruction. Nevertheless, the broad linear nature of the sand distribution probably represents a close approximation to the actual morphology.

Unfortunately, the sequence of sedimentary structures within the banks is unknown, and therefore the processes that shaped these sand bodies can only be inferred. Since the bank sands overlie muddy marine sediments, it is most likely that the morphology is due to normal marine processes near the present depth rather than the result of the earlier transgression. The sands may have been derived, in part, from preexisting sediments near the present shoreline to the north or, alternatively, from the westward drift of sand from the modern Sabine and Mississippi rivers. In both these river systems, the supply of sand is not abundant, and the present banks may have been shaped by longshore currents over a significant period of time. If sand accumulation was slow, it can be expected that benthic organisms occupied the bottom and probably reworked the shallow sediment. Consequently, primary sedimentary structures may have been largely destroyed.

The banks can be preserved in their present form if deposition of muddy sediment continues to blanket the area, thereby protecting the sands from future erosion. The resulting stratigraphic sequence could be readily interpreted if seen in a single cored section. The overall sequence is transgressive. The older fluvial channel might be expected to be dip-trending, and the local westward trend might not be anticipated. The overlying marine section might be interpreted with more confidence, and the overall discordance in trend between the nonmarine and marine sections probably could be predicted. The shallow-marine origin for the younger section might be indicated by bioturbation in both muds and sands. This environment is commonly dominated by longshore currents, and thus the elongation of sand in a strike trend is expected.

Locally, however, the sand is thin but extensive. Bioturbation, if present, would probably reduce permeability, especially if some interbedded muds were present. Shell fragments could result in patches of carbonate cement, also locally reducing permeability. The result might be a thin but widespread reservoir of low permeability. Therefore, Sabine and Heald banks form a possibly useful analog for the interpretation of some ancient shelf sandstones.

U.S. Atlantic Shelf

Sand ridges on the Atlantic shelf have been recognized for many years. Several hypotheses have been proposed for their origin, including relict sediments of either nonmarine or marine origin and as depositional features developed during Holocene transgression and by modern sedimentary processes (Duane et al. 1972; Swift et al. 1972). Recent study of the sand ridges off the New Jersey coast suggests that an interesting combination of constructional and degradational processes was responsible for their origin (Stubblefield, McGrail, and Kersey 1984).

Based on topography, there are three distinct areas of sand ridges (Figure 8–5): (1) Inner-shelf ridges are found in water depths of 5 to 15 m (16 to 49 ft) and occur in trends oblique to the modern shoreline. (2) A middle-shelf area shows poorly

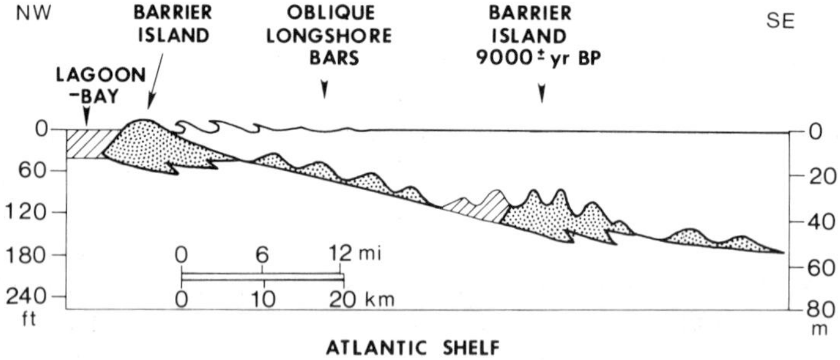

Figure 8–5 Diagrammatic cross section of the U.S. Atlantic shelf off New Jersey showing oblique longshore bars of the inner shelf, relict barrier-islands sands in the middle shelf, and oblique bars in the outer shelf. [Adapted from description by Stubblefield, McGrail, and Kersey 1984.]

oriented ridges in water depths of 25 to 45 m (82 to 148 ft). (3) Outer-shelf ridges are linear features similar to the inner-shelf ridges and are found in water depths greater than 45 m (148 ft).

The inner-shelf ridges are aligned in trends that range from 18° to 25° from the present coast. The shallow parts of these ridges are connected to the shoreface, but they extend in long, more or less continuous features into deeper water. The inner-shelf ridges may be designated, in general, as oblique longshore bars. The height of the ridges is from 4 to 10 m, and their widths are on the order of 2 to 4 km. They extend for 20 to 60 km and have wave lengths, or distance between ridges, on the order of 10 km. These ridges have a rather uniform texture of well-sorted, medium- to fine-grained sand, but samples from short cores suggest that the tops and flanks may be somewhat coarser grained (Field 1980; Stubblefield, McGrail, and Kersey 1984). Generally, a basal lag of coarse sand or gravel is present (Duane et al. 1972; Swift et al. 1972).

The middle-shelf ridges are parallel to the coast, but they are less continuous features and more ragged in outline. These ridges are also somewhat greater in height and are on the order of 10 to 15 m. The sands are fine grained, but they tend to coarsen upward. The crests of the ridges are abruptly coarser grained in the uppermost section of 1 m or less. Composition of the sand is high in quartz, which averages about 90 percent, with the balance of the composition being orthoclase and rock fragments.

The middle-shelf sand ridges appear to have been formed as barrier islands on an ancient shoreline developed during a stillstand in the Holocene transgression. This origin is consistent with textural grading and highly quartzose composition. Furthermore, cores show that the barrier-island deposits overlie older lagoonal sediments in a sequence similar to the modern barrier islands of the U.S. Atlantic Coast (Kraft 1971). The lagoonal sediments have a macrofauna that suggests an age as old as 10,950 years BP (before present). The macrofauna in the overlying sands suggests an age of deposition in the range of 4000 to 6000 years BP. Therefore, the deposition of the barrier-island deposits began about 9000 years BP and continued to about 4000 years BP.

At the time of barrier-island deposition, the outer-shelf sand ridges were probably formed as shore-connected, oblique longshore bars similar to those that now are being formed in the inner-shelf area. These linear ridges are found at angles of about 20° to 25° from the trend of the middle-shelf ridges, similar to the trends exhibited by the inner-shelf ridges. The similarities in form and orientation suggest a similar origin for these features.

After deposition of the barrier-islands sequence, Holocene transgression continued to the present high stand of sea level. During this transgression, wave and current action modified the older barrier-island deposits and produced their present discontinuous and irregular topography.

The inner-shelf sand ridges are modern constructional features. Their origin was previously thought to be due to storm-generated helical flow patterns close to the shoreline (Duane et al. 1972). Storm waves approaching the coast were believed to be refracted, forming northeastward flowing currents that transported sand in the form of linear sand ridges.

A new hypothesis suggests that construction of the inner-shelf sand ridges is due to persistent processes rather than to storm-induced currents (Stubblefield, McGrail, and Kersey 1984). Satellite imagery shows turbid waters near the Atlantic shore, and patterns indicate the presence of large vortices or eddy currents. The normal longshore current is toward the southwest, and the nearshore eddy currents are deflected backward along the shore toward the northeast. These currents are believed to be the transporting agents that form the linear inner-shelf ridges.

In either case, the oblique longshore bars of the innershelf are constructional features at the present time. The ragged barrier-island trend of the middle shelf represents relict sediment that was abraded during transgression. The linear ridges of the outer shelf represent the ancient oblique longshore bars developed at the time of barrier-island deposition. Thus, the differing morphologies are the result of both constructional and erosional processes that acted during transgression. Of particular interest for the study of ancient sands is the modification of the barrier-island morphology so that narrow linear beach trends are no longer present.

North Sea Sand Ridges

Strong tidal currents are responsible for construction of sand ridges in the North Sea (Stride et al. 1982). The ridges form shallow banks on broad shelves bordering the coasts of England and the Netherlands. They are especially well shown on the shallow shelf off the northeast coast of England called the Humber Spur (Figure 8–6). Ridges are found in water depths up to 100 ft (30 m), and they rise, in some places, nearly to sea level. The area is one of tidal currents that flow parallel to shore and in northwest-southeast directions. Surface velocities of tidal currents range from about 0.5 to 1.0 m/sec.

The ridges rise above the sea floor to an average height of about 50 ft (15 m). They are long narrow features that range from 2 to 5 km in width and 15 to 65 km in length, and they are nearly linear or broadly curved features, asymmetric toward the northeast. Internal structures of the sand ridges are poorly known. The gentle southwest flanks are covered by sand waves, or *mega ripples,* up to 6 m in height. Seismic profiling suggests an internal structure with a dip toward the northeast.

Short cores of the ridges give some indication of sand character (Houbolt 1968). Sedimentary structures are laminations that have dips of 10° to 20°, presumably representing a dune phase of sand transport. The median grain size ranges from 0.2 to 0.3 mm (medium grained). Cores from the steep faces show common dark clay laminae interbedded with sand, and these denote deposition from suspension during tidal reversals. The sand contains some clay clasts and carbonate skeletal fragments in amounts up to 5 percent. Sand is derived from older deposits on the shelf, probably composed of glacial outwash. A coarse lag of gravel is present on the sea floor between the sand ridges.

Local transport of sand is oblique, up the gentle slope of the sand ridges and

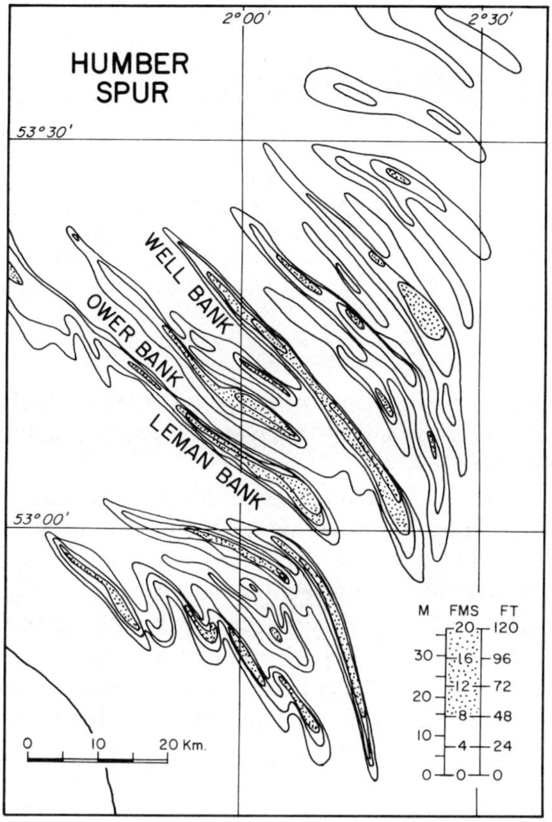

Figure 8–6 Sand ridges in the North Sea off the Suffolk coast of England. Height above bottom shown by form lines; contour interval four fathoms (7.3 m). [Adapted from Houbolt 1968.]

toward the northwest, in addition to sand transport parallel to the tidal currents. In this area, it is believed that the net transport direction is toward the northwest in the direction of local transport indicated by the sand waves (Figure 8–7). The vertical internal sequence of sand ridges must be inferred. A coarse-grained lag may be present at the base because of ridge migration across the interridge areas. The sequence is cross-laminated, probably representing a ripple to dune phase of flow regime. There appears to be no vertical change in grain size, and thin clay laminae may be present.

A somewhat different sequence has been postulated by McCave (1971); the deeper ridge flanks contain a benthic fauna, and if these flanks are preserved below the migrating ridges, a basal unit of bioturbated sand may be present. This possibility would seem unlikely because over time, the entire sequence would probably be reworked by tidal currents to produce a uniform sequence of cross-laminated sand. Another possibility is that the grain size increases upward within the sand ridges because maximum tidal velocities can be expected to be nearest to sea level.

Despite the uncertainties of internal sequence, the sand ridges provide an ex-

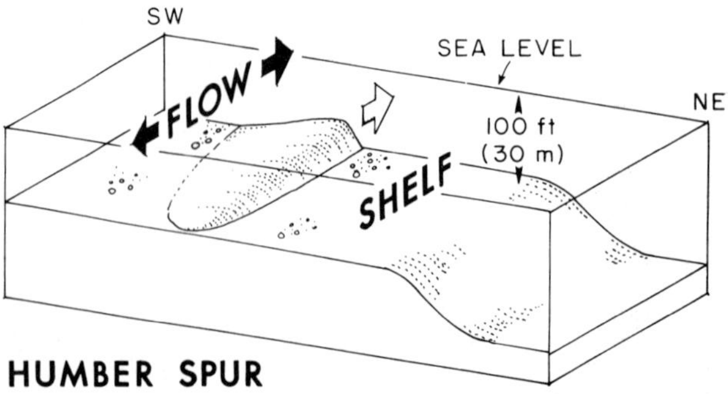

Figure 8–7 Diagram of sand ridges on the Humber Spur, North Sea, as viewed toward the southwest. Asymmetry suggests migration of ridges toward the northeast in addition to sand transport parallel to tidal currents. Open arrow shows net sand transport to the northwest. [Adapted from description by Houbolt 1968.]

ample of extensive sand bodies that might be formed on a shallow shelf. If these bodies were preserved as reservoirs, the sandstones would be well sorted and probably of high porosity and permeability.

A tidal origin has been suggested for sand ridges in many other shallow-marine areas of the world (Off 1963). Typically, these ridges are found in groups, parallel to tidal flow, and with more or less regular or "rhythmic" spacing. Most are 25 to 100 ft (8 to 30 m) in height, 5 to 40 mi (8 to 64 km) in length, and are spaced from 1 to 6 mi (1.6 to 10 km) apart. Their occurrence seems to be related to relatively wide shelves that have large tidal ranges of 10 ft (3 m) or greater (Off 1963, p. 332). Most ridges are found in shallow water of depths to 60 ft (18 m), and maximum water depths seem to be no greater than 100 ft (30 m), as in the North Sea.

Other Shelf Sands

The properties of shelf sands are not well known in many areas, but there is an important characteristic of shelf sediments that may aid in their identification in ancient deposits. Bioturbation, or the reworking of sediment by benthic organisms, is common, especially in the shallow parts of the shelf. In nearshore areas, bioturbation is abundant in fine-grained sands that lie in water depths of from 6 to 15 m (Reineck and Singh 1975), and bioturbation may extend as far as the shelf margin in muddy sediments.

Water depth is an important control on the distribution of benthic organisms and therefore on bioturbation. Inner neritic waters are in the photic zone of abundant plant and animal life, and benthic animals on, or within, the bottom sediments are numerous. In general, there is a decrease in the number of larger burrowing animals with an increase in water depth and hence a decrease in the biological disturbance of sediments. There are several types of burrowing organisms. The first type consists of those that burrow deeply into the bottom, passing the sediment through their digestive systems and extracting organic material as food. Wormlike animals affect this habit, and their feeding action can completely destroy primary sedimentary structures. A second type of burrower is the kind that forms an open tube or passageway into the sediment and uses the burrow primarily for shelter.

Tubes may be later filled by a contrasting type of sediment, thereby producing a distinctive and generally large burrow. Numerous crustacea possess this habit. A third type consists of animals that burrow only into the shallow sediment and feed on nutrients in the overlying water. Many pelecypods use this mode of feeding.

Water depth is not the only control on the distribution of benthic organisms. Other factors such as sediment type and available nutrients in the water also determine the abundance of bottom-dwelling animals. The importance of bottom sediment, or substrate, has been demonstrated by sediment cores along the continental shelf of the western Gulf of Mexico (Hill et al. 1982). Sandy bottoms extend far seaward in great lobes from the Rio Grande and Lavaca rivers, located in the southern and middle Texas coast, respectively. The intervening area is dominantly mud. Bioturbation is abundant in the sandy sediments almost to the edge of the shelf, whereas bioturbation is sparse in the muddy bottoms, even at water depths of 150 ft (45 m).

The control of nutrient supply on benthic fauna is expected but not well documented in relation to depth or sediment type. Thriving benthic faunas have been reported locally in excessive water depths, and apparently these faunas are maintained by cold nutrient-rich waters that flow from the shelf to greater depths.

The effects of bottom sediment and food supply may be related. Sandy sediment provides a suitable substrate for many animals and may owe its origin to sand transport by bottom currents, which, in turn, provide a constant flow of dissolved nutrients and plankton to support an abundant fauna. Furthermore, many marine species have swimming or floating larval forms that are easily dispersed by current flow to establish new benthic faunas. Therefore, currents may provide sand for proper substrate, food for sustenance, and a population-dispersal mechanism. It is not surprising, then, that bioturbated sands are common on marine shelves, and that these sands may be surrounded by nonbioturbated muds.

ANCIENT SHELF SANDS

Sandstones of marine-shelf origin are important reservoirs for oil and gas. In fact, these sandstones commonly form large stratigraphic traps because they are enclosed in marine shales. The characteristics of typical marine sandstones are shown by cored sections that display their sedimentary structures, texture, and composition in vertical sequence. The water depths in which these sandstones were deposited is not always certain. Stratigraphic position and the relation to known shoreline deposits indicate a shallow-water origin for some sandstones. Other sandstones appear to have been deposited in deeper-marine environments, but in these cases, estimates of depths are not precise. The shallow-water sandstones are characterized by a high degree of bioturbation, whereas those believed to be deposited in deeper water may show little or no bioturbation. In addition, depth may be inferred also from the nature of the surrounding shales and from the presence or absence of bioturbation in the enclosing sediments.

For some shallow-water sandstones, intense bioturbation may aid in establishing an inferred water depth, but evidence for the transport mechanism was destroyed by organic reworking. When primary sedimentary structures are preserved, it appears that low flow-regime currents were largely responsible for the transport of sand because ripple structure is common. More rarely, a somewhat higher flow-regime is indicated by the presence of inclined laminae that suggest a dune phase of transport. For the nonbioturbated sands of deeper-water origin, the dominance of ripple structure indicates transport by persistent submarine currents.

The morphologies of shelf sandstones appear to be of three general types (Figure 8–8). The first is that of broadly ovate bodies of wide areal extent that are characterized by a relatively high degree of bioturbation, and these sand bodies may have risen in moundlike structures above the surrounding sea-floor bottom. An example of this type of shoal mound is the Upper Cretaceous Woodbine sandstone in the Kurten field, southeast Texas.

A second type of shelf sandstone is narrow, linear, and shows a prograding vertical sequence. These sandstones are quite unlike those described on the modern shelf because they seem to have resulted from the concentration of currents along the seaward sides of gentle terracelike breaks. An example of these shelf-terrace sandstones is the Upper Cretaceous Sussex sandstone in the House Creek field, northeastern Wyoming.

A third type of sandstone occurs in elongate to broadly ovate bodies that may be of deeper-water origin. These sandstones contain ripple structures throughout, and bioturbation is essentially lacking. An example of this type is the Middle Pennsylvanian Upper Marchand sandstone in the East Binger field, central Oklahoma.

These shelf sandstones, then, display a large range in inferred water depth. The reservoir properties also range from relatively low permeabilities and porosities in the highly bioturbated shallow-water sands to moderately high values in some nonbioturbated sandstones.

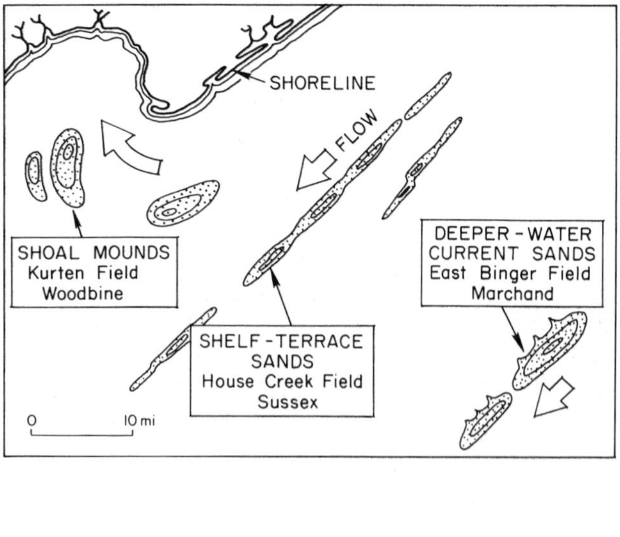

Figure 8–8 Diagram showing relative locations and morphologies of marine-shelf reservoirs described in this chapter. Open arrows indicate direction of sand transport. Scale in miles refers to the size of the sandstone bodies rather than to the distance from shore.

WOODBINE SANDSTONE, KURTEN FIELD, SOUTHEAST TEXAS

The Upper Cretaceous Woodbine Sandstone is the most famous reservoir in the Gulf Coast province. More than 4.8 billion barrels (763 million mm^3) of oil have been recovered from the Woodbine in the giant East Texas field discovered in 1930. Many other fields in the East Texas basin also produce oil from Woodbine reservoirs. More recently, the Woodbine equivalent of south Louisiana, the Tuscaloosa Sandstone, has been found to produce gas at depths of about 20,000 ft (6090 m), and the Woodbine-Tuscaloosa trend appears to have significant future potential.

In East Texas, the older Woodbine fields are located within areas of fluvial, deltaic, and shoreline sandstones in northeastern Texas. The Woodbine section passes abruptly to marine shale southward toward the Cretaceous shelf margin in southeast Texas. In this section of shelf deposits, thin lenticular sandstones of the Woodbine have been found to be oil-productive. The reservoirs have low porosities and permeabilities, typical of many shelf sandstones, and locally have wide areal extents. Characteristics of the Kurten Woodbine sandstones illustrate the nature of these reservoirs.

Geologic Setting

The East Texas basin is a major north-trending syncline in northeastern Texas (Figure 8–9). The basin is bounded on the east by the Sabine uplift. Across this struc-

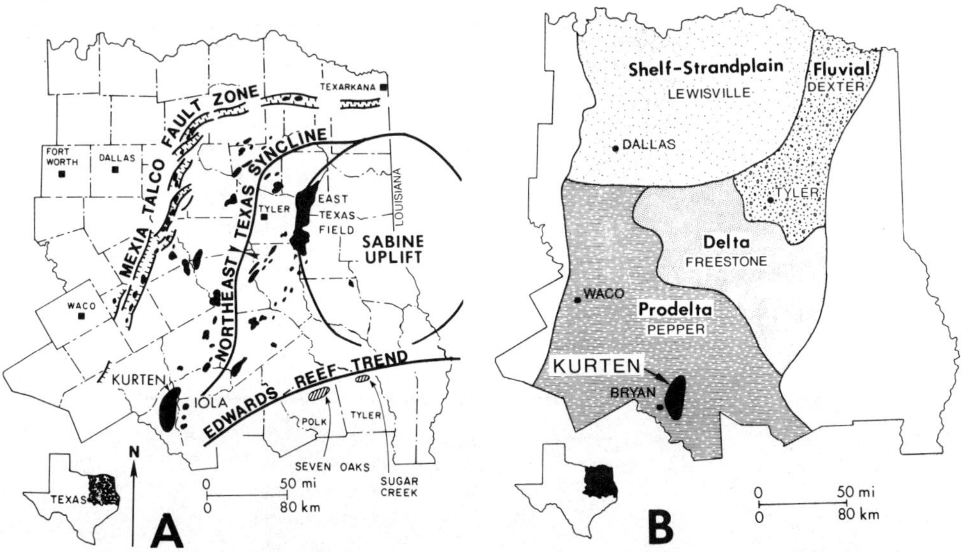

Figure 8–9 Maps of the East Texas basin showing relation to Kurten field to (A) major structural features, and (B) distribution of Woodbine facies and Lower Cretaceous shelf margin. [Modified from Oliver 1971 and Turner and Conger 1981.]

ture, Woodbine sandstones are truncated and overlain unconformably by the Austin Chalk, which resulted in an unconformity trap for oil in the East Texas field. Other fields in the east Texas basin are largely structural traps, either by normal faulting or by folds over salt domes.

The Woodbine sandstones were the earliest deposits of the Upper Cretaceous Gulfian Series (Table 8–2). In northeast Texas, the Woodbine Group is divided into a lower Dexter Formation, composed largely of thick fluvial sandstones, and the overlying Lewisville Formation of fluvial and shoreline origin (Oliver 1971). The Woodbine is overlain by the Eagle Ford Group, which is composed of shale with thin sandstones. The next overlying unit is the Austin Chalk, a prominent limestone unit that is present throughout the area. Below the Woodbine Group, the Lower Cretaceous Commanchean Series contains in its upper part a section of interbedded shales and limestones, which, in turn, overlies a thick limestone called the Edwards Formation.

In the Kurten field area, the stratigraphic units in the Woodbine and adjacent sections cannot be determined accurately because the Woodbine and Eagle Ford groups grade laterally into a thick shale section that contains isolated lenticular sandstones. Therefore, the entire section from the base of the Austin Chalk to the Buda Limestone is referred to informally as the Woodbine interval (Table 8–2). The

TABLE 8–2. STRATIGRAPHIC SECTION OF THE WOODBINE GROUP AND ADJACENT UNITS, EAST TEXAS

Series	Group	Formation	Kurten field
Gulfian	Navarro Taylor Austin	Austin Chalk	Austin
	Eagle Ford	Sub-Clarksville Sandstone Coker Standstone Harris Sandstone	Sub-Clarksville Woodbine interval
	Woodbine	Lewisville Sandstone Dexter Sandstone Pepper Shale	
Commanchean	Washita	Maness Shale Buda Limestone Grayson Shale Georgetown Limestone	 Buda
	Fredericksburg	Edwards Limestone	Edwards

Sub-Clarksville Sandstone is not everywhere present, but where it is found, the Sub-Clarksville appears to be the basal part of the overlying Austin Chalk and is not easily separated except where samples and cores are available. Therefore, the Woodbine interval is the section below the Austin or Sub-Clarksville.

In Kurten field, the Woodbine section contains five distinct sandstones (Turner and Conger 1981), which are designated, in descending order, the A through E sandstones (Figure 8–10). All of these sandy zones may produce oil in the area, but the principal reservoirs are the thicker B and C sandstones. The sandstones can be correlated widely throughout the field area, but toward the south and southeast, the upper Woodbine section is missing and appears to have been truncated by erosion below the Austin Chalk. On the unconformity surface, the Sub-Clarksville sandstone was deposited only locally and was succeeded by Austin limestone deposition.

East of the Kurten field area, thicker sands are found within the Woodbine interval (Figure 8–10, AA′). These sandstones have been referred to as the Harris delta (Nichols 1964; Oliver 1971). The Harris sandstones (Figure 8–11) reach a gross thickness of more than 400 ft (122 m) about 20 mi east of the Kurten field (Bell 1980; Turner and Conger 1981).

The location of the Kurten sandstones suggests that they were deposited in a marine shelf [Figure 8–9 (B)]. The Dexter and Lewisville sandstones were formed in fluviodeltaic and shoreline systems to the north of Kurten field, and these sandstones were replaced to the south by normal-marine deposits of the Pepper Shale (Oliver 1971). At a somewhat later stage, the supply of sand decreased, and the entire area of previous sand deposition was replaced by marine shales and sandstones of the Eagle Ford. During this period, a more limited amount of sand was supplied from the east, perhaps by erosion at the southern end of the Sabine uplift. The sands were distributed along the outer shelf and parallel to the shelf margin that was formed by the underlying Edwards reef trend. At this time, the Kurten Woodbine sands were deposited, probably by currents that flowed southwestward along the shelf margin. Thus, the Kurten Woodbine sandstones may be of Eagle Ford age (Anderson 1979, p. 19) and represent a succession of sands and shales that were not related to the older type Woodbine Formation of northeast Texas.

Kurten field is located about 30 mi (48 km) north of the Edwards reef trend. This Lower Cretaceous shelf margin also influenced deposition of Upper Cretaceous sediments. At Seven Oaks and Sugar Creek fields [Figure 8-9 (A)], the Woodbine interval contains thin turbidites that were deposited at the fill of slope channels (Siemers 1978; Foss 1979). The channel turbidites are succeeded by thin bioturbated sandstones that were outer-shelf deposits that prograded over the previous shelf margin.

The Kurten Woodbine sandstones also are most likely of shelf origin and were deposited relatively far from shore and close to the shelf edge. Sand was probably supplied to the Kurten area from the east by means of the Harris delta system (Figure 8–11), which prograded toward the shelf margin and provided sand to the adjacent shallow-marine environment.

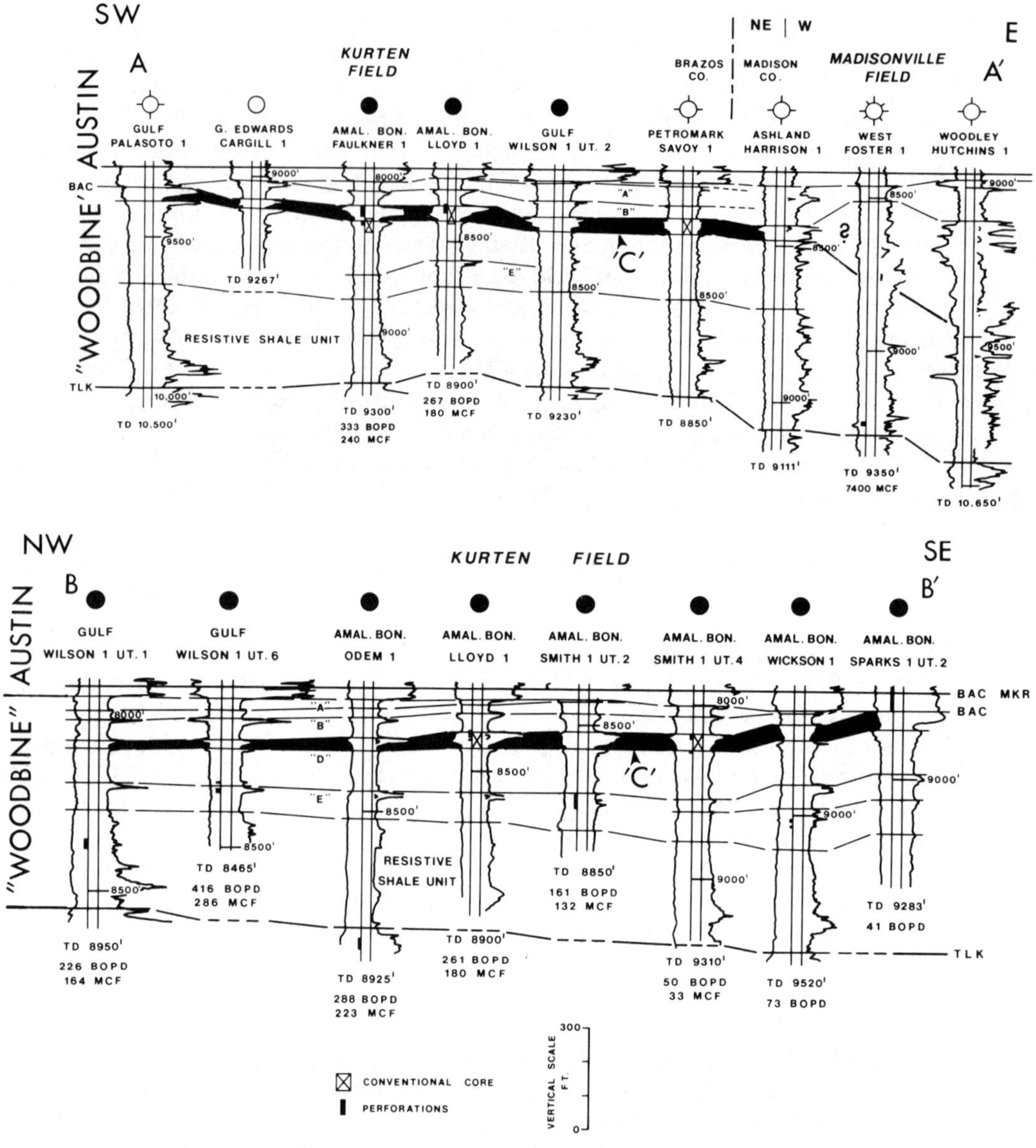

Figure 8-10 Stratigraphic cross sections of the Woodbine interval across Kurten field showing correlation of sandstones. Location of profiles AA′ and BB′ shown in Figure 8-15. Note truncation of Woodbine interval below the Austin Chalk. No horizontal scale. [Adapted from Turner and Conger 1981.] Location of sections shown in Figure 8-15.

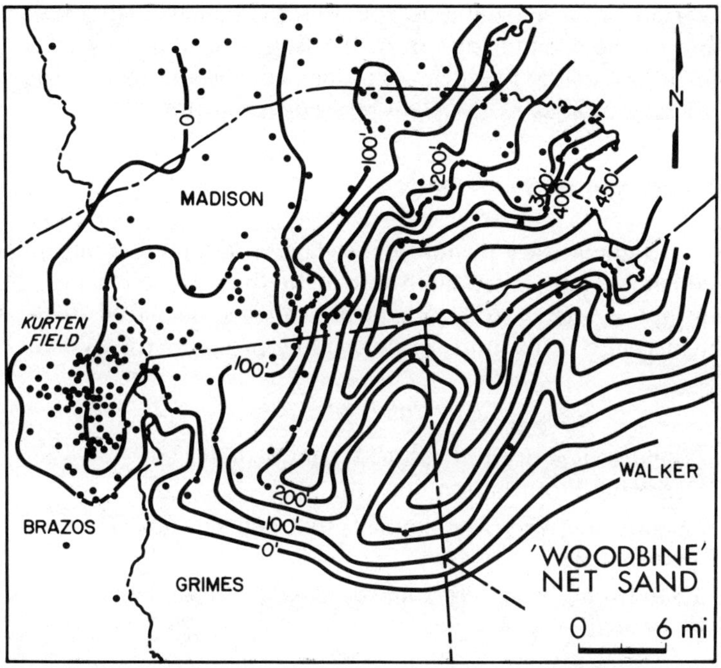

Figure 8–11 Total thickness of sandstone in the Woodbine interval. Thicker sandstones east of Kurten have been called the Harris delta. [Adapted from Turner and Conger 1981.]

Drilling History

Kurten field was discovered in June 1976 in a well drilled by Amalgamated Bonanza to test the possibility of oil in sandstones of the Harris delta. Oil production from these sandstones had been established previously in several small fields about 30 mi (48 km) east of Kurten. By 1980, there were more than 100 wells in the field over an area of 19,000 acres (46,930 ha). Initial production rates ranged from 60 to 380 bbls per day, and although flow rates generally decline about 30 percent during the first year of production, the wells were put on pump, and decline rates decreased (Bell 1980, p. 946). Total proven reserves have been estimated as 64 million bbls (10.2 million m^3) of oil and 60 billion cu ft (1.7 billion m^3) of gas, including primary and secondary recoveries (Bell 1980, p. 948).

All of the reservoir sandstones are relatively thin lenticular bodies, but the principal reservoir is the Woodbine C sandstone (Figure 8–10), which has an average thickness of about 30 ft (10 m) and a maximum thickness of 50 ft (15 m). The C sandstone is well developed in a core from the Amalgamated Bonanza Lloyd 1 well, located in the central part of the field, and its character has been described previously (Bell 1980; Turner and Conger 1981). Sedimentary structures are illustrated by a similar core from a nearby well, the Almagamated Bonanza Smith 2. Another core of the C sandstone from the Buttes Resources Wilson 2 well, located in the north part of the field, shows a thick contrasting facies. These cores, then, serve as typical examples of the main reservoir sandstone.

Description of Sandstones

Sedimentary structures. The Woodbine C sandstone shows a gradational sequence from clayey sandstone at the base to increasing sandstone with less clay upward in the section. A typical sequence displayed in the Lloyd 1 core can be divided into three units, in descending order, as follows:

Unit	*Thickness, ft (m)*
1. Sandstone, very fine grained, thin bedded, increasing bioturbation upward.	8 (2.4)
2. Sandstone, very fine grained, thin bedded, nonbioturbated.	11 (3.4)
3. Sandstone, very fine grained, clayey, intensely bioturbated.	19 (5.8)
Total	38 (11.6)

The nature of these units is illustrated by a similar core from a nearby well, the Amalgamated Bonanza Smith 2 (Figure 8–12). The lowermost unit (3) is nearly devoid of primary sedimentary structures because of the high degree of bioturbation [Figure 8–12 (E)–(H)]. Burrowing is generally of small size and essentially horizontal, although a few larger and sometimes vertical burrow tubes are present. The clay content is high, either intimately mixed with sand by bioturbation [Figure 8–12 (E)] or in thin discontinuous shale laminae [Figure 8–12 (F) and (G)]. The relict shale laminae suggest that the original sandstone beds were very thin and on the order of only a few centimeters in thickness. The base of the unit (3) grades upward from dark shale below through a thin zone of interlaminated sandstone and shale [Figure 8–12(H)].

The middle sandstone unit (2) is dominantly sandstone with thin shale laminae. The sandstone occurs in thin lenses that range from about 2 to 10 cm in thickness, and the lenses are outlined by laminae of black shale [Figure 8–12 (C) and (D)]. Laminations within the lenses appear to dip at relatively high angles of 10° to 15°, but the laminae are indistinct and are difficult to distinguish because of the highly quartzose nature of the sandstone. Some thicker beds appear to be more massive and range up to 15 cm thick.

The general character of the section appears to represent ripple lenses of sandstone separated by thin shale laminae, called *flaser bedding* (Reineck and Singh 1975), which is characteristic of periods of transport by low flow-regime currents

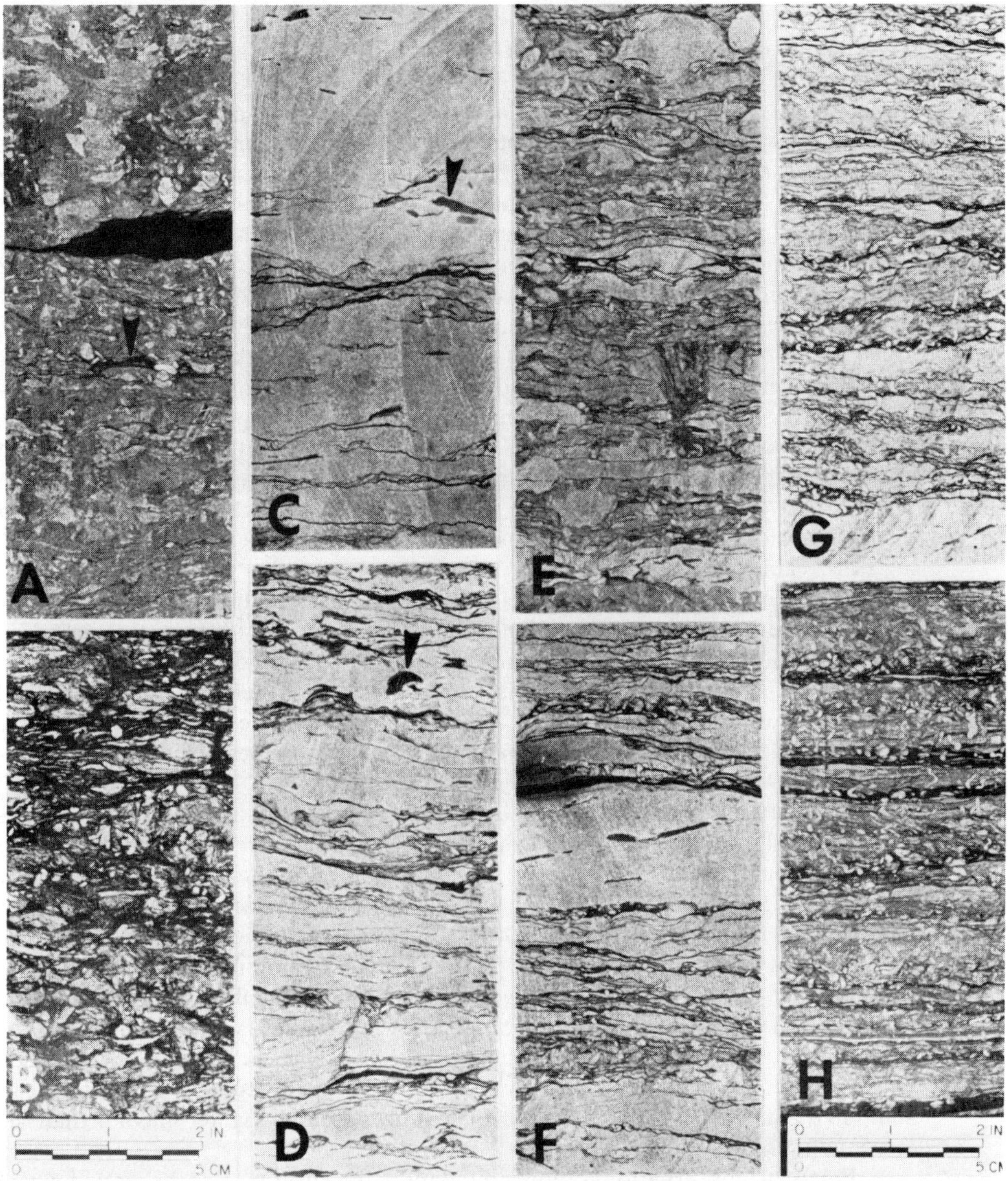

Figure 8–12 Sedimentary structures in Woodbine sandstones, Amalgamated Bonanza Smith 2, Kurten field. Note (A) and (B) are Woodbine B sandstone; (C)–(H) are Woodbine C sandstone. (A) highly bioturbated churned sandstone with remnants of shale laminae (arrow)—8474 ft; (B) highly bioturbated interbedded shale and sandstone; note small round burrows—8482 ft; (C) and (D) ripple-bedded sandstone with thin shale laminae and small shale clasts (arrows)—8527 ft, 8534 ft; (E) and (F) highly bioturbated sandstone with interlaminated shale but with some well-preserved ripple lenses of sandstone with a few shale clasts (F)—8546 ft, 8548 ft; (G) and (H) thinly interbedded sandstone and shale with increasing amounts of shale downward; note fine scale of bioturbation with small spiral burrow tubes—8552 ft, 8558 ft.

separated by periods of deposition of clay from suspension. Hence, these beds were characterized as flaser crossbeds by Turner and Conger (1981, p. 220). The sandstones contain scattered clasts of black to gray shale that are angular, thin, and range up to 2 cm in length. A few rounded pebbles of siltstone are also found, and these range up to 1 cm in diameter. Some of the thicker clasts and pebbles appear to have been deformed during compaction, suggesting that their origin was from fragments of partly consolidated mud.

The upper sandstone unit (1) contains thin ripple lenses of sandstone and shale laminae, similar to those of the middle unit (2), but the amount of interlaminated shale increases upward as does the bioturbation, which is generally of a fine scale. Thus, the upper unit (1) is gradational to the overlying black shale.

The overall aspect of the C sandstone sequence is one of progradation, showing increasing sand and decreasing shale and bioturbation upward with a more abrupt return to marine shale above. The ripple lenses of sandstone represent intermittent transport of sand under low flow-regime conditions. Although the exact nature of the lower unit (3) may be partly obscured by bioturbation, it appears that the thickness of sand beds increases upward, which also indicates a slightly increasing flow regime.

Other cores of the C sandstone in the central part of the field show the same general sequence with little variation. Turner and Conger (1981) distinguished an additional unit of thinly laminated sandstone and shale with decreased bioturbation that forms a lower transition zone in some cores. Nevertheless, the same general character is present through a wide area of C sandstone development. The three major units defined above are essentially the same as those described by D. K. Davies (in Bell 1980, p. 936) for the Lloyd 1 core.

A second major facies of the C sandstone is present in a core from the Buttes Resources Wilson 2 well in the north part of the field. Here, the sandstone has a maximum thickness of 56 ft (17 m), but the section is completely and intensely bioturbated. Bioturbation is of two types: an indistinct coarser bioturbation that produced a churned appearance, and a finer scale bioturbation in the form of thin vertical burrow tubes. Coarser bioturbation is most common in the upper part of the sandstone and produced a homogenized sand that has only indistinct convolute laminae. Generally, this sandstone lacks distinct burrow tubes, and the evidence for bioturbation remains in the totally churned aspect of the sandstone. Superimposed on the coarser bioturbation are a few small vertical burrow tubes that are only 2 to 3 mm in diameter. The tubes are sand-filled and commonly lighter in color than the surrounding matrix.

Fine-scale bioturbation is more abundant in the lower part of the section where an increasing amount of shale is present. The narrow burrow tubes are closely spaced and become increasingly more crowded until the entire section seems to be composed only of tubes. Still the burrows remain distinct and light gray in color against the darker and more clayey matrix. The character of the bioturbated facies resembles that of the clayey bioturbated sandstone unit (3) of the Almalgamated Bonanza Smith 2 core [Figure 8–12 (E)].

Despite the intense bioturbation of this facies, the Buttes Resources Wilson 2 core retains some evidence of bedding. The lower part has an indistinct horizontally laminated aspect, and the thinly bedded character is preserved by discontinuous and widely spaced laminae of shale. The upper part contains a few relict lenses of sandstone that suggest the entire interval was composed originally of thinly interbedded rippled sands and shale. Therefore, the bioturbated facies has the general aspect of the prograding facies found in the Lloyd 1 core, and the total section appears to have been the result of intermittent low flow-regime currents with increasing amounts of sand upward. The major difference is that the bioturbated facies was deposited more slowly, so that benthic organisms were able to survive the periodic influxes of sand.

Other sandstones in the upper Woodbine are most similar to the bioturbated facies of the C sandstone. For example, the B sandstone [Figure 8–12 (A) and (B)] is intensely churned, and all traces of primary sedimentary structures have been destroyed. Shale content, however, appears to decrease upward in each sandstone. Only the lowermost E unit has a distinctly different character. Thin sandstones are interbedded with black shale, and many beds show a sequence of massive to laminated or laminated to rippled and resemble the *AB* and *BC* divisions of turbidites (Turner and Conger 1981, pp. 219–20) as described in Chapter 9. These beds are about 1 ft (0.3 m) in thickness and have sharp basal contacts. Furthermore, bed thickness decreases upward, and the thin turbidites are replaced by ripple lenses of sandstone.

The Woodbine shales are black and generally massive but have a platy fracture. Thin laminae of fine silt are widely spaced at about 1 cm intervals through much of the shale. Scattered burrows are present, and these are small ovate tubes of silt that are 1 cm or less in diameter. The silt laminae and burrows are not common, however, and the gross aspect of the shale is structureless, largely a suspension deposit, and with no evidence for current transport.

The massive aspect of the shale indicates a quiet-water environment, and the rare bioturbation suggests an environment that was hostile to benthic organisms. Judging from these characteristics, and from geographic location and stratigraphic position, the nonsand environment was probably that of relatively deep water, perhaps outer shelf.

Petrography. In the Lloyd 1 core, the C sandstone has an average grain size of 0.14 mm (fine grained), an average maximum size of 0.54 mm (coarse grained), and a standard deviation of 0.05 mm (0.58 ϕ), which indicates that the quartz grains are well sorted. Grain size increases upward (Figure 8–13), which suggests that the transporting currents were persistent and of increasing velocity with time.

The sandstone has an average detrital composition of 73 percent monocrystalline quartz, 4 percent feldspar, 2 percent rock fragments, 1 percent other grains, and 15 percent matrix (Table 8–3). The feldspar grains are all orthoclase, and no plagioclase feldspars were recorded. Rock fragments are dominantly polycrystalline

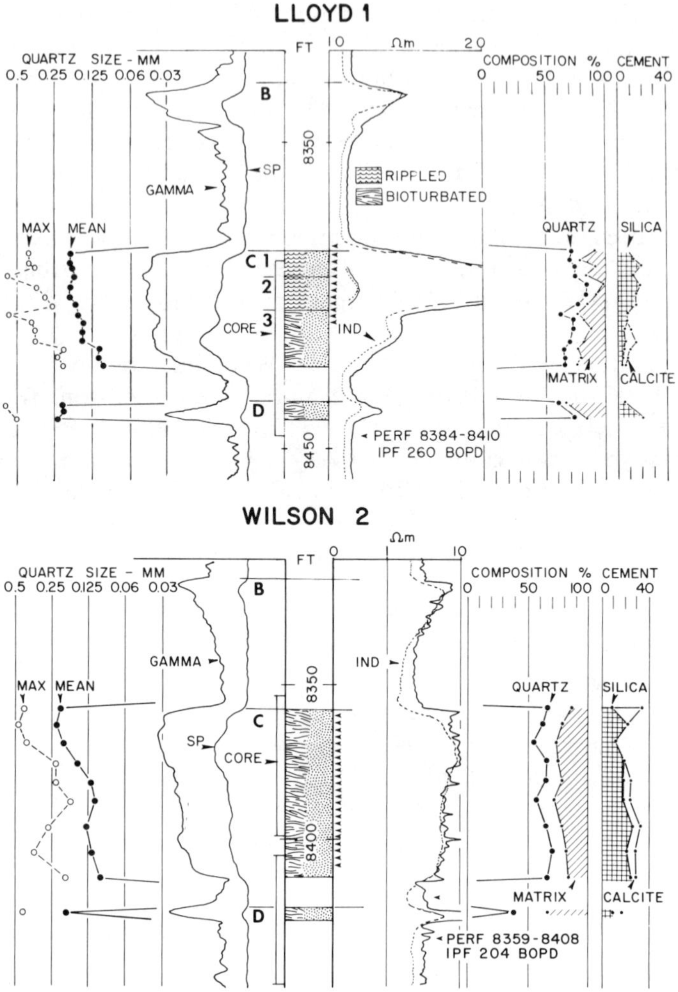

Figure 8–13 Log character, texture, and composition of Woodbine C sandstone. [Petrography in Amalgamated Bonanza Lloyd 1 from Turner and Conger 1981; petrography in Buttes Resources Wilson 2 by D. G. Kersey.] Location of wells shown in Figure 8–7.

quartz with only a trace of other rock types. Other grains include mica and rare opaque grains.

The matrix is composed largely of clay minerals and is found in thin discontinuous laminae and concentrations that apparently represent burrow fillings. Where the clay matrix is most abundant, it was probably introduced after sand deposition by bioturbation, as in the lower sandstone unit (3). The matrix tends to decrease upward but is somewhat variable. Small clasts are conspicuous in thin sections,

TABLE 8-3. PROPERTIES OF WOODBINE C SANDSTONES, KURTEN FIELD

				Quartz size[a]			Detrital composition[b]					Cement[c]			
Well	Sand	Unit	Number of samples	Mean (mm)	Max (mm)	σ (mm)	Qz (%)	F (%)	Rx (%)	Oth (%)	Mx (%)	Sil (% of total)	Cal	Permeability (md)	Porosity (%)
Lloyd 1	C	1	4	0.17	0.70	0.06	76	—	2	9	13	13	—	2.6	7.4
Lloyd 1	C	2	4	0.16	0.60	0.06	77	tr	2	9	12	15	—	2.5	7.7
Lloyd 1	C	3	6	0.12	0.40	0.05	69	tr	2	10	19	4	4	0.37	10.6
			Means	0.14	0.54	0.05	73	tr	2	10	15	10	1		
Smith 2	C	2	2	0.20	0.31	0.06	86	3	6	1	4	11	20	0.04	9.1
Smith 2	C	3	4	0.20	0.43	0.07	71	5	4	2	8	16	2	0.08	7.6
Smith 2	C	4	2	0.18	0.41	0.08	82	5	2	2	9	14	8	0.37	6.8
			Means	0.20	0.40	0.07	82	5	4	2	8	15	8		
Wilson 2	C	3	9	0.15	0.30	0.06	62	3	9	3	23	18	6	0.14	12.3

SOURCE: Petrography from Turner and Conger 1981.

[a] Long-axis measurements; σ = standard deviation.

[b] Qz = monocrystalline quartz, F = feldspars, Rx = rock fragments including polycrystalline quartz, Mx = matrix, Oth = other minerals, tr = less than 1 percent.

[c] Sil = silica grain overgrowths, Cal = calcite.

especially in the middle sandstone unit (2). The clasts are very fine-grained silty shale, partly deformed by compaction.

Cements form a significant part of total volume and average 11 percent in the Lloyd 1 core. Silica overgrowths on quartz grains are abundant in the cleaner sandstones of the upper units (1 and 2) and range up to 15 percent. Silica is less common in the lower bioturbated unit (3), where a small amount of carbonate cement is also present. The C sandstone in the nearby Smith 2 core (Table 8–3) has a greater amount of cement, and silica overgrowths and calcite are present in a total amount of 23 percent.

In the Wilson 2 core (Figure 8–13), the bioturbated facies of the C sandstone has about the same range in grain size. Average size is 0.15 mm (fine grained), and average maximum size is 0.30 mm (medium grained). Again, grain size increases upward, and the uppermost part of the sandstone is somewhat coarser grained than in the Lloyd 2 core. Composition is similar, but quartz content is reduced and the matrix is greater, a difference that is largely the result of bioturbation. The total content of cement is also high at 24 percent.

Other Woodbine sandstones have textures and compositions similar to the C sandstones. Where they are thin and bioturbated, their petrography resembles the bioturbated facies of the Wilson 2 core. Where they are thicker and contain cleaner rippled sandstone, their petrography is more like that of the Lloyd 1 core.

The overall composition of the sandstones is dominated by quartz, and even the rock fragments are mostly quartzose. The high quartz content probably indicates that the source sediment was derived largely from the weathering of older sedimentary rocks. High quartz is also consistent with transport by persistent low flow-regime currents.

Porosity and permeability. Generally low values of porosity and permeability appear to be the result of clay matrix, cements, and bioturbation (Figure 8–14). In the upper part of the Woodbine C sandstone, silica and carbonate cement are dominant, whereas in the lower part, carbonate is prominent. All of these components reduce permeability, but there seems to be be no systematic relationship between composition and permeability for individual samples. In the Wilson 2 core, bioturbation is probably most important in reducing the continuity of pores and in distributing clays more uniformly throughout the sandstone, thereby resulting in a lower average permeability.

Morphology

The Woodbine C sandstone is a large ovate body that has an overall north-south trend (Figure 8–15). Maximum thickness is somewhat greater than 50 ft (15 m) in the north part of the field as represented by the cored section in the Buttes Resources Wilson 2 well (Figure 8–15, BW2), but average thickness is about 30 ft (10 m). Within the body there is a northeast-southwest trend as based on net thickness of sandstone. The overlying B sandstone is a somewhat smaller body that has a max-

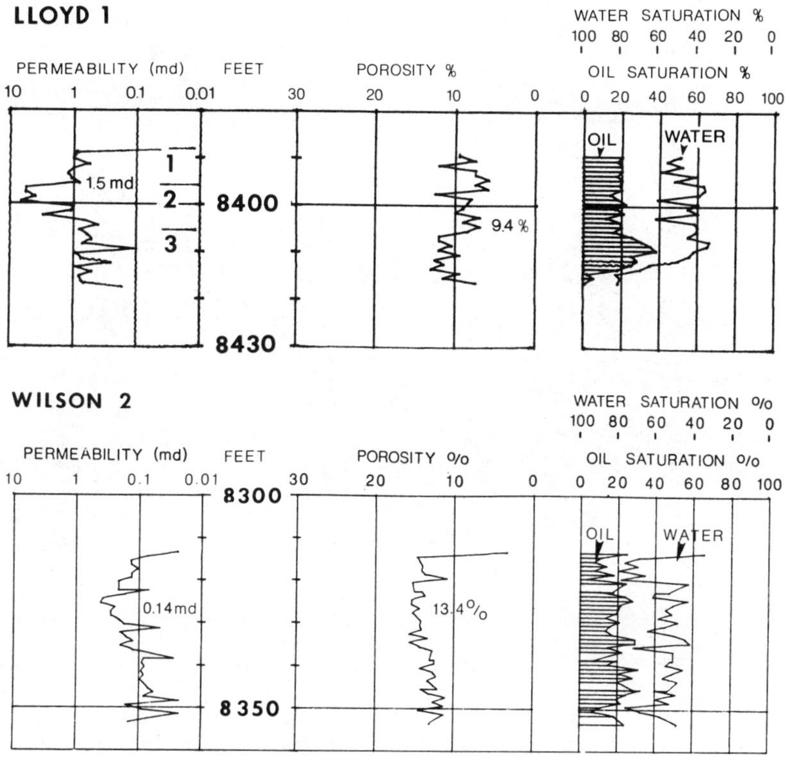

Figure 8–14 Porosities, permeabilities, and fluid saturations in the Woodbine C sandstone, Kurten field.

imum thickness of 25 ft (8 m), and internal northeast-southwest trends are well developed.

Other Woodbine sandstones have distributions that are similar to the B and C sandstones. The D sandstone is mostly restricted to the south part of the field, and the lowermost E sandstone is concentrated in the north part (Turner and Conger 1981, pp. 218–19). Both older sandstones also have internal northeast-southwest trends.

The upper Woodbine sandstones are truncated at the south and southeast margins of the field by an unconformity at the base of the Austin Chalk (Figure 8–10). Post-Woodbine uplift was probably caused by a deep-seated salt structure that appears to be present to the south of the field (Figure 8–16). The large domal uplift has intruded the Jurassic section and raised the overlying Cretaceous beds into a broad structure that may have closure over a wide area. Uplift took place before deposition of the Austin Chalk and caused erosion of the upper Woodbine sands in the area of the Hill field, located southeast of Kurten, where gas is produced from the partly truncated section. Following uplift, the Hill structure remained high,

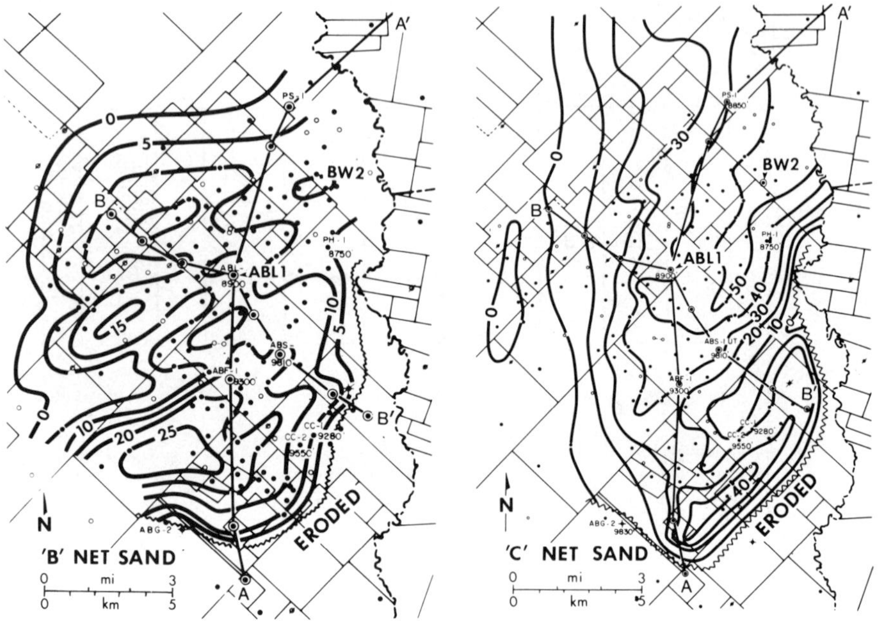

Figure 8–15 Net sandstone thickness of the Woodbine B and C sandstones, Kurten field. Contour interval 5 to 10 ft (1.5 to 3 m). [Modified from Turner and Conger 1981.] Cored sections described in text are indicated by letters ABL 1 (Amalgamated Bonanza Lloyd 1) and BW 2 (Buttes Resources Wilson 2). Correlation sections AA′ and BB′ are shown in Figure 8–10.

while coarse-grained sands of the Sub-Clarksville were deposited in adjacent low areas at the Iola and Martins Prairie fields (Barton 1982).

The Sub-Clarksville sandstones are quartzose and grade upward into bioturbated sands and shales. These thin sandstones are also gas reservoirs. Deposition of carbonate mud took place across the unconformity. There is no evidence of subaerial exposure, either in the truncated Woodbine or in the succeeding Sub-Clarksville, so erosion may have taken place under submerged conditions.

Early structure, therefore, played a role in the preservation of Kurten sandstones. Shelf processes were responsible for sand trends and morphology, and it is possible that contemporaneous salt uplift also influenced the patterns of sand deposition.

Interpretation

The Kurten Woodbine sandstones are believed to have formed as offshore bars in relatively shallow-marine waters (Bell 1980; Turner and Conger 1981). Sediment was derived from the Harris delta (Figure 8–11), which extends westward from the edge

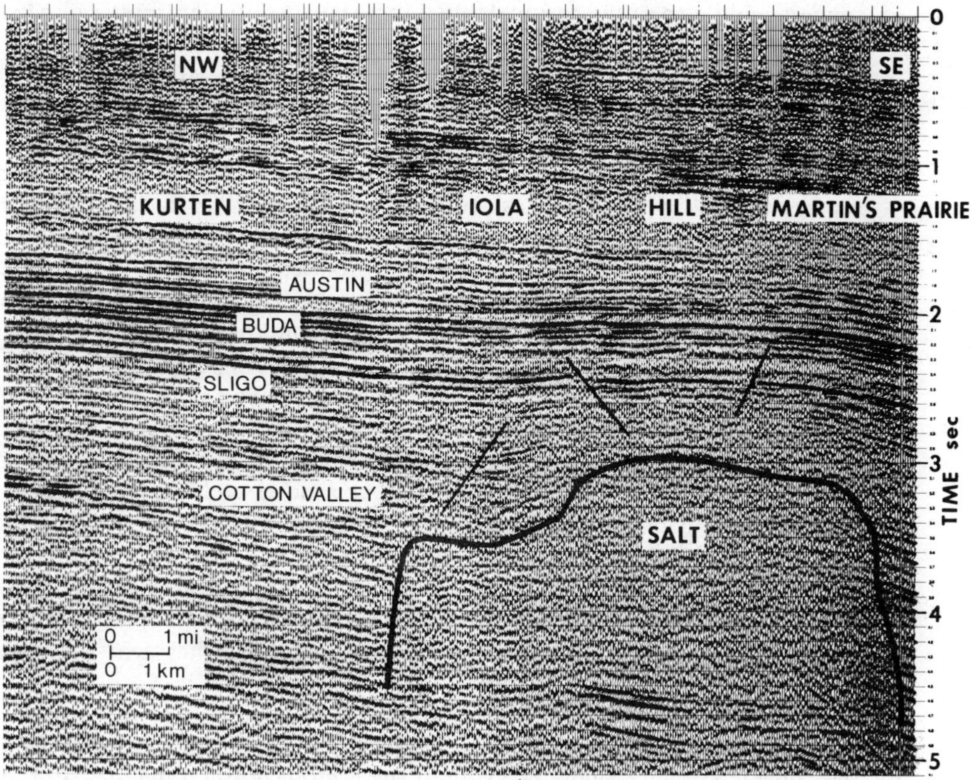

Figure 8–16 Seismic profile across the south end of Kurten field showing the outline of a deep salt structure and broad uplift of the overlying Cretaceous beds. [Seismic line provided by Seiscom Delta, Houston, Texas.]

of the Sabine uplift to within about 20 mi (32 km) of the Kurten field. Longshore currents flowed southwestward, parallel to the shelf margin, and then swept northward around the distal end of the delta, carrying sand to the Kurten area. Sand accumulated slowly, permitting an abundant benthic fauna to rework the bottom sediments.

A shallow-marine origin is supported by trace fossils. Common forms have been identified as *Chondrites, Teichichnus, Planolites,* and *Thalassinoides,* with the rarer forms *Arenicolites* and *Diplocraterion* (Bell 1980, p. 938). Other trace fossils reported are a small *Ophiomorpha, Asterosoma, Rhizocorallium,* and *Zoophycus* (Turner and Conger 1981, pp. 227, 228). This assemblage is undoubtedly of shallow shelf origin (Chamberlain 1978). Microfossils have been reported as absent or rare in the Kurten field area (Bell 1980; Turner and Conger 1981). In contrast, Stehli et al. (1972) reported on foraminifers from the Woodbine interval and showed an increase in the planktonic/benthic ratio from 4 to 8 across the area. These ratios suggest middle to outer neritic depths, increasing southward in the direction of the Cretaceous shelf margin.

The processes that constructed the offshore bars are not known. Previous interpretations have agreed that the bars rose in moundlike form above the sea bottom. The tops of the bars were either shallow and reworked by daily wave and tidal action (Bell 1980, p. 944), or they were somewhat deeper and only occasionally were reworked by storm waves (Turner and Conger 1981, p. 228). These processes were responsible for the cleaner quartz-rich sands of the upper bars. It is true that both wave and current action could have produced the common thin rippled beds of the upper C sandstone. Rather than being clay-free, however, the thin beds contain many laminae of shale, which attest to the intermittent nature of wave or current action. Between times of transport and reworking, a quiet water environment permitted the deposition of clays from suspension. The shale laminae seem to support a somewhat deeper-marine setting.

Local structure may have influenced the nature and distribution of sand. The south parts of the C sandstone appear to be sandier and less bioturbated, whereas the north part is highly bioturbated. The deep salt structure observed on the seismic profile (Figure 8–16) may have formed a broad uplift during the deposition of sand as well as afterwards. Cleaner sands were then concentrated on the flank of this uplift. The effect of local structure would also explain the apparent shallow origin of the trace fossils, despite the location of the area rather near the shelf margin (Figure 8–9).

Local construction of the bars could have taken place largely by current flow. The C sandstone shows a marked increase upward in grain size where the complete sequence is known (Figure 8–13), and the grain size trend denotes a progradation or building into the area of deposition, a process that requires currents. Wave action may have modified the tops of bars, of course, but perhaps only in the shallower locations near the deep structure.

The source of sand was probably from the east in the direction of the Harris delta. Thicker Woodbine sandstones south of the Iola field (Figure 8–9) are significantly coarser grained than the Kurten sandstones (Barton 1982), and this fact adds support to source direction. The primary mechanism of sand transport may have been by storm-induced turbidity currents. It has been noted that the oldest Woodbine E sandstone is composed largely of thin beds that consist of rippled and laminated units that resemble a *CDE* turbidite sequence (Turner and Conger 1981, p. 229). A similar mechanism of transport may have been responsible for the younger sandstones, but definitive evidence was soon destroyed by bioturbation, waves, or currents.

Other Upper Cretaceous sandstones form reservoirs of similar morphology in south Texas. One of these is the San Miguel sandstone of the Taylor Group that apparently originated as a broadly ovate body on the shallow shelf (Lewis 1977). The San Miguel is highly bioturbated, however, and the sandstone retains no evidence of transport mechanisms, although a primary deltaic origin has been suggested (Weiss 1979).

Conclusions

The Kurten sandstones were formed as shoal sands on a shallow marine shelf. Sand was transported into the area by currents, probably by a combination of longshore currents and by density currents from the Harris delta to the east. At the site of deposition, the sands were reformed by storm-induced currents and waves into broad, ovate, and elongate bodies that rose above the sea floor as gentle moundlike features. Deposition was slow, however, so that an abundant benthic fauna populated the area and thoroughly reworked the sediment. Thin-bedded nonbioturbated sandstones are most common in the Woodbine C sandstone toward the south, where a deep-seated salt structure is present. Salt uplift resulted in truncation of the upper Woodbine sandstones before deposition of the Austin Chalk. It is likely that earlier doming of the salt also influenced Woodbine sand deposition and produced a shallowing near which there was more persistent wave action that resulted in cleaner sands. Thus, the character of the Kurten reservoirs was determined by marine-shelf processes as well as by local contemporaneous structure.

SUSSEX SANDSTONE, HOUSE CREEK FIELD, NORTHEASTERN WYOMING

Upper Cretaceous marine sandstones have become important oil reservoirs in the Powder River basin of Wyoming. These sandstones are found as isolated lenses within thick shales of the Cody and Lewis formations, and several of these reservoirs display unusual linear morphologies. An understanding of depositional environment for these sandstones may be important in predicting their occurrence in future exploration and field development in the Powder River basin and elsewhere.

Geologic Setting

The Sussex and Shannon sandstones are present in the marine Cody Shale (Table 8–4), and large oil accumulations were discovered in these sandstones in 1949 at the Sussex and Meadow Creek fields in the southwestern Powder River basin (Figure 8–17). Limited drilling in the central part of the basin resulted in the discovery of oil in the Parkman sandstone in 1956 at Deadhorse Creek and Barber Creek fields, but further exploration was not successful. Then in 1968, oil was discovered in the Sussex Sandstone at House Creek field and in 1969 at Payne field, largely as a result of basinward extension of drilling to the Lower Cretaceous Muddy sandstones. At the same time, oil was discovered in the Teapot Sandstone at the Kaye field and later in the Teckla Sandstone at Poison Draw field. More recent discoveries at Well Draw and Hartzog Draw fields suggest that Upper Cretaceous sandstones have significant potential in the Powder River basin.

The origins for Upper Cretaceous sandstones have been variously given as offshore bar or delta-associated bar. At Deadhorse Creek, the Ferguson sandstone

TABLE 8-4. UPPER CRETACEOUS, POST-NIOBRARA MARINE ROCK UNITS IN THE CENTRAL POWDER RIVER BASIN, WYOMING

Formation	Member	Thickness ft	(m)
Lewis Shale	Unnamed Shale	400	(122)
	Teckla Sandstone	100	(30)
	Unnamed Shale	300	(92)
Mesaverde Formation	Teapot Sandstone	100	(30)
	Unnamed Shale	100	(30)
	Parkman Sandstone	300	(92)
Cody Shale[a] (part)	Mitten Shale	300	(92)
	Sharon Springs Shale	300	(92)
	Sussex Sandstone	50	(15)
	Unnamed Shale	300	(92)
	Shannon Sandstone	50	(15)
	Gammon Shale (lower)	800	(244)

[a]Member names adapted in part from Gill and Cobban 1973, pp. 24–25.

of the Parkman Member has been called a *bar sand* (Sabins and Peterson 1961), but it is coarser grained and distinct from the typical Parkman sandstones that are more clearly delta-associated and in part prodelta in origin (Hubert, Butera, and Rice 1973). The Teckla Sandstone has been called *deltaic* (Runge, Wicker, and Eckelberg 1973), although it is completely isolated in a thick normal-marine shale. At the outcrop in the southwest Powder River basin, the Shannon Sandstone has been interpreted as a marine shelf sand that was formed by current transport parallel to the shoreline (Spearing 1976). Because the total Upper Cretaceous section is composed of about 98 percent shale and only 2 percent sandstone, a shelf origin for these lenticular sandstones is a more likely choice.

The Sussex Sandstone at House Creek field is only one of several Upper Cretaceous sandstones of probable shelf origin, but its character is similar to other reservoir sandstones in the central Powder River basin. For example, the Teckla Sandstone (Table 8–4) displays an identical sequence at Poison Draw field, and the Shannon Sandstone has similar rock types in what appears to be a more complexly interbedded sequence at Hartzog Draw field (Martinsen and Tillman 1978). The following section describes the properties of the Sussex Sandstone in vertical sequence within a limited area of House Creek field (Figure 8–17), and it is taken in large part from a previous paper (Berg 1975).

Drilling History

Oil was discovered in the Sussex Sandstone at House Creek field in 1968, and initial production was 90 bbls/day (14 m^3/day) on pump from a depth of 8250 ft (2500

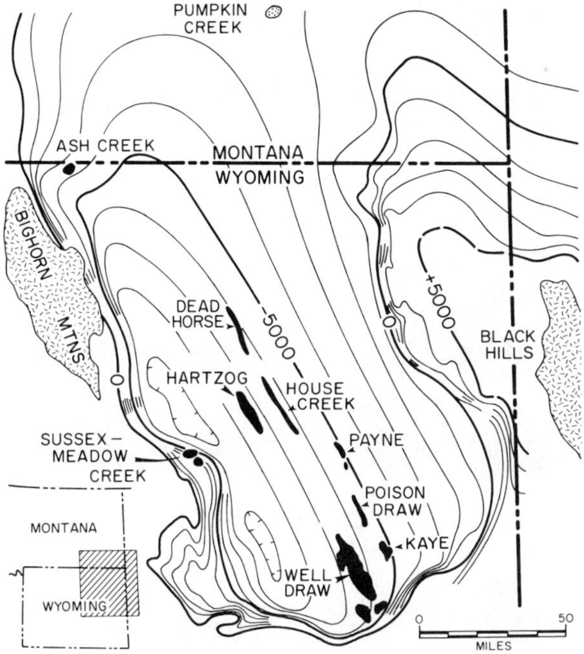

Figure 8–17 Structure of the Powder River basin on top of Lower Cretaceous Dakota Sandstone. Contour interval 1000 ft (303 m). Oil fields shown in black produce from Upper Cretaceous, post-Niobara sandstones. [Modified from Berg 1975.]

m). Because of the relatively poor yield at a time of low crude oil prices, further development was delayed until 1971 when better production was established. The field soon had 75 wells on 160-acre (60-ha) spacing and extended from Sec. 16–43N–72W to Sec. 6–46N–74W, a total distance of 25 mi (40 km) along the strike. Initial production rates averaged 400 bbls/day (64 m^3/day) per well, and total primary recovery of about 20 million bbls (3 million m^3) of oil is expected.

The Sussex reservoir is remarkably narrow compared with its length. The producing area is only 1-mi (1.6-km) wide, and the average net-pay section is 15 ft (5 m) in thickness (Figure 8–18). The reservoir is nearly parallel with the structural strike, and regional dip is about 100 ft/mi (19 km/m) to the southwest.

Total thickness of the Sussex Sandstone is about 40 ft (12 m), and porous reservoir sandstone is found only at the updip edge of a more extensive unit. Correlation of electric logs across the field (Figure 8–19) shows that a thick section of relatively low-permeability sandstone or siltstone occurs in dry holes downdip from the field to the southwest. A narrow zone of low-permeability sandstone occurs updip from the field to the northeast, but within a short distance the Sussex Sandstone is replaced laterally by a thin shale.

Units adjacent to the Sussex can be correlated by details of electric logs, and cores indicate that the overlying unit is a black platy shale, whereas the underlying unit is a dark gray massive claystone. There is no significant variation in these bounding units to suggest channeling or other stratigraphic anomalies in the section.

In order to determine the character of the reservoir, complete 4-in. (10-cm)

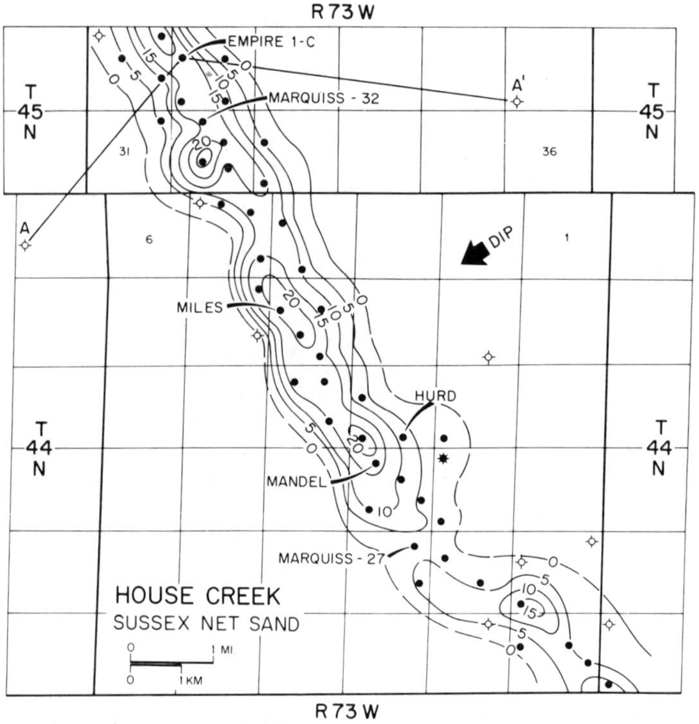

Figure 8–18 Net sandstone isolith of Sussex Sandstone in part of the House Creek field. Contour interval 5 ft (1.5 m). Total length of reservoir is 25 mi (40 km). Cores were examined from the named wells. [Map compiled by Barlow & Haun, Inc., Casper, Wyoming, and republished from Berg 1975 with the permission of the American Association of Petroleum Geologists.]

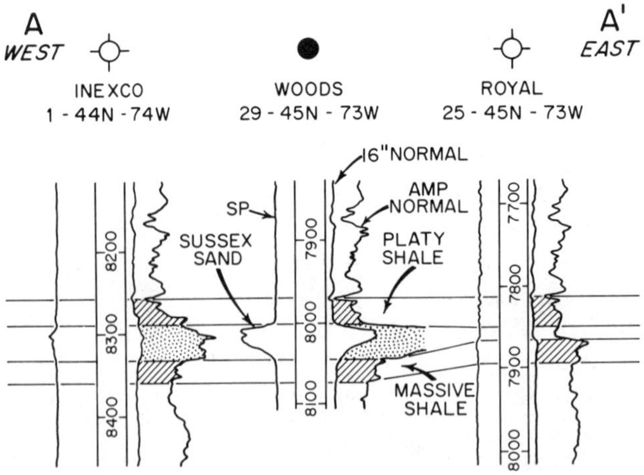

Figure 8–19 Correlation of electric logs across the House Creek field. Location of wells (AA′) given in Figure 8–18. [Reprinted from Berg 1975 with the permission of the American Association of Petroleum Geologists.]

full-diameter cores were examined from six wells drilled by Woods Petroleum, and five cores represent the entire Sussex section. The cored wells are shown on the map (Figure 8–18) and are designated by the royalty owner's name.

Description of Sandstone

Sedimentary structures. The Sussex sandstone has a sequence of beds that can be characterized by composition, texture, and sedimentary structures, and where complete, the sequence consists of the following beds in descending order:

Units	*Thickness, ft (m)*
1. Mudstone, bioturbated, with scattered coarse grains of chert.	5–8 (1.5–2.4)
2. Sandstone, medium to coarse grained, pebbly.	1–3 (0.3–0.9)
3. Sandstone, medium to fine grained, crossbedded, with deformed shale clasts.	5–18 (1.5–5.5)
4. Sandstone, fine grained, ripple-bedded with thin shale laminae.	6–16 (1.8–4.9)
5. Shale, with ripple lenses of fine-grained sandstone.	3–13 (0.9–4.0)

These distinctive units are gradational, but generally a change in bedding and rock type occurs within a few inches. The complete bed sequence is present in the Hurd core (Figure 8–20).

The underlying shale or claystone [Figure 8–20 (J)] is massively bedded, has a finely hackly fracture, and contains thin wisps or blebs of very fine silt; it is distinctly different from the smooth platy black shale that overlies the Sussex interval. Silt blebs appear to represent thin horizontal burrow tubes, but the massive and uniform character of the rock is not obviously disturbed by bioturbation.

The shale with interbedded ripple lenses of sandstone (unit 5) commonly occurs in very thin beds that are 1- to 5- cm thick, and sandstone lenses are of comparable thickness and may show a complete ripple form [Figure 8–20 (G)]. Minor bioturbation occurs throughout the unit in the form of discrete tubes or blebs of sand about 5 mm in diameter, and less commonly bioturbation is abundant enough to destroy the form of ripple lenses. Shale tends to increase in abundance downward, but the change to underlying shale is rather abrupt with only a few inches of gradation [Figure 8–20 (H)].

The characteristics of bedding types in units 3 and 4 are not mutually exclusive. A thin bed of cross-laminated sandstone may occur in the ripple-bedded sandstone with, perhaps, a large shale pebble, whereas the crossbedded sandstone above may contain a rare, but distinct, ripple lens. Generally, however, the two units are immediately distinguishable by the dominance of their characteristic bedding type.

The rippled sandstone (unit 4) is fine grained and commonly shows small-scale cross-laminae. Typically, gently curved laminae of shale outline the ripple form in

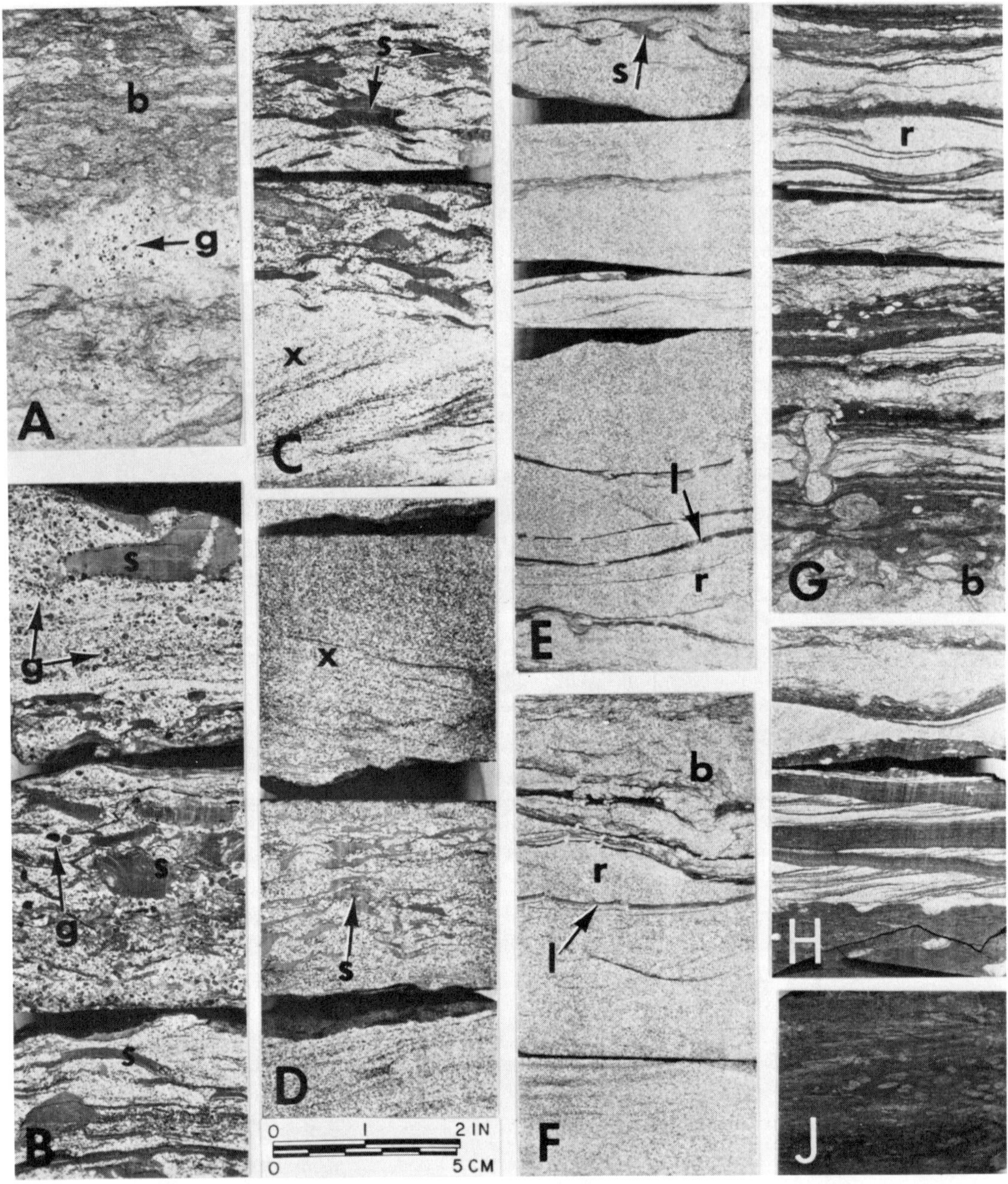

Figure 8–20 Sedimentary structures in Sussex Sandstone, Woods Petroleum Hurd-Federal 1, House Creek field: (A) bioturbated mudstone (b) with scattered coarse grains (g) of chert in thin sandstone bed—8163 ft (2514 m); (B) pebbly sandstone with abundant coarse grains and granules (g) of dark chert and large dark gray clasts (s) of shale—8166 ft (2515 m); (C) and (D) crossbedded sandstone (x) alternating wth thin beds that contain abundant deformed shale clasts (s)—8169 ft (2516 m) and 8171 ft (2517 m); (E) and (F) rippled sandstone with curved laminae (1) of shale that outline ripple lenses (r), few squeezed clasts (s, top E), and minor bioturbation (b, top F)—8176 ft (2518 m) and 8177 ft (2518.5 m); (G) and (H) shale with ripple lenses (r) of sandstone, minor bioturbation (b), and gradation to shale below (H)—8182 ft (2520 m) and 8184 ft (2521 m); (J) massive dark gray shale below Sussex with wisps and blebs of fine silt—8185 ft (2521 m).

the adjacent sandstone [Figure 8–20 (E)], and shale laminae are more abundant in the lower part of the unit. Rarely, bioturbation has destroyed the ripple bedding [Figure 8–20 (F)].

The crossbedded sandstone (unit 3) is medium to fine grained and is characterized by inclined laminae and by flat pebbles of dark gray shale [Figure 8–20 (C) and (D)]. The pebbles are large, up to 6 cm in length, but they are commonly thin and distorted or squeezed by compaction until in some cases only thin wisps of shale remain that are distinguishable as clasts only with difficulty [Figure 8–20 (D)]. The sandstone matrix may be massive or horizontally laminated, but it may also be crossbedded in thin sets only a few inches thick, in which laminae dip at 10° to 25°. Contorted laminae were observed in a few beds.

The pebbly sandstone (unit 2) contains pebbles and granules of chert in a medium- to coarse-grained matrix, but large pebbles of medium gray claystone are common [Figure 8–20 (B)]. In most cases, chert pebbles are concentrated in the uppermost part of the unit. An unusual type of clast occurs in the Mandel core, where yellow claystone cobbles are 2-cm thick and greater than 5 cm in length. The pebbly sandstone is a distinctive unit in the Hurd, Mandel, and Marquiss 32 cores, but it is missing in the Empire and Miles cores (Figure 8–18).

The bioturbated mudstone (unit 1) is highly disturbed by organic reworking, but there is an indistinct horizontal appearance to the bedding. Rounded coarse grains and small pebbles of chert up to 6 mm in diameter are found scattered through the shaly and sandy matrix, and more rarely the granules are concentrated in disturbed lenses of sandstone [Figure 8–20 (A)], which suggest an original thinly interbedded sand and shale before bioturbation. The bioturbated unit is somewhat more sandy in the Hurd core; commonly a higher content of clay gives the unit a dark gray color. Except for the coarse chert grains, the bioturbated unit has the characteristics of a relatively shallow normal-marine environment.

The sequence of bedding from unit 5 upward through unit 2 suggests a prograding section that was deposited by currents of increasing flow regime. Ripple lenses of units 4 and 5 display asymmetry in dip of laminae, and the thickness of ripple lenses generally increases upward [Figure 8–20 (E)–(H)]. The overlying thin beds of cross-laminated sandstone (unit 3) may represent small dunes, although their complete form cannot be discerned in the narrow core sections. The coarser pebbly sandstone (unit 2) probably represents the highest flow-regime, and in some places laminae dip as much as 25°.

Petrography. The sandstone has an average grain size of 0.22 mm (fine grained) as shown in the Mandel core (Figure 8–21). Mean grain size of quartz increases upward from 0.13 mm at the base to 0.28 mm in the pebbly sandstone and decreases to 0.20 mm in sandier parts of the bioturbated unit. Grain size is slightly finer in the Empire core, which lacks the pebbly sandstone, but the domi-

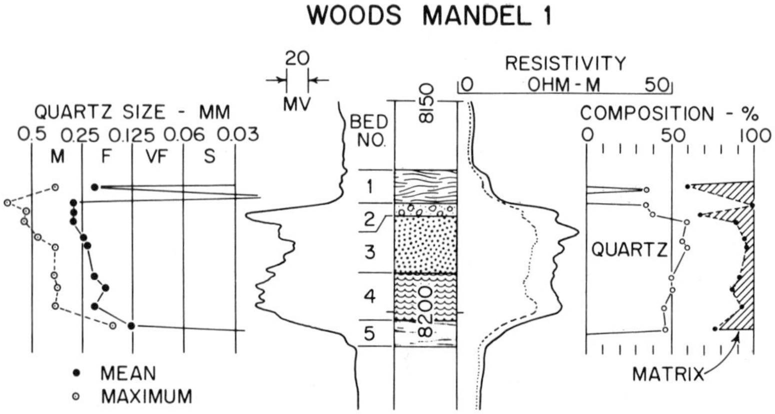

Figure 8–21 Grain size and composition of the Sussex Sandstone in the Woods Petroleum Mandel-Federal 1, House Creek field. Numbered units refer to bedding types described in text. [Reprinted from Berg 1975 with the permission of the American Association of Petroleum Geologists.]

nant sandstones (units 3 and 4) have mean grain sizes that are consistently finer than in the Mandel core.

In detrital composition, the sandstone has an average of 48 percent quartz, 11 percent feldspar, 23 percent rock fragments, and 15 percent matrix (Table 8–5). Quartz consists only of monocrystalline grains, and rock fragments include about equal amounts of polycrystalline quartz and chert; igneous and metamorphic rock fragments are rare. Matrix is chiefly kaolinite but also includes glauconite with minor chlorite. Glauconite grains average 4 percent and tend to be deformed and are not easily distinguished from the surrounding matrix.

Cement is mainly calcite and averages 11 percent of the total rock composition. Some quartz grains have overgrowths of silica, but these are present only in trace amounts.

Sussex sand was probably derived largely from older sedimentary rocks because of the dominance of monocrystalline quartz and chert, but significant amounts of polycrystalline quartz and feldspar suggest possible contributions from metamorphic or igneous terrains. The composition can be characterized as lithic graywacke and is not unlike some fluvial sandstones. It can be inferred that sediment was transported directly to its depositional site without an extensive period of reworking under shoreline conditions that might have increased its content of monocrystalline quartz. The presence of glauconite probably indicates a marine origin, but the chlorite is the result of diagenetic alteration of clay minerals.

Porosity and permeability. Composition and texture influence porosity and permeability, and therefore the different bedding units have distinctive reservoir

TABLE 8-5. PETROGRAPHIC SUMMARY OF SUSSEX SANDSTONE, WOODS MANDEL CORE, HOUSE CREEK FIELD, WYOMING

Bedding unit	Number of samples	Thickness ft (m)	Quartz size[a] Mean (mm)	Max (mm)	σ (mm)	Detrital composition[b] Qz (%)	F (%)	Rx (%)	Mx (%)	Oth (%)	Cement[c] (% of total)	Permeability (%)	Porosity (%)
1	1	8 (2.5)	0.21	0.36	0.08	36	6	16	40	2	—	1.2	6.1
2	2	3 (0.9)	0.28	0.68	0.10	38	10	35	16	1	12	97.	19.1
3	3	13 (3.9)	0.26	0.55	0.08	59	12	20	8	1	5	30.	14.7
4	3	11 (3.4)	0.20	0.38	0.06	49	15	19	10	7	6	13.	17.2
5	1	6 (1.8)	0.13	0.19	0.07	46	7	13	24	9	16	0.8	8.8
Means			0.22		0.07	48	11	23	15	3	11		

[a] Long-axis measurements; σ = standard deviation.

[b] Qz = monocrystalline quartz, F = feldspars, Rx = rock fragments, largely polycrystalline and chert, Mx = matrix includes clay, chlorite, and glauconite, Oth = other minerals, mainly heavies.

[c] Cement is largely calcite.

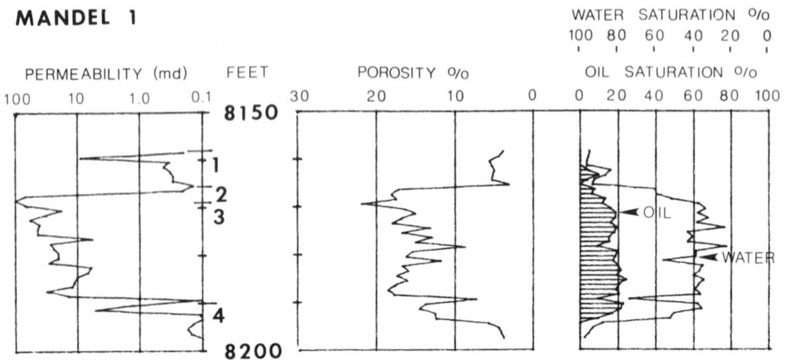

Figure 8–22 Permeability and porosity of the Sussex Sandstone, Woods Petroleum Mandel-Federal 1, House Creek field. Numbered units refer to bedding types described in text.

characteristics (Figure 8–22). The pebbly sandstone (unit 2) has the highest average permeability of 97 md and an average porosity of 19 percent in the Mandel core (Table 8–5), but in other wells, permeability is greatly reduced by clay. The crossbedded sandstone (unit 3) has the next highest average permeability of 30 md and an average porosity of 14.7 percent. The permeability increases upward, reflecting the increase upward in quartz content and grain size. The rippled sandstone (unit 4) has an average permeability of only 13 md and a porosity of 17 percent, reflecting its higher content of clay matrix and finer grain size.

Both the bioturbated mudstone (unit 1) and the ripple lenses of sandstone in the shale (unit 5) have low permeabilities that range from a few tenths to only a few millidarcys and porosities that range from 5 to 13 percent. Neither unit can contribute much fluid to a well bore because bedding characteristics tend to inhibit continuity of pore space.

Facies distribution. The House Creek cores present a section of the Sussex for 6 mi (10 km) along the structural strike, but they also present a limited view of changes in a dip direction. The Mandel and Hurd cores are 0.5 mi (0.8 km) apart along the dip, and the remaining cores are more or less centrally located within the sandstone along the strike (Figure 8–18). The Empire and Marquiss 32 cores, however, may be projected to a dip section in which it can be assumed that the Empire core is located a short distance updip from the Marquiss-32 core. The Miles core is located centrally in the sandstone, and the Marquiss 27 core to the south, although incomplete, appears to have a sequence much like the Hurd core.

A cross section may be constructed, then, that shows the strike changes in bedding units, and by exchanging the positions of the Marquiss 32 and Empire cores, the section may also indicate the dip changes within the Sussex (Figure 8–23). This arrangement shows that the crossbedded sandstone (unit 3) and rippled

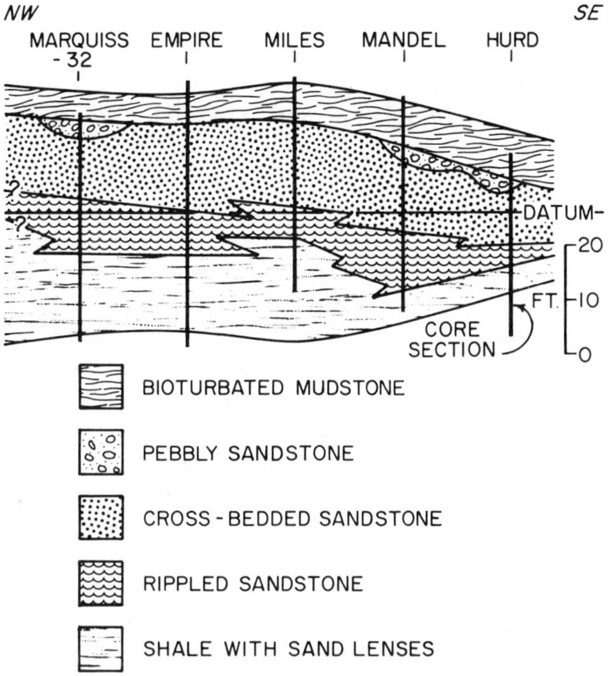

Figure 8–23 Relation of bedding units in the Sussex Sandstone, House Creek field, showing changes in thickness along the structural strike. This section may also be interpreted as representing change from downdip (left) to updip (right). Location of wells given in Figure 8–18. [Reprinted from Berg 1975 with the permission of the American Association of Petroleum Geologists.]

sandstone (unit 4) are the dominant sandstone units, but the rippled sandstone thins in an updip direction from the Mandel to the Hurd wells. If the relative locations of the Empire and Marquiss 32 cores are correct, then the rippled sandstone (unit 4) also thins slightly between these wells in an updip direction. The pebbly sandstone (unit 3) is variable in location, but it is thicker in the assumed updip direction at the Mandel and Hurd wells.

This reconstruction of facies suggests that farther downdip, the thicker Sussex may be represented entirely by shale with ripple lenses (unit 5), and this unit may form a single relatively impermeable section at the Inexco well (Figure 8–19). Furthermore, this distribution of bedding units reflects the presence of more permeable units (3 and 4) at the updip edge of the sandstone body before the entire section grades abruptly into shale (Figure 8–19).

This distribution of facies is partly confirmed by cores examined subsequently to the initial field study. For example, a dry hole drilled 12 mi (19 km) updip from the north end of House Creek field showed that bioturbated mudstone, identical to unit 1, occupied the entire Sussex interval (Shirley 1977). The presence of this facies updip from the field suggests that the units of the Sussex were deposited as successive overlapping lenses. In other words, during progradation, the several units appear to have formed by accretion from west to east. Recent interpretation of logs (Hobson, Fowler, and Beaumont 1982) also confirms this arrangement of facies.

Interpretation

The House Creek sandstone displays a gradation upward from fine grained to coarse grained, and thinly bedded to more thickly bedded and cross-laminated. This sequence is indicative of prograding clastic sediment that was transported by currents of increasing competence and flow regime. Deposition began with the transport of sand by intermittent low flow-regime currents that formed ripple lenses of sand with interlaminated shale (unit 5). The section contains rare, but typically marine, bioturbation. Overlying units show transport of larger amounts of sand, still by lowest flow-regime currents (unit 4), but later by stronger currents that deposited crossbedded sands in the form of small dunes (unit 3). Still stronger currents followed and deposited the pebbly sandstone (unit 2), perhaps even eroding parts of the previous section. When the strong currents ceased, the entire area was covered by an environment of quiet water into which there was an occasional influx of coarse sand, perhaps as thin sheets (unit 1). This environment contained a benthonic fauna that thoroughly reworked the bottom muds, and finally, there was a return to a deeper-water environment of black mud deposition that was generally hostile to a bottom fauna.

The total sequence is not unlike a delta front, except that there is no evidence in the cores for subaerial delta-plain deposition. Furthermore, the section is located in a thick shale sequence and far distant from any known shoreline deposits. The general depositional environment, then, can be inferred to have been marine, but the water depths and processes of transport and deposition are not apparent from the cored sequence.

There are few paleontologic data at House Creek to aid in the environmental reconstruction, but some inferences may be made from microfossils of the underlying Shannon Sandstone. Abundant foraminifers have been found in the Shannon section at Pumpkin Creek field in southeast Montana, about 130 mi north of House Creek field. The foraminifers represent both benthic and planktonic forms and were interpreted to have a depth range of less than 90 ft (27 m) to about 400 ft (122 m) (Parker 1958, p. 215). In this area, the foraminifers suggested a shallowing upward within the Shannon section. Although these foraminifers are from the Shannon Sandstone and are located many miles away, it may be inferred that the House Creek sandstone had a similar depth range at the time of deposition.

Transport direction. Evidence for transport direction may also be inferred from the Shannon sandstone, which has been studied at its outcrop in the southwest Powder River basin (Spearing 1976). The Shannon is composed of two sandstones, each about 50 ft (15 m) thick, consisting of a lower bioturbated shale and sandstone that grades upward to rippled sandstone in thin beds and to crossbedded sandstone that contains clay clasts. The entire sequence coarsens upward from silt to medium-sand size, and there is a sharp contact at the top with overlying marine shale but no evidence of erosion or emergence. On the basis of directional measurements from cross-laminae and ripple crests, the mean direction of transport was found to be southward. It was concluded that Shannon sands were transported parallel to the shoreline and that they accumulated in an offshore midshelf location.

The Shannon outcrop sequence is similar to the bedding sequence in the Sussex of House Creek field and indicates a prograding section deposited by low flow-regime currents. The Sussex sandstones may have had a similar southward transport direction, but in the case of House Creek sandstones, generally southeastward along the trend of the reservoir unit. The House Creek sandstone, however, is located about 40 mi to the east of the outcropping Shannon and Sussex sandstones and, presumably, was deposited in somewhat deeper water.

Depositional framework. The environmental setting may be related to the depositional framework established by correlation and mapping within thick Upper Cretaceous shale units of Wyoming (Asquith 1970). Thin beds of bentonite, or altered ash falls, are common in the section and can be readily identified on electric logs. By means of detailed correlations of bentonites, time-rock units were traced from areas of shoreline deposition to the deeper shelf and beyond (Figure 8–24). The correlations show that the bentonite time lines are parallel over long distances and enclose relatively thin sections of shale. The intervals expand rapidly at some distance from the old shorelines, and these thicker intervals contain increased amounts of interbedded siltstone and sandstone. The time lines then converge farther offshore toward a basin area of thinner deposits. These correlations establish a depositional configuration of shelf, slope, and basin, but the terms used for this depositional framework indicate only a relative change from shallower to deeper water. Most likely, the shelf-slope breaks represent a terracelike configuration for the broad Upper Cretaceous shelf.

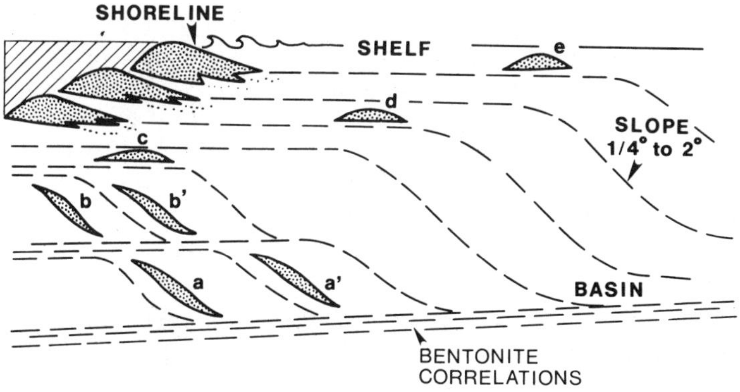

Figure 8–24 Diagram of bentonite correlations in thick sections of Upper Cretaceous shale, Powder River basin, Wyoming. Correlations established a depositional framework of thinner shelf and basin sediment with thicker slope accumulations. Slope sands may have been deposited as discontinuous, noncontemporaneous bodies (a, a′, b, b′); some sands may have been deposited on the shelf (c, d, e). [Adapted from Asquith 1970.]

These relationships demonstrate that ancient shelves were largely areas of sediment bypassing and that the slopes were sites of major sediment accumulation. In other words, shelf sedimentation was characterized by prograding slopes composed of fine-grained sediment. Based on stratigraphic relationships, the shelf margin may have been in water depths on the order of 200 ft (61 m) at the time of deposition. Thickening of slope sequences ranges from 200 ft to as much as 500 ft (61 to 155 m). Thus, it may be inferred that water depths from shelf margin to basin increased by approximately the same amount. Rates of thickening also indicate that inclination of the slopes may have been on the order of $\frac{1}{4}^\circ$ to 2°.

It is likely that the Sussex and similar sands were deposited within the areas of prograding slopes. The sands would have been discrete bodies, neither contemporaneous nor continuous, which were included within the slope sections by periodic transport of sand into the slope setting. Such sandstones would then form individual reservoir bodies that would appear to be at about the same stratigraphic level (see a, a′, b, and b′ in Figure 8–24). It is also possible that some sands may have been deposited on the shelf (Brenner 1978) as barlike features (see c, d, and e in Figure 8–24), but a shelf mound would appear to be unlikely because the depositional configuration indicates the shelf was an area of sediment bypassing rather than deposition. Therefore, the Sussex, and similar sands, were probably deposited in slope locations.

The slope configuration would appear to be difficult to establish for thin sandstones such as the Sussex. However, detailed correlations within the Sussex interval have established that a local slope configuration probably existed (Crews, Barlow, and Haun 1976). An interval selected for mapping was that between two bentonites that bound the Sussex sandstone (Figure 8–25). In the western part of the map area, two closely spaced bentonites diverge to enclose an interval that thickens from 50 to 75 ft (15 to 23 m) across the area. Then the interval thins again and disappears toward the east beyond the mapped area. Within the Sussex time-rock unit, there are several lenticular sandstones, discontinuous and strike-trending. These lenses appear to be partly overlapping and successively younger toward the east, as might be expected from the deposition of sand within a gentle prograding slope. Some lenses are reservoirs, as the House Creek Sussex sandstone and a more limited area of oil accumulation at the Triangle-U field (Anderman 1976), but at the time the map was constructed, most of the other lenses of sandstone were not reservoirs. Thus it appears that separate Sussex sandstones were deposited on slopes of low relief, perhaps not greater than the maximum thickness of the time-rock unit. The progression of separate lenses toward the east resembles the pattern of imbricate sandstone and shale units described for the Viking Formation of Saskatchewan (Evans 1970).

The depositional setting for the Sussex Sandstone resembles that postulated for the Upper Cretaceous Gallup Sandstone at Bisti field, northwestern New Mexico (Campbell 1971). For the Bisti reservoir, it was concluded that sand was transported by current flow parallel to the distant shoreline and that deposition was controlled by a gentle break in slope, or terrace, on a shallow marine shelf. Stratigraphic re-

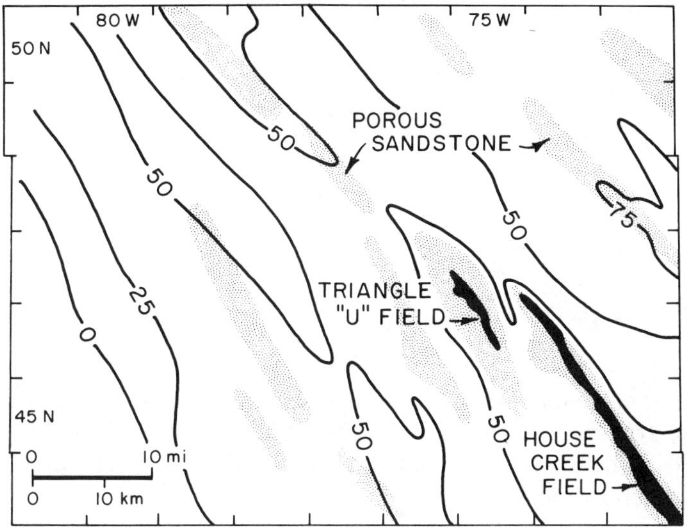

Figure 8–25 Thickness of the Sussex time-rock unit based on the correlation of bentonite beds in the central Powder River basin, Wyoming. The unit encloses discontinuous lenses of strike-trending sandstones (stippled). [Adapted from Crews, Barlow, and Haun 1976.]

lationships and intense bioturbation indicate that the Bisti sands were deposited in shallow water. The Sussex may also be termed a shelf-terrace sand, but bioturbation is relatively sparse and is associated only with current flow and sand transport. The surrounding shales are essentially nonbioturbated and perhaps of deeper-water origin.

The origin of Upper Cretaceous shelf currents is unknown. It has been suggested that the Sussex, Gallup, and similar sands were transported by tidal currents in the relatively narrow seaway of the Western Interior of the U.S. (Bridges 1982). In fact, tidal currents may have been the only mechanism that could have produced the linear morphology of these relatively coarse-grained sandstones.

Conclusions

The Sussex Sandstone at House Creek field was deposited in a middle- to outer-shelf location in water depths that were in the range of 100 to 200 ft (31 to 62 m) or even deeper. The nature of the section, however, is decidedly different from that which might be expected in such a marine environment. The bedding and textural sequences indicate a prograding section that was deposited by currents of increasing flow regime. Sand invaded an environment of dark shale deposition that was generally hostile to a benthonic fauna, and it was deposited in increasing amounts by currents of increasing velocity that flowed southeastward across the shelf, perhaps concentrating at the base of a prograding slope (Figure 8–26). At a later stage,

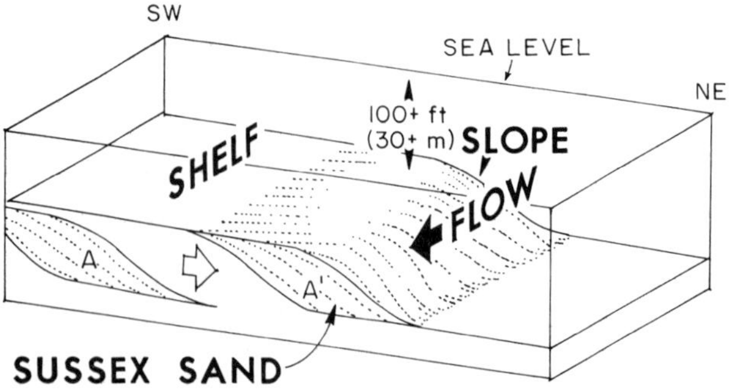

Figure 8–26 Diagram of transport and deposition of the Sussex sands at a shallow prograding shelf break, far from shore. View is toward the northwest.

strong currents carried large shale clasts and chert pebbles into the area. Finally, sand transport declined, and a benthonic fauna thoroughly reworked the bottom muds. The Sussex sand was later covered by a quiet-water environment of black mud deposition.

MARCHAND SANDSTONE, EAST BINGER FIELD, OKLAHOMA

The Pennsylvanian upper Marchand Sandstone is an oil reservoir in the deeper part of the Anadarko basin of central Oklahoma. These sands apparently were deposited far from shore, but probably in an outer-shelf, rather than basinal, location. Like other sands deposited in deeper-water environments, the Marchand has an unusual morphology, which, in this case, is composed of both strike- and dip-trending elements. In contrast to other shelf sandstones, composition, texture, and sedimentary structures are uniform through sections of 100 ft (30 m) or more of thickness, and they are undisturbed by bioturbation. These characteristics indicate transport and deposition by persistent low flow-regime currents in a deep-marine setting.

Geologic Setting

The Marchand sandstones are located in the deep southeast part of the Anadarko basin where the Pennsylvanian Missourian Series reaches a thickness of over 3000 ft (914 m) (Figure 8–27). The Missourian is composed largely of dark gray shales that have interbedded sandstones and limestones, and the limestones form marker beds that permit correlations within the Skiatook Group of central Oklahoma to the Hoxbar Group of the deeper basin and Witchita uplift of southwest Oklahoma (Table 8–6).

Throughout the Pennsylvanian period, the Anadarko basin was the site of clastic deposition in a section that has a maximum thickness of about 15,000 ft (4572 m) (Rascoe 1962). At the same time, the adjacent Wichita uplift was a source of coarse detritus that now forms an apron of generally arkosic sandstone on the faulted and folded northeast flank of the uplift. Apparently, most of the coarser clastics

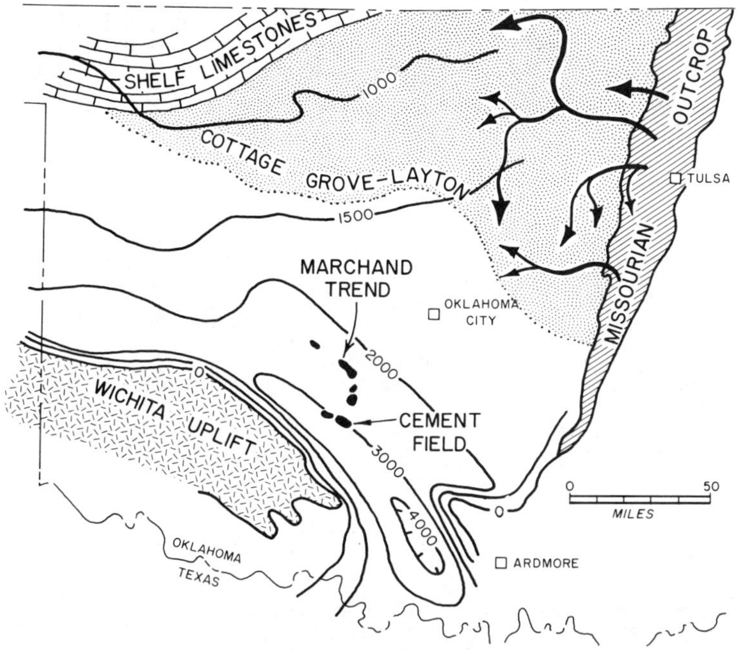

Figure 8–27 Thickness of Missourian Series in the southeast Anadarko and Ardmore basins, southwest Oklahoma. [Modified from Rascoe 1962.] Arrows on north-central Oklahoma shelf indicate transport directions for Osage-Layton fluvial-deltaic sandstones. [From Visher, Ekebafe, and Rennison 1975.]

were confined to the present deep basin, and their greatest extent in the southeast part of the Anadarko basin is at the Cement field. Here the name Marchand was originally applied to conglomeratic sandstones that have a maximum thickness of 380 ft (116 m). Oil production from the Marchand conglomerate was first established in 1943 at the Cement field (Figure 8–27), and cumulative production was 25 million bbls (4 million m^3) by 1960. The Marchand reservoir is not uniform across the Cement structure but thins northward and eastward (Herrmann 1961).

These conglomeratic sandstones contain granules and pebbles of chert that commonly range from 2 mm and 4 mm and have a maximum size of 17 mm (Eisner 1955, p. 10). The occurrence of pebbles is erratic, however, so that the mean grain size apparently is in the sand class. Composition of grains includes quartz and feldspar, but chert is the dominant component and ranges from 40 percent in the fine-sand fraction to 80 percent in the coarser fractions. The chert probably was derived from older Paleozoic rocks that were exposed to weathering and erosion in the Ouachita uplift during Missourian time (B. Rascoe, personal communication, 1975).

In 1967, oil was discovered at Northeast Verden, about 15 mi northeast of Cement field, and the producing sandstones occupy a stratigraphic position that is similar to the Marchand conglomerate. Consequently, the sandstones were also called

TABLE 8-6. STRATIGRAPHIC UNITS OF THE DESMOINESIAN AND MISSOURIAN SERIES, OKLAHOMA

	Northwest oklahoma shelf			Anadarko basin		
Series	Group	Formation	Member	Group	Formation	Member
Missourian	Lansing	Stanton Limestone Plattsburg Limestone Wyandotte Limestone Iola Limestone	Tonkawa Sandstone Cottage Grove Sandstone	Ochelata		Perry Sandstone Osage-Layton Sandstone
	Kansas City	Drum Dennis Swope Hertha		Skiatook	Dewey Limestone Hogshooter Limestone Checkerboard Limestone	Black Shale Marchand Sandstone
	Pleasanton					
Desmoinesian	Maramaton	Lenapah Altamont Pawnee Ft. Scott			"Big Lime" Oswego Limestone	

SOURCE: After Rascoe 1962.

Marchand, although in texture and composition they are considerably different. Exploratory drilling soon established significant oil production in several fields to the north and northwest of Chickasha, and the developing oil fields became known as the Marchand trend (Graff 1971). As of 1971, cumulative production for these Marchand fields was 3.8 million bbls (604 thousand m^3). Recent drilling has extended the trend to the East Binger area, about 30 mi northwest of the original development. However, the sandstones at East Binger appear to be at a stratigraphic level that is about 150 ft (46 m) above the original Marchand horizon (Table 8–6) and are referred to as Upper Marchand. Both sandstones are enclosed in a thick section of dominantly dark gray to black shale.

A major part of Missourian sediment in the basin may have been derived from deltaic systems that were present to the north and northeast (Figure 8–27). The Layton sandstones, which lie above the Marchand section, form channels of fluvial origin at the outcrop in northeast Oklahoma. These sandstones overlie marine prodelta mudstones and are associated with carbonaceous shales of delta-plain origin (Visher, Ekabafe, and Rennison 1975). The adjacent subsurface sections show that narrow channellike sandstone bodies extend westward into a dominantly shaly section that is bordered toward the northwest by shallow-marine shelf limestones. The maximum extent of these deltaic channels has not been mapped, but the Layton sandstones have thinned to only about 20 ft (6 m) where they are still over 60 mi (96 km) from the deeper part of the Anadarko basin. Slightly younger sandstones of the Cottage Grove Member are found across the Northeast Oklahoma shelf but do not extend toward the deeper basin.

Judging from their basinal position, it is likely that the upper Marchand sandstones were deposited in relatively deep-marine water. These isolated sandstone bodies reach a gross thickness of over 200 ft (61 m) and suggest that transport processes were effective in carrying sands a significant distance from the shoreline.

Description of Sandstone

Sedimentary structures. The Marchand section in a cored well at East Binger field contains only 30 ft (9 m) of sandstone in the interval 9920—9950 ft (3024–3032 m) (Figure 8–28). The sandstone consists of three bedding units as follows, in descending order:

Unit	*Thickness ft*	*(m)*
1. Sandstone, ripple-laminated and cross-laminated, with shale laminae and shale clasts	7.0	(2.1)
2. Sandstone, cross-laminated	7.5	(2.3)
3. Sandstone, ripple-laminated.	8.5	(2.6)

The lowermost unit (3) has small-scale cross-laminae throughout. The laminae are horizontal to inclined at 10° and are wavy, nonparallel, and continuous. The

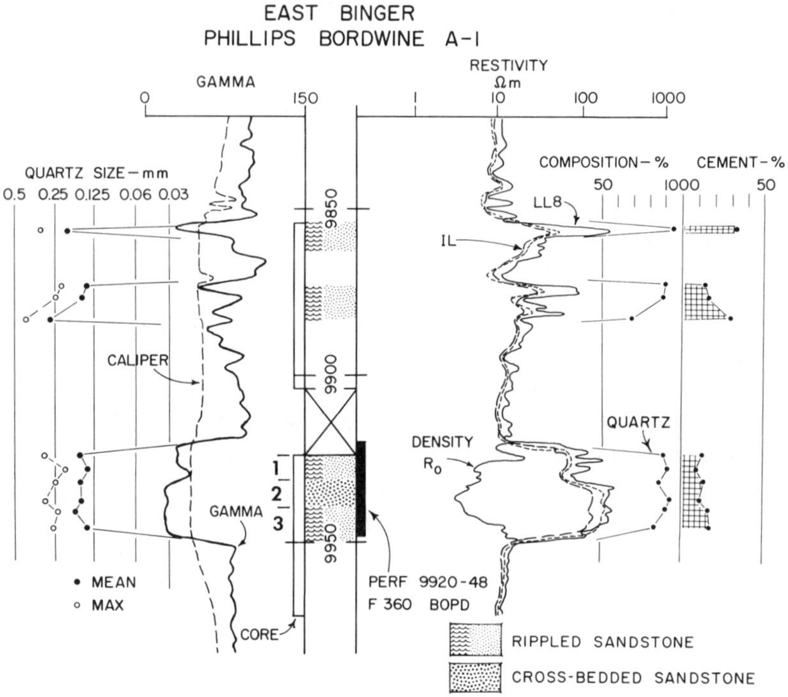

Figure 8–28 Characteristics of upper Marchand sandstones in the Phillips Bordwine A-1, East Binger field, Oklahoma.

laminae form thin ripple sets of about 1 to 2 cm in thickness. There is an increase upward from thinner sets below [Figure 8–28 (I)] to somewhat thicker sets above [Figure 8–28 (H)], which suggests slightly increasing low flow-regime currents. The unit also contains rare scattered small burrows and a few clay clasts up to 5 mm in length.

The overlying unit (2) consists of sandstone that has higher-angle cross-laminae that dip at 15° to 20°. The laminae are even, parallel, and continuous, and the beds appear to be on the order of 0.5 to 1 ft (0.15 to 0.3 m) in thickness [Figure 8–29 (F) and (G)]. This crossbedded unit represents dune lamination that was produced by currents of increased velocity but still within the low flow-regime. The sandstone also contains a few small burrows and scattered small clay clasts. Within this unit some of the laminae dip as much as 30°, and when steep dip is observed, the laminae tend to be irregularly wavy or crinkled in appearance, probably due to penecontemporaneous slumping [Figure 8–29 (G)].

The uppermost unit (1) also shows ripple structure as its dominant bedding, similar to the lowermost unit (3), but the ripple sets are commonly outlined by thin laminae of dark shale [Figure 8–29 (E)]. In other places, small-scale cross-laminae are present [Figure 8–29 (D)], and this ripple structure is similar to that which develops in the upper part of the lowermost unit (1) [Figure 8–29 (H)].

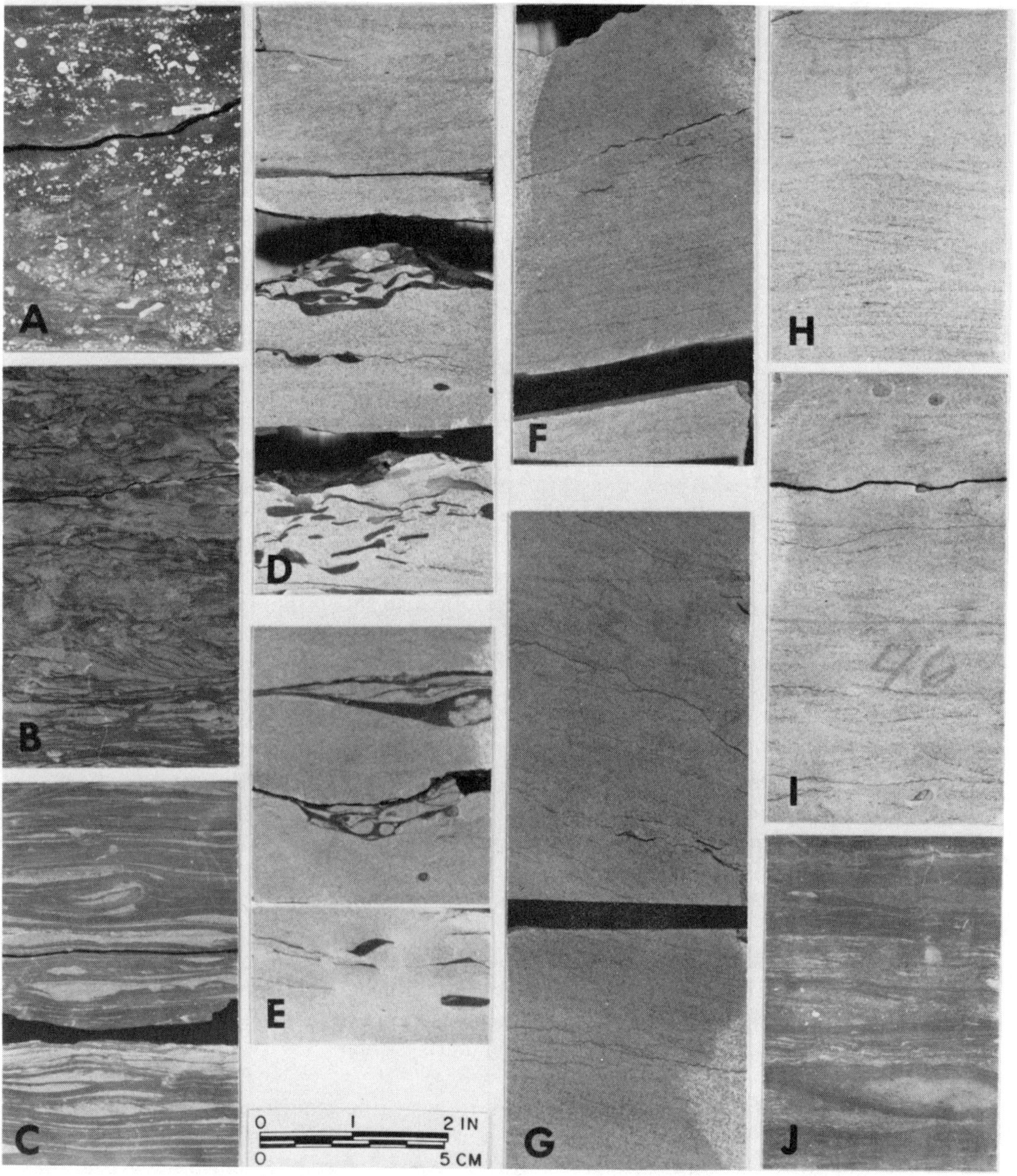

Figure 8-29 Sedimentary structures in upper Marchand Sandstone and associated shales, Phillips Bordwine A-1, East Binger field: (A) black shale with fossil fragments—9886 ft (3013 m); (B) bioturbated mudstone—9887 ft (3014 m); (C) black shale with ripple lenses of silt—9898 ft (3017 m); (D) sandstone with small-scale crossbedding and shale clasts—9926 ft (3025 m); (E) ripple lenses of sandstone with shale laminae and shale clasts—9928 ft (3026 m); (F) and (G) crossbedded sandstone—9932 ft (3027 m) and 9934 ft (3028 m); (H) sandstone with small-scale crossbedding—9942 ft (3030 m); (I) sandstone with ripple bedding—9946 ft (3032 m); (J) black shale with fine silt laminae and ironstone lens (f)—9947 ft (3032 m).

The upper unit also contains a conspicuous number of larger shale clasts that range up to about 8 mm in length [Figure 8–29 (D) and (E)]. Most of the clasts appear to be deformed or "squeezed," indicating that they were semiconsolidated during deposition and suffered plastic flow after burial.

The succession of bedding units from rippled to crossbedded and again to rippled suggests that the sandstone was deposited by low flow-regime currents that increased in intensity during deposition and then decreased during the final stages of sand deposition. The entire sequence appears to represent progradation and deposition by currents that were channelized, perhaps by sea-bottom topography.

The enclosing dark gray to black shales are generally massive but may contain thin laminae of very fine-grained sand and silt. The underlying shale [Figure 8–29 (J)] contains thin laminae of silt and at least one lenslike nodule that is stained by an iron mineral [Figure 8–29 (J)]. The overlying shale [Figure 8–29 (C)] contains thicker lens-shaped laminae of very fine-grained sand in thin beds up to 1 cm in thickness. These lenses appear to have very fine-scale ripple structure. The massive shales enclosing the Marchand sandstone show no evidence of bioturbation, and consequently they suggest deposition in deeper-marine waters that were generally unfavorable for benthonic organisms.

Thin sandstones higher in the section also show ripple structure and abundant dark shale laminae, similar to the uppermost unit (1) of the main sandstone section. One shale section shows relatively abundant bioturbation [Figure 8–29 (B)] as compared to the dominant aspect of the massive shales. This unit (9886—9892 ft in Figure 8–28) is considerably more sandy and appears to have originally contained thin ripple lenses of sand.

Another minor rock type in the section is a fossiliferous mudstone (9883.5–9886 ft in Figure 8–28) that contains abundant crinoid fragments in a black shale matrix [Figure 8–29 (A)]. This thin unit is somewhat sandy at the base and contains abundant fossil fragments in its basal portion; the number of fragments decrease upward in the section. In other words, the shale unit appears to be graded and may represent a submarine mudflow.

In summary, the upper Marchand sandstone probably was deposited by low flow-regime currents. Although bioturbation occurs in moderate abundance at one horizon, the dominant aspect of the enclosing massive shales suggests that the sandstone was deposited in relatively deepmarine waters.

Petrography. The sandstones are very fine grained and have a mean quartz size of 0.15 mm, an average maximum size of 0.25 mm, and an average standard deviation of 0.04 mm (Table 8–7). In vertical sequence, the sandstones are uniformly grained, and there appears to be no significant change in texture that might correspond with bedding units. The only exception is one sample that occurs immediately overlying the crinoid-bearing mudstone, and this sample (9883 ft in Figure 8–28) has a mean grain size of 0.26 mm and a maximum size of 0.41 mm, significantly greater than other sandstones in the same section. This medium-grained sandstone also contains a significant number of fossil fragments.

TABLE 8-7. AVERAGE PROPERTIES OF UPPER MARCHAND SANDSTONE, EAST BINGER FIELD

	Quartz size[a]			Detrital composition[b]				Cement[c]			
Depth (ft)	Mean (mm)	Max (mm)	σ (mm)	Qz (%)	F (%)	Rx (%)	Mx (%)	Sil (% of total)	Cal (% of total)	Permeability[d] (md)	Porosity[d] (%)
9857–9884	0.19	0.30	0.05	85	—	13	2	5	17	0.2	6.5
9924–9945	0.15	0.25	0.04	89	1	11	—	8	6	0.3	7.0

SOURCE: Petrography by Sonnenberg 1975.

[a] Long-axis measurements; σ = standard deviation.

[b] Qz = monocrystalline quartz, F = feldspars, Rx = rock fragments including polycrystalline quartz, Mx = matrix.

[c] Sil = silica as grain overgrowths, Cal = calcite.

[d] Whole-core analysis.

The sandstones are sublitharenites and consist on the average of 89 percent quartz, 11 percent rock fragments, and only 1 percent feldspar in their detrital fraction. The upper sandstones are similar but contain a minor amount of 2 percent clay matrix.

The quartz content is that of monocrystalline grains, and polycrystalline quartz grains are very rare. Most of the rock fragments have a groundmass that is very finely crystalline and has a reddish to orange color. These grains occur in distinct fragments, and many contain lathlike crystals that are scattered throughout the groundmass and appear to be composed of quartz. Rarely, the groundmass is altered to a micaceouslike mineral, probably chlorite.

The rock fragments are of uncertain origin. They may be either mudstones that contain very fine-grained detrital fragments or highly altered finely crystalline rocks such as those of volcanic origin. It is interesting that these rock fragments are the only other important detrital component besides the dominant quartz grains. Because of the negligible feldspar content and the lack of other evidence for igneous source rocks, it is presumed that the rock fragments represent altered sedimentary rocks.

The cementing minerals, calcite and silica, compose 14 percent of the total composition in the main sandstone unit, and these cements occur in subequal amounts. Calcite is found as a pervasive groundmass of clear sparry calcite, while silica occurs entirely as overgrowths on quartz grains. The relatively abundant cement has apparently filled most available pore space between the tightly packed detrital grains.

Porosity and permeability. The sandstone is characterized by an average low porosity of 7 percent and an average permeability of 0.3 md (Figure 8–30). Porosity and permeability increase upward and reach maximum values in the cross-laminated sandstone of the middle unit (2). Therefore, it appears that bedding type may control porosity and permeability. On the other hand, porosity and permeability do not seem to be greatly influenced by texture or composition. Even the cement variations are not reflected in changes of porosity and permeability. Some of the fine-grained rock fragments appear to have been deformed during compaction, and these may have flowed into adjacent pore spaces, reducing permeability.

Core analysis included two samples at the base of the sandstone that had unusually high permeabilities of 38 md and 98 md, and these values may be due to fractures in the samples. The high production rate of 360 bbl/day (57 m^3/day) from this low permeability section also indicates that natural fractures may contribute significantly to the permeability of the entire section.

Interpretation

Sedimentary structures show that sand was transported and deposited by currents of low velocity. This interpretation also implies that the currents may have been concentrated along a flow path or channel, but there is no evidence that such a

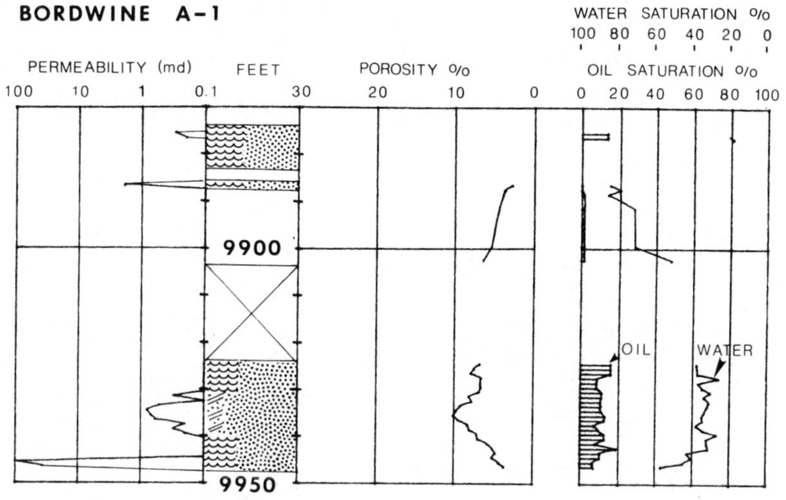

Figure 8–30 Porosity, permeability, and fluid saturation in the upper Marchand Sandstone, Phillips Bordwine A-1, East Binger Field, Oklahoma.

channel may have been erosional in origin. Texture and composition of the sandstone show no significant vertical changes, thereby suggesting a constant supply of fine-grained sediment and no abrupt change in the sedimentary processes during deposition. The highly quartzose nature of the sandstone and especially its lack of chert and feldspar further suggest that the sediment was derived from the Missourian shelf area to the north and northeast rather than from the tectonic uplifts to the south and southeast.

The general scarcity of bioturbation in adjacent shales indicates that the sand was deposited in a deep-marine environment, and this interpretation is reinforced by the location of the producing area near the deepest part of the Anadarko basin (Figure 8–27). It is concluded, therefore, that sand was carried by submarine currents that flowed from the central Oklahoma shelf to the deep part of the Anadarko basin. These currents were perhaps density flows that may have been routed by bottom topography. The sequence of sedimentary structures indicates that sand was carried relatively slowly but constantly to the site of deposition, as shown by the progradational nature upward from a ripple phase to a low dune phase. Once the channels received a certain amount of sediment, however, current velocities again decreased, probably as the main flow was diverted into other areas.

Morphology

Thickness of the upper Marchand sandstone shows that it has a general elongation parallel to the depositional strike (Figure 8–31). Maximum thickness is 86 ft (28 m), and the unit thins abruptly downdip to form an essentially linear boundary on the

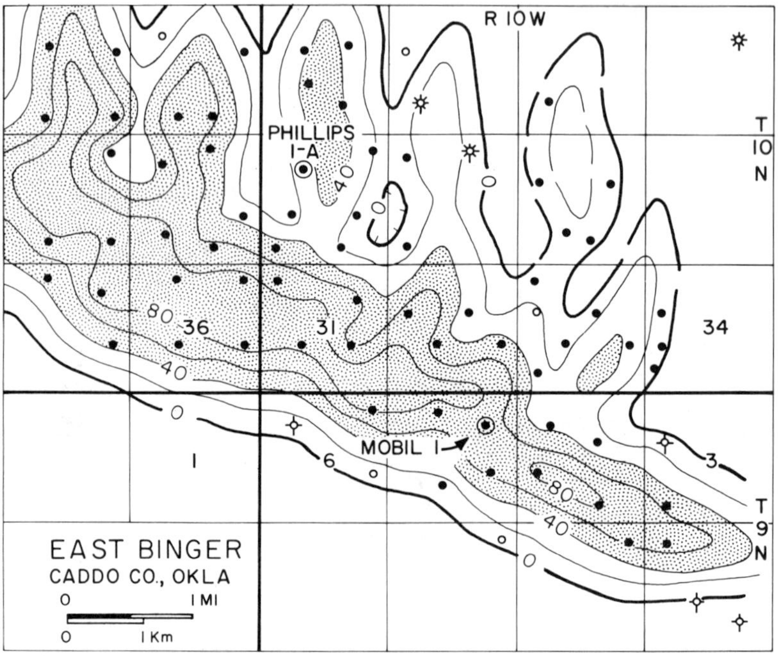

Figure 8–31 Gross thickness of the upper Marchand Sandstone, East Binger field, showing a strike trend of thick sandstone and possible dip trends of thin sandstones on the updip side of the field. Contour interval 20 ft (6 m).

southwest side of the field. Thinning in the updip direction is more gradual with common thicknesses of 20 to 40 ft (6 to 13 m) over an area of about 2 mi (2.2 km). In this area the map shows elongate trends that are oriented in a dip direction and at right angles to the trend of the main body of sandstone. In the updip area of thinner sandstones, more continuous sandstones are concentrated in the lower part of the Marchand interval as in the Phillips Bordwine A–1 well (Figure 8–31, Sec. 30-10N-10W). In the thicker downdip section, however, continuous sandstone appears to build upward until the maximum thickness of more than 80 ft (26 m) occupies the interval above that in the Bordwine A–1 well. A similar section is encountered in the thicker strike-trending sandstone downdip. For example, sedimentary structures in the Mobil Cheatsey 1 well (Figure 8–31, NE Sec. 5-9N-10W) showed a section of dominant ripple-laminated sandstone throughout the entire upper Marchand interval. Ripple structure is of low amplitude and identical to that of the Bordwine core [Figure 8-29 (H) and (I)], and the section contains only rare thin laminae of shale.

The peculiar pattern of both dip and strike trends in the East Binger field is confirmed by mapping in other Marchand oil fields. The overall distribution of the Marchand sandstones shows that they appear in a narrow belt that is about 30 mi (48 km) long, and the general trend of the sandstone cuts across the present structural strike at a low angle from the deeper part of the basin on the southeast to the East Binger area on the northwest (Figure 8–27). Within this trend, gross sandstone thickness reaches a maximum of 200 ft (62 m) in the Chickasha area on the southeast (Graff 1971), and there are local areas of thickening across the main trend of gross sandstone. These cross-trends are well developed in the Dutton-Verden field area (Figure 8–32) and suggest a series of dip-trending sandstone bodies that have coalesced within the main Marchand trend (Graff 1971, pp. 48, 51).

Subsurface control in the Dutton-Verden fields is sufficient to indicate that the dip trends are real, especially in the southeast part of the area (Figure 8–32). Maximum thickness is greater than 80 ft (26 m), and at least 60 ft (18 m) is present within both the dip- and strike-trending elements. Therefore, the dip trends within the East Binger field (Figure 8–31) are similar to those in the Dutton-Verden area, although of lesser thickness.

Differing interpretations have been made for the depositional environment of the Marchand. Based on the general strike trend, the Marchand sandstones were suggested to be offshore bars (Sawyerr 1972), whereas the prominent cross-trends of individual sandstone bodies were interpreted as a deltaic pattern of deposition (Graff 1971). Both of these interpretations were apparently made on the basis of subsurface mapping alone. Another interpretation based on cores designated the sandstone as channel deposits in water depths of at least 300 ft (91 m) (Shelton and Wilson 1978).

None of these interpretations can be supported by the sequence of sedimentary structures observed in the East Binger area. The dominance of low flow-regime

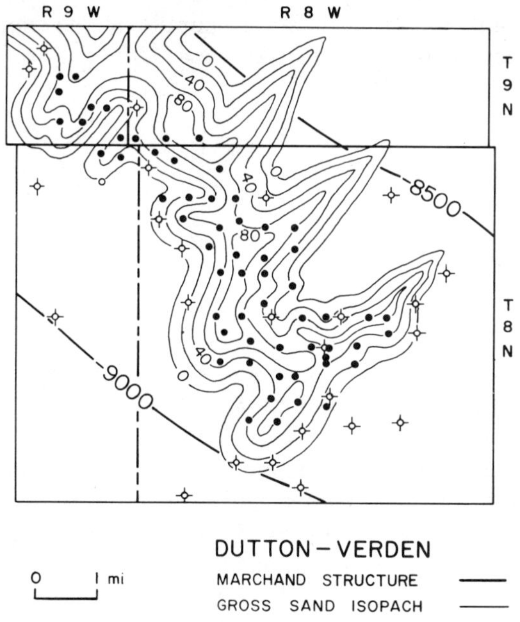

Figure 8–32 Thickness of gross sandstone in the Dutton-Verden fields southeast of East Binger field. [Map adapted from Graff 1971.] Contour interval 20 ft (6 m).

structures without textural gradation precludes a deltaic or channel sequence of deposits. The sands were probably transported in the traction mode by currents that flowed across the shelf and concentrated sands in deeper water, perhaps at a local break in slope, and basinal currents that existed between periods of dip transport probably reworked the sands into a general strike trend.

A recent study of the East Binger field (Baker 1979) has also concluded that the sands were deposited on the shelf. Transport of sand was inferred to be by tidal currents because some bidirectional ripple structure was observed in cores. Similar deeper-shelf current deposits may be more common than now recognized. One well-documented example is in the Pennsylvanian Strawn section of North Texas in which rippled sandstones reach a maximum thickness of 100 ft (30 m) in elongate ovate bodies that are generally parallel to the basin strike (Gunn 1979).

Conclusions

The Marchand sands were transported by low flow-regime currents into a relatively deep-marine environment. Current deposition is indicated by ripple structure and thin sets of high-angle cross-lamination. Relatively deep water is suggested by the enclosing black shales, which are uniform and generally undisturbed by bioturbation. Furthermore, the location of the Marchand sandstone near the deep axis of the Anadarko basin and on the order of 50 mi (80 km) or more from equivalent sandstones of known deltaic origin supports the conclusion that sand was transported by deep-marine currents. The nature of the currents is unknown. They may have been density currents in which flow was induced by high salinity or low temperature. A periodic tidal or storm origin is also possible. Between times of sand transport, contouring currents in the basin flowed parallel to the shoreline and reworked the sands to produce the prominent strike trend. Until a more specific mode of origin can be determined, these deposits might be referred to by the general term of *submarine-current sand.*

SUMMARY

Sandstones of marine-shelf origin are of three morphologic types: (1) broadly ovate bodies of wide areal extent, (2) narrow linear bodies, and (3) elongate ovate bodies (Table 8–8). In all types, sand transport appears to have been largely by low flow-regime currents that carried increasing amounts of sand into the site of deposition. This dominant process commonly resulted in a prograding sequence, with increasing upward grain size, increasing sand content, and increasing scale of sedimentary structures from small ripples to larger ripples and possibly to small dunes. Depositional rates were generally slow with periods of current flow that alternated with periods of clays from suspension as shale laminae between thin beds of otherwise clean sand. Periodic wave action may have been responsible for reworking some

TABLE 8-8. SUMMARY OF MARINE-SHELF RESERVOIR SANDSTONES

Reservoir	Evidence for progradation: Texture (upward)	Evidence for progradation: Bedding (upward)	Bioturbation	Morphology	Reference
San Miguel	Uniform?	Missing	Abundant	Broadly ovate	Lewis 1977
Woodbine	Coarsens	Ripples	Abundant	Broadly ovate	This volume
Bisti	Coarsens	Dunes?	Abundant	Narrow linear	Sabins 1963 and Campbell 1971
Sussex	Coarsens	Ripples to dune	Common	Narrow linear	This volume
Marchand	Uniform	Ripples to dune	Rare	Elongate ovate	This volume
Strawn	Coarsens	Ripples	None	Elongate ovate	Gunn 1979

marine sands, but its effects are not separable from the more common ripple structure produced by current flow.

In most sands, the slow rates of deposition permitted a prolific benthic fauna to populate the area, burrow deeply into the bottom, disrupt the bedding, and even homogenize the sediment. Consequently, primary sedimentary structures were altered or destroyed. In other sands, burrowing organisms were essentially absent, either because of more rapid deposition or because of a deeper-water more hostile environment.

Porosities and permeabilities are variable from generally low values in the highly bioturbated mudstones of mixed sands and clay to relatively high values in clean nonbioturbated sands. Many shelf sandstones can be characterized as "shaly" reservoirs because of intermittent current flow and common to abundant bioturbation. However, the wide areal extent of some bodies and the enclosure in thick shale sections provide conditions for large-scale accumulation of oil in stratigraphic traps.

REFERENCES CITED

ANDERMAN, G. G., 1976, Sussex Sandstone production, Triangle-U field, Campbell County, Wyoming: Wyoming Geological Association Guidebook, p. 107–113.

ANDERSON, E. G., 1979, Basic Mesozoic study in Louisiana, the Northern Coast Region and the Gulf basin province: Louisiana Geological Survey Folio series 3, v. 1, 58 p.

ASQUITH, D. O., 1970, Depositional topography and major marine environments, Late Cretaceous, Wyoming: American Association of Petroleum Geologists Bulletin, v. 54, p. 1184–1224.

BAKER, R. K., 1979, The depositional environment of the Pennsylvanian upper Marchand sandstones, northern Caddo County, Oklahoma, *in* N. J. Hyne, editor, *Pennsylvanian sandstones of the Mid-Continent:* Tulsa Geological Society Special Publication 1, p. 221–234.

BARTON, R. A., 1982, Contrasting depositional processes of Sub-Clarksville and Woodbine reservoir Sandstones: Gulf Coast Association of Geological Societies Transactions, v. 32, p. 121–136.

BEA, R. G., 1975, Gulf of Mexico hurricane wave heights: Journal of Petroleum Technology (September), p. 1160–1172.

BELL, W. A., 1980, The Kurten "Woodbine" field, Brazos County, Texas, *in* Facts and Principles of World Petroleum Occurrence: Canadian Society of Petroleum Geologists Memoir 6, p. 913–950.

BERG, R. R., 1975, Depositional Environment of Upper Cretaceous Sussex Sandstone, House Creek field, Wyoming: American Association of Petroleum Geologists Bulletin, v. 59, p. 2099–2110.

———, and R. R. POWELL, 1976, Density-flow origin for Frio reservoir sandstones, Nine Mile Point field, Aransas County, Texas: Gulf Coast Association of Geological Societies Transactions, v. 26, p. 310–319.

BRENNER, R. L., 1978, Sussex Sandstone of Wyoming—example of Cretaceous offshore sedimentation: American Association of Petroleum Geologists Bulletin, v. 62, p. 181–200.

BRIDGES, P. H., 1982, Ancient offshore tidal deposits, *in* A. H. Stride, editor, *Offshore tidal sands:* London, Chapman and Hall, p. 172–192.

CAMPBELL, C. V., 1971, Depositional model—Upper Cretaceous Gallup beach shoreline, Shiprock area, northwestern New Mexico: Journal of Sedimentary Petrology, v. 41, p. 395–409.

CHAMBERLAIN, C. K., 1978, Recognition of trace fossils in cores: *in* P. B. Basan, editor, *Trace fossil concepts;* Society of Economic Paleontologists and Mineralogists Short Course #5, Oklahoma City, p. 133–184.

CREWS, G. C., J. A. BARLOW, JR., and J. D. HAUN, 1976, Upper Cretaceous Gammon, Shannon, and Sussex sandstones, central Powder River basin, Wyoming: Wyoming Geological Association Guidebook, p. 9–20.

DUANE, D. B., M. E. FIELD, E. P. MEISBURGER, D. J. P. SWIFT, and S. J. WILLIAMS, 1972, Linear Shoals on the Atlantic inner shelf, Florida to Long Island, *in* D. J. P. Swift, D. B. Duane, and O. H. Pilkey, editors, *Shelf sediment transport, process and pattern:* Stroudsburg, Pa., Dowden, Hutchinson, and Ross, Inc., p. 447–449.

EISNER, S. M., 1955, Geology of the "Marchand" Conglomerate: Oklahoma City Geological Society Shale Shaker, v. 6, no. 4, p. 9–27.

EVANS, W. E., 1970, Inbricate linear sandstone bodies of Viking Formation in Dodsland-Hoosier area of southwestern Saskatchewan, Canada: American Association of Petroleum Geologists Bulletin, v. 54, p. 469–486.

FIELD, M. E., 1980, Sand bodies on coastal plain shelves: Holocene record of the U.S. Atlantic inner shelf off Maryland: Journal of Sedimentary Petrology, v. 50, p. 505–528.

FOSS, D. C., 1979, Depositional environment of Woodbine sandstones, Polk County, Texas: Gulf Coast Association of Geological Societies Transactions, v. 29, p. 83–94.

GILL, J. R., and W. A. COBBAN, 1973, Stratigraphy and geologic history of the Montana

Group and equivalent rocks, Montana, Wyoming, and North and South Dakota: U.S. Geological Survey Professional Paper 776, 37 p.

GRAFF, TOM, 1971, The Marchand trend: Oklahoma City Geological Society Shale Shaker, v. 22, no. 2, p. 40–53.

GUNN, R. D., 1979, Desmoinesian depositional systems in the Knox-Baylor trough, *in* N. J. Hyne, editor, *Pennsylvanian sandstones of the Mid-continent:* Tulsa Geological Society Special Publication 1, p. 221–234.

HAYES, M. O., 1967, Hurricanes as geological agents; case studies of hurricanes *Carla,* 1961, and *Cindy,* 1963; University of Texas Bureau of Economic Geology Report of Investigations 61, 56 p.

HERRMANN, L. A., 1961, Structural geology of Cement-Chickasha area, Caddo and Grady Counties, Oklahoma: American Association of Petroleum Geologists Bulletin, v. 45, no. 12, p. 1971–1993.

HILL, G. W., K. A. ROBERTS, J. L. KINDINGER, and F. D. WILEY, 1982, Geobiologic study of the South Texas Outer Continental shelf: U.S. Geological Survey Professional Paper 1238, 36 p.

HOBSON, J. P., JR., M. L. FOWLER, and E. A. BEAUMONT, 1982, Depositional and statistical exploration models, Upper Cretaceous Sandstone complex, Sussex Member, House Creek field, Wyoming: American Association of Petroleum Geologists Bulletin, v. 66, p. 689–707.

HOUBOLT, J. J. H. C., 1968, Recent sediments in the southern bright of the North Sea: Geologie en Mijnbouw, v. 47, no. 4, p. 245–273.

HUBERT, J. F., J. G. BUTERA, and R. F. RICE, 1973, Sedimentology of Upper Cretaceous Cody-Packman delta, Southwestern Powder River basin, Wyoming: Geological Society of America Bulletin, v. 83, p. 1649–1670.

KRAFT, H. C., 1971, Sedimentary facies patterns and geologic history of a Holocene marine transgression: Geological Society of American Bulletin, v. 82, p. 2131–2158.

LEWIS, J. O., 1977, Stratigraphy and entrapment of hydrocarbons in the San Miguel sands of southwest Texas: Gulf Coast Association of Geological Societies Transactions, v. 27, p. 90–98.

MARTINSEN, R. S., and R. W. TILLMAN, 1978, Hartzog Draw, new giant oil field (abstract): American Association of Petroleum Geologists Bulletin, v. 62, no. 3, p. 540.

MCCAVE, I. N., 1971, Sand waves in the North Sea off the coast of Holland: Marine Geology, v. 10, p. 199–225.

MURRAY, S. P., 1970, Bottom currents near the coast during Hurricane Camille: Journal of Geophysical Research, v. 75, no. 24, p. 4579–4582.

———, 1972, Wind, tidal, and density-driven currents in the vicinity of the Mississippi River Delta, *in* D. J. P. Swift, D. B. Duane, and O. H. Pilkey, editors, *Shelf sediment transport, process and pattern:* Stroudsburg, Pa., Dowden, Hutchinson, and Ross, Inc., p. 127–142.

NELSON, H. F., and E. E. BRAY, 1970, Stratigraphy and history of the Holocene sediments in the Sabine-High Island area, Gulf of Mexico, *in Deltaic sedimentation, modern and ancient:* Society of Economic Paleontologists and Mineralogists Special Publication 15, p. 48–77.

NICHOLS, P. H., 1964, The remaining frontiers for exploration in northeast Texas: Gulf Coast Association of Geological Societies Transactions, v. 14, p. 7–22.

OFF, THEODORE, 1963, Rhythmic linear sand bodies caused by tidal currents: American Association of Petroleum Geologists Bulletin, v. 47, p. 324–341.

OLIVER, W. B., 1971, Depositional systems in the Woodbine Formation (Upper Cretaceous) northeast Texas: University of Texas Bureau of Economic Geology Report of Investigations 73, 28 p.

PARKER, J. M., 1958, Stratigraphy of the Shannon Member of the Eagle Formation and its relationship to other units in the Montana Group in the Powder River Basin, Wyoming and Montana: Wyoming Geological Association 13th Annual Field Conference Guidebook, p. 90–102.

RASCOE, BAILEY, JR., 1962, Regional stratigraphic analysis of Pennsylvanian and Permian rocks in western Mid-Continent, Colorado, Kansas, Oklahoma, Texas: American Association of Petroleum Geologists Bulletin, v. 46, no. 8, p. 1345–1370.

REINECK, H. E., and J. B. SINGH, 1975, *Depositional sedimentary environments:* New York, Springer Verlag, p. 98–101.

RUNGE, J. S., W. L. WICKER, and D. J. ECKELBERG, 1973, A subsurface type section of the Teckla Sand Member of the Lewis Shale Formation: Wyoming Geological Association Earth Science Bulletin, v. 6, no. 3, p. 3–18.

SABINS, F. F., JR., 1963, Anatomy of a stratigraphic trap, Bisti field, New Mexico: American Association of Petroleum Geologists Bulletin, v. 47, no. 2, p. 193–228.

———, and F. A. PETERSEN, 1961, Geology and petrography of Dead Horse Creek and Barber Creek fields, *in Symposium on Late Cretaceous rocks:* Wyoming Geological Association Guidebook, p. 301–309.

SAWYERR, O. A., 1972, Subsurface stratigraphic analysis, Lower Hoxbar Group (Pennsylvanian), Dutton-Verden-Norge trend, Caddo and Grady Counties, Oklahoma: Oklahoma City Geological Society Shale Shaker, v. 28, p. 72–97.

SHELTON, J. W., and T. M. WILSON, 1978, Depositional environment of Pennsylvanian Marchand sandstone, northwest Norge and northwest Chickasha fields, Oklahoma (abstract): Geological Society of America South-Central Section Annual Meeting, Abstracts and Program, Tulsa, Oklahoma.

SHIRLEY, R. H., JR., 1977, Petrography and prediction of reservoir rock properties in the Sussex Sandstone, Powder River basin, Wyoming: Texas A&M University M. S. thesis, 91 p.

SIEMERS, C. T., 1978, Submarine fan deposition of the Woodbine-Eagleford interval (Upper Cretaceous), Tyler County, Texas: Gulf Coast Association of Geological Societies Transactions, v. 28, p. 493–533.

SNEDDEN, J. W., and D. G. KERSEY, 1982, Depositional environments and gas production trends, Olmos Sandstone, Upper Cretaceous, Webb County, Texas: Gulf Coast Association of Geological Societies Transactions, v. 32, p. 497–518.

SPEARING, D. R., 1976, Upper Cretaceous Shannon Sandstone: an offshore, shallow-marine sand body: Wyoming Geological Association Guidebook, p. 65–72.

STEHLI, F. G., W. B. CREATH, C. F. UPSHAW, and J. M. FORGOTSON, JR., 1972, Depositional history of Gulfian Cretaceous of East Texas Embayment: American Association of Petroleum Geologists Bulletin, v. 56, p. 38–67.

STRIDE, A. H., R. H. BELDERSON, N. H. KENGON, and M. A. JOHNSON, 1982, Offshore tidal deposits: sand sheet and sand bank facies, *in* A. H. Stride, editor, *Offshore tidal sands, processes and deposits:* London, Chapman and Hall, p. 95–125.

STUBBLEFIELD, W. L., 1980, Genesis and modification of the linear sand ridges, inner and middle New Jersey shelf, U.S.A.: Texas A&M University Ph.D. dissertation, 169 p.

———, D. W. MCGRAIL, and D. G. KERSEY, 1984, Recognition of transgressive and post-transgressive sand ridges on the New Jersey Continental shelf: *in* R. W. Tillman and C. T. Siemers, editors, *Siliciclastic shelf sediments:* Society of Economic Paleontologists and Mineralogists Special Publication 34, p. 37–42.

SWIFT, D. J. P., J. W. KOFOED, F. P. SAULSBURY, and P. SEARS, 1972, Holocene evolution of the shelf surface, central and southern Atlantic shelf of North America, *in* D. J. P. Swift, D. B. Duane, and O. H. Pilkey, editors, *Shelf sediment transport, process and pattern:* Stroudsburg, Pa., Dowden, Hutchinson, and Ross, Inc., p. 499–575.

TURNER, J. R., and S. J. CONGER, 1981, Environment of deposition and reservoir properties of the Woodbine sandstone at Kurten field, Brazos County, Texas: Gulf Coast Association of Geological Societies Transactions, v. 31, p. 213–232.

VISHER, G. S., S. B. EKEBAFE, and JAMES RENNISON, 1975, The Coffeyville format (Pennsylvanian) of northern Oklahoma, a model for an epeiric sea delta, *in* M. L. Broussard, editor, *Deltas, models for exploration:* Houston, Texas, Houston Geological Society, p. 381–397.

WEGGEL, J. R., 1972, Water motion and process of sediment entrainment, *in* D. J. P. Swift, D. B. Duane, and O. H. Pilkey, editors, *Shelf sediment transport, process and pattern:* Stroudsburg, Pa., Dowden, Hutchinson and Ross, Inc., p. 1–19.

WEISS, B. R., 1979, Wave-dominated deltaic systems of the Upper Cretaceous San Miguel Formation, Maverick basin, South Texas: Gulf Coast Association of Geological Societies Transactions, v. 29, p. 202–214.

9

Basin Sandstones

Clastic sediment is carried into deep-marine waters by currents, principally by density currents that flow downslope beneath the ambient sea water. If the currents are more dense than sea water because of suspended sediment, they are called *turbidity currents,* and their resulting deposits are called *turbidites.* These deposits are recognized by their ordered sequence of beds, and they form the most common sedimentary rock types, both of sandstone and carbonate, in deep-marine environments.

Turbidites are an important class of reservoir sandstones in ancient basinal settings, but turbidites are not the only sandstones of basinal origin, nor are they restricted to deep-marine environments. In fact, sediment transport by turbidity currents can take place in shallow waters of lakes as well as on the marine shelf. Nevertheless, turbidites are the dominant type of reservoir sandstones in deep-marine settings, and their character and morphologies determine the occurrence of oil and gas in many areas.

BASIN PROCESSES

The deep-marine environment is that below the maximum depth of wave action. Although the surface layers of the ocean basins are in continual motion, deeper waters are subject only to persistent currents of low velocity and occasional currents of higher velocity that are capable of transporting large amounts of sand. The only mechanism capable of transporting a significant amount of sediment is that of turbidity currents.

Turbidity flows are a type of density flow in which the motive force is that of gravity, which acts on the dense fluid containing suspended sediment. The fluid of greater density flows as a bottom-following current, and it can carry sediment great distances into the deep-water environment. Turbidity flows are formed by any mechanism that suspends sediment in an otherwise clear body of water. Suspension of sediment may be induced by storm waves, submarine slumping, or river discharge. The return flow of storm surge has been documented as a method by which turbidity flows can be produced (Hayes 1967). Hurricanes generate a great wave of water that piles against the shoreline during the approach and passage of major storms, and the return flow may entrain a sufficient amount of sediment that it becomes a turbidity flow. The products of such flow have been found on the shelf of the Gulf of Mexico following a major storm (see Chapter 8, Figure 8–3).

Submarine slumping takes place where rapidly deposited sediment forms an unstable slope such as on an active delta front. Oversteepening of the slope and the pounding of storm waves may cause sediment instability and abrupt mass movement of sediment, in which case a plume of suspended sediment could be formed on which the force of gravity then acts. Or the discharge of sediment-laden waters from the mouths of rivers in flood may produce a cloud of suspended sediment that could move as a turbidity flow. Thus, several mechanisms can be visualized as causing a turbidity flow that may then intrude the deep-marine environment.

The earliest modern studies of turbidity flow as a transport agent had a practical basis. The formation of Lake Mead behind the newly constructed Boulder Dam in the 1930s showed that silting of the lake was a serious problem. Subsequent studies showed that the Colorado River deposited its coarser load at the head of Lake Mead as a delta but that the suspended load formed a turbidity flow that could transport sediment for great distances down the axis of the lake (Bell 1942; Gould 1951). Experiments proved that these density currents could be formed if the density difference between the lake and river water was as low as 1×10^{-4} g/cm^3. It was soon recognized that turbidity flows may have been the principal agent whereby ancient sediment was transported to deep-marine basins.

Early experiments with turbidity flows in the laboratory showed that these flows could produce graded beds of silt and sand (Kuenen 1951). Later experiments attempted to scale the natural conditions to the flow of turbidity currents in flumes (Middleton 1966*a,* 1966*b*). Plastic beads that had a density of 1.52 g/cm^3 and a median diameter of 0.18 mm were used as sediment. The experiments showed that the turbidity flows developed a characteristic morphology that consisted of a rounded head, a thinner area of fluid or neck, a main body of flow, and a following tail (see Chapter 2, Figure 2–11). Motion pictures of the flows provided a means to determine their velocity and the factors that governed their movement.

Other experiments with saline density currents appear to have established that a steady state of flow may be achieved that will allow the turbidity flow to travel great distances on bottoms of low slope. The equation that governs the flow, as determined by the scaled experiments, may be that of the body velocity as given by the equation

$$v_b = (8\Delta\rho g d S/\rho_w)^{1/2} \tag{9–1}$$

where v_b is the body velocity, $\Delta\rho$ is the density difference between the flow and the ambient water, g is acceleration due to gravity, d is the flow thickness, and S is the slope or the sine of the angle of inclination of the bottom. This equation is similar to that which governs the flow of saline density currents [Eq. (2–56)], and it also resembles the Chezy equation [Eq. (2–10)] for flow in stream channels.

Density flows can travel great distances even where the slope of the bottom is at a very small inclination. Thus, it is possible that turbidity currents may carry sediment for many miles into the deep ocean basins and across the abyssal plains. Deposition from turbidity flows takes place where the velocity is reduced such that the flow can no longer transport sediment. These conditions may be imagined from Eq. 9–1. If the slope is reduced to a very low angle, velocity may be decreased such that deposition takes place. If mixing occurs between the flow and the ambient fluids such that the density difference is reduced, deposition may also take place. If the body thickness is reduced, as by spreading or overbank flow, velocity may also be reduced and deposition takes place.

Other conclusions of experiments are important to the study of both modern and ancient sediment that may have been transported by turbidity flows. The flows may carry a traction load as well as a suspended load, and the resulting sediment may be poorly sorted. Deposition from turbidity flows also produces a sediment that is graded or fines upward. The resulting sediment appeared to be massive or structureless and without lamination. This type of sedimentation is thought to be the result of high flow-regime (Figure 2–9) or deposition in the bed form of antidunes and plane beds (Harms and Fahnestock 1965). An alternative explanation for lack of structures is that deposition is rapid, and bed forms are not completely developed because of their transient nature.

Supplementary experiments have shown that the turbidity flows may override a thin layer of ambient fluid that is channeled beneath the flow (Simpson 1969). This observation may be important in explaining why the coarser deposits of turbidity flows commonly rest on unconsolidated basinal muds. The underlying sediment may have been protected by the thin layer of ambient fluid that separated it from the overlying flow, at least at the site of deposition. Therefore, the flows may not be erosive, especially when velocity is low and deposition is taking place.

TURBIDITE SEDIMENTS

Even the earliest oceanographic studies noted that coarse-grained deposits can exist on the abyssal plains of the ocean. More recent studies have also shown that leveed channels may extend for hundreds of kilometers across the abyssal plains. These facts suggest that turbidity flows are an important mechanism for sediment transport into the deep ocean basins.

Turbidites are defined as the deposits of turbidity flows. This genetic definition poses certain problems of application to both modern and ancient sediment. Therefore, a more objective definition should be based on the characteristics of those sediments believed to have been deposited by turbidity flows. Such a definition has been derived from the study of ancient turbidites. These sandstones show certain characteristics such as thin beds, interbedded shales, graded texture, and most important, a regular sequence of sedimentary structures (Bouma 1962). This sequence, often referred to as the "Bouma sequence," is different from other types of current-transported sediment in that the turbidites commonly lack the well-developed cross-lamination of the dune phase of transport. Single beds may have other characteristics such as sharp basal contacts, gradational tops, contorted bedding, and in some cases, only one bedding type. Despite this variation, turbidites are a distinctive class of sandstones, easily recognized from other categories.

The complete, or ideal, bedding sequence consists of a basal massive *A* division, a succeeding laminated *B,* a rippled *C,* thinly interlaminated silt and shale *D,* and an overlying *E* pelagic shale (Figure 9–1). The *A* and *B* divisions are believed to result from higher flow-regime deposition (Harms and Fahnestock 1965). The *C* is the lowest flow-regime, and *D* and *E* are deposits from suspension. This sequence does not show cross-laminations as in the dune phase of transport, except as indistinct lamination in otherwise massive *A* divisions. Laminae of the *B* division may be inclined only if the bed has been disturbed by soft-sediment deformation.

The complete sequence is least common. Most commonly observed are (1) thicker beds dominated by massive *A* or *AB* divisions, (2) thinner, more or less complete sequences, and (3) thin beds that lack the *A* division. These types represent, respectively, the deposits of channel, adjacent overbank, and distant overbank flow (Figure 9–1). The older designations of *proximal* and *distal,* in the sense of distribution along the transport system, are also applicable but in a more general context.

Bedding observed in turbidites may vary from the ideal sequence. The massive

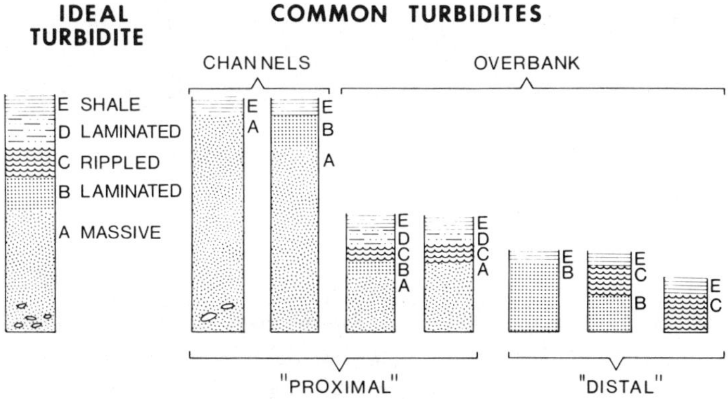

Figure 9–1 The complete or ideal turbidite sequence compared with commonly observed sequences. [From Bouma 1962.] Thicker beds dominated by massive *A* divisions are channel deposits; thinner sequences are overbank deposits. [Diagram from Berg 1978, p. 7–3.]

A division is not always structureless but may show indistinct laminae that are widely spaced at 1 to 2 cm and horizontal or inclined at low angles. Weathering at the outcrop commonly enhances this poorly developed lamination. More rarely within the *A* division are thin sets of converging laminae that have also been found in outcrop (Kruit et al. 1975). The laminated *B* division may also be inclined, but the dip is commonly the result of soft-sediment deformation. The contact between the laminated *B* and rippled *C* divisions is abrupt, and in fact, thin shale laminae may be present between them, which denotes that they were the products of two different depositional episodes separated by deposition of clay from suspension. The laminated *D* division is commonly thin and negligible in thickness as compared with the other divisions of the sequence. The shaly *E* division may have the appearance of a true basinal deposit of dark finely laminated pelagic clay, or it may contain a significant amount of silt and fine sand, which suggests that it too was the product of low-velocity current transport.

The base of beds is commonly in sharp contact with the underlying *E* shale, which suggests that at the site of deposition the turbidity flow did not erode the previously deposited sediment. The base may also contain deformation structures that resulted from rapidly deposited sand loading the underlying mud, or alternatively, the soft clay intruded upward into the sand because of density differences to produce wisps of shale or ''flame'' structure. Other basal features or *sole marks,* such as grooves and flutes seen in outcrop, are large-scale structures and seldom can be recognized in cores.

Finally, soft-sediment deformation may be present in single beds as well as throughout thick sections of multiple beds, denoting the instability of rapidly deposited sediment. The *A* division may have large folds of contorted sediment, and deformation of thick units may increase downward, indicating that large masses of sediment had slumped and deformed, possibly by block rotation, shortly after deposition.

Despite these variations in structure, the turbidite sequence remains unique as compared to those of other depositional processes. The generally thin beds, the repetition of bedding types, and the ''rhythmic'' nature of the sequence are distinctive features.

MODERN BASIN SANDS

Sands in modern ocean basins commonly form large fanlike bodies at the base of the continental slope. These *submarine fans* have certain bed associations that are characteristic of their location within the depositional system (Haner 1971). The fan can be divided into three parts [Figure 9–2(A)]. The inner fan consists of incised channels on the continental slope, the middle fan contains meandering channels, and the outer fan contains braided or distributary channels. The terms *inner, mid-*

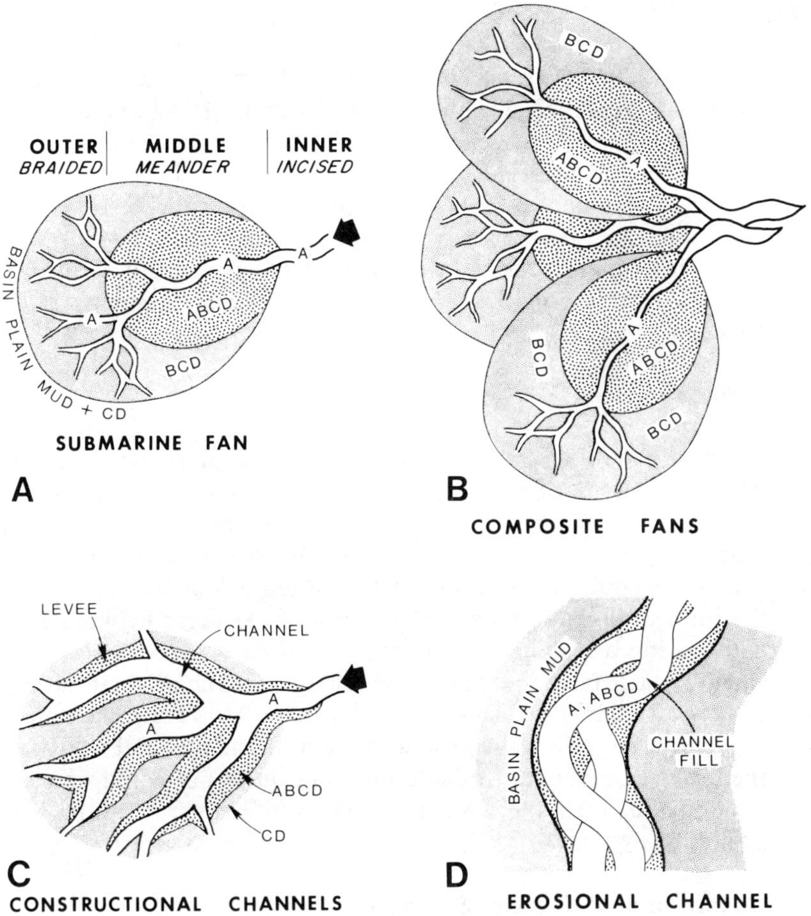

Figure 9–2 Turbidite morphologies and bed associations for (A) modern fans, (B) ancient composite fans, (C) constructional channels, and (D) erosional-channel fill. [Diagram A from Haner 1971 and B, C, and D from Berg 1978, p. 7–4.]

dle, and *outer* are best applied to ancient submarine fans. The modern-fan terms *upper, middle,* and *lower* denote relative elevation and can be confused with stratigraphic position. In modern fans, channels are filled by massive *A* sandstones, overbank deposits of the middle fan are generally dominated by more complete *ABCD* sequences, and overbank deposits of the outer fan have less complete *BCD* sequences. Other systems of fan terminology have been proposed (Mutti and Ricci-Lucchi 1972; Normark 1978; Walker 1978), but the scheme proposed by Haner (1971) appears to be best suited for ancient submarine-fan deposits.

ANCIENT TURBIDITES

The study of ancient turbidites, especially those that occur in reservoir sections, requires a prediction of morphology. From a vertical sequence of beds observed in cores, or inferred from well logs, an estimate may be made of the thickness and distribution of sandstone. Studies of modern turbidites give only a two-dimensional picture of sediment distribution, and the deposition of turbidites through time is best established by the study of reservoir sandstones.

Three distinct reservoir morphologies have been described (Berg 1981). These are called *submarine fans, constructional-channel* systems, and *erosional-channel* systems.

Submarine Fans

Submarine fans are relatively large bodies of interbedded sandstone and shale, in which the distribution of turbidites is similar to that described in modern fans. However, the vertical sequence of beds may be more heterogeneous than in the modern fans because of a complex depositional history. Ancient fans can be expected to be composite bodies of overlapping fans [Figure 9–2(B)]. An initial fan probably had the form of a gentle mound flanked by lower areas between the continental slope and the fan axis. Succeeding fans may then have occupied these low areas, and finally, a later fan could have prograded across the original fan to form a composite system of interbedded sandstone and shale. However, these ancient fans can also be expected to retain the general pattern of channel and overbank deposits so that the terms of inner fan, middle fan, and outer fan could be used to designate associations of turbidite sequences and to infer their locations relative to the composite fan system.

Constructional-Channel Systems

Another type of morphology has a channel-form pattern of sandstone distribution, and this has been called a constructional channel system [Figure 9–2(C)]. The sandstones are dominantly channel deposits that are closely bounded by more complete levee sequences which, in turn, are abruptly replaced away from the channel by basin-plain muds. This system commonly is formed by thin sandstones that are isolated by shale. These sections may contain fewer sandstones than submarine fans, and this system could be called a "sand-poor" fan. However, the term constructional-channel system better denotes the expected morphology of the sandstones.

Erosional-Channel Fill

Another type of channel-form pattern is found in channels that were eroded in basin plain sediments, and this deposit has been called erosional-channel fill [Figure 9–2(D)]. Channels were cut as deeply as 100 ft (30 m) and were subsequently filled by turbidites that were dominated by massive *A* sandstones, and which later may have been succeeded by other turbidite divisions. However, the massive sandstones are the most prominent divisions of the sequence. The channel-fill turbidites are similar to inner-fan deposits but occur farther basinward rather than at the slope.

Reservoir Examples

Many sandstones in basinal areas are the products of turbidity flow, and these turbidites commonly form reservoirs for oil and gas. The water depths in ancient basins were not, in many cases, of excessive depth, and stratigraphic relationships indicate that water depths may have been in the range of 1000 to 2000 ft (305 to 620 m) rather than of abyssal depths. Regardless of depths, the chief mechanism of sediment transport to these deep-water areas was turbidity flow, and the turbidites commonly formed a major part of sandstones in the deep-marine areas. The following sections concentrate on the different morphologies of turbidite reservoirs.

Late Pennsylvanian and Early Permian Canyon sandstones of the Val Verde basin, West Texas, form complex bodies of interbedded sandstones and shales that appear to have originated as parts of submarine-fan systems. A different mode of occurrence is postulated for Late Pennsylvanian sandstones of the Midland basin, West Texas, where narrow linear reservoir bodies were probably of constructional-channel origin. Also the Miocene Stevens sandstones of the San Joaquin basin, California, have long been called submarine fans, although these sandstones appear to have originated largely as channel-fill deposits. Finally, the Permian Delaware Mountain sandstones of southeastern New Mexico were deposited as the fill of erosional channels that were cut into the basin plain of the ancient Delaware basin.

CANYON SANDSTONES, VAL VERDE BASIN, WEST TEXAS

Turbidites are widespread reservoirs for natural gas in the northern Val Verde basin, and the basin may contain reserves of giant proportions. The thin-bedded sandstones are found in a variety of types from more complete to less complete sequences of turbidite divisions. Thicker beds of the *AE* and *ABE* types form the better reservoir sections, and thinner, more complete sequences of the *ABCDE* and *BCDE* types are most common in less-productive sections. The sandstones form groups of beds that differ markedly in bed thickness and in types of sequence present, and these bedding associations agree with the observed sequences of submarine fans.

Geologic Setting

The Val Verde basin is located on the southeastern margin of the greater Permian basin of West Texas. The Val Verde basin is bounded on the north by the Midland basin and Eastern shelf and on the south by the Marathon fold belt (Figure 9–3). The total section dips southwestward and thickens toward the Marathon fold belt,

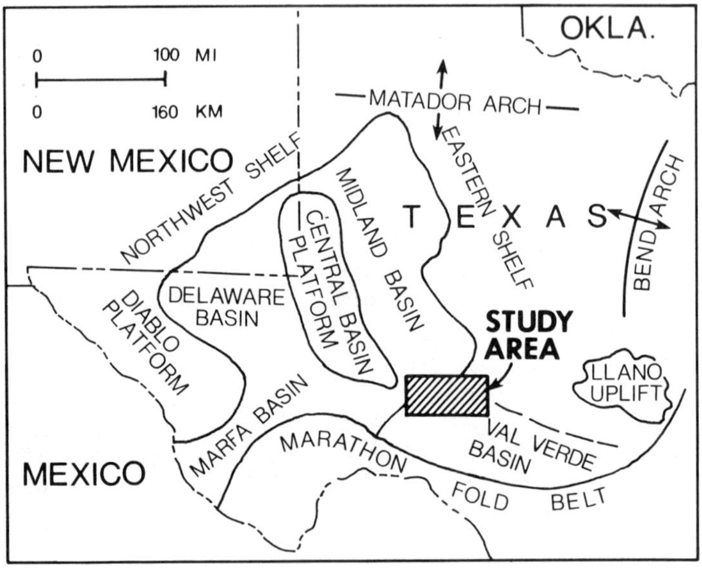

Figure 9–3 Map of west Texas showing major structural features and the location of the Val Verde basin study area (see Figure 9–4).

which is composed of Early Pennsylvanian and older Paleozoic rocks thrust northward across the deep trough of the Val Verde basin.

Late Pennsylvanian and Permian sandstones have produced gas for many years in limited areas across the north part of the Val Verde basin. During the early 1970s, increased exploratory drilling and the availability of cores from some wells permitted the first interpretation of depositional setting for the reservoir sandstones (Mitchell 1975). The work was concentrated along the north part of the basin designated as the study area (Figure 9–3).

Structure of the Val Verde basin on the Early Pennsylvanian Strawn limestone shows generally a low dip toward the south and southwest at the rate of about 150 ft/mi (30 m/km) from a broad arch that separates the southeast end of the Central Basin platform from the Eastern shelf of the Midland basin (Figure 9–4). Total sedimentary fill of the Val Verde basin is on the order of 18,000 ft (5486 m) in thickness, and Late Pennsylvanian and Permian sediments form a major part of this section. The Marathon fold belt to the south was uplifted during the Early Pennsylvanian, exposing older Paleozoic rocks to erosion and forming the main source for clastic sediments in the southern part of the basin (Vertrees, Atchison, and Evans 1959). The northern part of the basin received sediment that was derived, in part, from other sources to the northwest and northeast.

During Pennsylvanian and Permian deposition, the Eastern shelf prograded westward (Figure 9–5). Sediment was carried across the shelf by fluvial-deltaic trans-

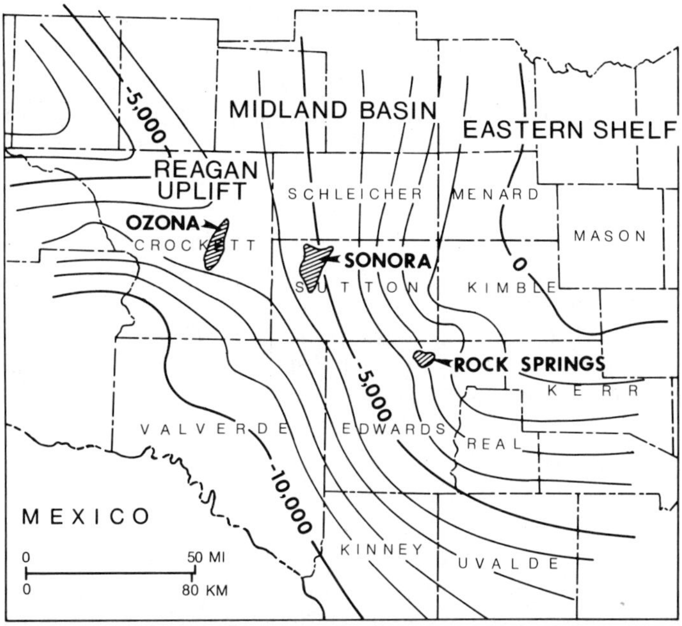

Figure 9–4 Structure of the Val Verde Basin on top of the lower Pennsylvanian Strawn limestone and location of selected Canyon sandstone gas fields. Contour interval 100 ft (305 m). Location of area shown in Figure 9–3. [Map adapted from Rall and Rall 1958.]

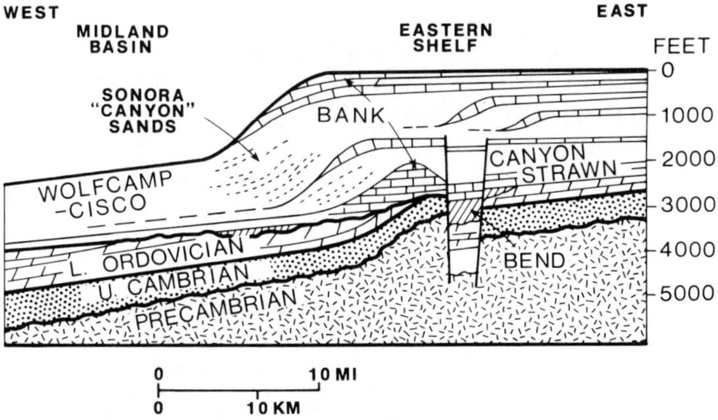

Figure 9–5 Diagrammatic cross section from the Eastern shelf to the Midland basin in the vincinity of Sonora field showing stratigraphic relationship of the Canyon sandstones and adjacent sections. [From Rall and Rall 1958.]

port systems, and thick sections of shales with interbedded sands were deposited as prograding slopes. Lesser amounts of sediment reached the deeper Midland basin, and thus the basin was "starved of sediment," and the stratigraphic units thin basinward (Adams 1962). Consequently, the gas-producing sandstones, generally designated "Canyon" were not the time equivalents of the Canyon Group recognized on the Eastern shelf but were considerably younger. Correlations from shelf to basin, and rare fossils, have established that the Canyon sandstones are Late Pennsylvanian (Cisco) to Early Permian (Wolfcamp) in age (Adams et al. 1951).

Thus, two major sources of sediment are possible for the Val Verde basin—the Eastern shelf to the northeast and the rising Marathon belt to the south. A possible third source may have been the Central Basin uplift to the northwest. These sources account for more than 10,000 ft (3050 m) of sediment deposited in the deeper parts of the Val Verde basin.

Drilling History

Canyon sandstones in West Texas produced oil and gas for many years, but only recently have these reservoirs been developed over wide areas in the northern Val Verde basin (Figure 9–4). At Sonora field, 50 gas wells were completed by 1962, and the producing area covered 56,000 acres (145,600 km^2) (Hills 1968, p. 1422). At Ozona field, 130 producing wells were completed by 1965, and the producing area covered more than 40,000 acres (104,000 km^2) (McWilliams 1966, p. 283). Reservoir depths range from 5000 to 6600 ft (1524 to 2112 m). Although production rates per well were modest at about 250,000 cu ft/day (7000 m^3/day), these fields suggest that important gas reserves could be established in Canyon sandstones.

During 1974, increased demand for gas stimulated exploration, and production was extended for many miles along the trend established by Sonora and Ozona. One of the recent discoveries was southeast of Sonora at Rock Springs field (Figure 9–4). By 1975, 23 wells had been completed over an area of 26,000 acres (66,000 km^2). Development has continued across the northern Val Verde basin and has resulted in a nearly continuous producing area from Rock Springs to Ozona. Reserves are conservatively estimated to be in excess of 1.5 trillion cu ft (42 billion m^3), and further drilling may establish this area as having gas reserves of giant proportions (Valder 1977).

Reservoir Section

During the period of recent field development, cores were obtained from widely separated stratigraphic intervals in the Sonora, Ozona, and Rock Springs fields. Core study established the sandstones as turbidites, and furthermore, the cored sections showed distinct associations of beds that can be interpreted as representing different locations on a submarine fan (Mitchell 1975). The general character of the section can also be recognized on borehole logs (Figure 9–6).

Recent wells have been drilled using air or gas as a drilling fluid, and therefore

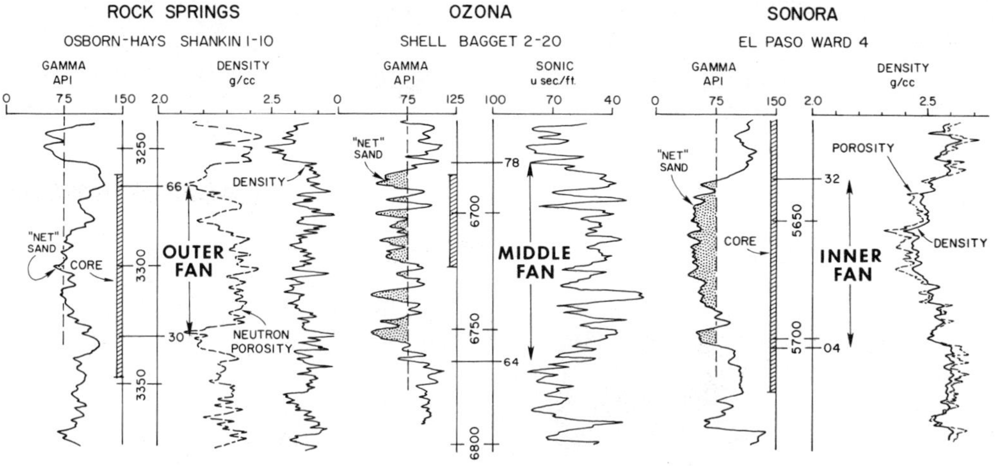

Figure 9–6 Well logs through Canyon sandstones showing character of submarine-fan facies as established in cores. An arbitrary cutoff at 75 API units of gamma-log response is used to designate net sand within the sections.

the common well logs are gamma-ray and density logs (see Chapter 1, Table 1-6). The most useful log for stratigraphic interpretation is the gamma log because of its response to variable amounts of sandstone and shale in the section. The inner-fan facies is shown in the El Paso Ward 4 well where the cored section is dominated by beds *AE* and *ABE* types. Sandstones are relatively thick, and the section contains little interbedded shale. Gamma logs indicate that this section has about 80 percent of net sandstone.

The middle-fan facies is shown in the Shell Baggett 2–20 well in the Ozona field. The core of the section shows a large number of *AE* and *ABE* turbidites with some complete *ABCDE* sequences. Gamma logs suggest about 50 percent net sandstone for this section.

The outer-fan facies is shown in the Osborne-Hayes Shanklin 1–10 well located near the Rock Springs field. The core of this section shows incomplete turbidites of the *BCE* and *BCDE* types in thin beds with significant amounts of interbedded shale. Gamma logs indicate a negligible amount of net sandstone within this section.

In all of the sections, the gamma logs indicate groups or "packages" of sandstone and shale that have thicknesses ranging from about 60 to 70 ft (18 to 21 m). Furthermore, the gamma log has an erratic or "nervous" character reflecting the thinly interbedded sandstone and shale, but the amounts of net sandstone within these sections differ according to bed thickness and type.

The gamma response shows increasing "shaliness" from the inner-fan, through the middle-fan, to the outer-fan facies within sections that have approximately the same thickness. The decreasing amount of sandstone can be expressed quantitatively, and therefore the gamma-ray log gives an estimate of net sandstone thickness as well as of the general character of the section within the submarine-fan deposits.

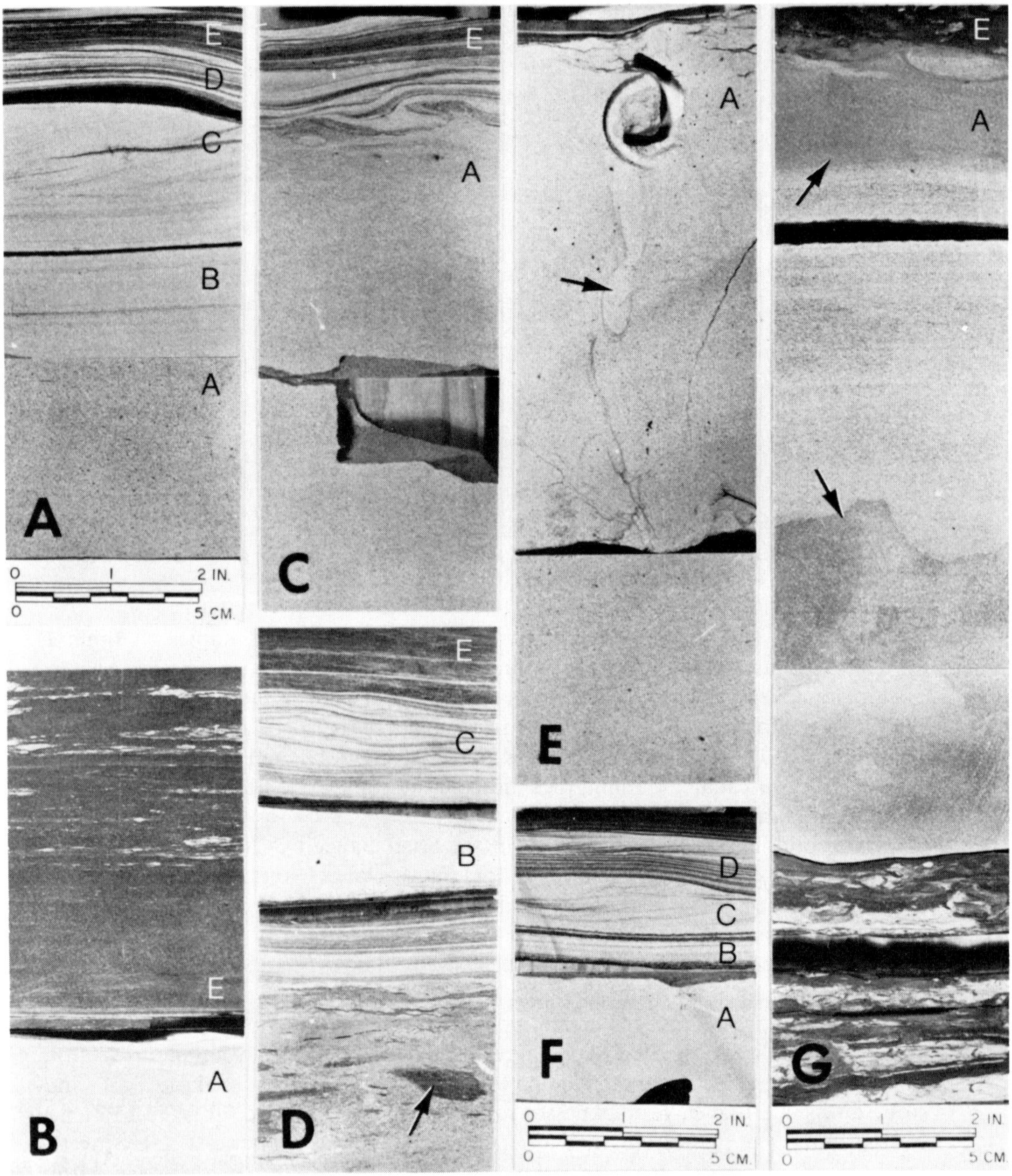

Figure 9–7 Sedimentary structures of middle-fan deposits in Canyon sandstones, Val Verde basin. Photographs are identified by boldface letters in lower left corner; turbidite divisions are designated by small capital letters on right side of photographs: (A) complete turbidite sequence of *ABCDE* divisions—6688 ft (2038 m); (B) massive *A* division overlain by *E* black shale—6698 ft (2042 m); (C) massive *A* division overlain by contorted bedding that may represent *B* or *C* divisions—6700 ft (2042 m); (D) thin turbidite of the *BCE* type overlying sandy

Sedimentary structures. The core from the El Paso Ward 4 well (Figure 9–6) shows sandstones that are generally massive and on the order of 1 to 3 ft (0 to 1 m) in thickness with a maximum of about 5 ft (1.5 m). The sandstones are dominantly of the *AE* or *ABE* type, and in all beds, the massive *A* division is the major part of bed thickness. The interbedded shales are very thin, inconspicuous, or absent, and there are also only a very small number of more complete bed sequences. The dominance of massive *A* sandstones suggests that the section is composed of stacked channel-fill sandstones, possibly of the inner fan [Figure 9–2(A)].

A contrasting section is shown in the core from the Shell Baggett 2–20 well in the Ozona field (Figure 9–6); here both complete and incomplete bedding sequences are found. Sandstones commonly range from 0.5 to 1 ft (0.15 to 0.3 m) in thickness. Several thin beds show a complete sequence of *ABCDE* divisions [Figure 9–7(A) and (F)], and the massive *A* division comprises about one-half of the bed thickness. Other sandstones consist mainly of a massive *A* division [Figure 9–7(B), (C), and (E)], and these beds are somewhat thicker and on the order of 1–2 ft (0.32–0.7 m). Some beds show an abrupt change from massive *A* to overlying *E* shale, but in other beds, the top of the *A* division may be contorted so that the character of succeeding divisions is in doubt [Figure 9–7(C)]. Most of the thicker beds are of the *AE* type, although a significant number also have horizontal laminae forming beds of the *ABE* type.

Only a few thin beds of the incomplete *BCE* types are present in the section [Figure 9–7(D)]. Other thin beds are composed of small to large shale clasts that are suspended in a matrix of sandy muds [Figure 9–7(D)], and these beds probably represent muddy debris flows of high viscosity.

Most of the sandstones are light gray, but some massive *A* sandstones have mottled areas that are tan or brown [Figure 9–7(G)]. These darker areas may occur at any level within the massive sandstone, and they indicate the presence of siderite cement in irregular patches.

Interbedded shales are dark gray to black and commonly contain thin laminae or wisps of fine silt and sand. The shales are massive and structureless [Figure 9–7(B)] and denote deposition of clays from suspension during times of no bottom current. In some sections the shales are somewhat sandy and may show intense burrowing by benthic organisms. Bioturbation is generally of a fine scale and may disrupt bedding so that laminations are indistinct or destroyed [Figure 9–7(G)].

The character of the section is distinctly different in the core from the Osborn-Hayes Shanklin 1–10 well in the Rock Springs field (Figure 9–6). Beds are generally thin and 0.5 ft (0.15 m) or less in thickness and are interbedded with shales that

---

mudstone with shale clasts (arrow)—6702 ft (2043 m); (E) massive *A* division with small shale clasts (dark spots) and slightly bioturbated top; note long vertical burrow (arrow)—6707 ft (2044 m); (F) top part of complete turbidite with *ABCDE* divisions— 6716 ft (2047 m); (G) massive *A* division with indistinct laminae, sharp basal contact on bioturbated shale; note darker areas (arrow) that indicate presence of siderite cement—2517 ft (767 m). (A)–(F) from Shell Baggett 2–20; (G) from Osborn-Hayes Shanklin 1–10.

have about the same thickness. Within the total section, beds of the *BC* type are dominant. These are characterized by a thick laminated *B* division, which is succeeded by thinner ripple-laminated *C* divisions [Figure 9–8(B)–(E)]. The *B* laminae are distinct and are marked by fine carbonaceous particles. The rippled *C* division is generally a single layer less than 1 in. (2.5 cm) in thickness, but rarely, the *C* division is somewhat thicker and may be composed of two distinct rippled sets [Figure 9–8(D)]. In a few thick beds, massive *A* sandstone is found at the base [Figure 9–8(E)], but the massive sandstone is generally thin.

The interbedded dark gray to black shale contains laminae and thin beds of sand that form sequences of the *BC* and *CD* types [Figure 9–8(F) and (G)]. At some levels, the interbedded sands and shales are more intensely bioturbated so that the original lamination of the sandstone is destroyed [Figure 9–8(A)]. Some bioturbation extends into the tops of underlying beds and partly destroys lamination [Figure 9–8(E)]. Masses of red-brown claystone (ironstone) are also found in the section. These occur as thin lenses up to several inches in thickness or as irregular replacements in sandstones [Figure 9–8(B), see f].

Petrography. Texture and composition reveal the detailed character of the turbidites (Table 9–1). Mean grain size is 0.17 mm (fine grained) and average maximum size is 0.54 mm (coarse grained). In general, grain size tends to decrease upward within bed sets. The massive *A* divisions are typically fine grained, whereas the *B* and *C* divisions tend to be very fine grained, and the *D* division is coarse silt. Thus, single beds that consist of two or more divisions show a fining upward sequence, but massive *A* sandstones may not show textural gradation [Figure 9–9(A)]. Beds of the *AE* and *ABE* types tend to be somewhat coarser grained as compared with similar divisions in more complete sequences. This relationship is especially well shown in those sections where more complete and less complete bed sequences occur together [Figure 9–9(B)].

Similar textural relationships are also shown in outer-fan sandstones (Figure 9–10). Thinner overbank sequences of the *ABCDE* and *BCE* types tend to be somewhat finer grained, and all show textural gradation within bed sets [Figure 9–10(A)]. A few thicker sandstones are also present in the outer-fan section, and these beds of the *AB* type are somewhat coarser grained in the dominant massive *A* division [Figure 9–10(B)].

Average detrital composition of the sandstones is 64 percent monocrystalline quartz, 7 percent feldspar, and 7 percent rock fragments with 20 percent matrix (Table 9–1). Orthoclase is the dominant feldspar, and rock fragments are primarily granitic but include grains of polycrystalline quartz and chert. Other grains constitute only 2 percent and include biotite, opaque minerals including pyrite, carbonaceous material, and fossil fragments. Matrix is largely clay and mostly detrital, although part of the matrix appears to be authigenic by alteration of feldspar grains.

Some compositional differences may be related to bedding type. For example, quartz content averages about 75 percent in samples from the inner-fan channel sandstones (Table 9–1), whereas quartz is lower and matrix is higher in the other

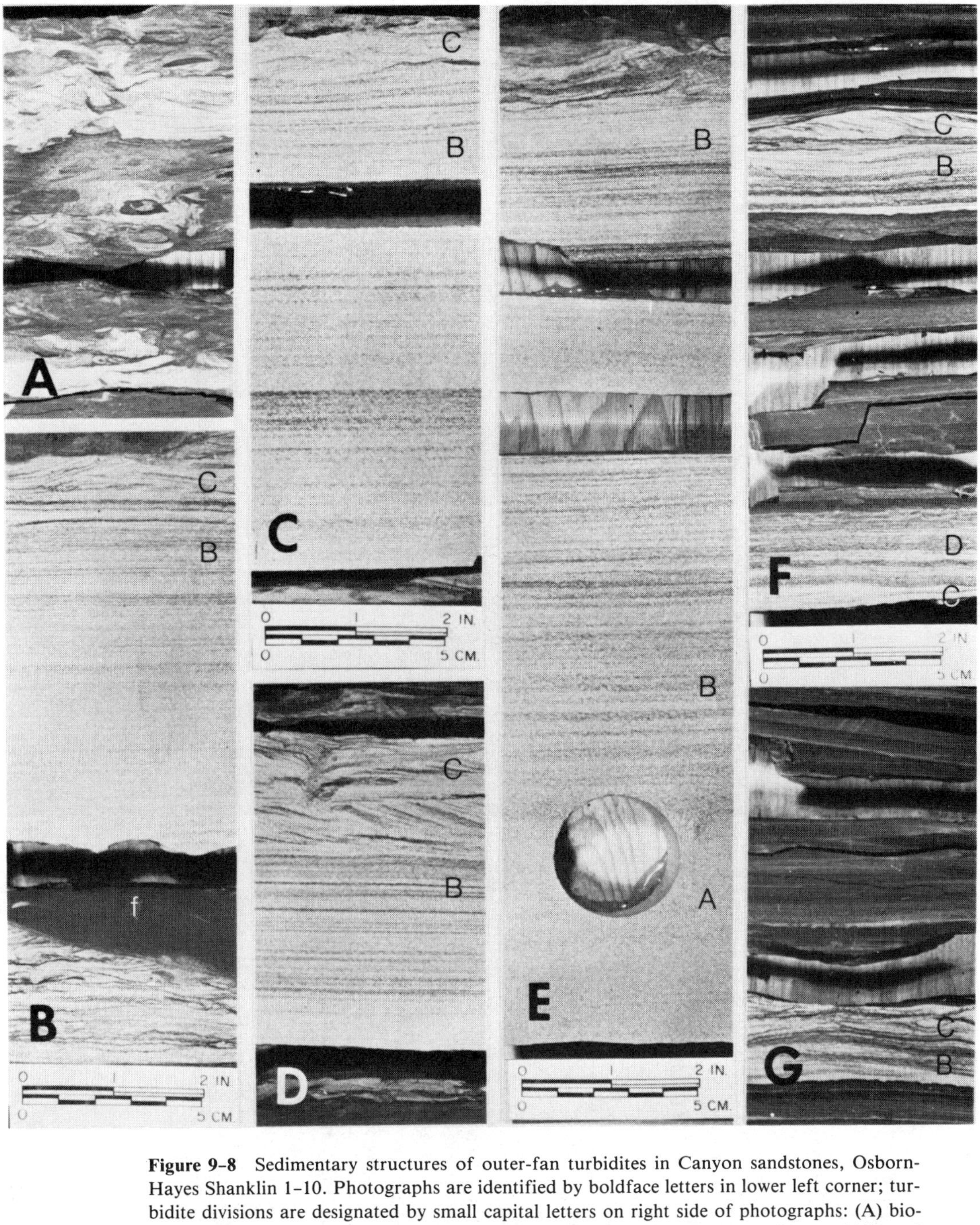

Figure 9–8 Sedimentary structures of outer-fan turbidites in Canyon sandstones, Osborn-Hayes Shanklin 1–10. Photographs are identified by boldface letters in lower left corner; turbidite divisions are designated by small capital letters on right side of photographs: (A) bioturbated ripple lenses of sand in black shale—3292 ft (1003 m); (B), (C) and (D) turbidites of the *BCE* type; note large "ironstone" mass (B, f)—3300, 3305, and 3308 ft (1006, 1007, and 1008 m). (E) turbidite of the *ABE* type; note thin *A* division and contorted bedding at top of *B* division—3311 ft (1009 m); (F) and (G) thin turbidites of the *BCE* and *CDE* types in black shale—4103 and 4104 ft (1251 m).

TABLE 9-1. SUMMARY OF ROCK PROPERTIES FOR CANYON SANDSTONES, VAL VERDE BASIN

					Quartz size[a]			Detrital composition[b]					Cement[c]				
Facies	Well	Number of samples	Bedding type	Depth (avg) (ft)	Mean (mm)	Max (mm)	σ (mm)	Qz (%)	F (%)	Rx (%)	Oth (%)	Mx (%)	Sil (% of total)	Sid	Cal	Permeability (md)	Porosity (%)
Inner Fan	Ward 4	7	Channel	5650	0.18	0.56	0.07	73	6	9	2	10	8	2	1	0.2	8.6
	Berger 3	7	Channel	6200	0.18	0.58	0.06	78	6	9	tr	6	11	2	tr	0.8	8.5
Middle Fan	Baggett 2	8	Channel	6700	0.15	0.50	0.06	63	5	6	2	24	1	—	9	<0.1	7.4
	Baggett 2	7	Overbank	6700	0.09	0.30	0.07	55	6	6	1	31	4	—	6	<0.1	7.4
Outer Fan	Whitworth 4	14	Channel	3200	0.26	0.93	0.10	61	9	11	1	18	4	7	1		
	Shanklin 1	15	Channel	2500	0.14	0.57	0.05	55	8	7	3	27	3	9	—		
	Shanklin 1	30	Overbank	3300	0.16	0.48	0.05	62	6	7	3	22	5	5	1	<0.1	8.0
	Shanklin 1	37	Channel	4120	0.18	0.83	0.06	69	6	6	2	17	6	2	—	0.3	9.0
			Means		0.17	0.54		64	7	7	2	20	5	4	1		

SOURCE: Petrography by Mitchell 1975.

[a] Long-axis measurements; σ = standard deviation.

[b] Qz = monocrystalline quartz, F = feldspars, Rx = rock fragments including polycrystalline quartz, Mx = matrix, Oth = other minerals.

[c] Sil = silica as grain overgrowths, Sid = siderite, Cal = calcite.

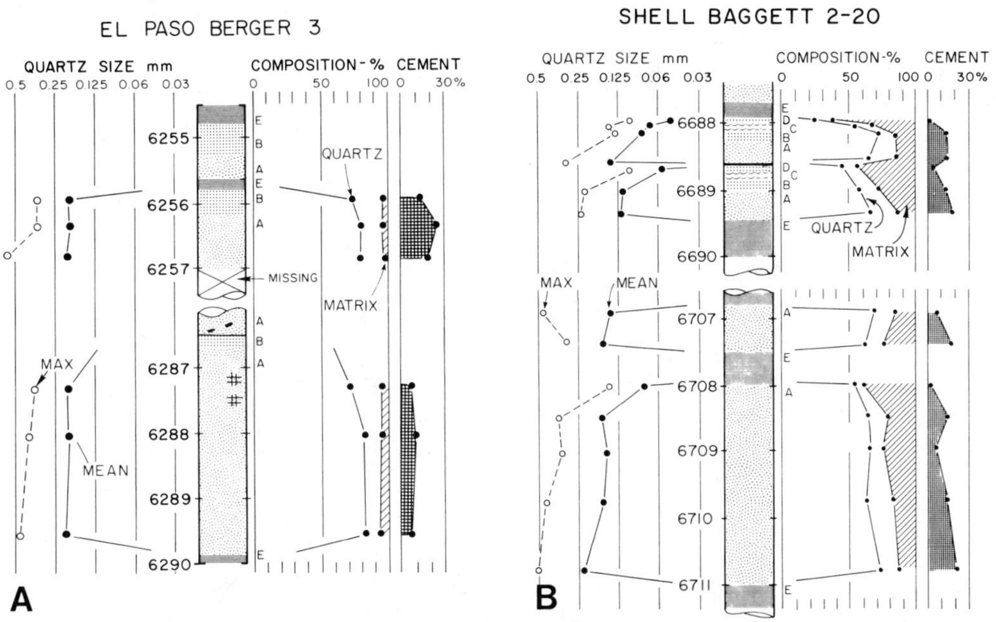

Figure 9–9 Texture, composition, and sedimentary structures of turbidites: (A) inner-fan sandstones, El Paso Berger 3, Sonora field; (B) middle-fan sandstones in the Osborn-Hayes Shanklin 1–10, Rock Springs field. Small capital letters to the right of the column indicate turbidite divisions.

sandstones. These differences, however, may be due to differences in source sediment rather than in bedding type.

Cements form an important part of the bulk composition of the sandstones. Total cement averages 10 percent and consists of silica as distinct overgrowths on quartz grains and carbonate cement of two types—siderite and calcite. Siderite is common as small euhedral crystals, and most commonly, the siderite is found as coatings on quartz grains beneath silica overgrowths. Calcite is an important cement only in the Shell Baggett core, where it averages about 8 percent (Table 9–1). This pattern of cement types suggests that the source sediment for the Ozona sandstones may have been different from the source for the other sandstones.

Porosity and permeability. The sandstones are characterized by porosities of about 8 percent and permeabilities in the range of 0.1 to 1 md. These low values appear to be due to the high content of matrix and cement (Table 9–1). In the Shell Baggett 2–20 core, permeability is reduced by matrix and calcite cement, whereas in the other cores, permeability is reduced by significant amounts of matrix and silica overgrowths. Siderite cement produces anomalously high values of permeability in some beds. Siderite that formed by replacement of calcite would result in reduction of mineral volume and increased porosity much the same as that which

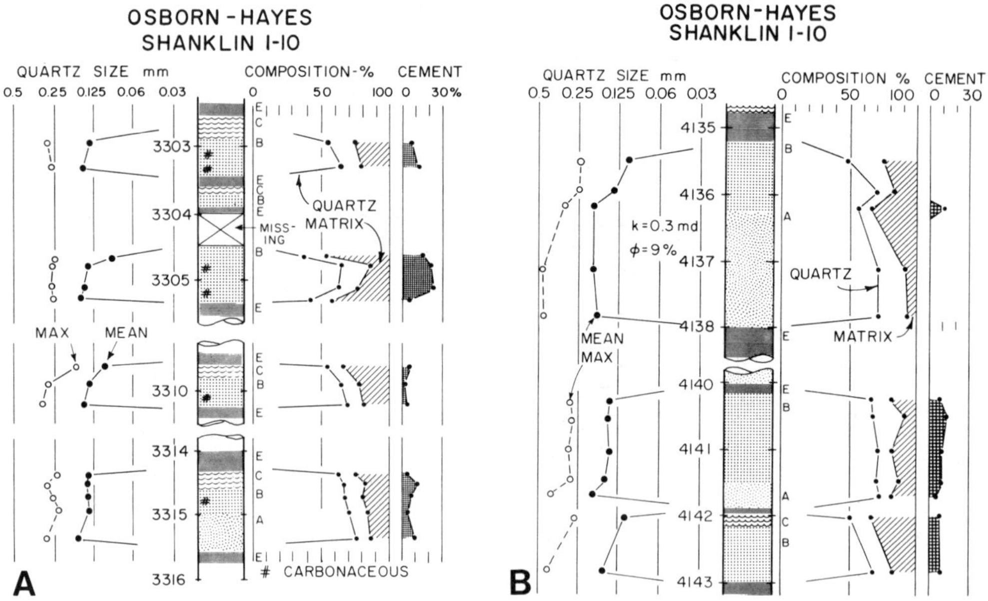

Figure 9–10 Texture, composition, and sedimentary structures of outer-fan sandstones, Osborn-Hayes Shanklin 1–10, Rock Springs field. Small capital letters to the right of the center column indicate turbidite divisions.

occurs by dolomitization. In addition, siderite coatings on quartz grains may have inhibited the addition of silica overgrowths.

The inner-fan channel sandstones have the highest porosities and permeabilities [Figure 9–11(A)]. These sandstones have the least amount of clay matrix and calcite cement. Many beds show permeabilities in the range of 0.1 to 0.5 md, but a few have permeabilities greater than 10 md. For sandstones that have permeabilities of 0.1 md or more, the average permeability is 7.1 md or 264 md·ft in 37 ft (11.3 m) of section. In contrast, outer-fan sandstones have an average permeability of 0.25 md and a total of 4.6 md·ft [Figure 9–11(B)]. Middle-fan sandstones have permeabilities that are within the inner-fan and outer-fan range.

Interpretation

The cored sections show three different sequences of turbidites as shown by a tabulation of bed types and frequency of occurrence (Table 9–2). These bed associations may be classified by their depositional location on a submarine fan according to the patterns defined by Haner (1971) [Figure 9–2(A)]. Cores from the Sonora field (El Paso Ward 4, El Paso Berger 3) contain a dominance of thicker beds of the *AE* and *ABE* types. These sandstones have a dominant massive *A* division and range from 2 to 8 ft (0.6 to 1.5 m) in thickness with little or no interbedded shale. This section represents the inner-fan sequence of channel-fill sandstone. The core

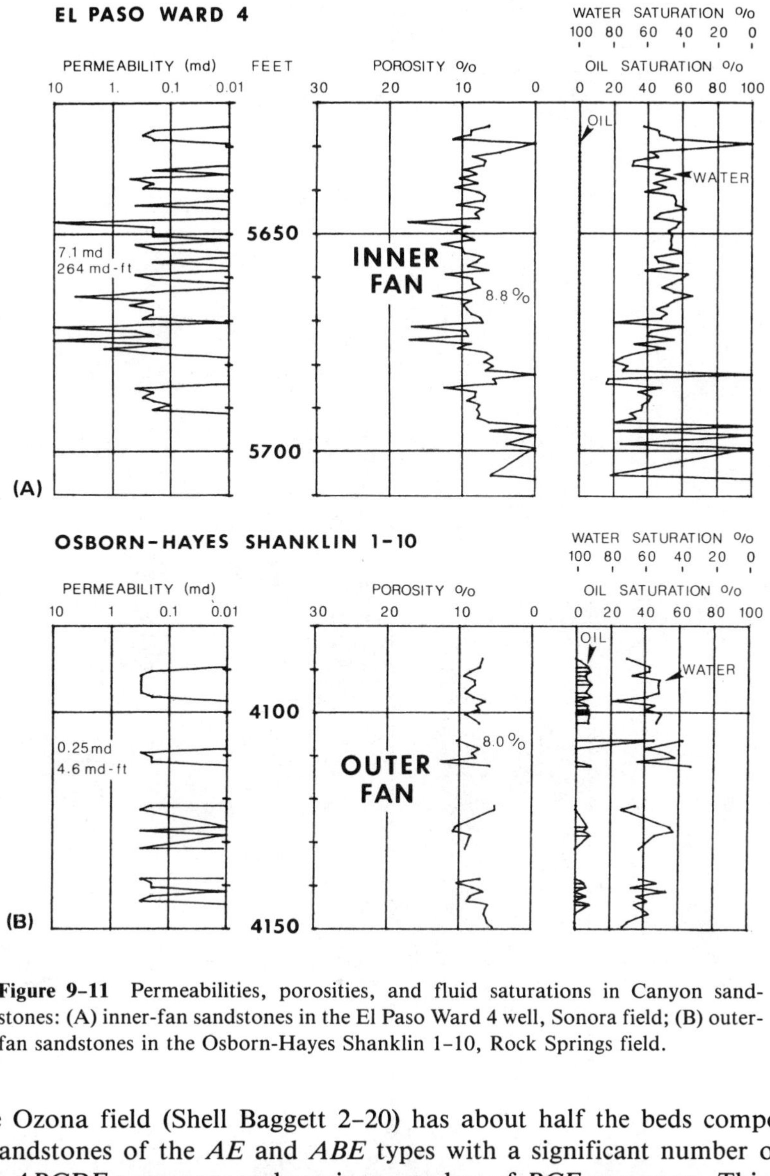

Figure 9-11 Permeabilities, porosities, and fluid saturations in Canyon sandstones: (A) inner-fan sandstones in the El Paso Ward 4 well, Sonora field; (B) outer-fan sandstones in the Osborn-Hayes Shanklin 1-10, Rock Springs field.

from the Ozona field (Shell Baggett 2-20) has about half the beds composed of thicker sandstones of the *AE* and *ABE* types with a significant number of more complete *ABCDE* sequences and a minor number of *BCE* sequences. This association of beds represents the middle-fan sequence. Cores from the vicinity of Rock Springs field (Midwest Witworth 4 and Osborn-Hayes Shanklin 1-10) are dominated by thin beds of *BCE* and *BE* types.

Average thickness of beds decreases from inner-fan through middle-fan to the outer-fan facies, which reflects the decrease in the number of thicker channel-fill sandstones (Table 9-2). The sandstone/shale ratio decreases from about 8 in the

TABLE 9-2. CLASSIFICATION OF TURBIDITE SANDSTONES IN CANYON GROUP, VAL VERDE BASIN

		Inner fan				Middle fan		Outer fan			
		El Paso Ward 4		El Paso Berger 3		Shell Baggett 2–20		Midwest Whitworth 4		Osborn-Hayes Shanklin 1–10	
Facies	Sequence	Number of beds	(%)	Number of beds	(%)	Number of beds	(%)	Number of beds	(%)	Number of beds	(%)
Channel	AE	21	70	15	47	10	40	6	17	3	4
	ABE	7	24	13	40	2	8	4	11	7	9
Overbank	ABCDE	—	—	—	—	3	12	—	—	2	3
	ABCE	1	3	—	—	5	20	—	—	6	8
	ACDE	—	—	—	—	3	12	—	—	—	—
	BCDE	—	—	—	—	—	—	3	9	2	3
	BCE	1	3	—	—	2	8	6	17	37	50
	BE	—	—	4	13	—	—	11	32	7	9
	CE,CDE	—	—	—	—	—	—	5	14	10	14
Totals		30	100	32	100	25	100	35	100	74	100
Average thickness (ft)		2.0		2.1		0.4		0.3		0.2	
Sand/Shale ratio		8.3		7.7		1.7		2.1		1.7	

SOURCE: From Mitchell 1975.

inner-fan facies to 2 or less in the middle and outer fan as the sections become increasingly shaly in the overbank deposits removed from the channels.

Morphology

The interpretation of the bed associations representing submarine-fan facies is confirmed, in part, by the location of the cored sections in relation to sandstone morphology. Maps of the older Ozona and Sonora fields show the distribution of net sandstone thickness within the reservoir section (Hills 1968). In the Sonora field, net sandstone thickness reaches a maximum of 70 ft (21 m) along a relatively narrow trend of sandstone that is normal to the adjacent margin of the eastern shelf [Figure 9–12(B)]. Cores that showed the stacked channel-fill sandstones of the inner fan are located within this dip-trending sequence. In the Ozona field, net reservoir thickness reaches a maximum of 100 ft (30 m), and the elongate trend of sandstone is composed of at least three overlapping lenses of reservoir sandstone. The broadly ovate configuration of these lenses suggests coalescing submarine fans. The core that showed a middle-fan bed association came from the northeastern end of this reservoir body.

For the Ozona field, a more recent map of sandstone thickness (McWilliams 1966) suggests that at least two narrow fingers of sandstone extend northwestward at right angles to the main reservoir body. These extensions may represent channel-

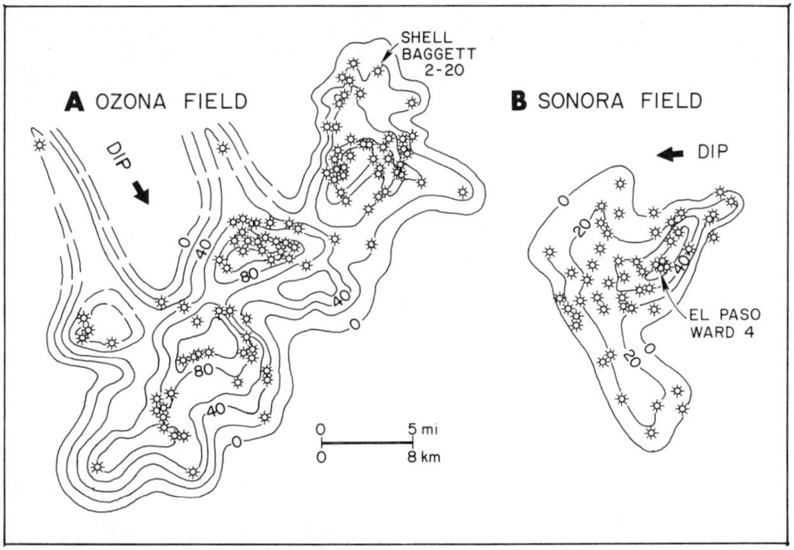

Figure 9–12 Net thickness of Canyon reservoir sandstones in the older producing fields: (A) Ozona field, Crockett County; (B) Sonora field, Sutton County. Location of fields shown in Figure 9–4. [Both maps adapted from Hills 1968.]

fill sandstones of the inner fan that were the feeder channels for the main standstone body downdip. In this case, the source of sediment was the Central Basin platform to the northwest.

It should be noted that for both fields the producing areas are large, and well control for the maps was widely spaced. However, recent infill drilling has consolidated production through the fields and has extended gas production over greater areas across the northern Val Verde basin. Much of the recent production has been established in Canyon sandstones at other stratigraphic levels that probably represent different submarine-fan systems.

During the development of the Rock Springs field, maps of major sandstone units were compiled (Mitchell 1975). The unit selected for mapping has an average depth of about 3000 ft (914 m). Within this section, interbedded sandstones and shales have a maximum gross thickness of 600 ft (183 m) and form a large ovate mass that is slightly elongated in a southwesterly, or dip, direction [Figure 9–13(A)]. Within this thick body of sandstone and shale, net sandstone was mapped using a 75 API cutoff on the gamma-ray logs [Figure 9–13(B)]. Net sandstone shows dominant dip trends that coalesce in a downdip direction to a maximum thickness of 250 ft (76 m). This thick body of sediment is a composite of several submarine fans, each of which is composed of inner-, middle- and outer-fan facies. Cores that represent a small part of this complex show middle- and outer-fan bed associations in

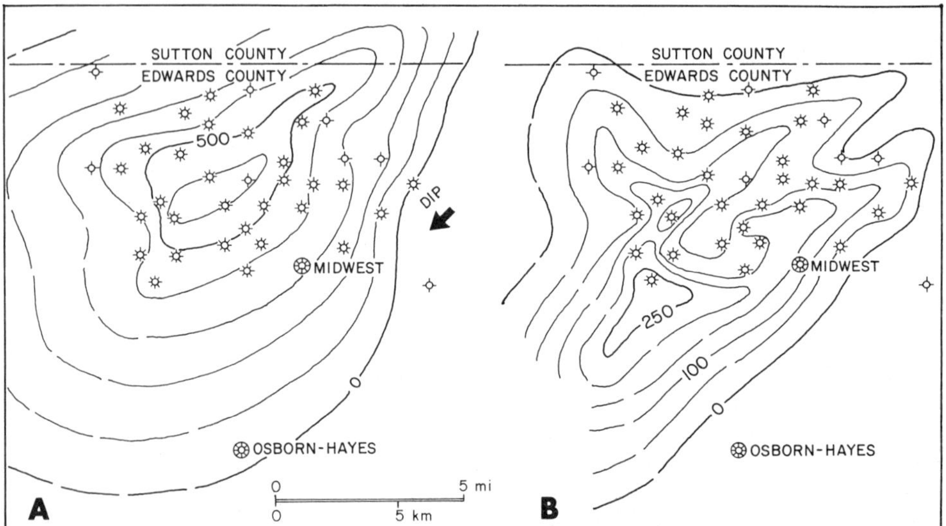

Figure 9–13 Thickness of Canyon sandstones in the Rock Springs field, Edwards County, Texas: (A) gross thickness of sandstones and interbedded shales at an average depth of 3000 ft (914 m); contour interval 100 ft (30 m); (B) net sandstone thickness within the same stratigraphic section. Contour interval 50 ft (15 m). Location of field shown in Figure 9–4. [From Mitchell 1975.]

the Midwest well, which is located within the area of net sand development. On the other hand, cores and logs from the Osborn-Hayes well show an outer-fan facies on the margin of the complex and outside the area of net sandstone. Thus, both cores and log interpretation tend to confirm the presence of a submarine fan. Concentrations of channel-fill sandstones in the middle-fan facies form the major part of the reservoir system.

Conclusions

The Canyon gas-reservoir sandstones of the northern Val Verde basin were deposited within large submarine-fan systems by turbidity flows that originated on the Eastern shelf to form the reservoirs at Sonora and Rock Springs fields, and possibly also from turbidity flows that originated on the Central Basin platform to the northwest of the Ozona field. Other gas reservoirs throughout the northern part of the Val Verde basin may have been derived, in part, from turbidity currents that flowed from the rising Marathon fold belt on the south. The submarine-fan complex contains multiple sandstone reservoirs of low porosity and permeability, but the reservoirs cover wide areas and contain gas reserves of major proportions. Bed associations in cores can be designated *inner-, middle-,* or *outer-fan* and correlated to gamma-log response for the purpose of mapping the fan facies in field areas.

CANYON SANDSTONES, MIDLAND BASIN, WEST TEXAS

Pennsylvanian sandstones are oil reservoirs in many small fields in the eastern Midland basin, west-central Texas. These sandstones are part of depositional systems that represent shelf, slope, and basin environments of deposition. The depositional systems prograded westward during Pennsylvanian and Permian time, gradually encroaching on and finally filling the basin by the end of the Permian.

Basinal sandstones of the Canyon and Cisco groups represent the fill of constructional channels near the base of the slope. The reservoir sandstones are thin and narrow bodies that have dip trends, and they form stratigraphic traps in and around older Pennsylvanian carbonate reefs. The channel-fill sandstones and adjacent overbank deposits have characteristic bed associations that are readily identified on well logs. Therefore, the details of these deep-water channel systems can be mapped in local areas (Berg 1982).

Geologic Setting

Pennsylvanian depositional patterns in north Texas were determined by shelf, slope, and basin topography (Figure 9–14). On the east, a broad shelf contained fluvial and deltaic systems that transported coarser sediment from the highlands in the Ouachita uplift to the deep basin on the west. In the Canyon Group, these areas of sandstone were the Perrin delta and Henrietta fan-delta systems (Erxleben 1975).

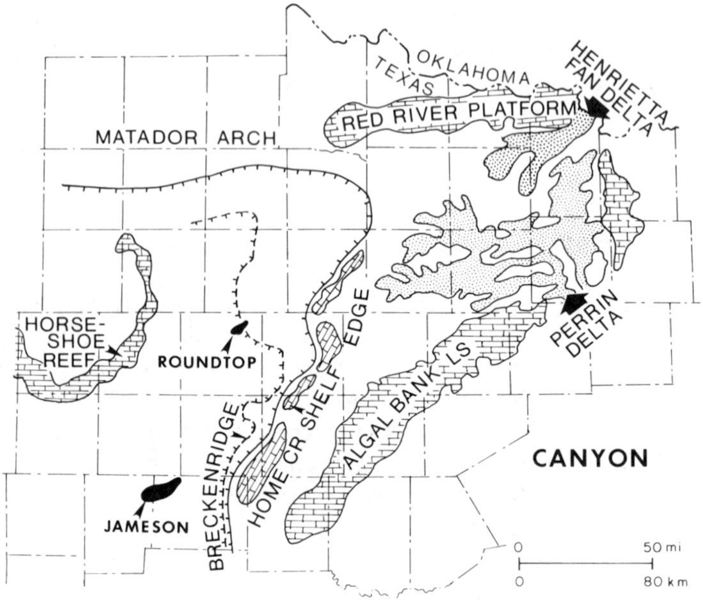

Figure 9–14 Generalized paleogeography of the Canyon Group in north-central Texas, showing shelf, shelf margin, and basin areas and location of Roundtop and Jameson fields. [Hachured lines represent successive shelf margins after Erxleben 1975.]

Bounding the fluvial systems on the north were areas of thick carbonate deposition that coincided with the Red River arch and with a broad shallow-marine shelf to the south. The marine shelf contained areas of carbonate deposits in the form of small banks composed largely of phylloid algal fragments and other biogenic debris (Wermund 1975). At the shelf margin were concentrations of reef and bank deposits, some of which persisted through extended periods and built thick sections of shelf-edge carbonate. Terrigenous sediment was supplied to the basin through gaps in the shelf-edge reefs.

Paleogeography at the end of Canyon Group deposition shows a northeastward-trending shelf margin of the Home Creek limestone (Figure 9–14). Succeeding deposits of the Cisco group display essentially the same distributions of sedimentary environments, but the shelf margin gradually prograded westward as shown by the Breckenridge Limestone (Figure 9–14). Fine sediment supplied to the basin caused lateral extension of the slope, gradually filling the basin area on the west. Earlier correlations did not recognize the thickening of section on the slope, and the basin sandstones were assigned to the Canyon Group or to the preceding Strawn Group.

Stratigraphic relationships from shelf to basin were finally established (Van Siclen 1958; Jackson 1964). For the Cisco Group, key beds of limestones were traced across the shelf to thickened sections of shelf-edge reef and bank limestones (Figure

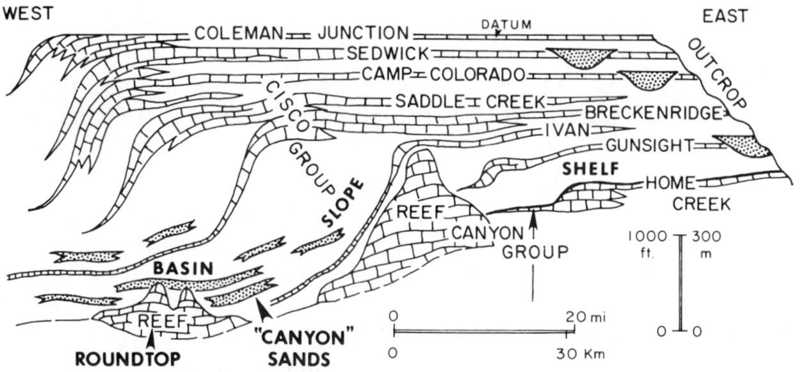

Figure 9–15 Stratigraphic diagram of the upper Canyon and Cisco groups showing position of basin sandstones in relation to shelf and slope deposits. [Modified from Jackson 1964 and Brown, Cleaves, and Erxleben 1973.]

9–15). Downslope, these limestones thin abruptly and grade to black calcareous shale or dense limestone toward the basin. The intervening shales thicken on the slope and then thin again into the basin. The resulting depositional pattern shows that terrigenous clastics were concentrated on the slope and rise, whereas the distant basin was "starved" of clastic sediment. The dominantly fine-grained slope deposits contain multiple thin sandstones that have dip trends. These sandstones have been identified as turbidites (Jackson 1964; Galloway and Brown 1972), and the several thick lenses of slope sediment have been termed the Sweetwater slope system.

The Roundtop sandstones lie above the Ivan Limestone and within the sublens called the Ivan slope sequence (Jackson 1964). The sands were deposited near the base of the slope and around limestone reefs that were established during the Strawn (Desmoinesian) and persisted through Canyon (Missourian) and into the Cisco (Virgilian). Progradation of the slope later covered the Roundtop reef, and a new sublens, the Breckenridge slope sequence, continued the basinward progradation (Figure 9–15). Other sands, such as those around the Jameson reef (Figure 9–14), were deposited within the Breckenridge sequence.

Thus, the Late Pennsylvanian shelf margin advanced westward across the thickened slope deposits, gradually restricting the Midland basin to a greatly limited area during the Early Permiam. The sandstones within slope sequences have been called *slope-trough* deposits (Galloway and Brown 1972), and in the Roundtop area they were believed to represent a submarine fan (Snyder 1977). The Roundtop sands were, in fact, deposited as turbidites in narrow channelways that supplied sediment to the deeper parts of the basin.

Drilling History

Oil was discovered at the Roundtop field in 1947 from a depth of 4500 ft (1372 m) in the Canyon Palo Pinto Limestone (Day and Galbraith 1950). Subsequent development showed reef buildup from an elevation of −2350 ft (−716 m) to an oil-water contact at approximately −3000 ft (−950 m) enclosing a productive area of about 5.6 sq mi (14.6 km^2) (Figure 9–16). The average well flowed 600 bbls/day (95 m^3/day) of 42.5° API gravity oil. Early drilling also found oil production in the overlying sandstones at a depth of 3900 ft (1190 m), but producing areas were small because of the lenticular nature of the reservoirs.

Canyon (Cisco) sandstones around and above the reef had shows of oil but were considered to have permeabilities too low for economic production. However, large areas of oil production were established after 1973 (Figure 9–16). Initial flows from Canyon reservoirs were on the order of 100 to 150 bbls/day (16 to 24 m^3/day) from net pay thicknesses of 10 to 20 ft (3 to 6 m).

During the period of recent drilling, cores were obtained from Tolar Canyon field near the north end of Roundtop reef field (Figure 9–16). Cores from the General Crude Oil Bowdry 12A and Bowdry 15A wells showed the character of nearly the entire thickness of the Canyon sandstone. A core from a later dry hole, General Crude Oil Stuart 1, located south of the Roundtop reef field, showed additional details of the Canyon section.

Reservoir Section

Canyon sandstones at Roundtop field are found at several levels (Snyder 1977). Lower sandstones are present updip and reach the east margin of the reef. A middle group completely surrounds the reef, partly overlapping the north end (Figure 9–16). Two upper groups are present, overlying the reef, but these sandstones are thin and not widely distributed. The total succession indicates a westward progradation

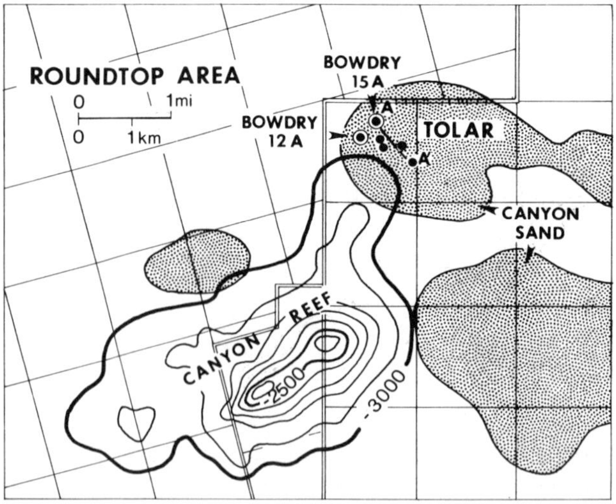

Figure 9–16 Roundtop field area, Fisher County, showing structure on top of Canyon reef, Canyon sandstone fields (stippled), and location of cored wells (circled). Contour interval 100 ft (30 m). Correlation section AA′ is shown in Figure 9–21.

of sands that gave way to dominant shale deposition. Thus, the sandstones probably represent basin-plain or rise deposits that were later covered by slope sediment.

Cores from the Roundtop area (Figure 9–17) are from the middle group of sandstones called the Canyon "C" unit by Snyder (1977). The section of thinly interbedded sandstone and shale has a maximum thickness of nearly 300 ft (92 m) in the Bowdry 15A well and thins to about 150 ft (46 m) in the Stuart 1 well.

The first core examined was from the Bowdry 12A well. Two groups of sandstones were present—an upper section 36-ft thick (4400–4436 ft, 1342–1353 m) and a lower section 90-ft thick (4450–4544 ft, 1557–1590 m). The upper section is characterized by thick channel sandstones of the *AE* and *ABE* types with some thin, more complete *ABCE* beds. Single beds are 1 to 2 ft (0.3 to 0.6 m) in thickness, but range up to 4 ft (1.2 m). The total aspect of this section is that of a middle-fan association, but channel bed types are dominant, so the section may be designated as being composed largely of "channels" (Figure 9–17).

The lower sandstone section consists of thinly interbedded sandstone and shale. The sandstones are more complete sequences of the *ABCDE* and *ABCE*. Beds average 0.5 ft (0.15 m) in thickness, but one complete set is 4-ft (1.2-m) thick. Interbedded shale is relatively abundant and contains thin ripple lenses of the *CE* and *CDE* types only 1 to 2 cm in thickness. The total aspect of this section is a middle-

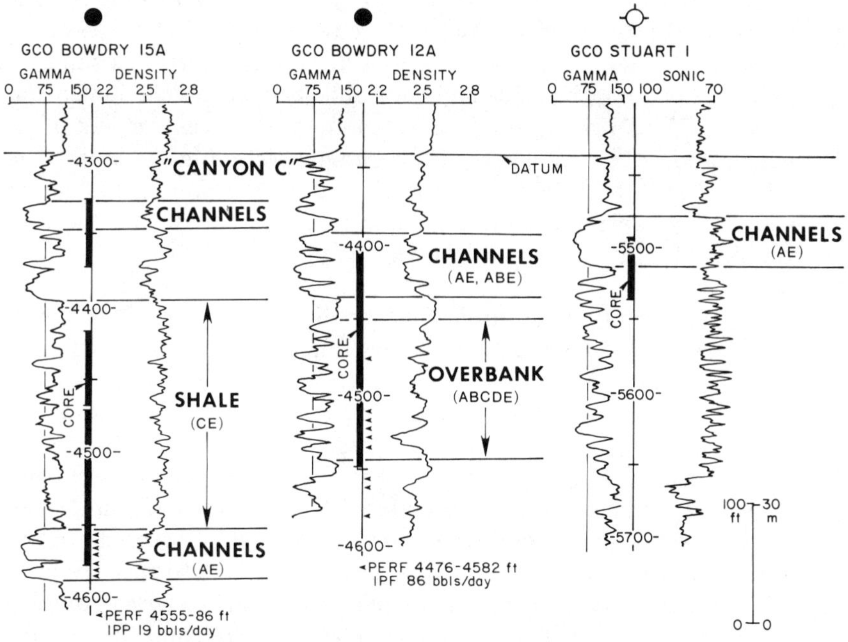

Figure 9–17 Comparison of well logs and cored sections of Canyon "C" sandstones, Roundtop area. Bowdry 12A and Bowdry 15A wells are located in the Tolar field, and Stuart 1 well is located south of Roundtop (Figure 9–16).

fan association, but it is highly shaly and completely lacking in channel-type beds. The section is designated "overbank" to denote its probable location as lateral to channel deposits.

The second core examined was from the Bowdry 15A well located only 1500 ft (458 m) northeast of Bowdry 12A (Figure 9-16). Groups of channel-dominated sections are found at two levels—4330–4347 ft (1321–1316 m) and 4552–4587 ft (1389–1399 m). The channel sections consist largely of *AE* beds that average only 0.5 ft (1.2 m) in thickness. A major part of the cored section consists largely of shale with thin ripple lenses of sandstone. This shale is at a level equivalent to the overbank section in the Bowdry 12A well and indicates an abrupt change in facies between the two locations (Figure 9–17).

A third core was examined from the Stuart 1 test, a dry hole located south of the Roundtop field. This core contains channel beds of the *AE* and *ABE* types (Figure 9–17) and is important in the interpretation of Canyon sandstones because it displays in detail the character of channel beds.

The Canyon cores, then, show three distinct bed associations: (1) generally thin multiple-channel sandstones, (2) interbedded shale and thin, more-complete bed sets that represent overbank deposits, and (3) dominantly shale sections that contain thin rippled sandstones. These facies are distinctly different in their response on the gamma-ray logs because of their differences in amounts of interbedded shale. These differences are shown in relation to the gamma reading of 75 API units (Figure 9–17). Stacked channel sandstones have readings that are less than 75 API units, and about 70 percent of these sections are indicated as "net sand." Overbank sections of thinly interbedded sandstone and shale have only 30 percent of the section less than 75 API units. Shale sections have 20 percent or less of net sand. Using these criteria, the major part of the Canyon section in the Stuart 1 test is indicated to be overbank because it contains an average of about 23 percent net sand (Figure 9–17). Only the upper part of the section was cored, and the presence of more complete *ABC* bed sets confirms the gamma-log criteria.

Sedimentary structures. Most oil is found in channel sandstones. Thicker bodies of multiple-channel sandstones range from 20 to 40 ft (6 to 12 m) in gross thickness, and bed sets commonly range from 1 to 6 ft (0.3 to 1.8 m) in thickness with a maximum of 12 ft (3.7 m) in the lower part of the Bowdry 15A core.

Beds are of the *AE* and *ABE* types, and the massive *A* sandstone is dominant within bed sets. The laminated *B* sandstones, when present, are commonly marked by abundant fine carbonaceous material along laminae [Figure 9–18(D)]. The basal contacts of massive *A* sandstones are sharp on thin underying *E* shales [Figure 9–18(C) and (E)], and in some cases, the *E* shale is deformed by loading [Figure 9–18(F)]. Shale clasts, either small angular fragments or larger rounded pebbles, may be found at the base of *A* sandstones [Figure 9–18(C) and (F)]. Small clasts or zones of coarse grains may also be present in the middle or upper parts of *A* sandstones [Figure 9–18(A) and (B)].

More complete sequences of the *ABCDE* type are present in overbank sections [Figure 9–18(G)]. These beds are commonly thin but may reach a maximum thickness of 4 ft (1.2 m). Some beds are of the *ADE* type in which the laminated *D* siltstone contains abundant fine carbonaceous material in discontinuous laminae [Figure 9–18(H)] similar to the *B* laminations. The *D* divisions, in these cases, are much finer grained and contain a greater clay content than the *B* divisions.

Petrography. Channel and overbank sandstones have an average mean grain size of 0.16 mm (fine grained), and maximum size ranges up to 0.75 mm (coarse grained) and averages 0.60 mm (Table 9–3). Within channel sections, individual beds commonly show textural gradation from fine-grained massive *A* sandstones to very fine-grained laminated *B* sandstones (Figure 9–19).

The sandstones have an average composition of 71 percent monocrystalline quartz, 3 percent feldspar, 18 percent rock fragments, 2 percent other grains, and 6 percent matrix (Table 9–3). Feldspars are mostly orthoclase, and many grains are highly altered and sericitized. Rock fragments are dominantly polycrystalline quartz and chert with lesser numbers of carbonate, siltstone, and shale fragments. Matrix is composed largely of clay minerals and is generally present in small amounts, except for the channel sandstones of the Stuart 1 well where the matrix averages 19 percent. Much of this matrix may be authigenic in origin. Although the carbonaceous laminae are conspicuous in some laminated beds, the total volume is small.

The detrital materials indicate that the sandstones are lithic quartzarenites. Rock fragments are dominantly of sedimentary origin, which suggests that the source terrain was composed largely of older sedimentary rocks with little or no exposure of igneous and metamorphic rocks.

Cements are an important part of the bulk volume and average 11 percent. Calcite is present in nearly all samples but commonly amounts to less than 10 percent. In a few samples, however, calcite is abundant and comprises about 40 percent of the bulk composition. Calcite was probably derived by solution of original detritus, either limestone or fossil fragments. Silica is present in most samples in the form of overgrowth on quartz grains, but its average amount is generally small.

Porosity and permeability. Reservoir properties are determined largely by bedding types, and sections of channel deposits that contain thick massive *A* sandstones tend to have the highest porosities and permeabilities (Figure 9–20). In the Bowdry 15A core, channel sandstones have an average permeability of 12.5 md and an average porosity of 16.4 percent. In the Bowdry 12A core, thicker channel sandstones have an average permeability of 8.7 md and an average porosity of 17.1 percent. In both cores, thinner channel sandstones of the *ABE* and *ADE* types have lower permeabilities. All of the channel sandstones have significant oil saturations.

Some channel sandstones have no oil saturation and show reduced permeabilities as, for example, in the Stuart 1 core where channel sandstones have an average permeability of 0.4 md and an average porosity of 9.6 percent (Table 9–3).

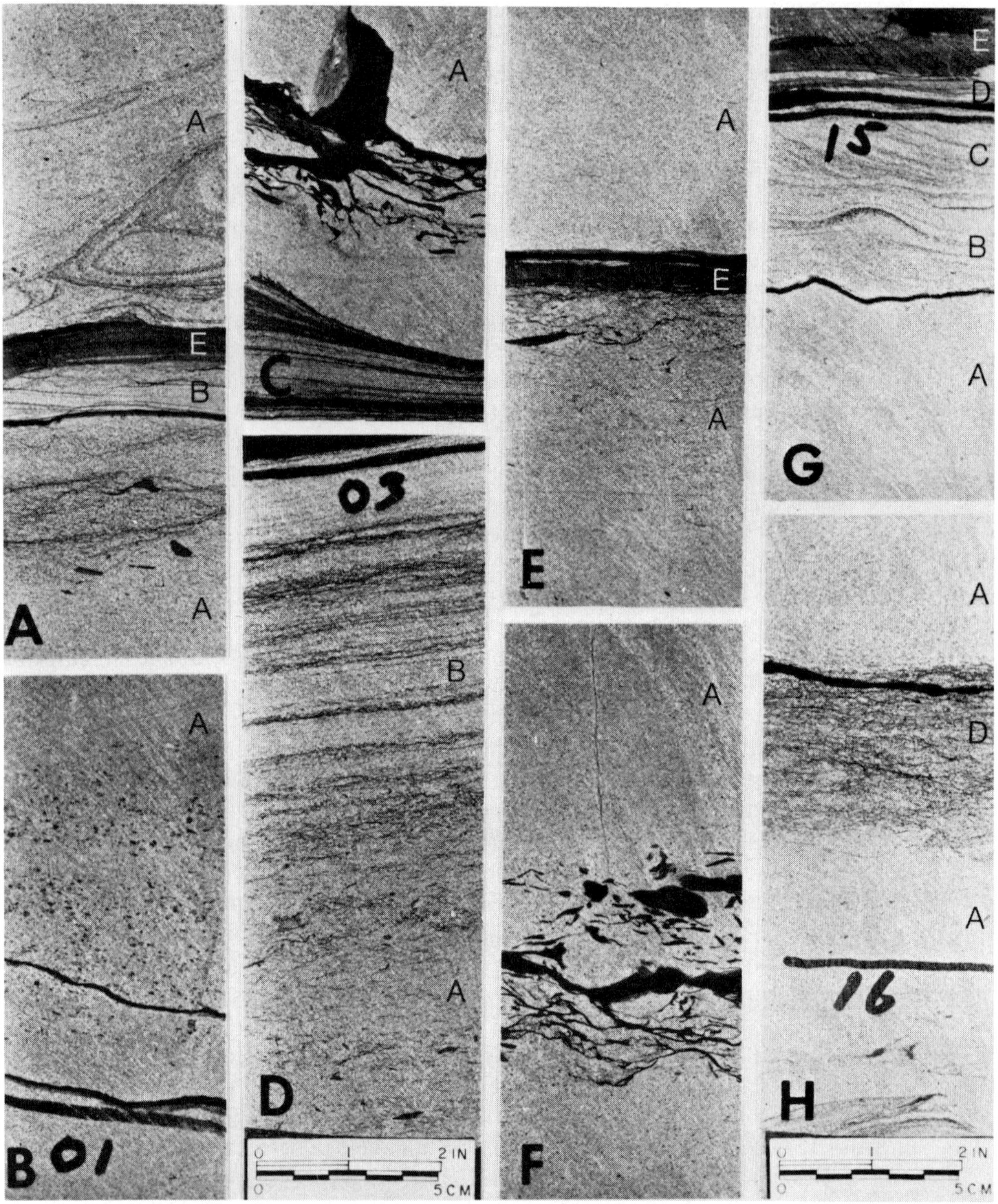

Figure 9–18 Sedimentary structures in Canyon sandstone, General Crude Stuart 1, Roundtop area. Photographs are identified by boldfaced letters at lower left; turbidite bed divisions are designated by small letters at right: (A) upper part of *ABE* turbidite set and overlying massive *A* sandstone showing "swirl" structure—5499 ft (1677.2 m); (B) middle part of massive *A* sandstone showing zone of coarse grains—5501 ft (1677.8 m); (C) base of massive *A* sandstone loaded on underlying shale and with contorted shale clasts—5502 ft (1678.1 m); (D) top part of *ABE* set showing abundant carbonaceous material marking laminated *B* division—5503 ft

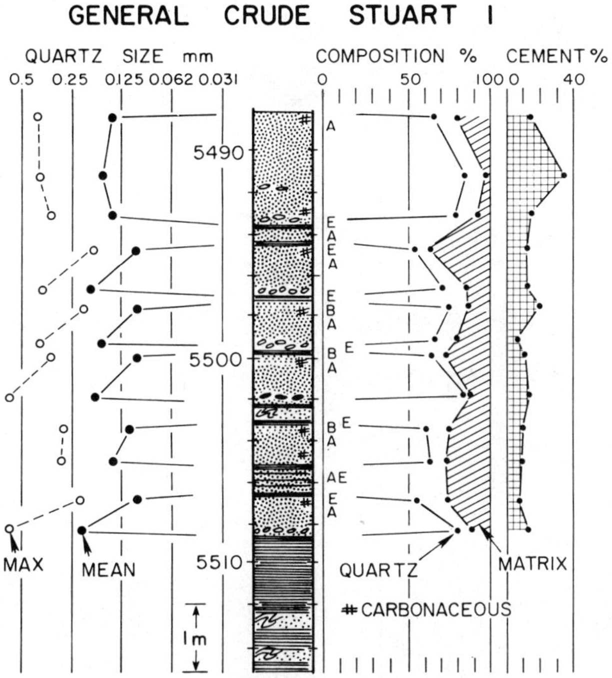

Figure 9–19 Texture, composition, and sedimentary structures of channel sandstones in the Stuart 1 well south of Roundtop field.

These sandstones have a relatively high matrix content of 19 percent, which probably accounts for the reduced porosity and permeability.

Thinner overbank sandstones generally have lower porosities and permeabilities than channel sandstones in the same section. The lowest values are found in thin ripple-laminated sandstones of the *CE* type where permeabilities of less than 0.1 md are probably the result of finer grain size and large content of clay matrix.

Morphology

The local distribution of facies can be mapped by the interpretation of well logs. The spontaneous potential (SP) curves of older wells reflect bed associations in the same way that gamma logs characterize sections in the cored wells (Figure 9–21). Stacked channel sandstones show the greatest departure from the shale base line of

(1678.4 m); (E) top of *AE* bed set with shale clasts and fine carbonaceous material in top of massive *A* sandstone—5506 ft (1679.3 m); (F) base of massive *A* sandstone with shale clasts and load structure on top of contorted sandstone and shale of underlying beds—5508 ft (1679.9 m); (G) complete turbidite bed set of the *ABCDE* type showing contorted laminations in the *B* division—5515 ft (1682.1 m); (H) turbidite of the *AD* type showing discontinuous wavy carbonaceous laminae of the *D* division; bed set truncated and overlain by massive *A* sandstone—5515 ft (1682.4 m).

TABLE 9-3. SUMMARY OF PROPERTIES FOR "CANYON" SANDSTONES, ROUNDTOP AREA

Well	Bed types	Depth (ft)	Number of samples	Quartz size[a] Mean (mm)	Max (mm)	σ (mm)	Detrital composition[b] Qz (%)	F (%)	Rx (%)	Oth (%)	Mx (%)	Cement[c] Sil (% of total)	Cal (% of total)	Permeability (md)	Porosity (%)
Bowdry 12A	Channels	4400–4436	8	0.16	0.46	0.07	65	5	25	2	3	1	3	2.6	16.2
Bowdry 15A	Channels	4333–4372	10	0.16	0.75	0.08	74	2	19	1	4	2	10	3.8	14.9
Bowdry 15A	Channels	4554–4574	7	0.19	0.68	0.08	77	2	19	—	2	4	7	11.0	15.4
Stuart 1	Channels	5488–5508	13	0.14	0.61	0.06	70	3	8	3	19	8	5	0.4	9.6
Bowdry 12A	Overbank	4474–4530	9	0.18	0.50	0.08	71	4	22	2	1	3	12	2.4	13.1
Bowdry 12A and 15A	Shale (*CE*)	—	3	0.08	0.25	0.04	66	3	9	1	21	1	4	<0.1	10.1
Means (channel and overbank)				0.16	0.60	0.07	71	3	18	2	6	4	7	3.5	13.4

SOURCE: Petrography by W. M. Ahr, D. G. Kersey, and T. Tieh.

[a] Long-axis measurements; σ = standard deviation.

[b] Qz = monocrystalline quartz, F = feldspars, Rx = rock fragments including polycrystalline quartz, Mx = matrix, Oth = other minerals.

[c] Sil = silica as overgrowths, Cal = calcite.

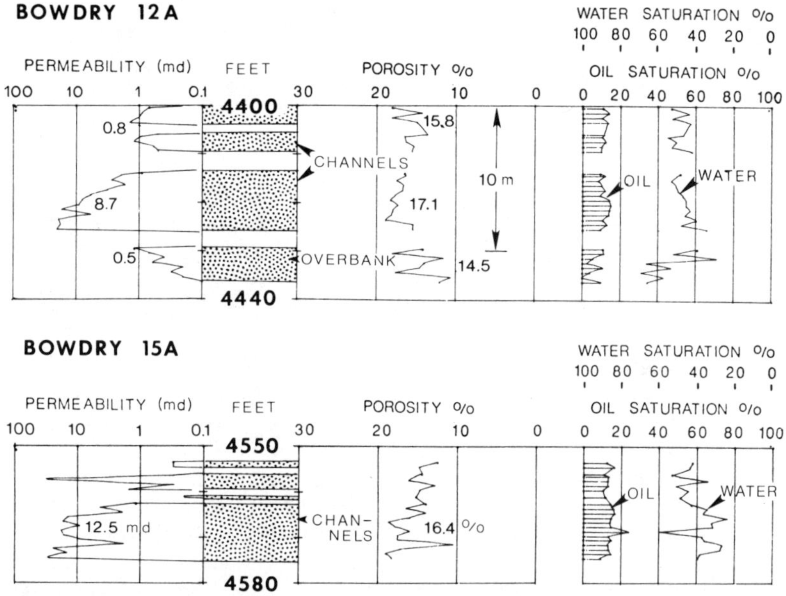

Figure 9–20 Permeabilities, porosities, and fluid saturations in selected sections of Canyon sandstones, Bowdry 12A and 15A wells, Tolar field.

about 60 mV or more; overbank sandstones and shales show a response of between 40 to 60 mV; and sandy shale units show 40 mV or less. These characteristic curves can then be identified in nearby wells (Figure 9–21).

The Canyon "C" section is divided into four zones numbered from the top downward. These units show rapid change laterally. For example, the lowermost zone 4 shows abundant channel sandstones that apparently are interbedded with thin overbank sandstones and shale. Zone 3 shows a lateral succession from shale in the Bowdry 15A to overbank in Bowdry 11A and then to stacked channels in Bowdry 8A (Figure 9–21). This change takes place within short distances between wells of only 800 and 400 ft (244 and 122 m), respectively. Zone 2 is again dominated by channels with interbedded shales in the Bowdry 15A and 11A wells, which are replaced by overbank deposits in Bowdry 8A before the zone grades to dominant shale to the southeast in the Bowdry 3A and 5A wells. The uppermost zone 1 is very thin and appears to consist largely of shale.

Abrupt lateral changes suggest that the sandy zones represent constructional channel systems in which the channelways were closely bounded by overbank deposits, probably as levees next to narrow channelways. Farther from the channels, sandy shales replace both facies, as in zone 3, or even enclose channel deposits, as in zone 2 (Figure 9–21). Between the sandy zones only clean shales were deposited when flow paths were temporarily abandoned.

Channel configurations are shown by maps of net porous sandstones as de-

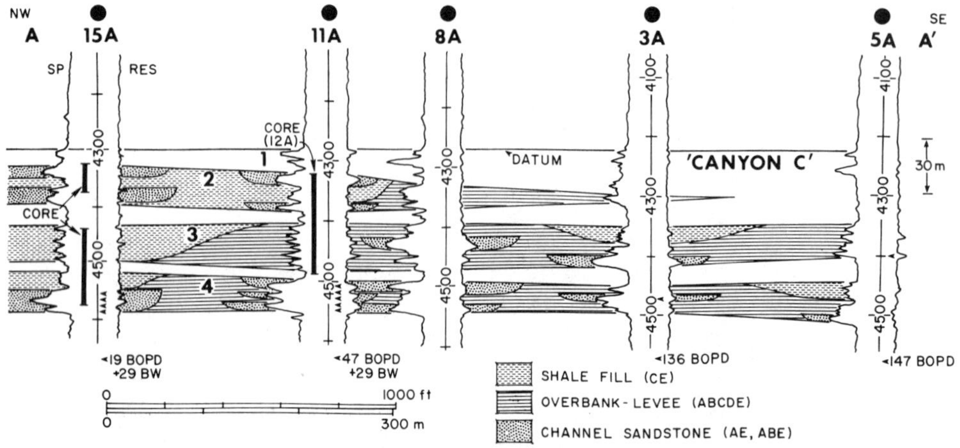

Figure 9–21 Correlation of electric logs and interpretation of constructional channel-fill and levee deposits in Tolar field. Cored section in Bowdry 12A is equivalent to section in Bowdry 11A. Location of profile shown in Figure 9–16.

termined from micrologs. Zone 4 has the greatest thickness of porous sandstone in this local area (Figure 9–22). The map is contoured to emphasize the presence of isolated thin channel-fill bodies as suggested by the correlation of logs (Figure 9–21). The exact configuration of individual channels is, of course, difficult to interpret from the existing well control, and single channels may be more extensive than shown. The higher zones have fewer porous sandstones. Zone 3, for example, also has channel-fill deposits at several levels (Figure 9–22), but their distribution is greatly restricted. The map of zone 3, however, again emphasizes that rapid lateral change from channel sandstone in Bowdry 8A to overbank deposits in Bowdry 11A and shale in Bowdry 15A (Figure 9–21).

The abrupt lateral changes in facies are unlike those of the submarine-fan system. The Canyon channel sandstones are, in gross aspect, thin isolated sandstone bodies within sections that are dominantly shale. These sandstones are highly variable in thickness and extent and disappear abruptly within short distances. In contrast, submarine fans show interbedded sandstones and shales in definite units that are more persistent laterally. Groups of sandstones within fan systems are generally correlatable for several miles.

Similar sandstones are present in the Cisco Group in many small fields on the east flank of the Midland basin, but the Jameson field shows an unusually large concentration of channel sandstones. In the Jameson field (Figure 9–23), the gross morphology of these sandstones has been mapped (Bloomer 1977). Net sandstone thickness shows a broadly lobate pattern that has a width of about 5 mi (8 km) and extends downdip for about 20 mi (32 km). Within this elongate body the Cisco sandstones are thin-bedded and composed largely of massive *AE* and *ABE* se-

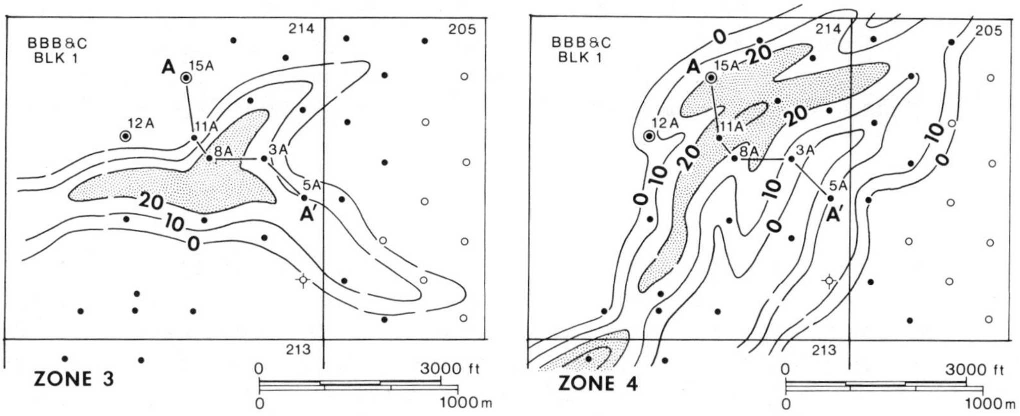

Figure 9–22 Net porous sandstone thickness in zones 3 and 4 of Canyon "C" unit, Tolar field. Thicker sections are multiple stacked channel sandstones closely bounded by overbank deposits. Contour interval 10 ft (3 m).

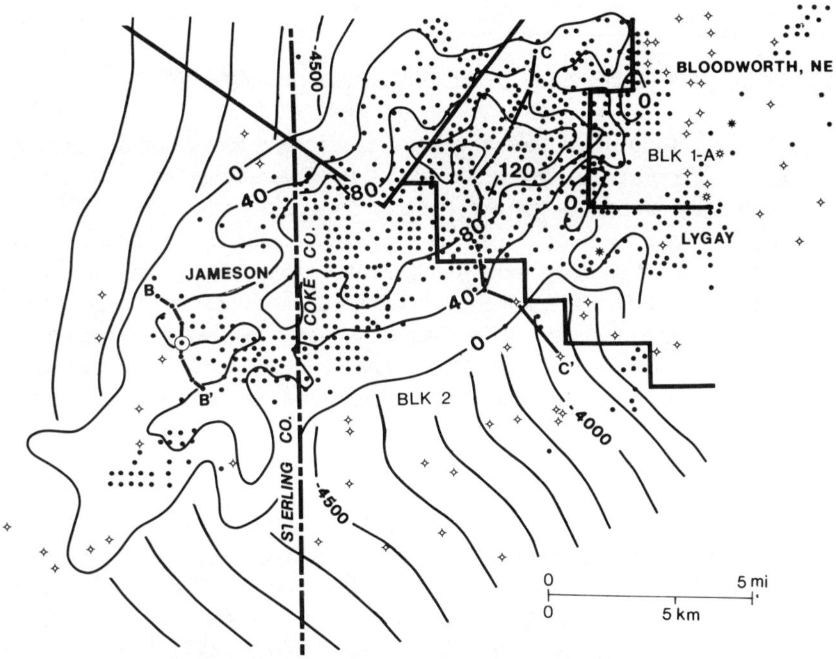

Figure 9–23 Net sandstone thickness of Canyon sandstones in the Jameson field. Contour interval 40 ft (12 m). [From Jones 1980 and Bloomer 1977.]

quences (Jones 1980, 1982). These individual channel sandstones form narrow sinuous trends within the composite lobate body, and the channels have local configurations similar to those in the Roundtop field (Figure 9–22). Thus, the channel sandstones at the Jameson field form a complex of thinly interbedded channel sandstones and shales.

Conclusions

Reservoir sandstones of the Cisco Group are largely channel turbidites. They are thinly bedded and occur as multiple stacked sandstones that are abruptly replaced laterally by overbank deposits and basin shales. These sandstones are interpreted as the fill of narrow channelways that were closely bounded by levees and are termed *constructional channel-fill sandstones.* The channelways may have been routes for transport of large volumes of sand carried from the slope to deeper parts of the basin by turbidity currents. The channel-fill deposits that remain, however, are generally sand-poor and are dominated by shale. The channel systems were later overridden by finer-grained sediments of the prograding slope that continued to advance toward the basin as broad aprons of shale. Recognition of constructional channel-fill deposits may provide evidence for larger masses of submarine fans in other basinal locations.

STEVENS SANDSTONE, SAN JOAQUIN BASIN, CALIFORNIA

The San Joaquin Valley of central California has large oil and gas resources that have been produced from rocks of Late Cretaceous to Pleistocene in age. A major producing section is the Stevens and equivalent sandstones of Upper Miocene age, which have long been known as deep-water sediments, and the reservoir sandstones produce oil primarily from stratigraphic traps. The primary source of sand was to the east, and the sands were carried from adjacent highlands across a narrow shelf and then into a deep-marine basin by turbidity flows to form a major part of the Stevens sandstones. Other source areas to the south supplied sands of differing compositions to more limited reservoir sections. In both cases, however, the reservoir sandstones have composite channel-form trends.

Geologic Setting

The southern end of San Joaquin basin is a deep narrow syncline. The basin is bounded by the San Andreas fault and parallel folds on the southwest, the Tehachapi Mountains and associated thrust faults on the southeast, and the gentle flank of the Sierra Nevada uplift on the northeast (Figure 9–24). The large anticlinal structures of Elk Hills, Buena Vista, and Paloma in the southwestern part of the basin are nearly parallel to the San Andreas fault. A broadly nosing structure called the Bakersfield arch trends southwest across the basin.

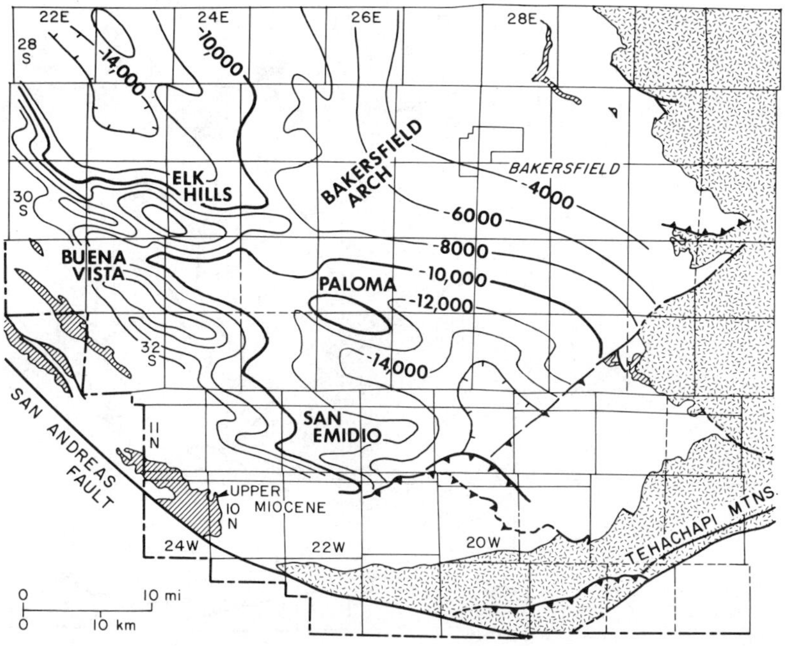

Figure 9–24 Generalized structure map on the "N" marker at the top of the Stevens sandstones and equivalent units. Contour interval 2000 ft (610 m). [Adapted from Webb 1981, p. 441.]

Upper Miocene rocks of the Mohnian Stage form a thick body of sandstone and shale across the entire basin. Maximum thickness is more than 4000 ft (1200 m), and the major part of this section is of deep-water origin. In the center of the basin, water depths are believed to have been greater than 2000 ft (610 m) (Bandy and Arnal 1969). On the eastern side of the basin, a narrow shelf facies is present in the shallower part of the section, and this facies comprises the Santa Margarita and Chanac formations. The basinal deep-water facies consists of the Stevens sandstone throughout much of the central area and its equivalent Monterey or Antelope Shale on the southwestern edge.

The Stevens sandstone and interbedded shale has a gross thickness of 3000 ft (915 m) and contains a net sandstone thickness of 2000 ft (610 m) according to Sullwold (1960). This great body of sandstone ends abruptly to the southwest along a line that crosses the east part of the Elk Hills anticline and extends southeastward down the present synclinal axis of the basin (Figure 9–25). The Stevens has been called a submarine-fan complex that had at least four different sources (MacPherson 1978). Three sources to the northeast provided sediment for the Rosedale, Fruitvale, and Bena fans, while a fourth source to the southeast provided sediment for the Valpredo fan. Sediment was supplied to the basin through narrow submarine canyons, or "gorges" in local terminology, which were cut deeply into the shelf

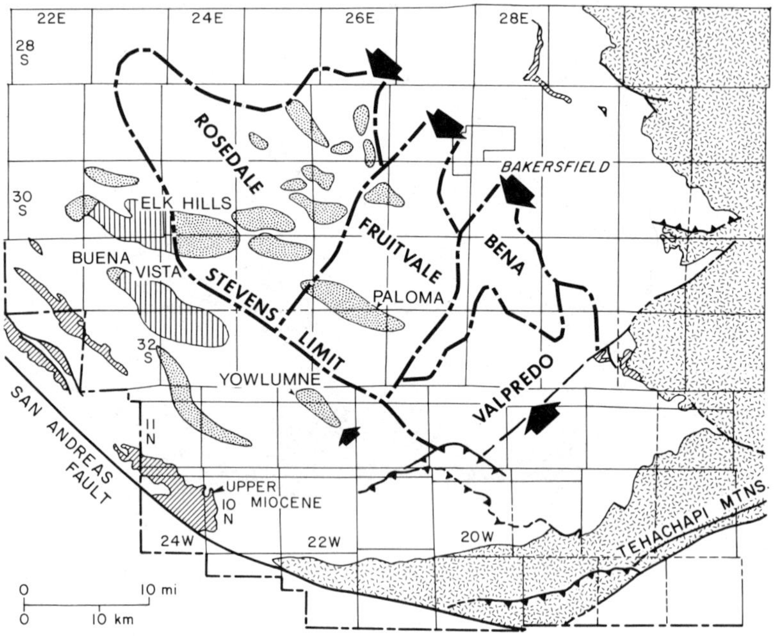

Figure 9–25 Distribution of Stevens sandstone in the southern San Joaquin basin showing oil fields in Stevens reservoirs (dotted) and in fractured Monterey shale (vertical lines) beyond the Stevens. [Adapted from MacPherson 1978.]

margin. Southwestward across the Bakersfield arch, the Stevens sandstone can be divided into four separate prograding units, each of which passes southwestward into siliceous shales of the basin facies. Furthermore, the fan facies were confined to the downthrown sides of major normal faults that trend to the northwest, and sediments across the faults show basinward thickening similar to the growth faults of the Gulf Coast province.

Beyond the limits of the Stevens sandstones are narrow channellike bodies that have different compositions. These sandstones have trends that generally parallel the southwestern margin of the main body of the Stevens such as the reservoir at Yowlumne field (Figure 9–25). At least some of these sandstones may have had their origin from a source area toward the south or southeast. Other isolated sandstones, such as those in the vicinity of Elk Hills, have compositions similar to the Stevens and may be extensions from an eastern source.

All the basinal sandstones were contemporaneous, and hence the stratigraphic relations among the sandstone units are complex. Correlations show lateral interfingering of sands from the different sources so that several "fans" have gradational boundaries. More important, compositional differences among the sandstones control porosity and permeability at depth. Therefore, an understanding of local depositional patterns, as well as sources, may be an aid to future exploration in the southern San Joaquin basin.

Stevens Oil Production

The Stevens sandstones and equivalent rocks have been estimated to contain reserves of more than 4 billion barrels (640 million m^3) of oil (Callaway 1971), and more than 2 billion barrels (320 million m^3) have already been recovered. Other horizons have contributed to the total of 4.3 billion barrels (680 million m^3) of oil that have been produced from more than 41 fields (MacPherson 1978, p. 2272) in the area. Thus, the Stevens and equivalent sections account for about half the total production from this area.

The Elk Hills structure accounts for a major part of the reserves from the Stevens sandstones on the eastern anticline, whereas fractured siliceous beds of the Monterey Shales have substantial reserves on the west. Smaller anticlines include the North and South Coles Levee fields, the Ten Section, and the Paloma fields. Other smaller fields on the Bakersfield arch are more subtle features, perhaps with structural control that originated largely by differential compaction over stacked reservoir sandstones (Dorman 1980). Other oil accumulations on the Bakersfield arch are either stratigraphic traps or partly controlled by faulting. Lower Stevens sandstones form reservoirs within a submarine canyon at Rosedale field (Martin 1963). Within the Stevens complex, only the large Paloma anticline has not been a major producing area.

Beyond the main body of Stevens sandstone, isolated sandstones produce oil at Elk Hills and in the Taft area (Callaway 1968). A similar isolated sandstone forms the principal reservoir in the recently developed Yowlumne field in the southwest part of the basin. At Yowlumne, the complex nature of these reservoirs is well known (Taylor 1978; Dorman 1980).

Yowlumne Field History

The development of the Yowlumne field in the southern San Joaquin basin revealed an isolated reservoir body composed of stacked channel sandstones. The reservoir, however, contains few volcanic rock fragments unlike the normal Stevens, and its source was to the south rather than from the Sierra Nevada uplift to the east. Nevertheless, the Yowlumne channel sandstones present a complex of channel-fill deposits within a single channel, probably similar to those that exist in the Stevens but which cannot be defined within the thick section of multiple sandstones. Therefore, the Yowlumne section presents an unusual view of channel reservoir sandstones that may be considered typical of other channel deposits in the San Joaquin basin.

The Yowlumne area has had a long history of exploration and development (Taylor 1978). The field is located on an eastward plunging structure known as the San Emidio nose (Figure 9–26). Geophysical exploration began in 1934, and the first dry hole was drilled in 1935. Later drilling resulted in a total of 10 dry holes. Three fields finally were discovered: San Emidio nose in 1958 (Bazeley 1972), Yowlumne in 1974, and Rio Viejo in 1975. The optimism and persistence of many geologists and geophysicists made these discoveries possible.

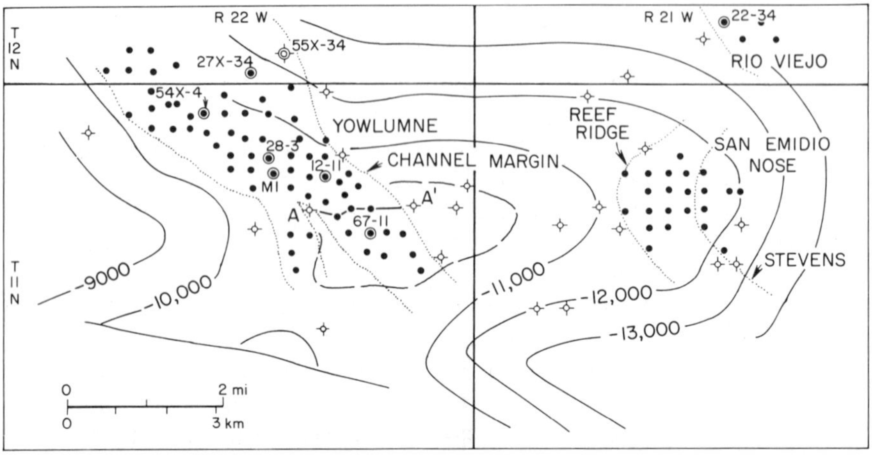

Figure 9-26 Generalized structure map of the San Emidio Nose showing the location of oil fields and trend of the Yowlumne channel reservoir. Circled wells are those in which Yowlumne sandstone was cored. Contour interval 1000 ft (303 m). [Adapted from Taylor 1978.]

The Yowlumne discovery well was completed for 800 bbls/day (127 m^3/d) from the zone now known as the Yowlumne sandstone at a depth below 11,000 ft (3353 m). Subsequent drilling established production in basal Pliocene sandstones at a depth of about 10,300 ft (3139 m) and in a lower zone called the "10–4" sandstone at a depth of about 11,700 ft (3566 m). By October 1977, a total of 58 wells had been completed in the field.

The Yowlumne reservoir is a lenticular channellike body that trends northwestward across the San Emidio structure. The channel has a maximum thickness of about 400 ft (122 m) and ranges in width from 1.0 to 2 mi (1.6 to 3.2 km). The channel-fill deposits do not appear to have occupied an eroded channel but seem to have been deposited in a topographic depression (Taylor 1978).

Details of the reservoir section were observed in cores from three wells. The cored sections ranged from 116 to 209 ft (35 to 64 m) in length and together represented nearly the entire Yowlumne sandstone. Typical sections are described from a core in the Tenneco 54X–4 well (Figure 9-27). In this well the total thickness of Yowlumne sandstone is about 200 ft (61 m), which is an average reservoir thickness for the field. The gamma log indicates the dominantly sandy nature of the sequence with prominent shales restricted to a thin unit in the upper part of the section.

Description of Sandstones

Sedimentary structures. The Yowlumne sandstone consists of turbidites that range from about $\frac{1}{2}$ to 9 ft (0.15 to 2.7 m) in thickness, and the average bed thickness is on the order of 2 ft (0.6 m). The sandstones consist largely of massive

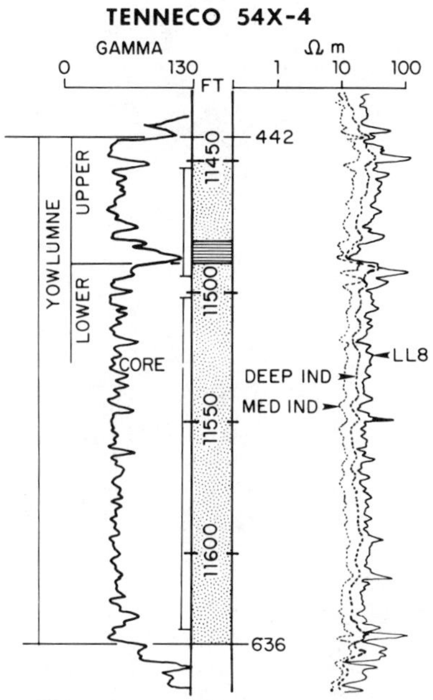

Figure 9–27 Well logs and cored interval in Yowlumne sandstone, Tenneco 54X–4. Location of well shown in Figure 9–26.

A divisions separated by thin beds of *E* shale. Less commonly, a massive *A* division is succeeded by laminated *B* and rippled *C* sandstones.

Most of the beds are of the *AE* type in which the massive *A* sandstones are succeeded by *E* shale, either abruptly or with a thin zone of gradation [Figure 9–28(A)]. The gradational zone may be only a few centimeters in thickness and is characterized by fine-grained angular fragments composed of carbonaceous or other organic material. Less commonly, the massive *A* sandstone is overlain by a thin section of fine-grained laminated *B* sandstone [Figure 9–28(C)]. The *B* division is characterized by dark laminae and also may contain small angular fragments of carbonaceous material. The *B* division may be succeeded abruptly by black shale, or more rarely, there appears to be a gradation upward to sandy shale. The fine-grained rippled *C* division is exceedingly rare and can be identified with certainty only in a few thin beds.

Many beds show contorted laminae in a thin zone immediately above the massive *A* sandstones. Contorted laminae may show simple folding, in which case the section retains the character of a laminated *B* sandstone [Figure 9–28(D)]. In other cases the contortion is somewhat more severe, and folded laminae are overturned, in which case the identification of a laminated *B* sandstone is less definite. In other beds, zones of more highly contorted laminae are found, and the original attitude of internal laminae is difficult to determine. The contorted zone may retain a relict character with internal steeply dipping micro laminae, which suggests that the original bedding may have been rippled *C* sandstone [Figure 9–28(F)]. The contorted beds may be succeeded by undisturbed black shale, and rarely there is an overlying

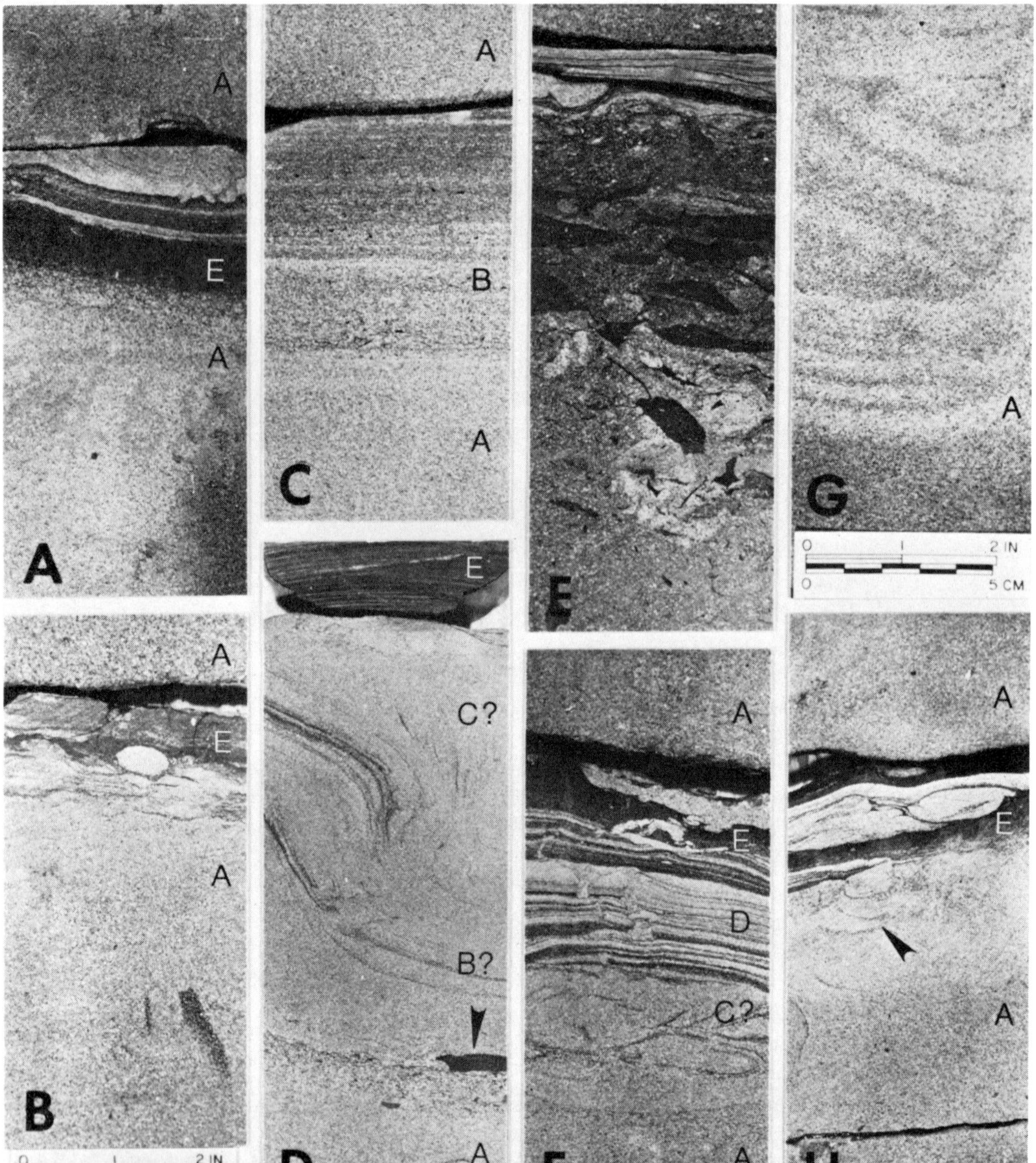

Figure 9–28 Sedimentary structures in Yowlumne sandstone, Tenneco 54X–4. Photographs designated by boldface letters at lower left; turbidite division identified by letters at right: (A) top of *AE* sequence showing gradation upward of massive sandstone to thin *E* shale loaded by overlying massive *A* sandstone—11,467.5 ft; (B) thin *AE* unit with disturbed bioturbated (?) top overlain by *E* shale and coarse-grained *A* sandstone— 11,507 ft; (C) massive *A* sandstone grading up to laminated *B* sandstone and thin *E* shale; note small black carbonaceous

thin section of laminated shale and sandstone that may be designated a *D* division [Figure 9–28(F)].

Most of the contorted bedding appears to have been the result of soft-sediment slumping. This conclusion is suggested by the fact that the contorted beds are overlain by a thin *E* shale that shows relatively little deformation. Although the *E* shale may have a dip up to 10°, it does not appear to be severely deformed as might be expected in the case of loading by overlying sediments. In some beds the massive *A* sandstones lack an overlying *E* shale, apparently due to truncation by overlying massive *A* sandstone.

In two places the massive *A* sandstones exhibits "dish structure" in which indistinct laminae are broadly curved, concave upward, and appear to be discontinous and broken [Figure 9–28(G)]. Dish structure is present in thin intervals about ½ ft in thickness in the middle of massive *A* sandstones and can be explained as the result of fluid escape on dewatering of the sediment, resulting in broken and upturned laminae.

Bioturbation is exceedingly rare in the cored section. Only one bed was found in which massive *A* sandstone is succeeded by finely disturbed shaly sandstone [Figure 9–28(B)]. This thin section also contains a distinct ovate burrow tube. Thus, the finely churned appearance of the bed may also be due to bioturbation.

Interbedded shales are black and massive, and thick shale units commonly show conchoidal fracture consistent with the massive character [Figure 9–28(A)]. The shales may contain very fine sand and silt, rarely in thin laminae and more commonly in distinct zones. These sandy zones are less than 0.5 ft (0.15 m) in thickness and have scattered coarse sand grains and large angular shale clasts [Figure 9–28(E)]. The sandy zones appear structureless or have contorted bedding or a finely "swirled" appearance. These sandy muds may have originated as highly viscous debris flows.

Petrography. The Yowlumne sandstone has a mean grain size of 0.29 mm (medium grained) and an average maximum size of 1.02 mm (very coarse grained). Most beds show normal textural grading, or fining upward, within massive *A* sets (Figure 9–29), and grading is commonly expressed in both the mean and maximum grain sizes. Laminated *B* and rippled *C* units are generally fine grained and, when present, emphasize the textural grading within sets. Some thick sandstones may show reverse grading, or coarsening upward, through most of the bed as in the section illustrated (Figure 9–29, right). In this case, a zone of scattered coarse grains up to 3 mm in diameter appears to be the base of a massive unit (11,595 ft) that overlies

---

clasts—11,529.5 ft; (D) contorted laminated fine sandstone overlying massive *A* sandstone; note large shale clast (arrow)—11,533 ft; (E) sandy mudstone representing viscous flow with large clasts of black shale—11,478 ft; (F) contorted laminated fine sandstone overlying massive *A* sandstone and succeeded by interlaminated shale and sandstone—11,571.5 ft; (G) dish structure in massive *A* sandstone—11,590 ft; (H) top of *AE* unit showing disturbed finely churned structure (arrow) and contorted ripple (?) lens of sandstone in *E* shale—11,615.5 ft.

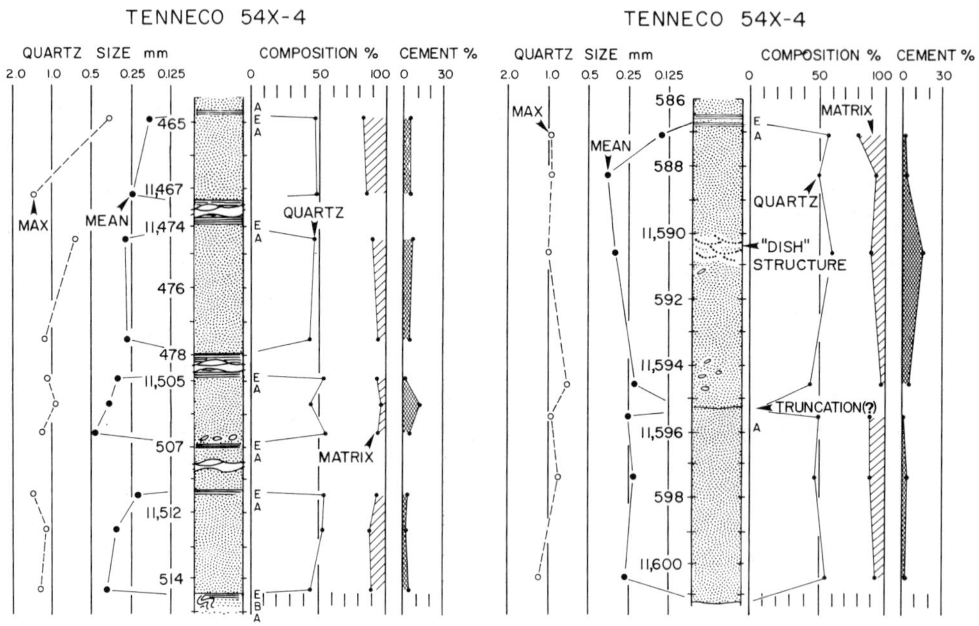

Figure 9–29 Texture and composition in massive bed sets of the Yowlumne sandstone, Tenneco 54X-4, showing normal graded beds (left) and thick beds with reverse grading (right).

another massive sandstone without an intervening *E* shale. Despite the coarse grains, the mean size is fine grained at the base and increases upward to medium before decreasing again at the top of the beds.

Composition is that of an arkose with total feldspar of 37 percent and quartz content of 50 percent (Table 9–4), and the rock may be termed an orthoclase arkose to denote that potassium feldspar grains are about twice as common as plagioclase. Rock fragments are only 3 percent and of granitic type. Calcite cement occurs in small patches or mottles and constitutes only 6 percent of bulk volume.

Composition of the Yowlumne sandstone contrasts with that of other sandstones in the area (Table 9–4). Immediately underlying the Yowlumne is the "10–4" sandstone, which has a similar feldspar content of 29 percent, but plagioclase grains are about three times as common as potassium feldspar grains. This sandstone may be termed a plagioclase graywacke because of a matrix content of 22 percent. To the east at Rio Viejo field (Figure 9–26), the reservoir sandstone is called Stevens, but it has a composition similar to that of the Yowlumne. These compositions are distinctly different from that of the main body of Stevens sandstone in the eastern part of the basin. At Gosford and Bowerbank fields on the Bakersfield arch, the typical Stevens has a low quartz content of 25 percent and a significant amount of rock fragments of both volcanic and granitic types. Based on this composition, the main Stevens sandstone can be termed a volcanic arenite. The volcanic

TABLE 9–4. AVERAGE PROPERTIES OF SELECTED STEVENS AND EQUIVALENT SANDSTONES, SAN JOAQUIN BASIN, CALIFORNIA

Field/Sandstone	Average depth (ft)	Number of samples	Grain size[a] Mean (mm)	Max (mm)	σ (mm)	Detrital composition[b] Qz (%)	Kf (%)	Pf (%)	VRx (%)	ORx (%)	Mx (%)	Cement[c] (% of total)	Permeability (md)	Porosity (%)
Yowlumne/"Yowlumne"	11,544	34	0.29	1.02	0.20	50	26	11	—	3	10	6	74	18.0
Yowlumne/"10-4"	11,630	6	0.24	0.96	0.17	48	7	22	—	2	22	4	6	14.6
Rio Viejo/"Stevens"	14,134	14	0.34	1.36	0.26	53	22	8	—	4	13	4	31	15.9
Gosford/Stevens	8,109	25	0.33	1.00	0.21	26	—	33	25	16	—	9	372	25.0
Bowerbank/Stevens	11,150	13	0.42	1.30	0.26	24	—	23	26	27	—	7	30	18.4

SOURCE: Petrography by R. L. Findley 1975, and T. T. Tieh.

[a] Long-axis measurements; σ = standard deviation.

[b] Qz = quartz, Kf = potassium feldspar, Pf = plagioclase feldspar, VRx = volcanic rock fragments, Orx = other rock fragments, Mx = matrix.

[c] Cement = calcite.

fragments are dominantly glass, partly devitrified, and highly altered to the clay minerals smectite, vermiculite, and chlorite.

Porosity and permeability. The Yowlumne sandstone has variable porosities and permeabilities, but higher values are present in the lower part of the section where the beds are generally thicker. Average porosity for the total section is 18.4 percent, and average permeability is 74 md (Table 9–4). Maximum permeabilities of about 300 md are common in the lower sandstones, and these values are significantly greater than those in sandstones of other compositions at similar depth.

The role of composition in determining both porosity and permeability in the southern San Joaquin basin is shown by a plot of porosity as a function of depth (Figure 9–30). The volcanic arenites of the main body of the Stevens sandstone maintain good porosities and permeabilities to depths of about 12,000 ft (3660 m), below which permeability is greatly reduced by the distortion of volcanic rock fragments between more competent grains. The volcanic fragments appear to have flowed into adjacent pores, effectively restricting the continuity among pores.

The orthoclase arkoses of the Yowlumne and Rio Viejo sandstones have a lower rate of porosity reduction and retain adequate permeabilities for oil production to depths of at least 15,000 ft (4560 m). The plagioclase graywackes of the "10–4" sandstone have greatly reduced porosities at the same depths, and permeabilities are reduced to low values. The reasons for these differences appear to be due to diagenesis of the feldspars (Tieh et al. 1981). The orthoclase grains are altered in part to kaolinite, which fills pores but does not greatly reduce porosity or perme-

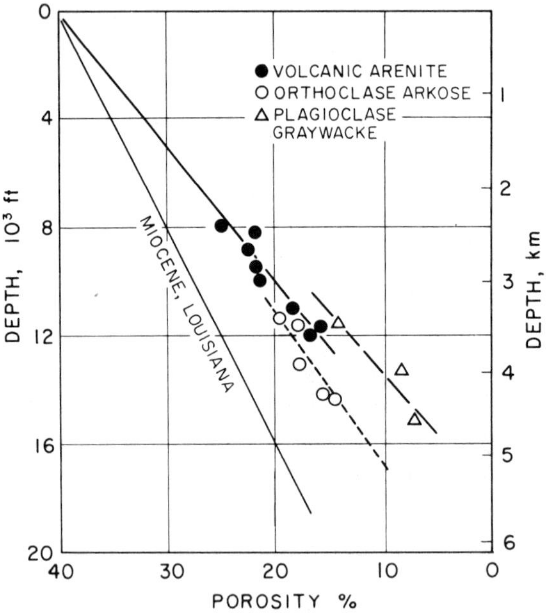

Figure 9–30 Porosity reduction in Late Miocene sandstones, southern San Joaquin basin, illustrating the effects of differing compositions. Reduction of porosity in Gulf Coast Miocene sandstones of south Louisiana are shown for comparison.

ability. In contrast, the plagioclase grains are more highly altered to the expandable clays—montmorillonite and interlayered smectite-illite, which have a larger relative volume in the pores and thereby a greater effect in reducing porosity and permeability.

Interpretation

The Yowlumne sandstone is composed of relatively thick beds of massive sandstone or massive and laminated sandstones succeeded by thin black shales. The bed sets are, therefore, channel-fill turbidites of the *AE* and *ABE* types. Only in the upper Yowlumne are there a small number of more complete tubidite sequences that represent overbank deposits (Table 9–5). Other cores from the Yowlumne sandstone show a similar dominance of massive channel turbidites.

The stacked channel-fill sandstones form a lenticular body (Figure 9–31) that has a maximum thickness of about 250 ft (76 m). The sandstones appear to be inserted in the section with no evidence for channel erosion. Log correlations are based on conspicuous cherty shales, the "N" and "P" cherts, and on distinctive marker beds, "M1" and "M2," and they show that the section thickens abruptly to accommodate the sandstones. The channel margins (Figure 9–31) outline the lateral extent of reservoir sandstones and show a channel about 1 mi (1.6 km) wide across the top of the structure, increasing to more than 2 mi (3.2 km) wide downdip toward the northwest. Within the channel reservoir section, the sandstones appear to form thick overlapping lenses, progressively younger toward the east (Dorman 1980).

Other cores examined in the field were composed of stacked channel sandstones, and there was no evidence of thick levee or channel-margin sediments in any of the reservoir section (Figure 9–26). Judging by the log response in adjacent

TABLE 9–5. BED TYPES, AVERAGE THICKNESS, AND FREQUENCY IN THE YOWLUMNE SANDSTONE, TENNECO 54x-4

		Upper yowlumne		Lower yowlumne				Total	
		(11,453–11,479 ft)		(11,491–11,520 ft)		(11,520–11,630 ft)			
Bed types		Number of beds	Thickness (ft)	Number of beds	Thickness (ft)	Number of beds	Thickness (ft)	Number of beds	(%)
Channel	AE	12	1.5	8	2.0	22	2.4	42	(72)
	A[a]					1	6.0	1	(2)
	ABE	2	2.0	1	1.5	9	3.1	12	(20)
Overbank	ABCDE	1	0.3					1	(2)
	ABCE	1	0.5					1	(2)
	ACE	1	1.1					1	(2)

NOTE: Well logs of cored section are shown in Figure 9–31.

[a]Truncated.

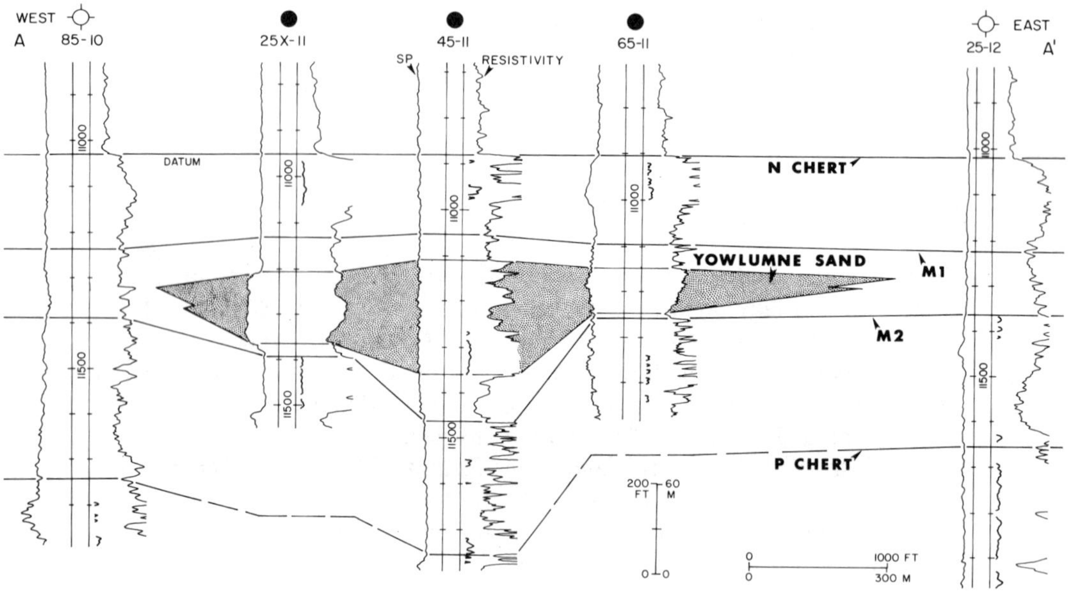

Figure 9–31 Correlation of well logs through the Yowlumne reservoir showing stacked channel-fill turbidites in a lenticular body 250-ft (76-m) thick. Location of correlation section shown in Figure 9–26. [Adapted from Taylor 1978.]

dry holes, the channel-fill sandstones pass abruptly to what appears to be basin-plain mud (Figure 9–31). Therefore, deposition of the channel sandstones may have been controlled by topography of the sea floor (Dorman 1980). The initial deposits were probably confined to the channel, but as additional sand was brought in, the flows may have been partly confined by overbank deposits. When the depression on the sea floor was filled, succeeding flows were diverted to other areas, and the Yowlumne sandstones were covered by basin muds.

Conclusions

The Yowlumne sandstone shows the character of channel turbidites that were deposited in an isolated preexisting channel that was largely the result of bottom topography. Similar channel turbidites are known in the main body of the Stevens sandstone that forms a thick sequence of sediments that have been referred to as "submarine fans" (Figure 9–25). However, an extensive study of many cores from the Stevens (Findley 1975) showed a dominance of channel turbidites of the *AE* and *ABE* types with no evidence for thinner overbank deposits except in the most distal part of the Stevens section. Although the main body of the Stevens sandstone had a different source and composition, the channel character is identical to the Yowlumne, and therefore the submarine fans of the Stevens were probably formed by a succession of stacked channel deposits.

BELL CANYON SANDSTONE, DELAWARE BASIN, NEW MEXICO

Sandstones of the Permian Delaware Mountain Group have long been recognized as deep-water deposits. This thick section of interbedded sandstone, siltstone, and shale occupies the Delaware basin, while correlative rocks of the shelf and shelf margin are largely carbonates and evaporites. The famous Capitan "reef" is exposed in the Guadalupe Mountains, plunges northeastward into the subsurface, and extends along the northern and eastern margins of the Delaware basin (Figure 9–32). Shelf carbonates and evaporites occupy the Central basin platform of West Texas and the Northwest shelf of southeastern New Mexico. Thus, the basinal clastics of the Delaware Mountain Group present a striking contrast in facies to the shallow-water carbonates of the shelf.

In the basin, massive sandstones form narrow sinuous "channels" that extend for many miles southwestward across the basin floor, and these sandstones contain oil in stratigraphic traps (Figure 9–32). Most of the oil fields are also aligned southwestward like beads on a string, obviously controlled by the prevailing orientation

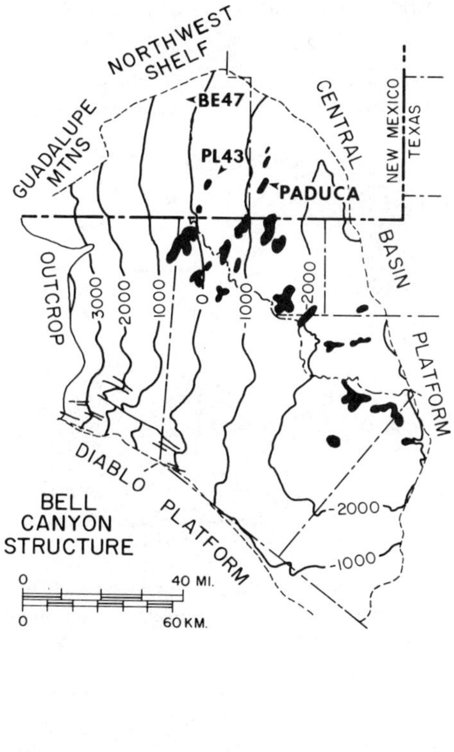

Figure 9–32 Structure map of the Delaware basin, west Texas and southeastern New Mexico, showing structure on top of the Delaware sandstone. Contour interval 1000 ft (300 m). Stratigraphic oil fields in the upper Delaware Mountain group are shown in black. Locations referred to are Big Eddy Unit 47 (BE 47), Poker Lake Unit 43 (PL 43), and Paduca field.

of the sandstone "channels." Thus, the oil accumulations pose questions regarding the origins of the reservoirs and the mechanisms of oil entrapment.

Geologic Setting

The general stratigraphic relationships of Guadalupian rocks (King 1942) are well known from surface and subsurface control (Figure 9–33). Major assemblages of rock types show the principal facies from the shallow shelf, through the shelf margin, and to the deep basin. These facies have been defined and named according to their common lithic characteristics (Meissner 1972). The Chalk Bluff facies is composed of interbedded evaporites, dolomites, and red sandstones of the Northwest shelf and designates the Artesia Group and its formations, in descending order, the Tansill, Yates, Seven Rivers, Queen, and the updip part of the Grayburg. These rocks denote the ancient platform of shallow subtidal to supratidal environments of deposition. The Chalk Bluff facies is replaced farther to the north by a thinner section of red beds designated as the Bernal facies.

The shelf margin is composed of two facies, the Carlsbad and Capitan "reef." The Carlsbad facies is composed of dolomite wackestone and clean sandstones that denote normal marine environments at the shelf edge. The sediments were deposited as a relatively narrow band of wave-washed sand with immediately adjacent areas of more restricted circulation in which algal muds were formed. The outer shelf margin is composed of the reef facies and includes basin-slope deposits of carbonate talus.

The basinal facies is composed of sandstones and shale of the Delaware Moun-

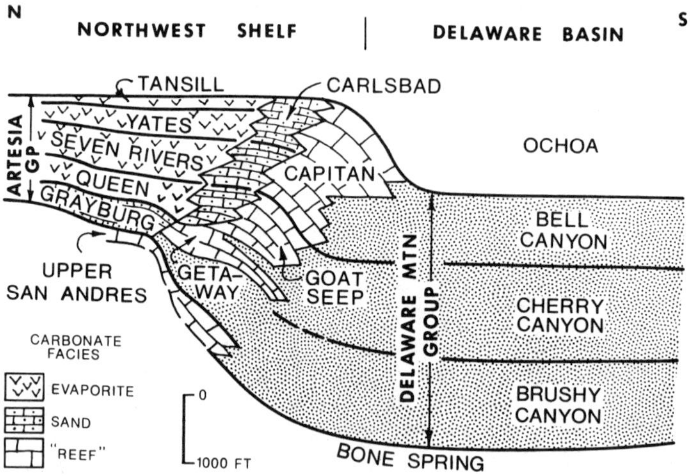

Figure 9–33 Diagrammatic cross section from the Northwest shelf (north) of the Delaware basin (south) showing stratigraphic relationships in the Middle Permian Guadalupian Series. [From Meissner 1972.]

tain Group and its three formations, in descending order, the Bell Canyon, Cherry Canyon, and Brushy Canyon. The basinal facies also includes thin beds of dark dolomite, but these are a minor part of the total section.

The Delaware Mountain Group reaches a maximum thickness of nearly 4000 ft (1219 m) in the central basin. This thick sequence of fine clastics contrasts with the thinner shelf and shelf-margin deposits that are dominantly carbonate. In order to explain this distribution of sediments, a cyclic pattern of deposition has been proposed (Silver and Todd 1969; Meissner 1972).

During high stands of sea level, the shoreline extended far onto the Northwest shelf. Near the shore were supratidal sabkhas and tidal flats, farther seaward were saline lagoons, near the shelf margin were lime sands, and at the shelf edge were organic reefs. At the same time, carbonate talus accumulated on the slope, and basin sediment was pelagic clay and lime muds, probably with some carbonate turbidites. When carbonate precipitation was active on the shelf, the basin was "starved" of sediment. Then the sea level was lowered, perhaps by glaciation, and clastic sediment prograded across the shelf. Massive amounts of very fine sand and coarse silt were carried to the shelf margin and, eventually, into the deep basin. At the same time, organic precipitation of carbonate ceased on the shelf, while thick sections of sand and silt were deposited in the basin. This alternation of shelf carbonate and basinal clastic cycles has been referred to as *reciprocal sedimentation* (Meissner 1972, Figure 8).

In the subsurface of the Delaware basin, deposition of the Delaware Mountain Group was preceded by dark micritic limestones called the Bone Springs (Figure 9-32). This formation represents the basinal equivalent of Leonardian San Andres dolomites of the Northwest shelf. The San Andres section shows shelf and shelf-margin facies similar to those of the Artesia Group (Meissner 1972, pp. 221–26). Delaware Mountain sandstones are succeeded by evaporites of the Ochoan Series, which represent the final phase of Permian deposition in the Delaware basin.

The Delaware Mountain Group is subdivided into the Brushy Canyon, Cherry Canyon, and Bell Canyon formations by means of thin but persistent limestones that extend across the basin. The uppermost Bell Canyon Formation is best known in the subsurface because it contains the principal shallow oil reservoirs.

Drilling History

Small areas of oil production were found in the upper Delaware Mountain Group in the 1930s (Adams 1936), but for many years exploratory drilling was concentrated on the Northwest shelf and Central Basin platform, which established major oil reserves in Paleozoic carbonates. Not until the late 1950s were many small fields discovered in the Delaware basin, largely as the result of drilling for deeper objectives. These accidental discoveries of stratigraphic traps led to an active drilling campaign that extended to about 1965. The fields were at shallow depths of 2500 and 5000 ft (762 to 1524 m), but thin lenticular reservoir sandstones and significant water production (Grauten 1965) made the play hazardous and not highly profitable.

A typical reservoir section is shown by logs from the Paduca field (Figure 9–34). The first "Delaware sand" represents a convenient marker bed for structural mapping. The succeeding section is divided into the Ramsey Sandstone, Ford Shale, and Olds Sandstone members (Nottingham 1960). The more porous sandstones of the Ramsey represent the common oil reservoirs in the central basin, and the porous sandstones of the Olds are less commonly productive. Older sandstones also produce some oil, but the most abundant reservoirs are those of the Ramsey Member. Generally, sandstones that have porosities greater than 20 percent produce oil and the best reservoirs for water-free production are those of 25 percent or more porosity. Thicknesses of reservoir sandstones are on the order of 30 ft, (9 m), and they form the linear "channel" sandstones of the basin.

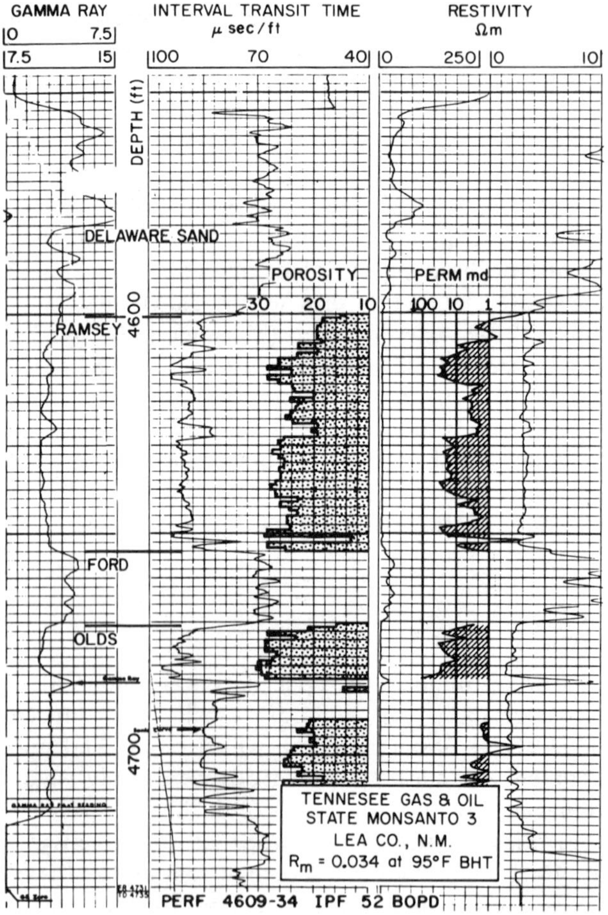

Figure 9–34 Well logs and core analysis of the upper Bell Canyon Formation in the Paduca field showing stratigraphic nomenclature of the subsurface section. Well is located in SW SW 16-25S-32E.

The reservoir sandstones and their equivalents have been studied at the outcrop and in the subsurface, but there is no agreement as to their origins. At the outcrop, the generally thin-bedded sandstones were interpreted to be either turbidites in submarine fans (Jacka et al. 1968) or traction deposits from density currents (Harms 1974; Williamson 1977). Cores from the subsurface were also subject to conflicting interpretations (Payne 1976; Bonzanich 1979). The reasons for differing interpretations is probably due to the effect of weathering. At the outcrop, weathering enhances an indistinct lamination in otherwise massive sandstones while it obscures the details of bedding in fine-grained rocks. Cores from the subsurface are friable, contain highly saline water, and deteriorate rapidly on exposure to air, thereby destroying important features of the sequence. Only in fresh continuous core slabs can the details of sedimentary structures be observed. Study of complete cores representing 1143 ft (349 m) of section revealed that the reservoir sandstones are channel turbidites interbedded with basinal suspension deposits (Berg 1979).

Description of Sandstones

Sedimentary structures. Bell Canyon sandstones are very fine grained and generally display bedding features that are of a very fine scale. Massive (structureless) sandstones may dominate a section, but interbedded sandstones and siltstones are thinly laminated to finely bioturbated. An ordered sequence of bedding types is commonly seen, and when complete, the succession is that of the standard turbidite sequence. In addition to the divisions *A* through *E* (Bouma 1962), a bioturbated sandstone is also found in the section. This rock is designated an *F* division for convenience in referring to the common succession, although the bioturbated *F* sandstone is a basinal unit and not part of the turbidite sequence.

Bell Canyon rock types are illustrated in cores from Poker Lake Unit 43 and Big Eddy Unit 47 wells (Figures 9–35 and 9–36). The massive *A* sandstone is always a prominent unit, and although generally structureless, it may show widely spaced and indistinct laminae that are even, parallel, and may dip at relatively high angles of up to 15° [Figure 9–36(F)]. Rarely, laminae are found in converging groups [Figures 9–35(G) and 9–36]. Basal contacts are sharp and may display loading and small flame structure [Figure 9–35(G)]. The lower parts of many *A* sandstones show indistinct, curved, and convolute laminae that give the rock a churned or "fluidized" appearance, as if turbulent streamlines had been frozen in the sediment [Figure 9–36(A) and (B)]. A common characteristic of the *A* division is the presence of tiny shale clasts that are only a few millimeters in diameter and are scattered through the rock [Figures 9–35 and 9–36(C)]. Less common are larger clasts [Figure 9–35(D)] up to several centimeters in diameter that may occur singly or be concentrated in zones, either at the top or base of the unit. More rarely, larger shale clasts form a basal conglomerate [Figure 9–35(F)]. In spite of the presence of indistinct laminae and zones of clasts, the overall aspect of the *A* division is a generally nonbedded appearance.

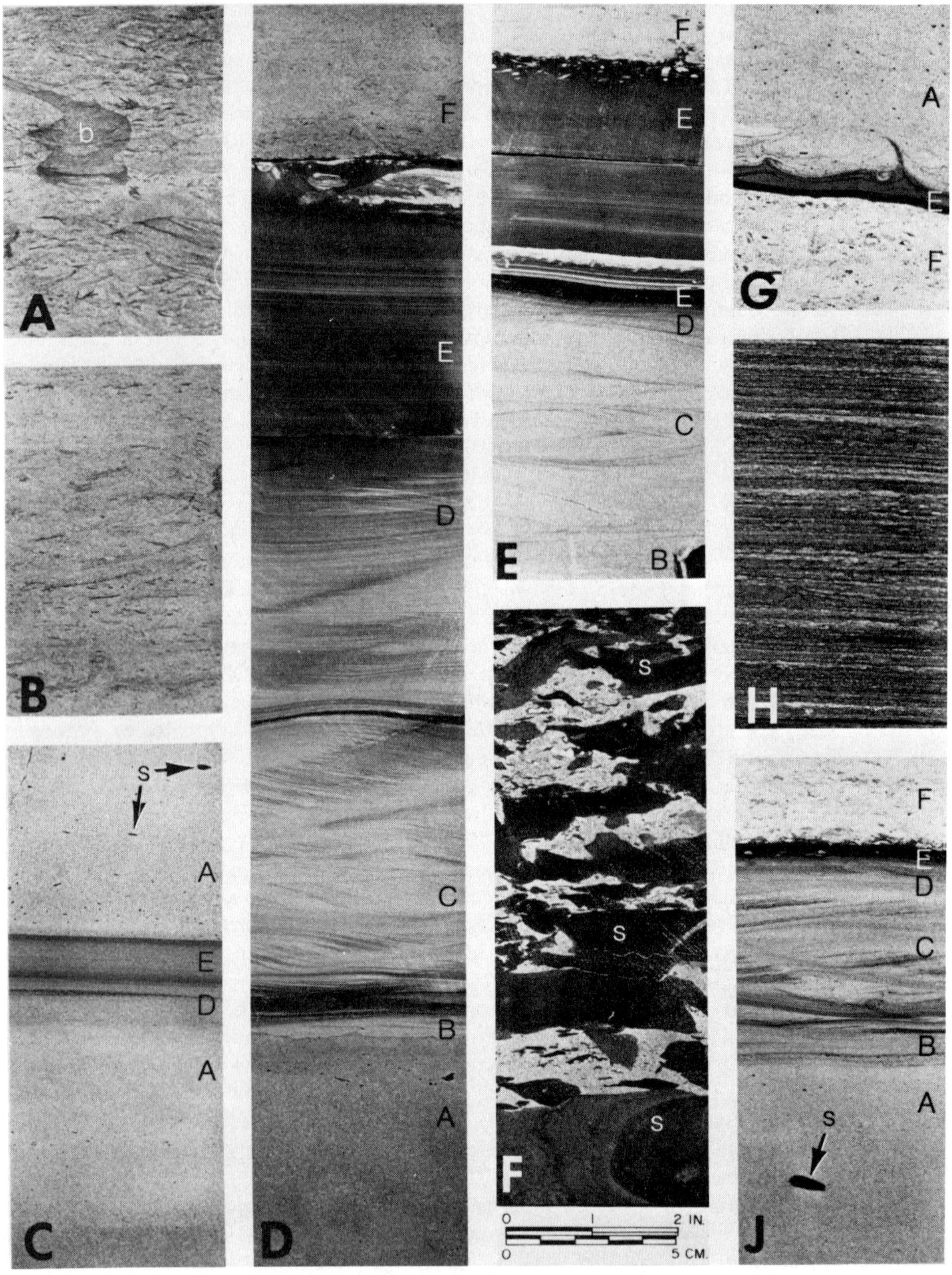

A
b
B
C
s
A
E
D
A
D
F
E
D
C
B
A
E
F
E
E
D
C
B
F
S
S
S
0 1 2 IN.
0 5 CM.
G
A
E
F
H
J
F
E
D
C
B
A
s

The *B* division is rather poorly developed and may be seen only as a few distinct laminae at the top of an *A* division [Figure 9-35(D) and (J)]. At some places, however, the *B* division attains a significant thickness of several feet. The laminae are even, parallel, and uniformly spaced at a few millimeters to about 1 cm [Figure 9-36(E)]. Because of its usual thinness, the *B* division can only be identified where the core is continuous and unbroken.

The *C* unit is thin but commonly present as well-developed ripples of low amplitudes on the order of 1 to 3 cm and of lengths of 10 cm or more [Figures 9-35(D) and (J) and 9-36(D)]. The common form is that of "climbing ripples," denoting sediment fallout from current flow. The amplitudes of ripples decrease upward. The base of the *C* unit is generally in sharp contact with the underlying *A* or *B*, perhaps with a thin lamina of black shale [Figure 9-35(D)]. The upper contact is gradational to *D* or *E* over a thin interval with increasing numbers of black shale laminae.

The *D* unit is generally quite thin, and it is often difficult to identify as a distinct bed because of the upward gradation of *C* laminae into overlying black shale [Figure 9-35(D)]. In complete cores, however, there appears to be always a few horizontal laminae at this level, even if only a few millimeters in thickness.

The *E* unit is a massive black shale that contains thin even laminae of very fine silt [Figure 9-35(D) and (E)]. The *E* unit is nearly always present but thin, and beds greater than a few centimeters are not common. Silt laminae may be gently curved, resembling ripple structure of very low amplitude, or the division may contain laminae of fine sand that show distinct load structures at their base [Figure 9-35(E)].

The *F* division is a common bedding type in the Bell Canyon section [Figure 9-35(A) and (B)]. It is finely bioturbated and is characterized by small dark spots or "flecks" that appear to be shale clasts. Most likely they represent concentrations of fecal material in fossil tracks or trails (Payne 1976). These flecks are irregular in outline, 1 to 5 mm in diameter, and give the rock a spotted appearance. Other particles appear to be shale clasts. These particles are elongate and gently curved, mostly concave upward, and may be up to 1 cm in diameter. These shale "flakes" apparently are remnants of shale laminae. Original bedding is marked by a few

Figure 9-35 Sedimentary structures in upper Bell Canyon sandstones, Bass Enterprises Unit 43, Poker Lake area. Photographs designated by boldface letters in lower left corner; turbidite divisions designated by small letters on right side of photographs: (A) coarsely bioturbated sandstone with large burrow (b)—4391 ft (1339 m); (B) finely bioturbated sandstone—4396 ft (1341 m); (C) turbidite divisions *ADE* overlain by *A* unit with small shale clasts (s)—4405 ft (1344 m); (D) turbidite divisions *ABCDE* overlain by bioturbated *F* sandstone—4418-4419 ft (1348 m); (E) turbidite divisions *BCDE* (*A* unit below not shown) overlain by bioturbated *F* sandstone—4439 ft (1354 m); (F) conglomerate of large shale clasts (s) in sandstone matrix—4443 ft (1355 m); (G) flame structure (f) at base of *A* unit overlying *E* shale and bioturbated *F* sandstone—4446 ft (1356 m); (H) shale with thin sandstone laminae—4460 ft (1360 m); (J) turbidite beds *ABCDE* overlain by bioturbated *F* sandstone; note large shale clast (s) in *A* unit—4465 ft (1362 m).

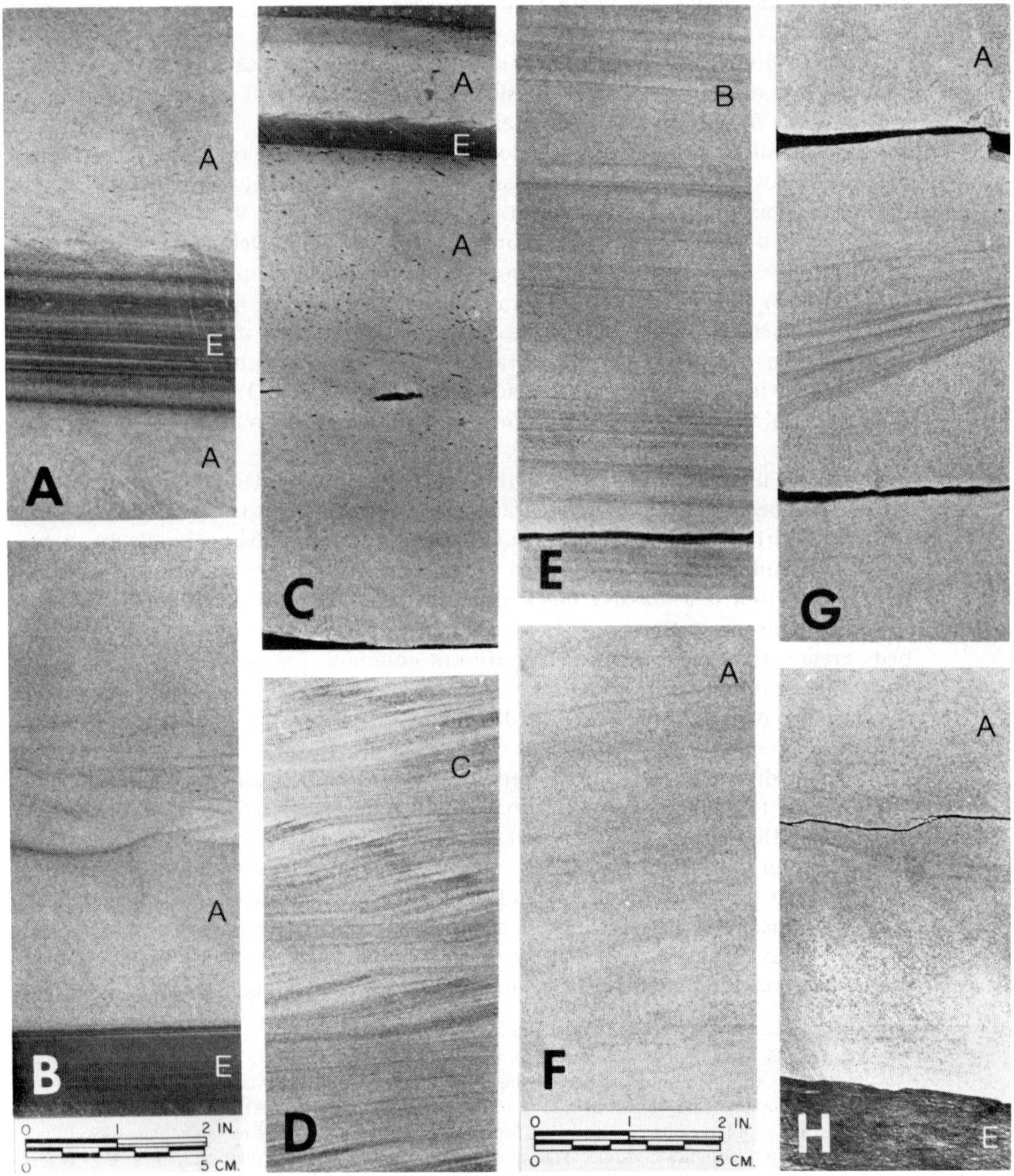

Figure 9–36 Sedimentary structures in upper Bell Canyon sandstones, Bass Enterprises Unit 47, Big Eddy area. Photographs designated by boldface letters in lower left corner; turbidite divisions designated by small letters on right side of photographs: (A) massive *A* sandstone (below) grading upward to *E* shale, which is overlain by massive *A* sandstone with basal flame structure—2786.5 ft (850 m); (B) massive *A* sandstone (base of lower sandstone in *A*, above) with indistinct curved laminae overlying *E* shale—2790 ft (851 m); (C) thin massive *A* sandstones with shale clasts—2807 ft (856 m); (D) thick rippled *C* sandstone forming upper part of sequence shown in (e), (f), (g), and (h) below—2816.5 ft (859 m); (E) thick laminated *B* sandstone—2819.5 ft (860 m); (F) indistinctly laminated *A* sandstone—2821.5 ft (860.5 m); (G) converging laminae in *A* sandstone—2825.5 ft (862 m); (H) massive *A* sandstone with sharp basal contact on underlying *E* shale—2828.5 ft (863 m).

distinct laminae of black shale, but the laminae are more commonly broken and discontinuous, or they may be completely destroyed by fine-scale bioturbation. A horizontal aspect to the bedding is, however, generally preserved, despite the finely churned nature of the rock. Bioturbation is largely confined to sandstone or siltstone and rarely extends into adjacent shales, as if the organisms found black mud unpalatable.

Distinct burrow tubes [Figure 9–35(A)] are not common in the *F* division, but they occur often enough so that the finely churned appearance of the rock is confirmed to be the result of organic reworking. Large burrow tubes are approximately circular in outline and vertical in orientation. Many tubes appear to have been filled by ingested sediment and contain tiny fecal pellets or undigested flakes of shale. Other tubes have collapsed interiors caused by the withdrawal of probing organisms.

Bioturbated *F* sandstones are abundant only in the Bell Canyon Formation and are rare or absent in older formations. Most common in the lower Delaware is a finely laminated sandstone and shale unit that is composed of alternating very fine sand in laminae about 2-mm thick and black shale laminae less than 1-mm thick. The laminated sandstone and shale are present in thick units that are gradational to black shale with sand laminae, like the Ford Shale [Figure 9–35(H)]. The remarkable fineness and uniformity of laminae suggest a cyclic pattern of deposition. It has been reported that laminae in the Lamar Shale at the top of the Delaware Group can be correlated over a distance of 15 mi (24 km), and it was concluded that the laminae probably represent annual layers of sediment (Anderson et al. 1972, p. 74). Apparently, thin sheets of sand were spread widely in the basin either by current flow or by settling from suspension. The bioturbated *F* sandstones of the Bell Canyon Formation could have been derived from such finely laminated sands that were reworked by a benthic fauna. This means that a general environmental change may have taken place in the Guadalupian basin from an earlier more hostile environment to a later more congenial environment for benthic organisms.

The complete sequence of bedding divisions from *A* through *E* is observed at many places and ranges in thickness from a few feet to as much as 15 ft (5 m). This common and ordered succession means that the sequences are turbidites. Thicker beds are the most complete. Generally, the massive *A* division is dominant, and succeeding *BCDE* divisions are 10 percent or less of the total thickness (Figure 9–37, Poker Lake 43). Only one sequence was observed in which *B* and *C* were significantly thick and together form about one-third of the total section (Figure 9–37, Big Eddy 47). In this sequence, the *A* division was also atypical in that indistinct laminae were more abundant.

Complete sequences are also found in thinner sections that range from about 3 to 7 ft (1 to 2 m) in thickness. These sections generally lose thickness in the massive *A* division, whereas the succeeding divisions are more prominent. The *B* division, however, tends to thin also in these sections and may be insignificant.

Some sequences are not complete, especially those about 1 ft (0.5 m) or less in thickness. The *B* and *C* divisions are commonly lacking in these sequences, and the *D* may be so thin as to be almost overlooked [Figures 9–35(C) and 9–36(A) and (C)]. Consequently, thin turbidites can be expected to be of the *AE* type, although ripples of small amplitude or thin silt laminae might be present. In only a few cases are the *AB* divisions absent, and the sequence is composed only of *CDE*. An unusual thickness of such a sequence is 3 ft (1 m) (Figure 9–37), Big Eddy 47), and others are only about 1 ft (0.3 m) or less in thickness.

Most commonly there is a sharp contact that separates the *AB* and *CDE* divisions. This break might not be conspicuous except for the fact that in some sections a distinct lamina or thin bed of dark silt or black shale intervenes between *B* and *C* [Figure 9–35(D)]. This separation confirms the interpretation made previ-

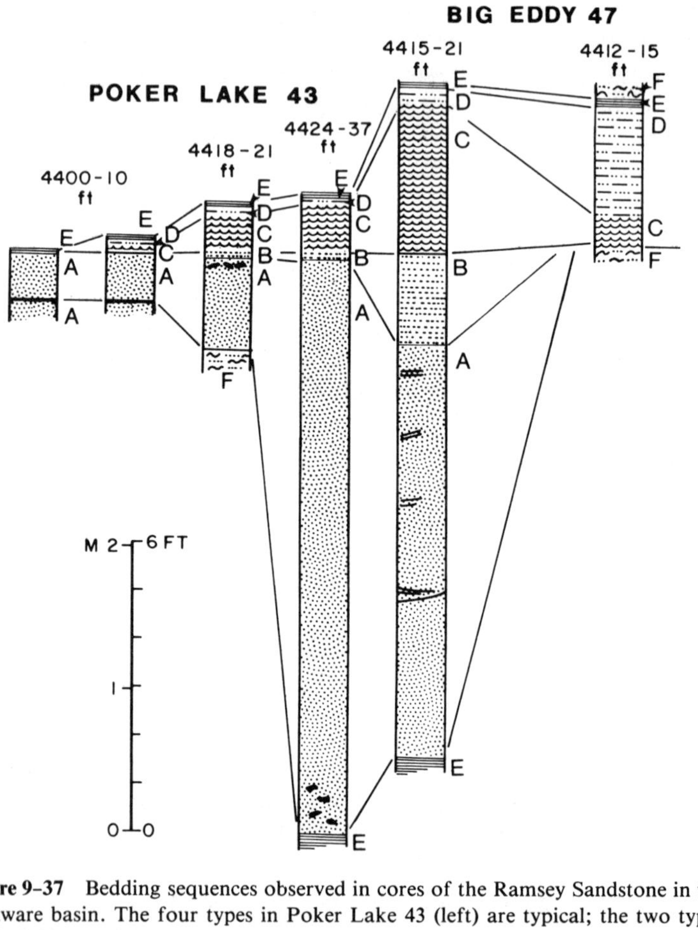

Figure 9–37 Bedding sequences observed in cores of the Ramsey Sandstone in the Delaware basin. The four types in Poker Lake 43 (left) are typical; the two types in Big Eddy 47 (right) show unusual thicknesses of the *B*, *C*, and *D* divisions. Sedimentary structures shown in Figures 9–35 and 9–36 illustrate these sequences.

ously that two processes are involved in the deposition of complete turbidite sequences. First, there was the turbidity flow that deposited the *A* and *B* divisions during abruptly decreasing flow velocity. Then, following currents or "entrained flow" reworked the sediment to produce the low flow-regime *C* division, which in turn was followed by *D* and *E* settling from suspension. Furthermore, the observed contacts suggest that a significant period may have elapsed between the passage of the turbidity flow and the "following" current.

Petrography. Bell Canyon sandstones have a mean quartz size of 0.09 mm (very fine grained), average maximum size is 0.18 mm, and standard deviation is only 0.03 mm. Textural grading is evident within bed sequences (Figure 9-38). Massive *A* divisions are generally in the range of 0.09 to 0.11 mm, *B* divisions are on the order of 0.06 to 0.08 mm, and other units have a minimum size of 0.05 to 0.07 mm (Table 9-6). The massive *A* sandstone itself commonly does not show textural grading. There may be some variation within the *A* division, possibly a slight increase upward to the middle or upper parts of the unit. Despite their overall fineness, quartz grains are rather well rounded, indicating that the quartz has been recycled and probably was derived largely from the weathering of older sedimentary rocks.

In detrital composition, the sandstones are feldspathic quartz arenites with 70 percent monocrystalline quartz, 11 percent feldspar, 10 percent rock fragments, 1

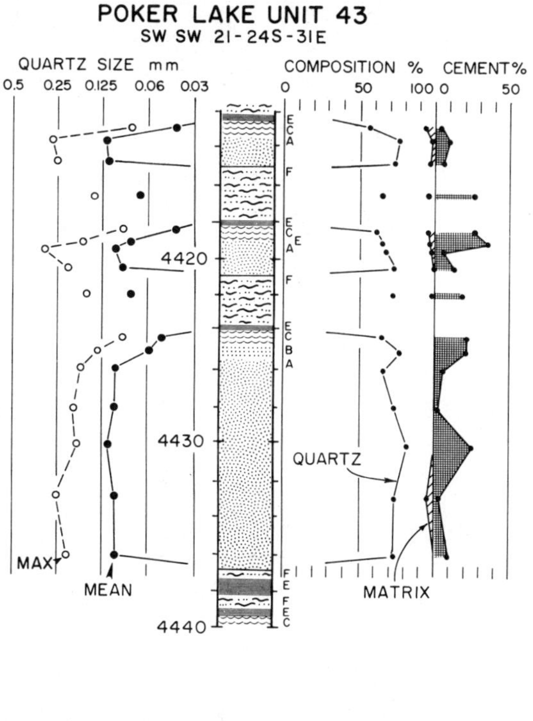

Figure 9-38 Texture, composition, and sedimentary structures in the Ramsey Sandstone. Letters to the right of central column designate turbidite divisions. Sedimentary structures shown in Figure 9-35 are from this cored section.

TABLE 9-6. SUMMARY OF ROCK PROPERTIES FOR UPPER BELL CANYON SANDSTONES IN POKER LAKE UNIT 43

Unit	samples	Quartz size[a]			Detrital composition[b]					Cement[c]		
		Mean (mm)	Max (mm)	σ (mm)	Qz (%)	F (%)	Rx (%)	Oth (%)	Mx (%)	Carb (%)	Permeability (md)	Porosity (%)
F	3	0.07	0.14	0.03	68	9	4	15	4	17	1.5	18.9
C	4	0.05	0.10	0.02	62	5	13	18	2	16	0.3	15.9
B	2	0.07	0.15	0.02	71	13	3	12	1	28	31.	23.0
A	18	0.10	0.20	0.04	72	11	12	4	1	10	45.	25.5
Means		0.09	0.18	0.03	70	10	11	8	1	13	33.	23.2

[a] Long-axis measurements; σ = standard deviation.

[b] Qz = monocrystalline quartz, F = feldspars, Rx = rock fragments including polycrystalline quartz, Mx = matrix, Oth = other minerals including mica.

[c] Carb = dolomite and calcite.

percent matrix, and 8 percent other minerals (Table 9–6). Feldspars are largely orthoclase and microcline but also include some plagioclase. Rock fragments are dominantly polycrystalline quartz with rare chert, but a significant number are composed of limestone. Matrix is unusually low and consists of detrital clasts with organic material. An important component is mica, both fresh and altered, which occurs most abundantly in the finer-grained *B, C,* and *F* units.

Carbonate cement is an important part of the bulk composition and averages 7 percent. Calcite commonly occurs in small discrete patches or "blebs" and may have been derived in large part from the dissolution of limestone fragments. Some samples have small dolomite rhombs that occupy intergranular spaces, but the volume of this authigenic carbonate is small. Silica occurs on many quartz grains as overgrowths, but its volume is negligible.

Porosity and permeability. Reservoir properties of Bell Canyon sandstones are clearly related to rock type. The channel sandstones, which are dominantly massive *A* units, have high porosity and permeability, whereas the basinal sandstones, the bioturbated *F* units, have low porosity and permeability (Table 9–6). Average permeabilities of channel sandstones in the Ramsey Member are 14 to 90 md; permeabilities of channel sandstones in the Olds Member are 16 to 26 md; and bioturbated sandstones are 1.5 md to 4 md (Figure 9–39).

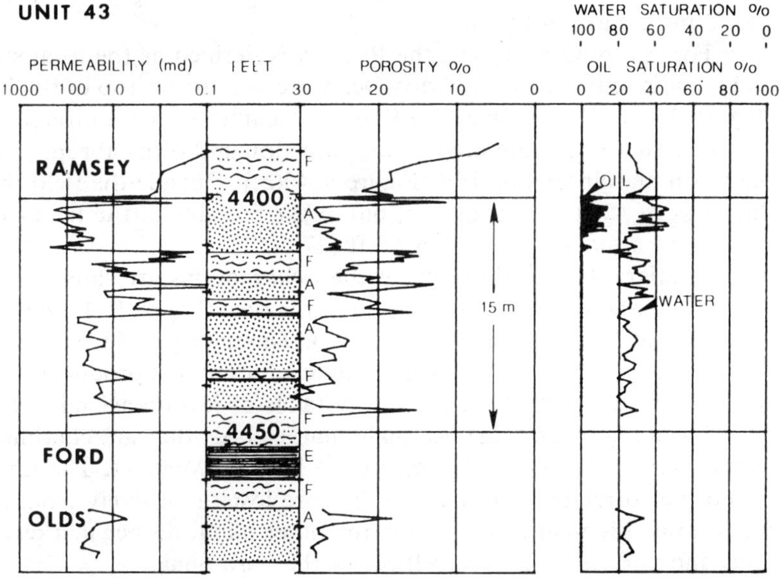

Figure 9–39 Permeabilities, porosities, and fluid saturation of Bell Canyon sandstones in Poker Lake 43.

The difference in permeabilities of channel and basinal sandstones is only partly due to grain size, and it is clear that compositional differences also affect porosity and permeability. The principal compositional differences between the rock types are the amounts of calcite cement and "other" minerals, especially mica (Table 9–6). It is also likely that bioturbation in the *F* units reduces permeability even when cement and mica are not abundant.

Morphology

In the subsurface, linear sandstone bodies of the Bell Canyon have been called ***channels,*** and massive porous and permeable sandstones within these channels form the principal oil reservoirs. The thickness and form of these producing sandstones probably reflect the basinal extension of eroded channels observed at the outcrop (Harms 1974) and inferred in the subsurface (Williamson 1977). Stratigraphic relationships in a typical Delaware field help to explain the origin and morphology of reservoir units (Berg 1979).

The Paduca field produces oil from the Ramsey Sandstone Member of the uppermost Bell Canyon Formation (Figure 9–40). The field is located along a trend of porous sandstones that extends southwestward, crossing structural strike at an angle of about 30° (Figure 9–40). Paduca was discovered in 1960, and development was slow because of low productivity, commonly less than 100 bbls/day (16 m^3/day) per well, as well as erratic distribution of oil and water within the reservoir similar to other Delaware fields (Grauten 1965). The structure on top of the Delaware Mountain Group shows a slight terrace in the regional dip of approximately 100 ft/mi [Figure 9–40(A)].

For mapping purposes, the Ramsey is defined as the total section between a highly radioactive shale overlying Delaware sand to the top of the Ford Shale below (Figure 9–41). Thickness of the Ramsey Member shows a broadly curved channel-like form that is arcuate in an updip direction and turns abruptly southeastward at the south margin of the field [Figure 9–40(b)]. Outside the field the Ramsey has a thickness of 40 ft (12 m) or less, but at the margins of the field there is an abrupt increase to a maximum of over 90 ft (27 m).

Stratigraphic relationships at the field margins are shown by two correlation profiles, one across the updip margin of the field (Figure 9–41, AA′) and one across the downdip margin (Figure 9–41, BB′). The updip profile shows a Ramsey thickness of less than 40 ft (12 m) in a dry hole adjacent to a producing well. The Ramsey thickens to nearly 80 ft (24 m) in a distance of less than 1 mi (1.6 km). Below the Ford Shale are several distinct shale marker beds that are continuous and parallel to the overlying shale at the top of the Ramsey Member. The upper marker beds cannot be correlated continuously but are lost successively from the top down as the Ramsey Member thickens. On the other hand, deeper marker beds are continuous and parallel at a level well below the Ford Shale.

Channel unconformity. The convergence of the Ford Shale with underlying marker beds suggests an unconformity below the Ford that marks the base of a

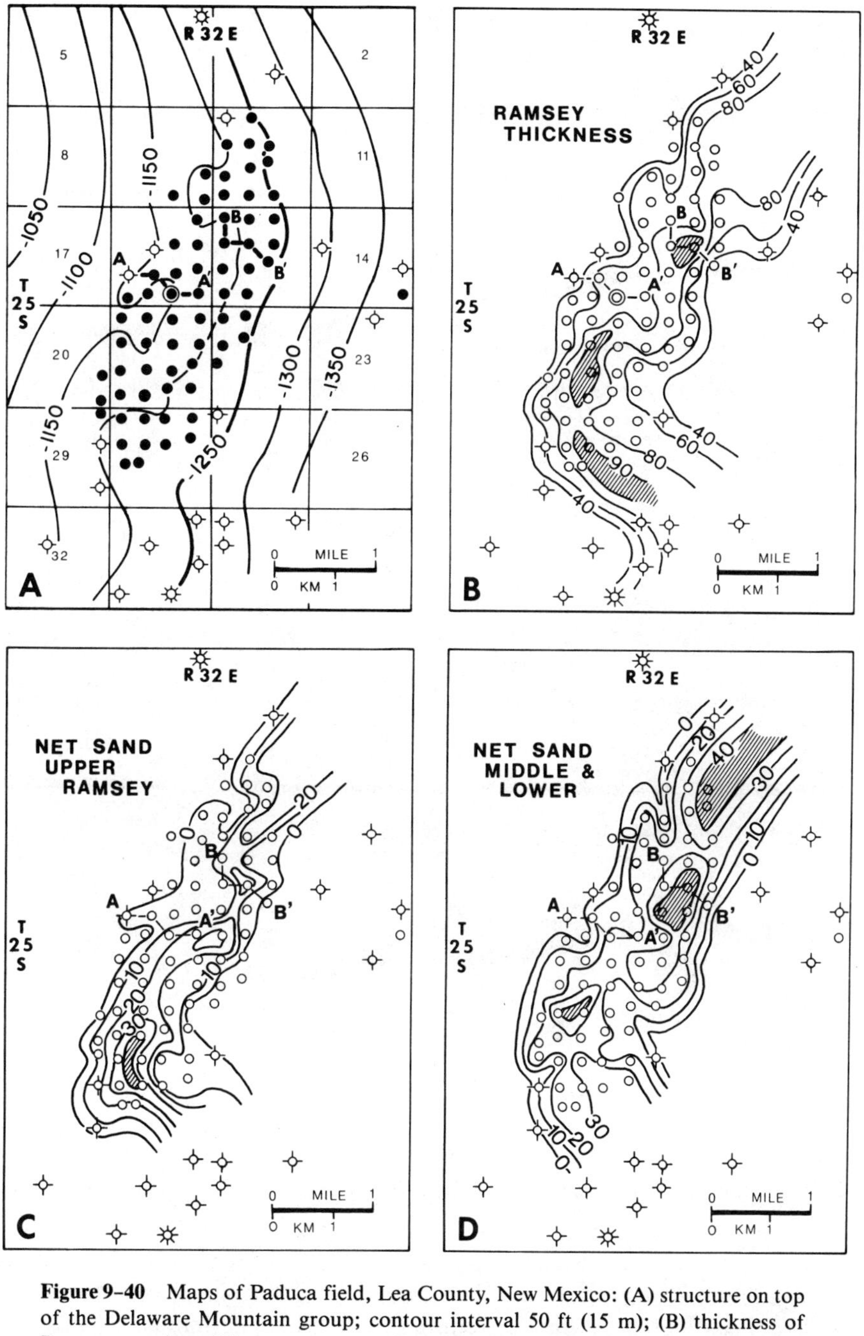

Figure 9–40 Maps of Paduca field, Lea County, New Mexico: (A) structure on top of the Delaware Mountain group; contour interval 50 ft (15 m); (B) thickness of Ramsey member; (C) net porous sandstone in the upper Ramsey; (D) net porous sandstone in the middle and lower Ramsey. Isopach contour interval 10 ft (3 m). Areas of maximum thickness are patterned. [Maps from Weinmeister 1978.]

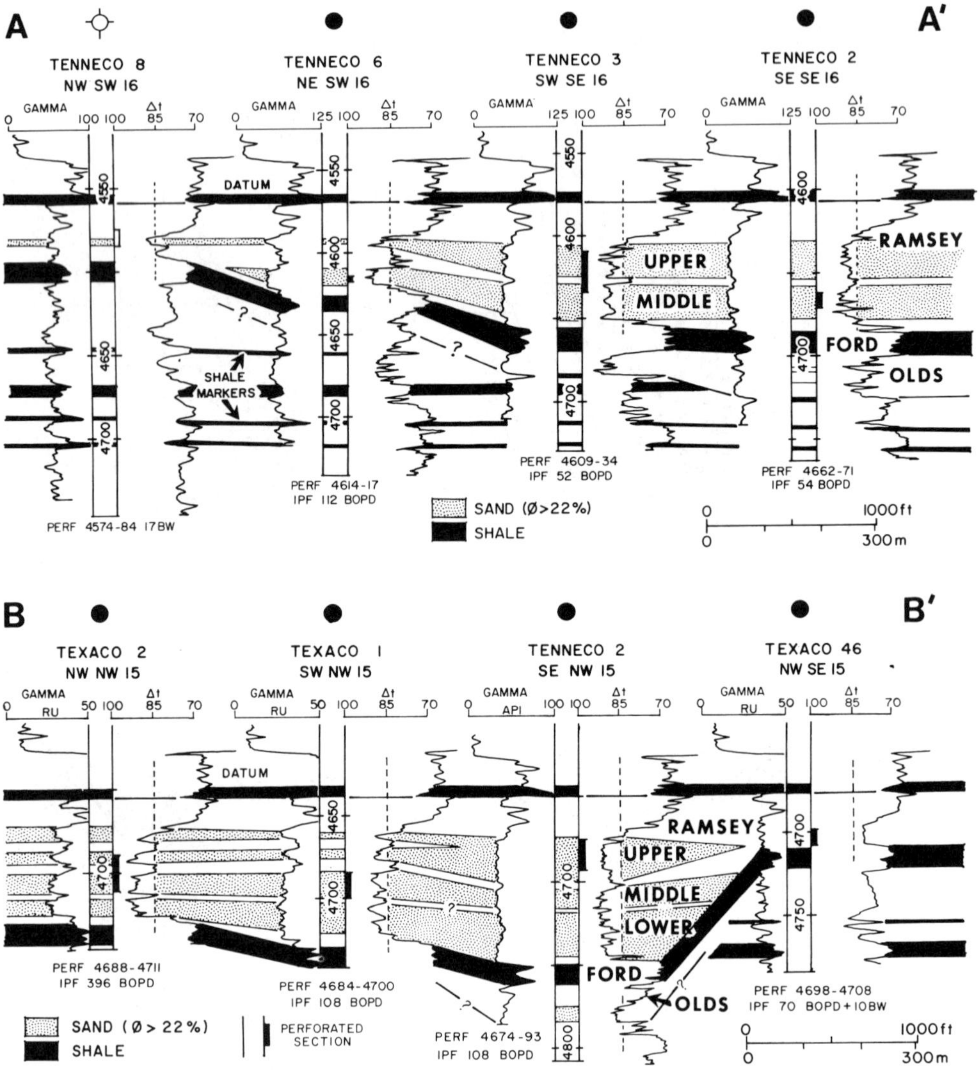

Figure 9–41 Correlations in the upper Bell Canyon formation at the updip (AA′) and downdip (BB′) margins of Paduca field. Locations of cross sections shown in Figure 9–40.

channel that was eroded in basin sediments. The porous Olds Sandstone lies immediately below the Ford Shale and probably is within the eroded channel, but correlation of an Olds section cannot be made throughout the field. Therefore, the maximum depth of the channel cannot be determined, and channel configuration is approximated by the thickening of the Ramsey Member (Figure 9–40).

Near the downdip margin of the field, the Ramsey Member reaches a maximum thickness of nearly 100 ft (30 m) but thins abruptly to less than 40 ft at the last producing well (Figure 9–41, BB′). Below this thin section of Ramsey at least two of the shale marker beds reappear in the section underlying the Ford Shale. In fact, the character of this section as expressed by well logs is nearly identical to that at the updip margin of the field (Figure 9–41, Tenneco 8, AA′). This relationship confirms a downdip edge for the channel and shows that it is somewhat irregular and asymmetric in cross section.

The Paduca channel was apparently eroded into basin-plain mud and sands to a depth of more than 55 ft (16 m) below the surrounding basin. Porous sands of the Olds Member were the first deposits in the deepest part of the eroded channel. The massive shale of the Ford Member was next deposited widely across the eroded channel as well as in the basin area. Succeeding sandstones of the Ramsey Member are sequences of the *AE, ABE,* and *ABCDE* type (Payne 1976; Weinmeister 1978) and were confined to the eroded channel until the channelway was filled to a level of the surrounding basin. The following deposits were largely interbedded black shale and bioturbated *F* sandstones, which again represent widespread basinal sediments. Finally, black shale was deposited over the basin area to form the excellent marker bed immediately above the Ramsey.

Channel patterns. The distributions of individual channel sandstones cannot be mapped within the field because of difficulties in correlating thin beds between wells. However, groups of porous sandstones can be differentiated by means of thin marker beds within the channel-fill sequence. These marker beds represent bioturbated *F* sandstones and divide the Ramsey Member into lower, middle, and upper parts (Figure 9–41). The lower part is confined largely to the areas of thicker total Ramsey, whereas the middle and upper parts are more extensive. Maps of net porous sandstone were prepared for the combined lower and middle units and for the upper unit [Figure 9–40(C) and (D)]. Net porous sandstone within the lower and middle units has an average thickness of about 25 ft (8 m) and a maximum thickness of 40 ft (12 m). The trend of the thick sandstones shows a pattern that is more highly sinuous than the larger eroded channel. Net porous sandstone in the upper unit has an average thickness somewhat less than 20 ft (6 m) and a maximum thickness of 40 ft (12 m). Again, the trend of maximum thickness is more highly sinuous than that of the eroded channel.

These relationships indicate a difference between those flows that eroded the major channel and those that followed and filled the channel. From a study of modern streams, channel patterns are related to discharge and gradient (Leopold and Wolman 1957). High discharge or high gradient results in a braided pattern of low sinuosity, whereas low discharge or low gradient results in a meander pattern of high sinuosity (see Chapter 5, Figure 5–5). These relationships may apply also to submarine flows. If it is assumed that low gradients in the basin were essentially constant, it may be concluded that flows that eroded the Paduca channel were those of high discharge and that they scoured a channel of low sinuosity. Subsequent flows, responsible for deposition of porous sandstones, were those of low discharge and resulted in the more highly sinuous patterns for the channel-fill deposits.

Prediction of channel-fill morphology. Although the gross morphology of channel-fill sandstones is well established by subsurface control (Figure 9–40), a general hypothesis is needed to predict the size and shape of individual sandstone units. Two assumptions may be made based on sandstone distribution (Figure 9–40). First, turbidity flows had a meandering pattern, and second, deposition took place laterally on the inside of meander loops. These are reasonable assumptions because meandering channels have been observed on modern submarine fans, and meandering is a common pattern of current flow, whether in stream channels or in unconfined density currents such as the modern Gulf Stream (see Chapter 5, Figure 5–11). Furthermore, deposition within meanders always takes place as point bars that prograde laterally within meander loops.

Based on these assumptions, patterns of sand deposition can be inferred. If meandering turbidity flows deposited sand on the inside of meander loops, a number of such flows could have produced a lateral succession of point-bar deposits composed largely of massive *A* or *AB* sands (Figure 9–42). These deposits A_1, A_2, . . . , A_4 built laterally to form a series of broadly curved lens-shaped bodies that tended to be displaced in a downstream direction, similar to patterns of point bars in modern streams. After each flow, succeeding currents of low flow-regime followed the same flow path and produced a layer of rippled sediment. After flow

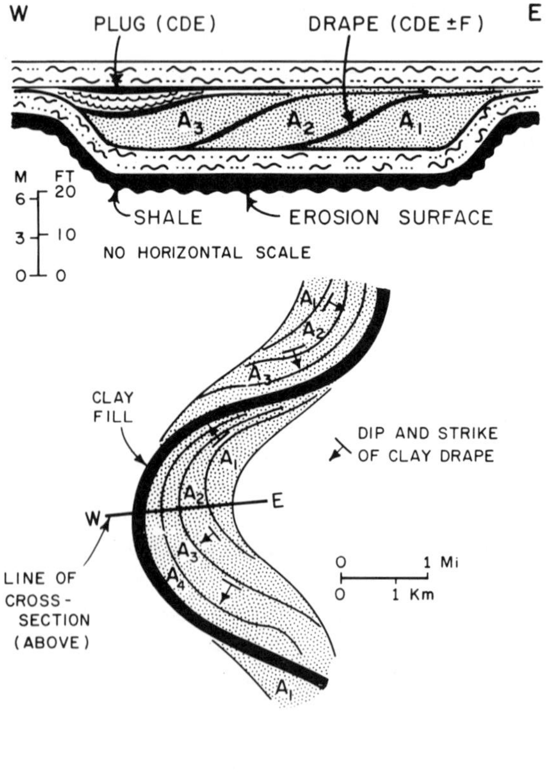

Figure 9–42 Diagrammatic cross section (above) and plan view (below) of turbidite deposits in a meandering channel.

ceased, very fine sediment was deposited from suspension to form a "clay drape" over the entire channel. After a number of such flows had nearly filled the channel with their deposits, succeeding flows were diverted to another location. The abandoned flow path was filled with rippled sand and then with pelagic clay to form a "clay plug."

This pattern of deposition explains the observed sequences that generally begin with one or more massive *A* or *AB* sandstones of variable thickness, each of which may be followed by a thinner section of *CD,* and nearly always by a thin *E* shale. The thicker sequences of *ABE* or *ABCDE* divisions are commonly overlain by thin beds of the *AE* type, each of which may represent a local overbank deposit of a thicker flow in the main channel. The least common sequences of the *CDE* type that occur without a massive *AB* unit may represent the final fill of an abandoned channel. The diagrammatic cross section (Figure 9–42) shows a lateral succession of these beds that have been observed in vertical succession. Some units are exaggerated in thickness, such as the final clay plug, but all the variable sequences of the cored sections are represented.

"Clay drapes" can be expected to conform with topography within channels after each period of point-bar deposition. Hence, the direction of dip for these fine-grained sediments should be normal to the trend of the channelway and should point toward the outside of the meander loop (Figure 9–42). The amount of dip can be expected to be on the order of 10° or less, and the base of eroded channels should be marked by a dip discordance. All such dips could be recorded by logs and could provide clues to the morphologies of channel-fill sandstones. A recent study of a Cherry Canyon field confirms that such dips are present and can aid in defining channel morphology (Phillips 1981).

The diagrammatic cross section (Figure 9–42) shows only a lateral succession of deposits in a single channel, and maximum thickness for such sequences is on the order of 15 ft (3 m). If these deposits did not completely fill the previously eroded channel, then other successions could have been deposited over the first until the original channel was filled and was no longer a path for turbidity flows. Thus, a deep eroded channel could have been filled by a succession of tabular sandstone bodies that formed a complex of lensing deposits and, in some places, a nearly continuous section of massive sandstones of aggregate thickness on the order of 30 ft (10 m) or more. The resulting channel sands would then comprise a composite body that closely resembles the variable thicknesses of channel-fill sandstones of the outcrop and subsurface.

Conclusions

Delaware sandstones display bedding sequences that are identical to those of turbidites. It can be concluded that very fine sand was transported by turbidity currents and was deposited as channel-fill sediment in sinuous topographic depressions that represent earlier erosional channels. Sand deposition was largely confined to channelways, and overbank deposits were not formed. Channel-fill sands were enclosed in laminated very fine sand and silt that were transported by wind and were deposited from suspension as widespread sheets. In the Bell Canyon section, the basinal sediment was subsequently finely bioturbated.

The channel-fill sandstones of the Delaware are dominated by thick massive *A* divisions commonly succeeded by *B, C,* and *D* divisions and separated by *E* shales. The channel-fill sandstones probably formed as submarine point bars within meandering flow paths.

SUMMARY

The common reservoir sandstones in deep basins were deposits of turbidity flows, and turbidites are recognized by objective criteria. The sandstones show an ordered sequence of sedimentary structures, which, when complete, consist of a basal massive *A* division, a succeeding laminated *B,* a rippled *C,* an interlaminated silt and shale *D,* and an overlying pelagic shale *E.* Turbidites generally show textural gradation and are thinly interbedded with shale, and the beds have sharp basal contacts. Thicker beds that consist of massive *A* or *AB* sequences are channel deposits. Thinner, more complete sequences represent overbank deposits.

Reservoir turbidites were deposited in three major morphologies: submarine fans, constructional-channel systems, and erosional-channel systems. Submarine fans can be further subdivided into inner-fan channels, middle-fan channel and overbank deposits, and outer-fan, distal sequences with thin channels. Submarine fans are characterized by decreasing frequency and thickness of channel sandstones from inner to outer fan, and therefore facies can be recognized by their characteristic well-log response, especially on gamma-ray logs.

In all three depositional settings, the better reservoir sandstones consist of thicker stacked channel-fill deposits. Porosity and permeability of turbidites are controlled, as in all other sandstones, by composition and depth of burial, but in general, turbidites represent reservoirs of low porosity and permeability and in many deep basins are major gas reservoirs.

REFERENCES CITED

Adams, J. E., 1936, Oil pool of open reservoir type: American Association of Petroleum Geologists Bulletin, v. 20, p. 780–796.

——, and others, 1951, Starved Pennsylvanian Midland basin: American Association of Petroleum Geologists Bulletin, v. 35, p. 2600–2607.

——, 1962, Foreland Pennsylvanian rocks of Texas and eastern New Mexico, *in* C. C. Branson, editor, *Pennsylvanian System in the United States:* Tulsa, Okla., American Association of Petroleum Geologists, p. 372–384.

ANDERSON, R. Y., W. E. DEAN, JR., D. W. KIRKLAND, and H. I. SNIDER, 1972, Permian Castile varved evaporite sequence, west Texas and New Mexico: Geological Society of America Bulletin, v. 83, p. 59–86.

BANDY, O. L., and R. E. ARNAL, 1969, Middle Tertiary basin development, San Joaquin valley, California: Geological Society of America Bulletin, v. 80, p. 783–820.

BAZELEY, WILLIAM, 1972, San Emidio Nose oil field, California, *in* R. E. King, editor, Stratigraphic oil and gas fields: American Association of Petroleum Geologists Memoir 16, p. 297–312.

BELL, H. S., 1942, Density currents as agents for transporting sediments: Journal of Geology, v. 50, p. 512–547.

BERG, R. R., 1975, Capillary pressures in stratigraphic traps: American Association of Petroleum Geologists Bulletin, v. 59, p. 939–956.

——, 1978, Exploration for sandstone stratigraphic traps: American Association of Petroleum Geologists Continuing Education Course Notes Series 3, revised edition.

——, 1979, Reservoir sandstones of the Delaware Mountain Group, southeast New Mexico, *in* N. M. Sullivan, editor, Guadalupian Delaware Mountain Group of West Texas and southeast New Mexico: Society of Economic Paleontologists and Mineralogists, Permian Basin Section, Publication 79–18, p. 75–95.

——, 1981, Deep-water reservoir sandstones of the Texas Gulf Coast: Gulf Coast Association of Geological Societies Transactions, v. 31, p. 31–39.

——, 1982, Turbidite channel reservoirs in Cisco sandstones, Roundtop area, Fisher County, Texas, *in* D. W. Cromwell, editor, Middle and Upper Pennsylvanian System of north-central and West Texas: Society of Economic Paleontologists and Mineralogists Permian Basin Section Guidebook, Publication 82–21, p. 215–225.

BLOOMER, R. R., 1977, Depositional environments of a reservoir sandstone in west-central Texas: American Association of Petroleum Geologists Bulletin, v. 61, no. 3, p. 344–359.

BOUMA, A. H., 1962, *Sedimentology of some flysch deposits: a graphic approach to facies interpretation:* Amsterdam, Elsevier Publishing Company, 168 p.

BOZANICH, R. G., 1979, The Bell Canyon and Cherry Canyon formations, eastern Delaware Basin, Texas: lithology, environments, and mechanisms of deposition, *in* N. M. Sullivan, editor, Guadalupian Delaware Mountain Group: Society of Economic Paleontologists and Mineralogists Permian Basin Section Publication 19–18, p. 121–141.

BROWN, L. F., JR., 1959, Problems of stratigraphic nomenclature and classification, Upper Pennsylvanian, north-central Texas: American Association of Petroleum Geologists Bulletin, v. 43, p. 2866–2871.

——, A. W. CLEAVES, and A. W. ERXLEBEN, 1973, Pennsylvanian depositional systems in north-central Texas: University of Texas Bureau of Economic Geology Guidebook 14, 122 p.

CALLAWAY, D. C., 1968, Habitat of oil on the west side, San Joaquin valley, California, *in* S. E. Kemp, editor, Geology and Oil Fields, West Side, Southern San Joaquin Valley: American Association of Petroleum Geologists-Society Economic Paleontologists and Mineralogists, Pacific Section Guidebook, p. 47–55.

——, 1971, Petroleum potential of San Joaquin basin, California, *in* I. H. Cram, editor, Future petroleum provinces of the United States—their geology and potential: American Association of Petroleum Geologists Memoir 15, v. 1, p. 239–253.

Day, Willard, and George Galbraith, 1950, Notes on Roundtop field, Fisher County, Texas: Abilene Geological Society Geological Contributions, p. 66–68.

Dorman, J. H., 1980, Oil and gas in submarine channels in the Great Valley, California: San Joaquin Geological Society Selected Papers, v. 5, p. 38–53.

Erxleben, A. W., 1975, Depositional systems, Canyon Group (Pennsylvanian System), north-central Texas: University of Texas Bureau of Economic Geology Report of Investigation 82, 75 p.

Findley, R. L., 1975, Facies relationships in the Stevens Sandstone, Kern County, California: Texas A&M University M.S. thesis, 218 p.

Galloway, W. E., and L. F. Brown, Jr., 1972, Depositional Systems and shelf-slope relationships in Upper Pennsylvanian rocks, north-central Texas: University of Texas Bureau of Economic Geology Report of Investigation 75, 62 p.

Gould, H. R., 1951, Some quantitative aspects of Lake Mead turbidity currents *in* J. L. Hough, editor, Turbidity currents and the transportation of coarse sediments to deep water: Society of Economic Paleontologists and Mineralogists Special Publication 2, p. 34–52.

Grauten, W. F., 1965, Fluid relationships in Delaware Mountain sandstone, *in* A. Young and J. E. Galley, editors, Fluids in subsurface environments: American Association of Petroleum Geologists Memoir 4, p. 294–307.

Haner, B. E., 1971, Morphology and sediments of Redondo submarine fan, southern California: Geological Society of America Bulletin, v. 82, p. 2413–2432.

Harms, J. C., and R. K. Fahnestock, 1965, Stratification, bed forms, and flow phenomena (with examples from the Rio Grande), *in* G. V. Middleton, editor, Primary sedimentary structures and their hydrodynamic interpretation: Society of Economic Paleontologists and Mineralogists Special Publication 12, p. 84–115.

Harms, J. C., 1974, Brushy Canyon Formation, Texas: a deep-water density current deposit: Geological Society of America Bulletin, v. 85, p. 1763–1784.

Hayes, M. O., 1967, Hurricanes as geological agents; case studies of hurricanes *Carla,* 1961, and *Cindy,* 1963: University of Texas Bureau of Economic Geology Report of Investigations 61, 56 p.

Hills, J. M., 1968, Gas in Delaware and Val Verde basins, West Texas and Southeastern New Mexico, *in* V. W. Beebe and B. F. Curtis, Natural Gases of North America: American Association of Petroleum Geologists Memoir 9, v. 2, p. 1394–1432.

Jacka, A. D., R. H. Beck, L. C. St. Germain, and S. C. Harrison, 1968, Permian deep-sea fans of the Delaware Mountain Group (Guadalupian), Delaware Basin, *in* Guadalupian facies, Apache Mountains area, West Texas: Society of Economic Paleontologists and Mineralogists Permian Basin Section, Publication 68-11, p. 49–90.

Jackson, W.E., 1964, Depositional topography and cyclic deposition in west-central Texas: American Association of Petroleum Geologists Bulletin, v. 48, p. 317–328.

Jones, J. W., 1980, Depositional environment and morphology of Canyon sandstone reservoirs, Central Midland basin, Texas: Texas A&M University M.S. thesis, 147 p.

———, 1982, Depositional environment and reservoir morphology of Canyon sandstones, Central Midland basin, Texas (abstract): American Association of Petroleum Geologists Bulletin, v. 66, p. 244–245.

King, P. B., 1942, Permian of West Texas and southeastern New Mexico: American Association of Petroleum Geologists Bulletin, v. 26, p. 535–763.

KRUIT, C., J. BROUWER, G. KNOX, W. SCHOLLNBERGER, and A. VANVLIET, 1975, Une excursion aux cones d'alluvions en eau profonde d'age Tertiaire pres de San Sebastian (province de Guipuzcoa, Espagne): 9th International Congress of Sedimentology, Nice, Excursion 23, 75 p.

KUENEN, P. H., 1951, Properties of turbidity currents of high density: *in* J. L. Hough, editor, Turbidity currents and the transportation of coarse sediments to deep water: Society of Economic Paleontologists and Mineralogists Special Publication 2, p. 14–33.

LEOPOLD, L. B., and M. G. WOLMAN, 1957, River channel patterns—braided, meandering, and straight: U.S. Geological Survey Professional Paper 282, p. 34–85.

MACPHERSON, B. A., 1978, Sedimentation and trapping mechanisms in Upper Miocene Stevens and older turbidite fans of southeastern San Joaquin Valley, California: American Association of Petroleum Geologists Bulletin, v. 62, p. 2243–2274.

MAHER, J. C., R. D. CARTER, and R. J. LANTZ, 1975, Petroleum geology of Naval Petroleum Reserve No. 1, Elk Hills, Kern County, California: U.S. Geological Survey Professional Paper 912, 109 p.

MARTIN, B. D., 1963, Rosedale channel—evidence for Late Miocene submarine erosion in Great Valley of California: American Association of Petroleum Geologists Bulletin, v. 47, p. 441–456.

MCWILLIAMS, B. H., 1966, Ozona, East, *in* Oil and Gas fields in West Texas: West Texas Geological Society Publication 66-52, p. 282–285.

MEISSNER, F. F., 1972, Cyclic sedimentation in Middle Permian strata of the Permian Basin West Texas and New Mexico, *in* Cyclic sedimentation in the Permian Basin, 2nd edition: Midland, Texas, West Texas Geological Society, p. 203–232.

MIDDLETON, G. V., 1966a, Experiments on density and turbidity currents—Part 1, Motion of the head: Canadian Journal of Earth Science, v. 3, p. 523–546.

——, 1966b, Experiments on density and turbidity currents—Part 2, Uniform flow of density currents: Canadian Journal of Earth Science, v. 3, p. 627–637.

MITCHELL, M. H., 1975, depositional environment and facies relationships of the Canyon sandstone, Val Verde basin, Texas: Texas A&M University M.S. thesis, 211 p.

MUTTI, E., and F. RICCI LUCCHI, 1972, Le torbiditi dell' Appennino settentrionale: introduzione all'analisi di facies: Mem. Soc. Geol. Italiana, v. 11, p. 161–199.

——, 1978, Turbidites of the northern Appenines: introduction to facies analysis, translated by T. H. Nilsen, International Geological Review, v. 20, p. 125–166.

NORMARK, W. R., 1978, Fan valleys, channels, and depositional lobes on modern submarine fans: characters for recognition of sandy turbidite environments: American Association of Petroleum Geologists Bulletin, v. 62, no. 6, p. 912–931.

NOTTINGHAM, M. W., 1960, Recent Bell Canyon exploration in the North Delaware basin (New Mexico-Texas), *in* Natural gas in the Southwest: Southwestern Federation of Geological Societies Transactions, v. 1, p. 139–153.

PAYNE, M. W., 1976, Basinal sandstone facies in the Delaware Mountain group, west Texas and southeast New Mexico: American Association of Petroleum Geologists Bulletin, v. 60, p. 517–527.

PHILLIPS, SANDRA, 1981, Depositional patterns and reservoir morphology of Guadalupian Cherry Canyon sandstones, Indian Draw field, Eddy County, New Mexico: Texas A&M University M.S. thesis, 173 p.

RALL, R. W., and E. P. RALL, 1958, Pennsylvanian subsurface geology of Sutton and Schleicher counties, Texas: American Association of Petroleum Geologists Bulletin, v. 42, p. 839–869.

SILVER, B. A., and TODD, R. G., 1969, Permian cyclic strata, northern Midland and Delaware Basins, West Texas and southeastern New Mexico: American Association of Petroleum Geologists Bulletin, v. 53, p. 2223–2251.

SIMPSON, J. E., 1969, A comparison between laboratory and atmospheric density currents: Royal Meteorological Society Quarterly Journal, v. 95, p. 758–765.

SNYDER, J. M., 1977, Petrography and depositional environment of Upper Pennsylvanian sandstones, subsurface of Fisher County, Texas: University of Texas at Arlington M.S. thesis, 144 p.

SULLWOLD, H. H., JR., 1960, Turbidites in oil exploration, *in* J. A. Peterson and J. C. Osmond, editors, *Geometry of sandstone bodies:* Tulsa, Okla., American Association of Petroleum Geologists, p. 63–81.

TAYLOR, D. S., 1978, California's Yowlumne field—from basics to barrels: Oil and Gas Journal, March 20, p. 192–200.

TIEH, T. T., R. R. BERG, J. D. PIKE, and J. D. BRASHER, 1981, Feldspar diagenesis in the Yowlumne sandstone—I. Textural and mineralogical observations and effects on reservoir rock properties (abstract): Geological Society of America Annual Meeting, Abstracts with Programs, p. 567.

VALDER, CLAYTON, 1977, Gas production from tight sandstones of Permo-Pennsylvanian age, Val Verde basin, Texas (abstract): American Association of Petroleum Geologists Southwest Section Meeting program, Abilene, Texas.

VAN SICLIN, D. C., 1958, Depositional topography—examples and theory: American Association of Petroleum Geologists Bulletin, v. 52, p. 1897–1913.

VERTREES, C., C. H. ATCHISON, and G. L. EVANS, 1959, Paleozoic geology of the Delaware and Val Verde basins, *in* Geology of the Val Verde basin and field trip guide: West Texas Geological Society Publication 59–43, p. 64–73.

WALKER, R. G., 1978, Deep-water sandstone and facies and ancient fans, models for exploration for stratigraphic traps: American Association of Petroleum Geologists Bulletin, v. 62, no. 6, p. 932–966.

WEBB, G. W., 1981, Stevens and earlier Miocene turbidite sandstones, southern San Joaquin Valley, California: American Association of Petroleum Geologists Bulletin, v. 65, p. 438–465.

WEINMEISTER, M. P., 1978, Origin of upper Bell Canyon reservoir sandstones (Guadalupian), El Mar and Paduca fields, southeast New Mexico and west Texas. Texas A&M University M.S. thesis, 96 p.

WERMUND, E. G., 1975, Upper Pennsylvanian limestone banks, north-central Texas: University of Texas Bureau of Economic Geology Geological Circular 75–3, 34 p.

WILLIAMSON, C. R., 1977, Deep-sea channels of the Bell Canyon Formation (Guadalupian), Delaware Basin, Texas-New Mexico, *in* Upper Guadalupian Facies, Permian Reef Complex, Guadalupe Mountains, New Mexico and Texas: Society of Economic Paleontologists and Mineralogists, Permian Basin Section, Publication 77–16, p. 409–431.

Index

A

B

C

D

E

Q

R

S

T

U

V

W

X

Y